Geologic Structures of the Arctic Basin

Alexey Piskarev • Victor Poselov
Valery Kaminsky

Editors

Geologic Structures of the Arctic Basin

 Springer

Editors
Alexey Piskarev
All-Russian Research Institute of Geology
and Mineral Resources of the World Ocean
(VNIIOkeangeologia)
Saint Petersburg, Russia

Saint Petersburg University
Saint Petersburg, Russia

Valery Kaminsky
All-Russian Research Institute of Geology
and Mineral Resources of the World Ocean
(VNIIOkeangeologia)
Saint Petersburg, Russia

Victor Poselov
All-Russian Research Institute of Geology
and Mineral Resources of the World Ocean
(VNIIOkeangeologia)
Saint Petersburg, Russia

ISBN 978-3-319-77741-2   ISBN 978-3-319-77742-9   (eBook)
https://doi.org/10.1007/978-3-319-77742-9

Library of Congress Control Number: 2018943128

Printed on acid-free paper

This Springer imprint is published by the registered company Springer International Publishing AG
part of Springer Nature.
The registered company address is: Gewerbestrasse 11, 6330 Cham, Switzerland

Preface

By the mid-1980s of the last century, the uniqueness of the Arctic Ocean became apparent to the researchers involved in its regional investigations. They found that nowhere on Earth so many diverse structures are confined to such limited space according to global standards. Perhaps, the proximity of the region to the rotational pole of the planet, where rotational forces control to some degree the orderliness of the tectonic movements to their minimum, may explain such diversity.

For the whole abyssal part of the Arctic Ocean, there is only one well drilled in 2004 under the project ACEX (**A**rctic **C**oring **Ex**pedition) TD-ed at 400 mbsf; only few deep seismic soundings (DSS) are available, leaving many structures with no deep data at all. Hence, the ambiguity of interpretation of the existing geological and geophysical information is limited until recently only to the potential field anomalies maps, bathymetry, and sporadic DSS.

Situation drastically improved at the beginning of the twenty-first century when Circum-Arctic States – Russia, Canada, USA, Denmark, and Norway – initiated active research programs to substantiate the claims enlarging the outer boundaries of the Arctic continental shelf. With this goals in sight, many scientific expeditions using specialized research vessels and even nuclear icebreakers studied abyssal parts of the Arctic Ocean.

Integrated geological and geophysical investigations were run on the Mendeleev and the Lomonosov Ridges and their corresponding junctions with Siberian shelves, on one side of the Arctic Ocean, and with the Greenland and the Canada Archipelago shelves, on the other. These efforts included DSS and multi-channel seismic (MCS) surveys, bottom samplings, and aeromagnetic and airborne gravity surveys. High quality MCS surveys covered most of the Amundsen and the Nansen Basins in Eurasian Arctic and Canada Basin – in Amerasian.

The results of these investigations were published worldwide. Nevertheless, there is still no universally accepted seismic stratigraphy scheme for the sedimentary cover of the Arctic Basin. Despite the progress in absolute geochronology, many questions regarding the evolution of magmatism in the Arctic region remain unanswered. Classification and evolution of the main structural elements are also far from being finalized.

The main purpose of this volume is to demonstrate the contemporary level of our understanding of the geology of the Arctic region and outline remaining unsolved topical geological problems.

Saint Petersburg, Russia   Valery D. Kaminsky
                                                                           Alexey L. Piskarev
                                                                            Victor A. Poselov

Acknowledgments

We would like to express our gratitude to many people and organizations who took part in the studies of the Arctic Basin and Polar Regions. This book could not have been translated into reality without their effort and field and research experience.

Special thanks go to the colleagues from VNIIOkeangeologia: Ekaterina G. Astafurova, Evgeny V. Bruj, Clara I. Bulatkina, Alexandr S. Zholondz, and Liubov V. Aplonova – for their help in preparing illustrations; and to Vera A. Kosheleva and Irina G. Dobretsova (Polar Marine Geological Expedition) for their significant contribution to the petrographic description of the sediment thin sections. We are also grateful to Sergey P. Pavlov and Tatiana A. Kirillova (Marine Arctic Geological Expedition) for their assistance during the different parts of our work and manuscript preparation.

The authors of Chap. 10 would like to thank the many individuals and organizations who have been involved in acquisition, processing, and interpretation of the new data upon which the chapter is based. There are far too many people to mention individually, but they especially wish to express appreciation to Dr. Gordon Oakey for the potential field data compilations, Mr. John Shimeld for his exceptional efforts in seismic data acquisition and processing, and Mr. Kai Boggild for helping compile the subbottom reflection data.

Contents

List of Contributors

Irina A. Andreeva All-Russian Research Institute of Geology and Mineral Resources of the World Ocean (VNIIOkeangeologia), Saint Petersburg, Russia

Daria E. Artem'eva All-Russian Research Institute of Geology and Mineral Resources of the World Ocean (VNIIOkeangeologia), Saint Petersburg, Russia

Georgy P. Avetisov All-Russian Research Institute of Geology and Mineral Resources of the World Ocean (VNIIOkeangeologia), Saint Petersburg, Russia

Dmitry V. Bezumov All-Russian Research Institute of Geology and Mineral Resources of the World Ocean (VNIIOkeangeologia), Saint Petersburg, Russia

Victor V. Butsenko All-Russian Research Institute of Geology and Mineral Resources of the World Ocean (VNIIOkeangeologia), Saint Petersburg, Russia

Andrey A. Chernykh All-Russian Research Institute of Geology and Mineral Resources of the World Ocean (VNIIOkeangeologia), Saint Petersburg, Russia
Saint Petersburg University, Saint Petersburg, Russia

Daria V. Elkina All-Russian Research Institute of Geology and Mineral Resources of the World Ocean (VNIIOkeangeologia), Saint Petersburg, Russia
Saint Petersburg University, Saint Petersburg, Russia

Yury G. Firsov All-Russian Research Institute of Geology and Mineral Resources of the World Ocean (VNIIOkeangeologia), Saint Petersburg, Russia

Vladimir Yu. Glebovsky All-Russian Research Institute of Geology and Mineral Resources of the World Ocean (VNIIOkeangeologia), Saint Petersburg, Russia

Evgeny A. Gusev All-Russian Research Institute of Geology and Mineral Resources of the World Ocean (VNIIOkeangeologia), Saint Petersburg, Russia

Deborah R. Hutchinson United States Geological Survey, Woods Hole Coastal and Marine Science Center, Woods Hole, MA, USA

Mikhail V. Ivanov All-Russian Research Institute of Geology and Mineral Resources of the World Ocean (VNIIOkeangeologia), Saint Petersburg, Russia

Sergey P. Kashubin A.P. Karpinsky Russian Geological Research Institute (VSEGEI), Saint Petersburg, Russia

Valery D. Kaminsky All-Russian Research Institute of Geology and Mineral Resources of the World Ocean (VNIIOkeangeologia), Saint Petersburg, Russia

Gennady S. Kazanin Marine Arctic Geological Expedition, Murmansk, Russia

Artem A. Kireev All-Russian Research Institute of Geology and Mineral Resources of the World Ocean (VNIIOkeangeologia), Saint Petersburg, Russia

Alexey A. Krylov All-Russian Research Institute of Geology and Mineral Resources of the World Ocean (VNIIOkeangeologia), Saint Petersburg, Russia
Saint Petersburg University, Saint Petersburg, Russia

Natalia E. Leonova All-Russian Research Institute of Geology and Mineral Resources of the World Ocean (VNIIOkeangeologia), Saint Petersburg, Russia

David C. Mosher Geological Survey of Canada, Natural Resources Canada, Bedford Institute of Oceanography, Dartmouth, NS, Canada

Anatoly D. Pavlenkin All-Russian Research Institute of Geology and Mineral Resources of the World Ocean (VNIIOkeangeologia), Saint Petersburg, Russia

Vera I. Petrova All-Russian Research Institute of Geology and Mineral Resources of the World Ocean (VNIIOkeangeologia), Saint Petersburg, Russia
Saint Petersburg University, Saint Petersburg, Russia

Alexey L. Piskarev All-Russian Research Institute of Geology and Mineral Resources of the World Ocean (VNIIOkeangeologia), Saint Petersburg, Russia
Saint Petersburg University, Saint Petersburg, Russia

Victor A. Poselov All-Russian Research Institute of Geology and Mineral Resources of the World Ocean (VNIIOkeangeologia), Saint Petersburg, Russia

Vasily A. Savin All-Russian Research Institute of Geology and Mineral Resources of the World Ocean (VNIIOkeangeologia), Saint Petersburg, Russia

Oleg E. Smirnov All-Russian Research Institute of Geology and Mineral Resources of the World Ocean (VNIIOkeangeologia), Saint Petersburg, Russia

Sergey M. Zholondz All-Russian Research Institute of Geology and Mineral Resources of the World Ocean (VNIIOkeangeologia), Saint Petersburg, Russia

Anna G. Zinchenko All-Russian Research Institute of Geology and Mineral Resources of the World Ocean (VNIIOkeangeologia), Saint Petersburg, Russia

Chapter 1
The Current State of the Arctic Basin Study

Georgy P. Avetisov, Victor V. Butsenko, Andrey A. Chernykh,
Yury G. Firsov, Vladimir Yu. Glebovsky, Evgeny A. Gusev,
Artem A. Kireev, Alexey A. Krylov, and Anna G. Zinchenko

Abstract The chronology of investigations and geomorphological analysis of the underwater terrain in the abyssal parts of the Arctic Ocean is presented. History of airborne and ship-borne gravity and magnetic surveys is accompanied by the detail technical, quantitative and qualitative analysis of existing datasets and regional potential fields anomalies maps.

The up-to date location maps show the MCS seismic coverage grid inside the Arctic Basin. The important geological structures are highlighted by detail fragments of both TWT and depth converted interpreted seismic sections. The composite velocity models calculated from DSS data along regional geotraverses are also present.

The progress of the seismological observations over several decades presents a telling picture of distribution of hypocenters and focal mechanisms of the modern seismicity related to the mid-ocean spreading zone and its continuation onto the Laptev Sea shelf.

The results of the Arctic Basin deep water seafloor sampling and drilling provide additional input to our knowledge base.

Keywords Arctic Basin · Morphology · Gravity · Magnetic anomalies · MCS · DSS · Seismology · Sampling and drilling

G. P. Avetisov (✉) · V. V. Butsenko · Y. G. Firsov · V. Y. Glebovsky · E. A. Gusev ·
A. A. Kireev · A. G. Zinchenko
All-Russian Research Institute of Geology and Mineral Resources of the World Ocean
(VNIIOkeangeologia), Saint Petersburg, Russia
e-mail: vicb@vniio.nw.ru; gleb@vniio.nw.ru; gpavet@mail.ru

A. A. Chernykh · A. A. Krylov
All-Russian Research Institute of Geology and Mineral Resources of the World Ocean
(VNIIOkeangeologia), Saint Petersburg, Russia

Saint Petersburg University, Saint Petersburg, Russia
e-mail: a.a.chernykh@vniio.ru

© Springer International Publishing AG, part of Springer Nature 2019
A. Piskarev et al. (eds.), *Geologic Structures of the Arctic Basin*,
https://doi.org/10.1007/978-3-319-77742-9_1

1

1.1   Ocean Floor Topography and Morphology

All modern offshore maps of different content and scale, used for variety of fundamental scientific or practical purposes, are based on enormous amount of bathymetric data collected to date all over the World Oceans. But, until recently, the Arctic
Ocean was an exception. The climate and logistics handicapped the scientific investigations and for a long time this region remained unexplored.

It is sufficed to say that, up to the middle of the last century, all the scientists had
at their disposal 11 (eleven) depth soundings made by F. Nansen during the famous
"Fram" drift. Therefore, the central part of Arctic Ocean was described as a large
single abyssal depression (Fig. 1.1).

The Soviet High-Latitude Airborne Expeditions (HLAE) of the late 40th of the
last century (under Gakkel's guidance), opened a new era of the Arctic Basin investigation. By 1956 more than 400 depth soundings were collected, the Lomonosov
Ridge with minimal depth of only 1005 m was discovered and several bathymetric
maps, which revolutionized the existing conceptions of subsea morphology, were
published (Gakkel 1957, 1959, 1960; Gakkel et al. 1968a,b; Kiselev 1979).

American polar researchers jointed the efforts in the 50th of last century putting
on maps Marvin Spur and Alpha Ridge (1957–1958) (Ostenso 1962; Weber 1985).
The Chukchi Plateau was mapped in details in 1959.

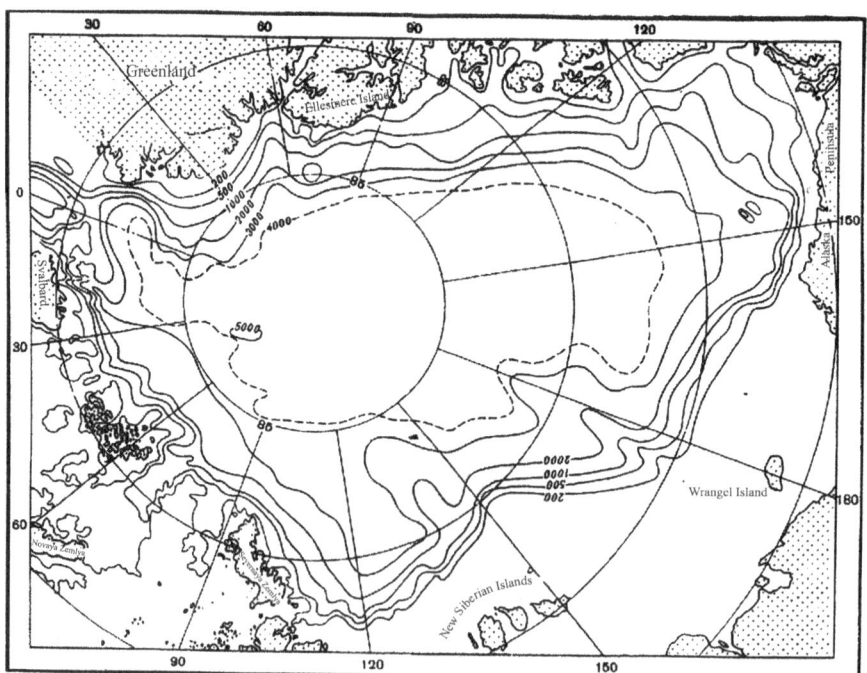

Fig. 1.1 The Arctic Basin bathymetry (Gakkel 1957)

US Navy started Arctic Basin investigations using nuclear submarines in the late 50th, but acquired information remained classified for a long time. From 1958 to 1960, nuclear submarines "Nautilus", "Skate" and "Sargo" were working mainly in the Eurasian Basin (Diets and Shumway 1961; Ostenso 1962; Weber and Roots 1990), collecting detail information along traverses from the Fram Strait to 86°N. Sonic soundings found the north- western fragment of the Mid-Arctic Ridge having a highly contrasting terrain (later, this area was named "Region of Seamounts" (Diets and Shumway 1961)). However, despite their quite accurate description of the seafloor topography, the information was not completely reliable due to low accuracy of submersible navigation. Presumably, these data were incorporated into later generation of bathymetric maps and international map GEBCO, Chart No. 5-17, (Canadian Hydrographic Service 1979). It must be noted, that information for the region between 80°N and 85°N is critical for understanding interrelations between the Gakkel Ridge and Greenland-Spitsbergen Fault, which connects the rift system of the Arctic Ocean and that of the North-Eastern Atlantic.

By 1960, the Russian and foreign expeditions collected up 20,000 depth sounding, widely scattered all over the Arctic Basin (Gakkel 1960; Ostenso 1962), sufficient enough for only general delineation of the major morphological units. It was established that the Lomonosov Ridge divides the Arctic Ocean on Eurasian and Amerasian Basins and the Gakkel Ridge split the Eurasian Basin into Nansen and Amundsen abyssal depressions. The Gakkel Ridge was traced from the Laptev Continental Margin (Sadko Trough) to the Greenland-Spitsbergen gap where it connects to the Norvegian-Greenland system of mid-ocean ridges. The continuity of the continental shelf edge around the entire ocean and ortogonality between the continental margins (including edge itself) and oceanic ridges was also noted. Confidential status of the shelf data made them unavailable for the detail morphological analysis of important junctions between oceanic and shelf structures.

Heezen and Ewing (1961) published several very informative papers on the Arctic Mid-Oceanic Ridge. After careful analysis, they selected about 4000 reliable depth sounding in the Arctic region (excluding submarine data) and compiled the bathymetric map of the whole Eurasian Basin, based on assumption that the Mid-Arctic Ridge joints the mid-oceanic ridges in the North-Eastern Atlantic (Ostenso 1962).

Some bathymetry data from the Soviet HRAE were also included in the database, as well as the results of RV "Ob" and "Lena" expeditions (1956–1957). The latter discovered the abyssal trough more than 400 km long at depth 3500–3900 m between Spitsbergen and Greenland, known as Lena Trough. B.C. Heezen and M. Ewing, according to their assumption of uninterrupted chain of ridges from the Arctic to the Atlantic, considered it as a mid-ridge rift valley of a hypothetical mid-oceanic ridge connecting the Arctic and Norvegian-Greenland rifting systems. The authors did not deny the subjectivity of their theory, insisting that the most plausible interpretation must based on correct hypothesis (Ostenso 1962). The latest information proved the Lena Trough to be an axial zone of the complex Greenland-Spitsbergen fracture zone, more than 350 km long.

Fig 1.2 Arctic basin seafloor Terrain Map (fragment) (Heezen and Tharp 1971)

Despite these categorical statements, the global bathymetric maps, compiled by B.C. Heezen, M. Ewing and M. Tharp, for many years served as a base for studying underwater morphology. Later, Heezen and Tharp (Heezen and Tharp 1971, 1975) included their information into seminal World Ocean Floor Map (Figs. 1.2, 1.3).

The bathymetric map was the first one showing the Mid-Arctic Ridge joining the Laptev Continental Margin at 79°30′ N. But the ridge itself was presented rather schematically as a regular system of more than 25 blocks separated by similarly regular system of transform faults displacing the axial rift valley. (Fig. 1.3).

Mid-ocean ridges in some other locations were presented in similar way. In general, these maps, often based on subjective idea of continuous chain of mid-ocean ridges, should be considered as mere schemes. But regardless of these shortcomings, maps published in the early 1970th turned out to be closer to the contemporary maps than maps published later (Ostenso 1962; Sobczak 1977; Sobczak and Sweeney 1978). For instance, the 1977 map (Sobczak 1977) shows the Mid-Arctic Ridge as a system of narrow, low-amplitude (300–500 m) ridges and troughs only up to 84°N, after which the Nansen and Amundsen Basin merging together.

Several subsequent expeditions contributed greatly to the knowledge base of the region. Among them – Canadian expeditions LOREX (Lomonosov Ridge, 1979) and CESAR (Alpha Ridge, 1983). Collected data improved quality of seafloor map sheets at the scale 1:250000 (Sweeney et al. 1989; Weber 1980; Weber and Jackson

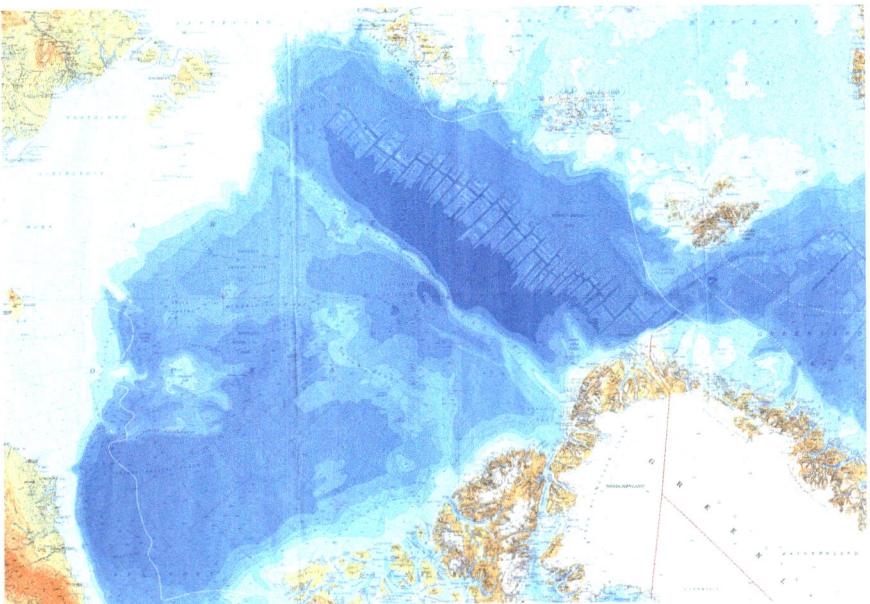

Fig. 1.3 Bathymetric map of the Arctic ocean (fragment) (Heezen and Tharp 1975)

1985; Weber and Sweeney 1990). The overview of these bathymetric works is given in (Weber and Roots 1990).

Beginning from 1961, continued native hydrographic investigations - airborne or from ice-based camps, ships and submarines – covered more than 80% of the Arctic Ocean with highly accurate depth measurements including 21,120 echosoundings, 17,246 ice-based seismic sounding and 92,000 km of submarine-based recording (Fridman 2007).

Currently, the majority of the information is open to public, but some. The Central Arctic Uplifts of the Amerasian Basin and Eurasian continental margin junction remain classified.

Introduction of ice-based seismic surveys into the Arctic region greatly increased amount of bathymetric information, because seismic, along with valuable geological information, provides an accurate estimates of seafloor depth. In total, the ice-based seismic investigations under the "Sever" program covered more than 4 million sq.km of the Arctic Ocean. Large amount of available information dictated need in unified database, which was created in 1997. The overall database positioning accuracy is ±900 m, depth - 0.5%.

By 1999, Russian hydrographic expeditions investigated more than 85% of the Arctic Basin. Despite the irregular spatial distribution of multidiscipline observations, the major structural and morphological forms and their interrelationship were established, e.g. the orthogonal junction of the abyssal ridges and plateaus with continental margins.

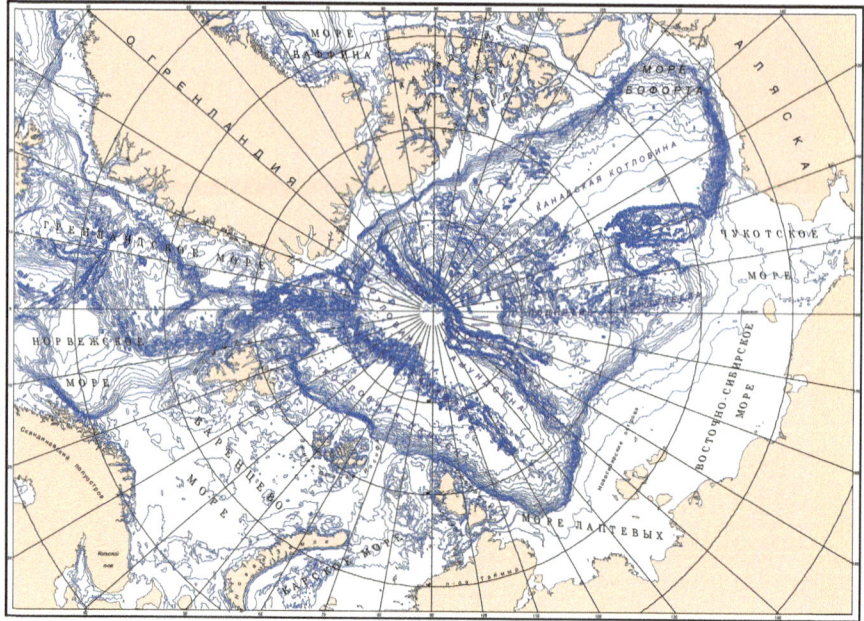

Fig. 1.4 Bathymetric map of the Arctic Ocean

It was found that the shelf edge line serves as a natural boundary of the Arctic Basin, completeley encompassing it. The depth of the edge varies from 100–200 m to 400 m, dipping to 600–800 m at the intersecting troughs. In average, depth soundings were located 2–5 km apart within shelf and upper continental slope, including junction zones with deeper uplifts; 5–7 km – in the central parts of the slope; 15–20 km – at continental rise and abyssal planes. At the western slopes of the Mendeleev Ridge and Podvodnikov Basin the distance grew to 27 km (this numbers are true only for the Russian part of the Arctic Basin).

The proper interpretation and imaging the bathymetric data became possible with development of certain principles of morphological and morphometric analysis. (Imaging of Underwater Terrain on Naval Maps 1973). The morphological analysis, or surface-mapping, starts with defining terrain structural lines which, in turn, form the structural wireframe of the terrain being investigated. The structural lines are drawn along elemental terrain forms (slopes, escarpments etc) fixing their configuration and combined to form a wireframe representing more complex combination of structural lines. The wireframe becomes an fixed image of the terrain form independent from variations of bathymetry and is combined with isobaths. This method creates a detail representation of seafloor surface, as well as size, shape and interrelations of morphological structures of different orders. The bathymetric map of the Arctic Ocean with 200 m contour interaval (Fig. 1.4) was compiled using this method (Seafloor Topography of the Arctic Ocean 1998).

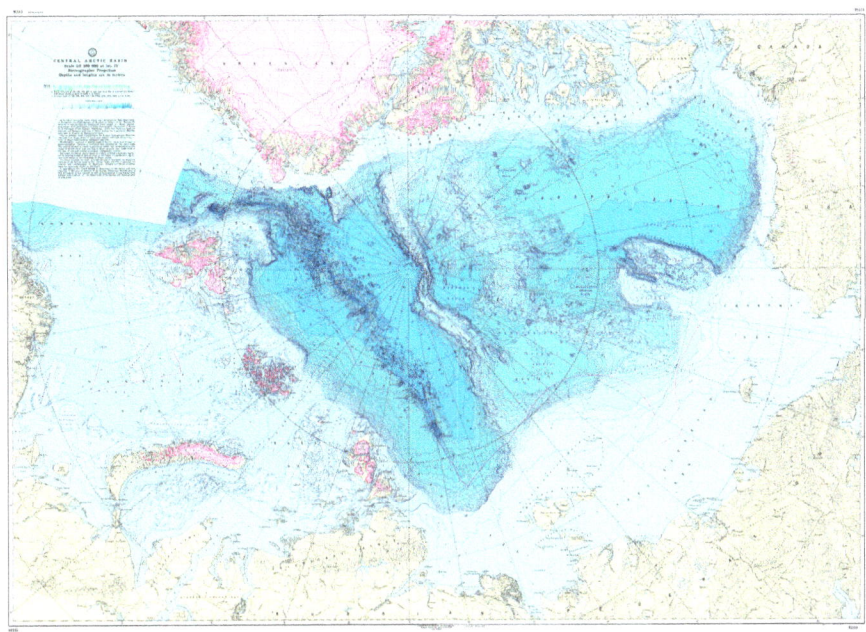

Fig. 1.5 Central Arctic Basin. (Central Arctic Basin 2002)

Later version of this map at the scale of 1:2500000 based on newly acquired data was published in 2002 (4 sheets, Admiralty ref., No. 91115) [Fig. 1.5], (Central Arctic Basin 2002).

The first version of The International Bathymetric Chart of the Arctic Ocean (IBCAO) in polar stereographic projection was presented to the American Geophysical Union in 1999 (Jakobsson et al. 2000). It utilized all bathymetry values and their coordinates available to date regardless their vintage, reliability and accuracy, including the data from the latest surveys with satellite navigation and multibeam sonar soundings. The database also included declassified depth information from American and British submarines, as well as multi-beam sonar data from icebreaker "Polarstern" (Alfred Wegener Institute, Germany).

For areas without actual data, the depths were digitized from navigation charts and other existing sources. Among them – paper version of the small-scale (1:5000000) bathymetry chart of the Arctic Ocean compiled by the Chief Directorat of Navigation and Oceanography under the Russian Ministry of Defence (Seafloor Topography of the Arctic Ocean 1998). In 2005, the digital graphic raster file of the Central Arctic Basin map (1:2500000) was sent to the IBCAO editors. In total, the IBCAO digital dataset included 1,643,477 actual values and 1,948,749 – digitized (Jakobsson et al. 2008). Detail analysis of the IBCAO map can be found in (Fridman 2007). The IBCAO map quickly became a widely used source of global oceanographic information and its interpretation.

The beginning of the XXI century marks wider use of modern and precise method of seafloor topography mapping – multi-beam sonar survey. Concentrated mostly in the Cenrtal Arctic Basin, the investigations were operated from the ice-breakers – "Polarstern" (Germany), "Healy" (USCG) and "Oden" (Sweden)".

The most interesting results were obtained by the joint American ("Healy") and Germany ("Polarstern") project AMORE (Arctic Mid Ocean Ridge Expedition) of 2001 on the Gakkel Ridge. USCG "Healy" later extended its operations to the Amundsen Basin, near-polar section of the Lomonosov Ridge (2005), Canada Basin and Chukchi Plateau (2003, 2004, 2007).

Open to the scientific community, new sonar data were included into second version of IBCAO map, this time compiled under the aegis of the International Hydrographic Committee and the International Arctic Science Committee. The map (2 × 2 km grid) was plotted in Polar Stereographic Projection with coordinates reduced to WGS-84 ellipsoid (75^0N central latitude) and bathymetry values - to mean sea level (m. s. l.) (Jakobsson et al. 2008).

The Alaska continental slope mapping benefited the most from new acoustic data ("Healy" and "Palmer"). The sonar data revealed numerous regularly spaced troughs, orthogonal to the slope, and depositional fans at their mouths, clearly visible on the map even after digitizing to 2 × 2 km grid (Jakobsson et al. 2008).

USCG "Healy" did extensive mapping of Nortwind Ridge and Chukchi Plateau from 2003 to 2007. It was found that the Northwind Ridge faces the Canada Abyssal Plane with very steep slope (>10°) and amplitude 3000 m.

American USCG "Healy" and Canadian "Louis S. St-Laurent" ice-breaker tandem did multi-beam sonar mapping in the Canada Basin from 2008 to l 2012, with "Louis S. St-Laurent" simultaneously running the multi-channels seismic (MCS) (Gardner et al. 2008). The ships covered 200,000 km^2, with 33,885 lineal km of traverses.

The regular mapping of the Lomonosov Ridge in the central part of the Arctic Ocean started in 2007 under the project LOMROG (Lomonosov Ridge of Greenland - Denmark, Sweden, Canada). Three expeditions: LOMROG-2007, LOMROG-2009 and LOMROG-2012 of Swedish icebreaker "Oden", with multi-beam sonar EM-122 on board, mapped the central sections of the Lomonosov Ridge with accuracy high enough for 100 m grid cell size. In the process, several disagreements with IBCAO 2.0 map were found.

New IBCAO 3.0. version (Jakobsson et al. 2012), was another step forward. After including the latest multi-beam sonar data, the share of the accurate data in the database increased from 6% in IBCAO 2.0 version to 11% in IBCAO 3.0. Using "remove-restore" algorithm, the 2 × 2 km grid was merged with 500 × 500 m grid, resulting in overall global grid of 500 m cell size, making the finest details of the underwater terrain visible. All versions of IBCAO bathymetric model are posted in *website of National Centers for Environmental Information* (www.ngdc.noaa.gov/ mgg/bathymetry/arctic/downloads.html).

Complete information on the Arctic bathymetry is concentrated in The Center for Coastal and Ocean Mapping/Joint Hydrographic Center (CCOM/JHC), the University of New Hampshire, USA. The Centre's website (www.ccom.unh.edu)

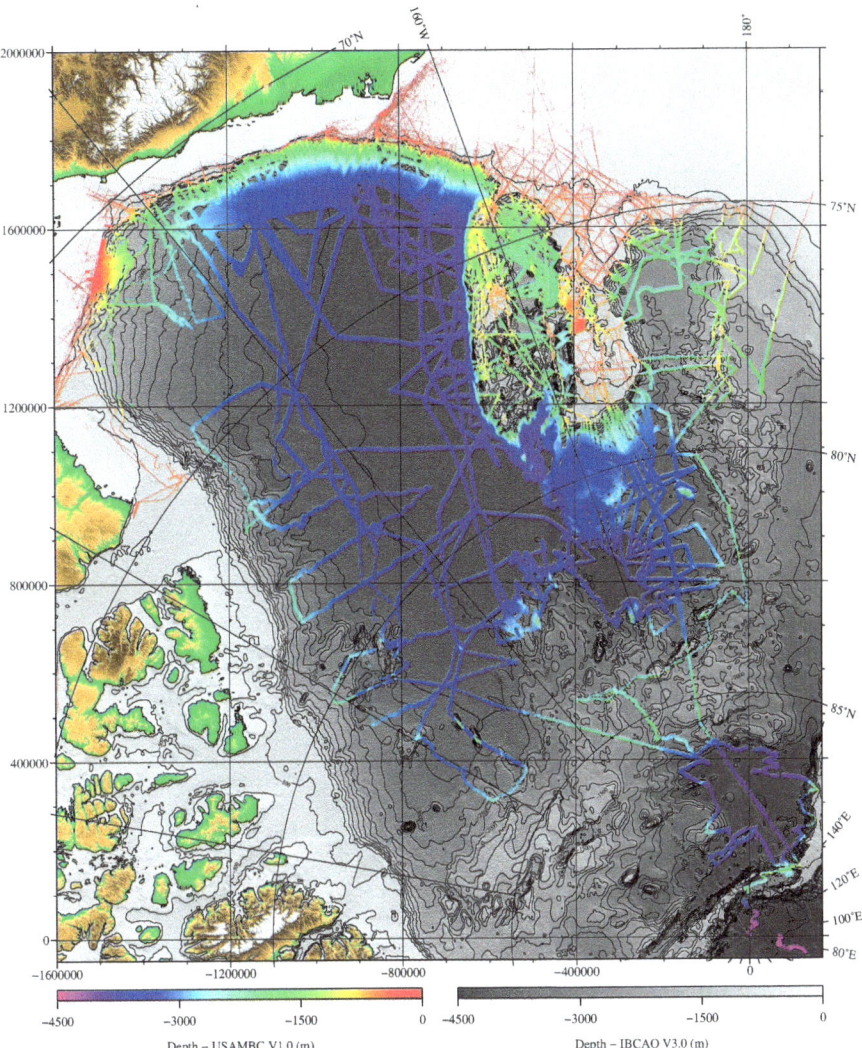

Fig. 1.6 Traverses of USCG "Healy" expeditions (2003, 2004, 2007–2012)

contains reports of all USCG "Healy" expeditions and digital bathymetric grids in three different formats. The resulting map of the American sector seafloor with 40 m cell size (USAMBC v.1.0) is also available on the University of New Hampshire website. Fig. 1.6. shows the locations of USCG "Healy" traverses of 2003–2012 surveys used for compilation of this map.

The map on Fig. 1.6. demonstrates uneven distribution of oceanographic information. The best studied are the following segments of the Arctic Ocean floor:

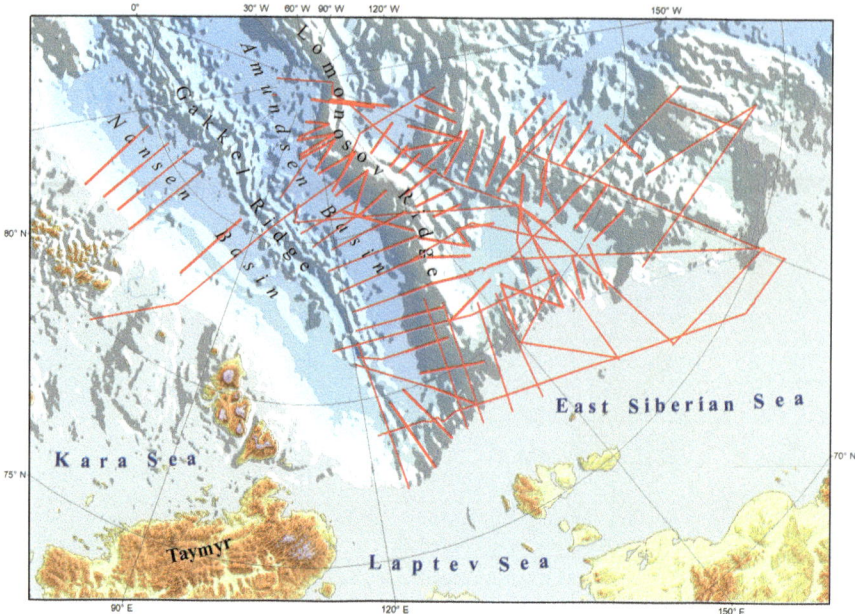

Fig. 1.7 Multi-beam sonar surveys of 2010—2014 carried out by the Russian expeditions in the Arctic Ocean

1) Alaska Continental Slope;
2) Abyssal part of the Mendeleev plain next to the Chukchi Plateau;
3) Southern section of the Northwind Ridge;
4) Northwind escarpment and adjacent part of the Canada Basin;
5) Southern section of the Alpha Ridge near Ryabov Seamounts.

Since 2002, the Russain Federation performed an unprecedented amount of observations on more than 35,000 km of traverses.

Fig. 1.7 demonstrates the locations of multi-beam sonar surveys of 2010—2014 carried out by the Russian expeditions in the Arctic Ocean. The 2010 expedition was specifically concentrated on defining the continental rise. The 2012 and 2014 operations combined sonar and MCS operations.

Surveys were carried out with the RV "Akademik Fedorov". The sonar equipment included multi-beam sonar EM-122, single–beam sonar EA-600 and profilograph TOPAS PS 18 (the latter optionally could operate as a single-beam sonar), manufactured by Kongsberg Maritime, Norway. Navigation system SEAPATH 330 was used for positioning, and hydrographic software package CARIS HIPS - for data processing.

Due to difficult ice condition, the RV "Akademik Fedorov" was accompanied by the icebreaker. The sweep was 3–4 km wide. With depth sounding accuracy ±7.6 m and antenna location ±3.4 m (post-processed).

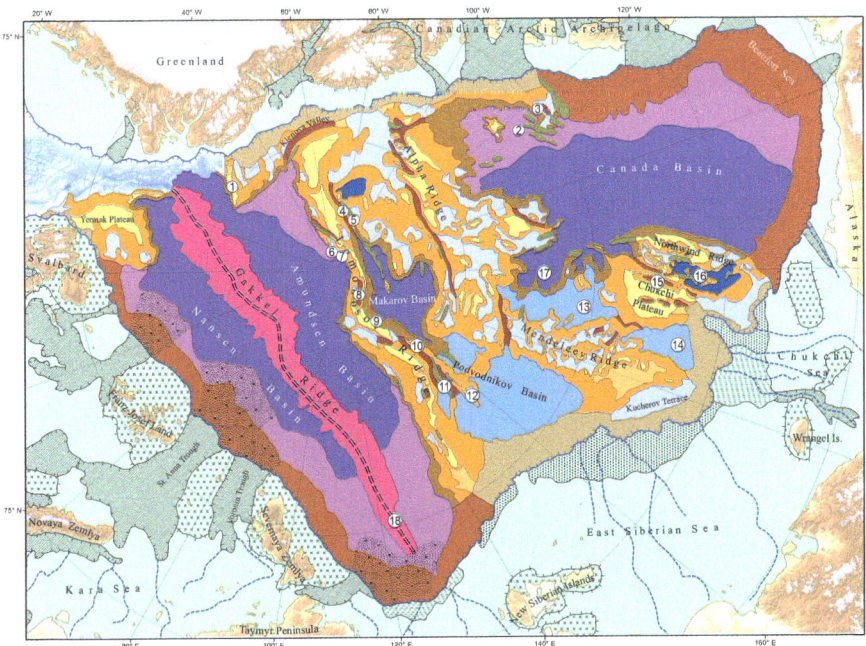

Fig. 1.8 Geomorphology of the Arctic Ocean – the legend

Geomorphology interpretations of constantly growing volume of the Arctic Basin bathymetric information is represented by multitude of small-scale regional maps (Dibner et al. 1965; Atlas of the Oceans 1980; Treshnikov 1985; The Arctic and Pacific Oceans 1985; Lastochkin and Naryshkin 1989; Orographic Map of the Arctic Basin 1995). Some information can also be found in some publications dealing with geology and geophysics of the region as a whole, or its parts. (Pogrebitsky 1984; Kiselev 1986; Jakobsson et al. 2008). The pressing issue of defining the outer limits of the Russian continental shelf gave additional impetus to geomorphological analysis at the end of XX and beginning of XXI centuries (Kulakov et al. 1986; Explanatory Notes to the Arctic Basin Maps: Orographic Map of the Arctic Basin 1999; Orographic Map of the Arctic Basin 1995; Geomorphological Aspects of the Russian Continental Shelf Exterior Boundary in the Arctic 2005; Alekseev et al. 2012).

The map on Fig. 1.8 demonstrates the major geomorphological elements of the Arctic Basin and adjacent shelves. The map (based mainly on the IBCAO v.3 database, 1:5000000 scale) was published in 2015 and later updated with new information. The legend (Fig. 1.8) is divided in two blocks: one for continental margin (including shelf, continental slope and rise) and another for ocean floor (including abyssal planes and mod-oceanic ridge).

The Arctic Ocean shelf consists of several segments: Barents-Kara, Greenland-Canada, Laptev, East-Siberian-Chukchi and Alaska, each with its own width, depth,

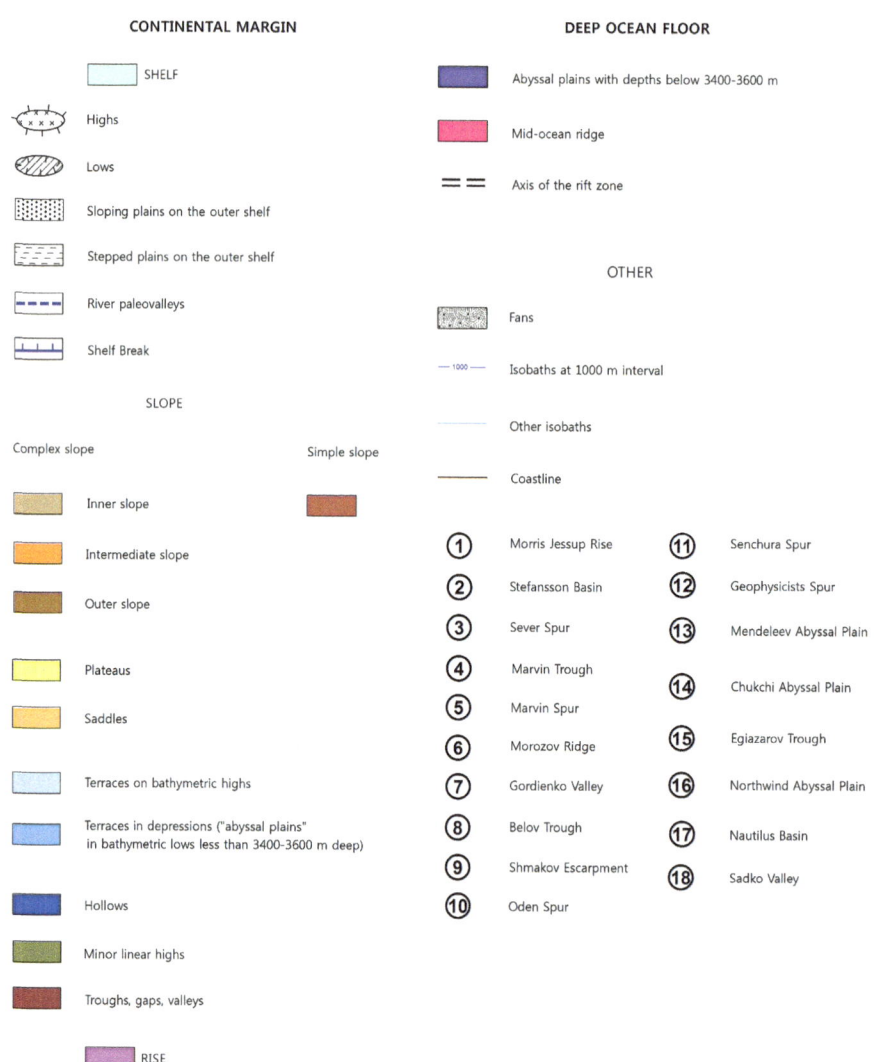

Fig. 1.8 (continued)

ruggedness and quantity of insular land. Local features inside shelves are divided into two groups: "positive" (plateaus, banks, mounts etc) and "negative" (trough, gaps, valleys, hollows etc). Some peculiar parts of Laptev and Chukchi shelves, described as "sloping" and "stepped" plains, were noted due to their proximity to the adjoining Amerasian uplifts. Most prominent paleo-river valleys formed during the last Neo-Pleistocene regression, and presently acting as conduits for currents and sediments (perhaps, indirectly controlled by geological structures), are marked by blue dashed lines.

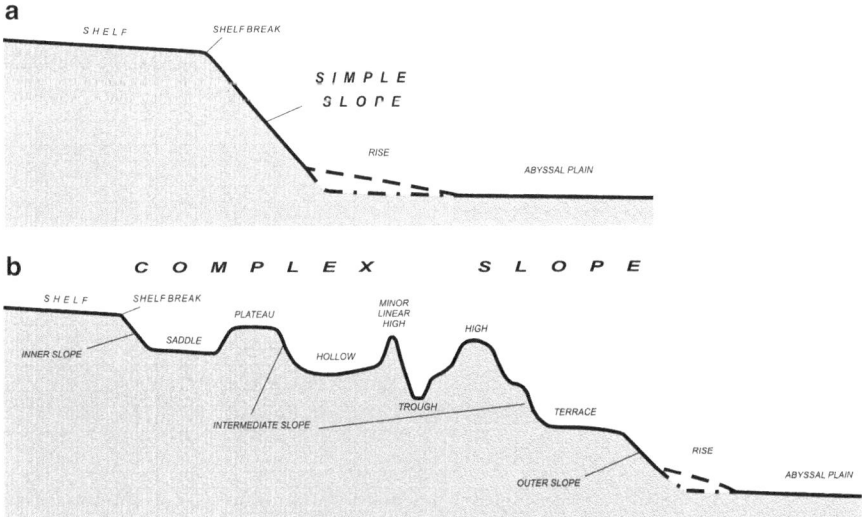

Fig. 1.9   The simple (**a**) and complex (**b**) types of continental slopes

Continental slopes are subdivided into simple and complex slopes. The simple slope starts right at shelf edge and continues down-slope (with gradual increase of dip angle) to the continental rise or, lacking one, directly to the abyssal plain (Fig. 1.9a). Presence of fans depositing sediments at the base makes the plain-slope transition even more gradual. Typically, the simple slopes are 70–150 km wide.

The problem of the continental shelf boundary delineation which lately acquired ever increasing political and economic importance, drove attention to the fine details of slopes morphology (Kulakov et al. 1986; Khain and Lobkovsky 2003; Zinchenko 2004; Geomorphological Aspects of the Russian Continental Shelf Exterior Boundary in the Arctic 2005; Pogrebitsky et al. 2005). Influx of new information reveal a new type of continental slopes with saddles, plateaus, terraces and mounds of different size, amplitude and steepness (Symonds et al. 2000), classified as a complex continental slopes. This classification of slopes was used in the geological maps of the shelves and adjacent regions (1:1000000 scale) (The Geological Map 2011).

The typical complex slope consist of: inner (upper) slope, just below shelf edge; outer (lower) slope, in contact with the rise or abyssal plain; intermediate slopes in-between sculpturing complicating elements cited above (Fig. 1.9b),

The simple slopes, dipping at first degrees or less and with amplitudes around 2500–3000 m, are typical for most of the Barents-Kara margin (except the area around the Ermak Plateau), Laptev, Beaufort and Alaska shelves.

The complex slopes, much wider that simple slopes, are common in the Central Arctic Uplifts (Lomonosov, Mendeleev, Northwind and Alpha Ridges, Chukchi and Morris Jessup Plateaus), bridging the Greenland-Canada and East Siberian – Chukchi shelves (Kiselev 1986; Kulakov et al. 1986; Lastochkin and Naryshkin

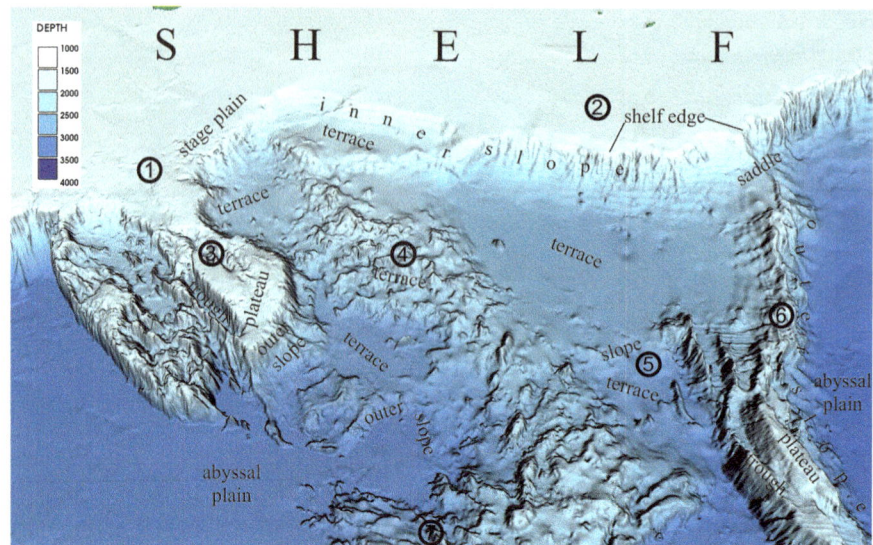

Fig. 1.10 Elements of bottom morphology in the Arctic Ocean. 3D image based on IBCAO grid v. 3.0 looking south from the Alpha Ridge. Numbers in circles: 1 – Chukchi Shelf; 2 – East Siberian Shelf; 3 – Chukchi Plateau; 4 – Mendeleev Ridge; 5 – Podvodnikov Basin; 6 – Lomonosov Ridge

1989; Orographic Map of the Arctic Basin 1995; Explanatory Notes to the Arctic Basin Maps: Orographic map of the Arctic Basin 1999).

The boundaries between the bathyal Mendeleev, Chukchi and Podvodnikov Basins and the adjacent Nansen, Amundsen, Makarov and Canada abyssal Basins are marked by gentle low amplitude simple slopes. Therefore, it can be stated that both positive and negative forms of this region have benched, or stepped, morphology. At the uplifts, however, this morphology is more contrasting with numerous troughs, gaps and valleys 200–1000 m deep (even more at the Marvin Gorge). The abundance of these forms may reflect extensional forces in action and subsequent normal faulting. The depth of uplifts gradually increases toward the central parts of the Arctic Ocean.

The inner (upper) slope at the hinge line is a permanent feature of all complex continental slopes. As a rule, its amplitude is higher at contacts with depressions, than with uplifts and, also as a rule, there are saddles at the foot of the inner slope separating it from adjacent uplift further down-slope (Fig. 1.10).

It was noted that the range of saddle depth is different:

1600—1800 m - at Eurasian termination of the Lomonosov Ridge;
1000—2000 m (up to 2600 m) – at Greenland-Canadian termination of the Lomonosov and Alpha ridges (here the saddle is almost obliterated by the trough opening to the Amundsen Basin);
1400—1700 m - at south end of the Mendeleev Ridge;
600—1000 m - at the Chikchi Plateau;
≈ 1200 m – at the Northwind Ridge.

In any circumstances, the saddles are much shallower than the surrounding abyssal plains and, as parts of continental margins, link the uplifts of the Central Arctic Basin with continental shelf.

The terraces are frequently observed at the transition from slope to depression. Some of them (Kucherov Terrace at the Mendeleev Ridge - East-Siberian Shelf junction) become a part of the uplifts stepped terrain and confirming internal integrity of the Central Arctic Basin.

The continental rise, a gently (0.1–3°) sloping surface between base of continental slope and abyssal plain, formed by wedge of clastic material delivered from upper parts of the continental margins. Its width depends on volume of deposited sediments and reaches maximum at the opening of underwater troughs and canyons, where multiple overlapping depositional fans are often observed. On the other hand, the rise is practically absent at both west and east bases of the near-polar segment of the Lomonosov Ridge, Northwind Ridge and the Alaska Continental Shelf.

The abyssal sub-horizontal (pitch 0,1° or less) Amundsen, Nansen, Canada and Makarov Basins with seafloor depth of about 4000 m represent the lowest tier of the Arctic Basin topography (www.gebco.net). All plains are completely, or almost completely, encircled by continental rises, lower slopes or flanks of mid-oceanic ridge.

The Lomonosov Ridge separates two principal components of the Arctic Ocean – Eurasian and Amerasian Basins (the ridge itself is considered as a part of the latter). The elongated Eurasian Basin includes Nansen and Amundsen Abyssal Plains separated by the Gakkel Ridge. It is the deepest part of the Arctic Ocean ($\geq$ 4000 m at the plains and 5000–5500 m at the Gakkel's Ridge central rift valley) with many simple slopes, continental rises and depositional fans at the perimeter. The largest fans were developed at outfalls of deep canyons on Barents-Kara shelf (Frantz-Victoria, St. Anna, Voronin). Opposite the Severnaya Zemlya Archipelago, the continental rise of Kara Sea shelf merges with that of Laptev Sea, because the slopes and flanks of terminating Gakkel Ridge created geomorphologic traps for sediments transported from the continental margins. This process flattened the lower slopes, widened the rises and determined a parabolic configuration of isobaths, especially in the Amundsen Basin (Orographic Map of the Arctic Basin 1995).

Located at both sides of the Gakkel Ridge, the outlines of the Nansen and Amundsen Basins roughly follow its axial line. Most of the Nansen Basin is 3800 m deep reaching and exceeding 4000 only in its western part, while the most of the Amundsen Basin is deeper than 4200 m. The Basins floor is mostly flat, with some gentle mounds at the foothills of the Gakkel Ridge. The Basins topography gets more contrasting in the western parts with some benches in the Amundsen Basin (Explanatory Notes to the Arctic Basin Maps: Orographic Map of the Arctic Basin 1999). The perimeters of the basins also differ: the Nansen Basin is surrounded mostly by simple continental slopes about 3000 m tall; most of the Amundsen Basins periphery consists of complex continental slopes of the Morris Jessup Plateau and slopes of the Lomonosov Ridge. The Lomonosov Ridge itself consist of several en-echelon displaced segments deepening toward the North Pole.

The Gakkel Ridge stands out as a large linear uplift with rugged topography and omnipresent, but not everywhere equally well developed, central rift valley. The ridge axial line has several "kinks", but does not show substantial transformal displacements. The axial rift valley is the most revealing in the western half of the ridge where it is more than 5000 m deep. Eastward, the topography becomes less contrasting and more asymmetric across the axial line and depth rarely exceeds 4000 m. The Laptev termination is marked by increasing depth and anomalous oval depression more than 5000 m deep.

Amerasian Basin has a near-triangular shape. Along with elongated Canada Basin approximately 3800 m deep, it includes the region of the Central Arctic Uplifts. It consists of several uplifts (Lomonosov, Alpha, Mendeleev Ridges, and Chukchi Plateau) and separating them depressions (Podvodnikov, Nautilus, Mendeleev and Chukchi Basins). This region is thought to represent a complex continental slope linking the Asian and North Americam continental margins. The presence of a relatively small (compare to other Arctic Basins) "oceanic" element (Makarov Basin, 3800 m deep) does not disturb the morphological coherence of this Trans-Arctic continental "bridge".

The geomorphology of other parts of the Arctic Ocean will be discussed in the appropriate Chapters.

1.2 Potential Fields Data

1.2.1 Russian Surveys

The Russian studies of the Arctic Ocean gravity field started in 1948—1956 by the Arctic Research Institute and the Earth Physics Institute of the Academy of Science, as a part of High-Latitude Airborne Program "North". Since 1963, the program continued by the Polar Expedition of then Arctic Research Institute (NIIGA) under the Navigation and Oceanography Directorate of the Ministry of Defense.

Until 90s of the last century, the most of gravity measurements were obtained during airborne, or continuous ship-borne, surveys. Later, the fully airborne surveys became the principle method of data acquisition.

1.2.1.1 On-ice Gravity Surveys

Gravity surveys in the abyssal part of the Arctic Ocean, eastern Kara, Laptev, East Siberian and Chukchi Seas were run using landing airborne method, with fixed-wing airplane landing at certain intervals and gravity measurements taken on ice using several models of land quarts gravity meters GDK and GAK. At every location, the seafloor depth was determined using either deep water lead or seismic echo-sounding; station geographical coordinates - by astronomical or triangulation methods (40% of total, ±0.1–0.8 km error), or by phase radio navigation (60% of total, ±0.2–1.5 km error).

Average density of gravity observation over the entire Arctic Shelf is close to 1 point per 100 sq. km. However, it decreases to 1 point per 625 sq.km in the East-Siberian and Chukotka Seas, dropping even lower over to 1 point per 1600 sq. km in the abyssal parts of the Arctic Ocean.

In average, the RMS errors of landing airborne data vary from 1 to 3 mGal, reaching up to 5 mGal in the deep-water parts of the Arctic Ocean. The raw gravity data and points locations of the Russian gravity surveys outside the Russian parts of the Artcic Shelf still remain confidential. At present, only very general compilations and digital models of Free Air gravity anomalies at the scales from 1:2500000 to 1:6000000 are declassified and available for the public (Glebovsky et al. 2002).

The results of landing airborne surveys in East-Siberian and Chukchi Seas are published as a sheets of "The State Map of Free Air Gravity anomalies of the USSR at the scale of 1:1 000 000".

1.2.1.2   Ship-Borne Gravity Surveys

The first area ship-borne gravity surveys and limited volume of depth sounding were conducted in Barents and Kara Seas in 1968–1969 on 30 × 30 km and 25 × 25 km grids. Combination of astronomical observations and radio-geodetic navigation brought positioning accuracy to less than 1 km, and gravity ±1–5 mGal. The data were integrated into the summary "Gravimetric map of the USSR Arctic Shelf and surrounding areas at the scale 1: 2 500 000" in both Free Air and Bouger (2.67 g/ cm^3) gravity anomalies.

More accurate and comprehensive gravity surveys of Barents and Kara Seas (1:1000000) started in mid 70th under the aegis of international program "World gravity mapping, II class", run by Marine Exploration Expedition (currently JSC "MAGE"). Gravity was recorded by specially designed onboard quartz gravity meter GMN, positioning – by radio-geodetic system "Poisk-S". Gravity RMS error was kept within 1–3 mGal, positioning – 0.25–0.4 km.

By the end of 80th the entire western of the Russian Arctic Shelf was covered by gravity survey at the scale 1:1000000 with average density of observation 1 point per 100 sq.km. These data, integrated with the data obtained earlier by landing air-borne surveys, were used for compilation of "The State Map of Free Air Gravity anomalies of the USSR at the scale of 1:1000000".

Generalized information regarding the gravity field datasets acquired in the Arctic Ocean prior to 2000th and available for furhre mapping and interpretation were cited in (Kaminsky et al. 2000; Maschenkov et al. 2001).

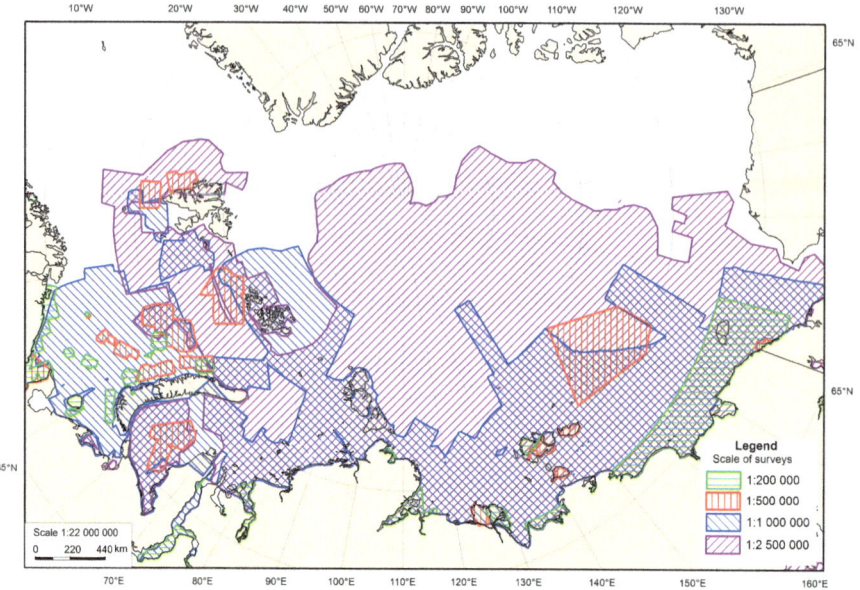

Fig. 1.11 Gravity data distribution over the Arctic Ocean, as of 2016. Generalized boundaries of surveys (scales 1:200000–1:2500000) are shown

1.2.1.3 Aerogravimetric Surveys

Implementation of true airborne gravimetric methods signified a real progress in investigation of Arctic Seas gravity field. In short period, from 1993 to 2000, the entire western part of the Russian continental shelf and "ocean-continent" transitional zone around the Franz-Joseph Archipelago, was covered by aerogeophysical (gravity and magnetic) survey using satellite navigation system MX-4400 GPS Navstar. The surveys were flown using IL-38 plane (400 m a.s.l over open water and 1000—1500 m a.s.l. overland), string airborne gravity meters GAMS and vertical accelerometers SIVS. Flight elevation was measured by ice radar MPI- 300. The accuracy of navigation was less than ±100 m in 1993 and ±20 m in 1998—2000. The RMS error of gravity values was ±5.0—5.5 mGal.

Later, the Russian geophysicists run similar investigation at geotraverses "Artica-2005" and "Artica-2007" (Kaminsky et al. 2014), located at the convergence of Mendeleev Rise and Lomonosov Ridge, accordingly, with the continental shelf. In 2005, gravity meters GAMS, GSD-M and SIEL-1300 were used, in 2007 – gravity meter "CHEKAN-AM". Both programs used GPS NAVSTAR for navigation. The positioning and gravity anomalies were accurate within ±50 m and ±3,5 mGal in 2005 and ±100 m and ±3 mGal in 2007, accordingly.

One of the recent, and most detalied so far (1:200000 scale), airborne gravity survey was run over coastal waters of East-Siberian and Chukchi Seas, using aerogravimeter GT-2A and in 2012–2013, achieving the high accuracy of gravity data (±0,76 mGal) and positioning (±0,2 m) – see Fig. 1.11. Another modern survey

(1:500000 scale) was run in 2014–2015 at the northern part of the East-Siberian Sea using GPS Navstar navigation. The positioning and gravity RMS errors were estimated at 3 m and 1 7 mGal.

In all, the Russian organizations accumulated about 62,500 discreet points of airborne gravity observations, run more than 200,000 km of ship-borne gravity observations along marine traverses and recorded the gravity data over more than 150,000 km of profiles. The raw data of all types of surveys mentioned above are kept in digital form in the proprietary catalogs of organizations and companies which have ran the surveys. Generalized geographical distribution of the existing gravity data is illustrated by the Fig. 1.11.

This data were repeatedly compiled into analog gravity maps at the scales form 1:1000000 to 1:6000000 (Glebovsky et al. 2008).

In the late 90th the available analog gravity datasets were entered into digital databases, unified, digitized and interpolated to a regular 10x10 km grid. This grid was used for compilation of the first Russian map of Free-Air gravity anomalies of the Arctic Ocean open for geological interpretation (Glebovsky et al. 2002, 2008). Lately, this grid was gradually updated by adding new gravity data from geotraverses "Artica-2005" and "Artica-2007". Several new digital datasets and gravity maps for selected sheets of the latest generation of the "State Geologiacl Map at 1:1000000 scale" (GTK-1000/2 and GTK-1000/3) were also utilized (Glebovsky et al. 2012; Kaminsky et al. 2014).

The significant volume of gravity data acquired prior to mid-2010th was utilized in several international projects dealing with compilation of unified digital gravity database and gravity maps for the Arctic Ocean and the Arctic Region as a whole. Brief description of these projects can be found below.

1.2.1.4 Aeromagnetic Surveys

Beginning from the end of 1950th, the Russian aeromagnetic surveys were flown over the greatest part of the Arctcic Ocean. Experimental research and development flights over widely spaced grid of traverses gradually transformed into small-scale areal surveys. They peaked in 1961–1964 and marked the beginning of systematic investigation of the Artcic Ocean. The Research Institute of the Geology of Arctic the (NIIGA) ran the operations in conjunction with the Navy's Polar Hydrographic Expedition, all as a part of High-Latitude Airborn Program "North" ("Sever"). The aeromagnetometers AM-13 and AEM-49 were used. Diurnal variations were recorded at stations located at the ice camps or on land. The state of technology could not produce high accuracy surveys at that time. Approximate astronomical observations, dead reckoning and, sporadically, aero photography, could provide positioning within several, or even tenth, kilometers. The first airborne radio geodetic system Rym, developed in 1949–1953, did not improved navigation accuracy. Nevertheless, it was possible to picture the major elements of magnetic field over almost entire conditional Soviet sector of the Arctic Ocean (around 2,500,000 sq.km), not only within shelf, but also in the deep-water parts.

The following years witnessed the sharp increase in numbers and size of aeromagnetic surveys over polar seas. As a rule, they were flown at 300–600 m altitude and later, in 1970th – 100-300 m (overland, or during high risk of icing, the plains climbed to 1–1.5 km). The surveys designs and specifications as a function of technical support and distances flown, were also constantly developed and improved. The original ferromagnetic magnetometers (AM-13 and AMM-13) were replaced first by proton (PPM, AMP-7 etc), then quantum (KAM-28, AKM-01 etc) magnetometers and, eventually, by specialized magnetometric stations (BMK-95 and others).

The initial positioning system RYM was augmented by several models of newly developed system "Poisk"; the number of diurnal registration station, located mostly on land, grew up to four. As a result, the accuracy of navigation grew from kilometer to fist hundreds of meters, and of survey – from 65–85 to 10–15 nT.

Further improvement came in late 1980th with introduction of satellite navigation reducing positioning errors to 50–100 m and magnetic measurements – to 3–10 nT.

The aeromagnetic surveys tied to the geotraverses "Transarktika 1989–1991", "Transarktika 1992", "Arctic 2005" and "Arctic 2007" are among the best high–resolution surveys flown over the abyssal parts of the Arctic Ocean. They cover swaths along of seismic profiles between De Long Islands and Makarov Basin, Amundsen and Makarov Basin, as well as junctions of abyssal Makarov Basin and Lomonosov Ridges with continental shelves of Siberia and Chukotka.

High-resolution areal magnetic on shelf surveys were flown over Franz-Joseph Land Archipelago (2000), in central Kara Sea (2000) and southern parts of Laptev, East-Siberian and Chukchi Seas (2012–2013). The aeromagnetic maps at the scale 1:200000 of the latter group was presented together with aerogravity maps of the same scale mentioned in the previous chapter (Fig. 1.12).

In total, more than 2 million kilometers of aeromagnetic data was acquired by the Russian geophysicists covering large part of the abyssal Arctic Ocean and, practically speaking, the entire continental shelf. However, the areal data distribution turned out to be very irregular (Fig. 1.12). The best studied are Barents Sea and south-west of Kara Sea, island archipelagoes an adjacent waters. Laptev. East-Siberian and Chukchi Seas are the worst. Here the density of observations are even less then in deep-water parts of the Arctic Ocean. North-eastern Kara Sea remains practically untouched.

The veracity of aeromagnetic data strongly depends on accuracy of data measurements and geographical positioning. Old data acquired prior to 1990th and displayed graphically mostly as a fence-diagrams of ΔT profiles, have low accuracy both of data readings (10–90 nT) and positioning (from one to tens kilometers). Characteristically, the worst data are concentrated in the least studied eastern parts of the Russian Arctic (Laptev. East-Siberian and Chukchi Seas).

The available data were periodically compiled, integrated and generalized into regional maps of magnetic anomalies, analog at first, and later – digital. One of the largest among them is unpublished "Generalized contour map of magnetic anomalies of the Arctic Ocean at the scale 1:6000 000", compiled by V.V. Verba and

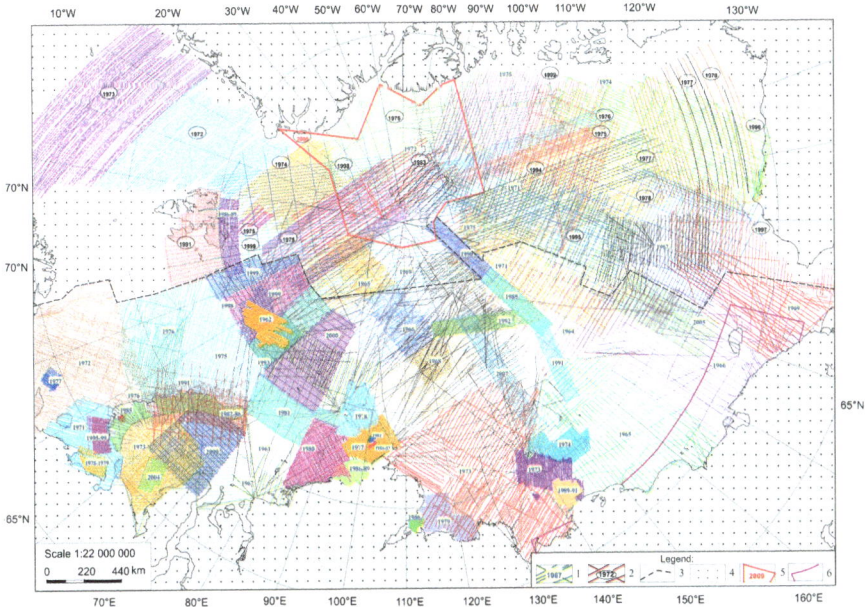

Fig. 1.12 Aeromagnetic data coverage of the Arctic Ocean (public domain). 1 - Russian sources; 2 - National sources; 3 - Overlaps of various sources; 4 - areas of CAMP-GM Project (Gaina et al. 2011) digital compilation; 5 - LOMGRAV - 2009 (Denmark-Canada) survey (Matzka et al. 2010; Døssing et al. 2013); 6 - "AeroGeophysica" aeromagnetic survey (2012–2013, 1:200000 scale), northern boundary

V.E. Volk in 1988. This map was later digitized and included in the first digital international model and map of magnetic anomalies of the Arctic, NorthAtlantic and adjacent territories, created by the Atlantic Division of the Geological Survey of Canada (Verhoef et al. 1996).

In the following years, the analog profile magnetic data were digitized, reconciled and reprocessed. The resulting database were included in number of contemporary international projects(see below) aimed at creating a new generation of regional digital models and maps of anomalous magnetic field of the Arctic Ocean and the Arctic region as a whole.

1.2.1.5   Hydromagneteic Surveys

Ship-borne hydromagnteic surveys were run using research vessel "Geophysik" in the second half of 1990th - early 2000th and helped to clarify the patterrn of magnetic anomalies in the Barents, Kara and Laptev Seas. The surveys were run in differential mode using the magnetometric station MAMP-01 with two automatic magnetometers, marine differential magnetometer MDM (later – magnetometer SeaSpy) and GPS navigation with positioning accuracy within 50 m (1990th) and

2–5 m later in 2000th. Analysis of tie-lines intersections demonstrated instrumental RMS error within 1.5–4.5 nT. Some of the marine magnetic data were included in the digital models and aeromagnetic maps inclided as attachments to the sheets of the attached to the new generations of the "State Geologiacl Map at 1:1000000 scale" (Glebovsky et al. 2012; Kaminsky et al. 2014).

1.2.2   National Survey Programs

1.2.2.1   Gravity

The various national organizations started systematic gravity works in the Arctic in the early 1960th and went through the same stages as their Russian counterparts: on land and ice at first, then switching to landing airborne and ship-borne methods later in the 70th and, in 1980th, uilizing submarines, planes and satellites.

Gravity observation on ice have produced remarkably valuable data. Among these semi-permanent stations the most productive were:

- ARLIS II (US Navy, 6800 points, 1961—1965);
- FLETCHER (USA, La Mont Observatory, 9000 points, 1962—1974);
- LOREX-1979 (Canada, Geological Survey, 260 points 1979);
- FRAM I-IV (Norway Polar Institute, USA, La Mont Observatory, 1979—1982);
- CESAR (Canada, Geological Survey, 1610 points, 1983).

The accuracy of gravity observations steadily increased from 4 mGal in the 1960th to 1–2 mGal in 1970—1980th. Positioning errors dropped off drastically depending on the distance from fixed base from hundreds meter to more than ten of kilometers.

One of the largest ship-borne gravity surveys was run by US Geological Survey offshore Alaska, in Chukchi and Beaufort Seas in 1970th acquiring close to 55,000 km of gravity profiles with RMS error of 2 mGal.

As a result, almost half of the Arctic region was covered gravity observations. Later, the generalized gravity map at the scale 1:6000000 was compiled from 57 separate 1:1000000-scale gravity contour maps produced by geophysicists of USA, Canada, Denmark, Finland and France (Sobczak et al. 1990). This gravity map (Free Air gravity offshore, Bouger - overland) became the most detailed representation of anomalous gravity field of the Arctic Ocen, compare to the previous works (Bowin et al. 1982).

The spatial distribution of gravity observations, on which this map is based, is rather uneven. Most of the data are concentrated in continental margins of Canada, Alaska and Beaufort Sea. Less data are available for the Alpha Ridge and near-pole part of the Lomonosov Ridge and basins adjacent to it, even less – for Canada Basin and Chukchi Plateau. Northern part of Mendeleev Ridge and Lincoln shelf remained totally uncovered.

One of the most important underwater submarine-based gravity and bathymetry survey in the Arctic Ocean was conducted by USA under the SCICEX (Science Ice Exercises) program in the late 1990th. The observation profiles of this program, stretching from the North American continental margin to the Nansen Basin, crossing the principal structures of the Amerasin Basin and Lomonosov Ridge. In 1996 and 1988, the submarines "Pogy" and "Hawkbill" concentrated on the Gakkel Ridge, covering its axial zone between meridians 5E and 75E and 5th magnetic anomalies across. Using sonars SCAMP and SeaMARC, the structure of 100–200 m-thick section of sedimentary layers was studied. Obtained gravity and bathymetry were used for estimates of density and thickness of the crust (Coakley and Cochran 1998). Similar investigation was conducted in 1999, covering all three segments of the Lomonosov Ridge (Coakley et al. 1999).

US Navy Research Laboratory (NRL) run simultaneous aerogravity and aeromagnetic surveys from 1992 till 1999 (Brozena et al. 2003). The program covered major part of western half of the Eurasian Basin with the Lomonosov Ridge and Amerasian Basin with Canada Basin and Chukchi Plateau. At the same time, the Danish agency KMS flew aerogravity in Lincoln Sea, north from Greenland coast, adjoining the American survey (Forsberg et al. 2001). The data, accurate within 2–3 mGal, were merged into unified database, interpolated to a regular grid and published as a contour Free Air gravity map (Brozena et al. 2003).

Further combined aerogravity and aeromagnetic investigations continued jointly by Denmark and Canada under the project LOMGRAV in 2009 covering around 550,000 sq. km over Lomonosov and Alpha Ridges, Morris Jessup Rise and Ellesmere Island shelves (Fig. 1.12). In all, 58,000 km of survey profiles, offset by 12–15 km (and tie-lines – by 255–300 km), were flown using DC3 Basler plane, equipped with SL-1 and Lacoste-Romberg S gravimeters, magnetometer Geometrics G-823 and dual-frequency GPS receivers, at altitude 600–700 m. RMS errors of aerogravity and aeromagnetic data were estimated at ±2,4 mGal and ±0,39—1,82 nT, accordingly (Døssing et al. 2013, 2014). This high-quality information was integrated with previous aerogeophysical works (Brozena et al. 2003), refining the existing digital datasets to a 2.5 × 2.5 km grid and revealing the finer details of structure and history of the region.

The satellite altimetry data, as an indirect contributor to the study of Polar region gravity field, also must be mentioned. The breakthrough in using these data came after development of algorithm for processing these data in ice-covered waters of the Arctic Ocean (Laxon and McAdoo 1994, 1998). Using the altitude information recorded by European Remote-sensing Satellites ERS-1 and ERS-2 launched in 1991 and 1995, accordingly, the authors produced the gravity anomaly map covering parts of the Arctic Ocean to latitude 82°N. The resolution of this map, especially for the long wave-length part of the spectrum (more than 35 km), turned out to be comparable with that of the aerogravity maps.

Launched in 2003, the American satellite ICESat (Ice, Cloud, and land Elevation Satellite) with high-resolution laser altimeter GLAS, designed for measuring ice thickness, ocean surface altitude, extended the coverage to 86°N. Even newer satellites (Jason-1 and 2, Cryosat-2 and Saral-AltiKa), designed mostly for environmental studies, also had a GPS–controlled system of geographical positioning of the recorded altitude data. It reduced the RMS error of altitude and gravity derived from it, thus refining existing models of gravity field in Polar regions. An important feature of the satellite Cryosat-2 - its ability to differentiate between the reflections from ice-covered and from open waters – made possible to use only the latter to transform them to gravity values.

The latest version of Artic Seas gravity database (DTU13) was created in Denmark State Cosmic Institute (Andersen et al. 2014). It is based on radar altimetry by satellites ERS-1, Jason-1 and 2, and Saral-AltiKa in 1 Hz interval and multichannel database from Cryosat-2, recorded along orbital trajectories at 6 km increment. This database was later reprocessed by "Nordic Geoscience Pty. Ltd" to a Free Air gravity model NORDIC13 at $1' \times 1'$ grid. This grid covers the Arctic Ocean between 65^0 and 90^0 N. The gap 88^0 and 90^0 N without any satellite data was filled with data from the Arctic Gravity Project (ArcGP).

It was shown that, compare to marine gravity data (Christensen and Andersen 2015), NORDIC13 model is more accurate with RMS error $\pm$ 1.0–2.0 mGal. The further analysis of this model (Christensen and Andersen 2016) demonstrates that NORDIC13 also is superior to ArcGP model. Therefore, it is obvious that the satellite data made a significant contribution to our knowledge and understanding of gravity field of the Arctic Ocea, especially of its abyssal part and, being more accurate, refined some older gravity data.

1.2.2.2 Magnetics

In the 1950th, US and Canada began experimental high-altutude (3–6 km a.s.l) aeromagnetic surveys in the Arctic, switching in the following decade to the regular investigations under large projects "Magnet" (1960–1963) and "Artcic Basin" 1961–1964). The attempts to increase accuracy of data by flying at lower altitude ($\approx$ 450 m) were thwarted by the huge positioning error reaching $\pm$10—15 km. Therefore, the only very general picture of the anomalous magnetic field was developed.

Thanks to technological progress (mainly in positioning) in 1970th, this decade witnessed a noticeable increase of both data quality and geographical reach of investigations. The surveys expanded into the abyssal part of the Arctic Ocean. Flown mostly by US Navy at 150 m a.s.l., with the positioning as good as $\pm$1–2 km (however, occasionally as bad as $\pm$5 km), these regional aeromagnetic surveys, for the period from 1972 to 1978, covered Norwegian and Greenland Basins in the Atlantic and adjacent half of Eurasian Basin, Lomonosov Ridge, Lincoln Shelf Canadian Basin and Alpha Ridge in the Arctic.

The data were processed, reprocessed and compiled into generalized maps of separate regions of the Arctic Ocean on the regular basis. Among them are:

- Composite map of residual magnetic anomalies of Arctic Canada and adjacent waters (Riddihough et al. 1973);
- Fence-diagram maps of ΔT profiles from the papers by (Vogt et al. 1979; Taylor et al. 1981);
- Fence-diagram maps of residual magnetic anomalies of the Arctic Ocean (Kovacs et al. 1984, 1987);
- Map of magnetic anomalies of Norway and adjacent oceanic basins (Olesen et al. 1997), etc.

Due to a constant influx of new data, the value and veracity of composite maps grew exponentially. However, all of them suffered from common flaws hampering the process of compilations of the different databases: erroneous orientation of flight paths and their uneven spacing; presence of false anomalies due to undocumented diurnal variations, etc.

Nevertheless, in the early 80th, the geologists and geophysicists started to understand the complex nature of the Arctic Basin where the structures of both oceanic and continental origin exist side by side, with specific patterns of magnetic anomalies peculiar to them.

The beginning of 1990th marked noticeable acceleration of development of new digital technologies of registration potential fields data and satellite-based navigational support of geophysical investigations. From 1992 to 1999, US Naval Research Laboratory (NRL) completed several carefully planned and executed aeromagentic and aerogravity programs in the Arctic Ocean. The on-board registration complex consisted of two modified marine Lacoste-Romberg gravity meter, proton-precession magnetometer and radar altimeter. GPS satellite navigation provided 3D coordinates of a flight pass accurate within ±1 m (Brozena et al. 2003).

The programs eliminated all remaining "white spots" from the data coverage and re-surveyed significant part of the western half of the Eurasian and Amerasian Basins including Lomonosov Ridge and Chukchi Plateau, considerably improving the reliability of newly compiled composite maps.

The next large project - already mentioned Canadian-Danish LOMGRAV, (Fig. 1.12) - was competed in 2009. It also re-surveyed the areas of older (1970–1990) investigations, producing high quality data and drastically refining the existing magnetic maps in one of the most disputable, from tectonic and evolutional point of view, region north from Greenland and Ellesmere Island, as well as the entire Arctic Ocean.

While the total mileage of all aforementioned national surveys is comparable to that of the Russian projects of 1960–1980 (Fig. 1.12), the latter are still mostly available only in analog form, while the majority larger part of the former are presented as digital databases of high-quality data.

All publicly available potential fields data sets were used by several international projects for compilation of regional composite digital maps of potential fields of the Arctic.

1.2.3   International Projects

Growing international cooperation in understanding the geology of the Arctic Ocean in the 1990th stressed the urgent need to integrate all available geological and geophysical information, especially gravity and magnetic, into unified digital database. As was mentioned above, the American and Canadian regional digital compilations covered mostly western parts of the Arctic Ocean, while the Russian data, mainly in analog form, were concentrated in the East. Both overlapped each other in the central abyssal part of the Arctic Basin (Figs. 1.11, 1.12).

1.2.3.1   Gravity

The first international digital 10 × 10 km grid an contour map of Free Air gravity anomalies was jointly compiled by VNIIOkeangeologia and NRL (Brozena et al. 1999), using the Russian, American and Canadian data. The additional gravity information from satellite altimetry (Laxon and MacAdoo 1998) was included as well. This compilation had a serious shortcomings due to poor positioning and was only partially utilized by subsequent international projects.

The following are the most fundamental international compilations of the Arctic gravity data made with Russian (chiefly VNIIOkeangeologia and VSEGEI) participation:

1. 5' × 5' grid of Free Air gravity by the Arctic Gravity Project, ArcGP (www.earth-info.nga.mil/GandG/wgs84/gravitymod/). This model was based on variety of sources: airborne, ship- and submrine-borne, aerogravity and satellite altimetry (Kenyon et al. 2008). This compilation, encompasses not only the all northern waters, but also the adjacent territories north from latitude 64°N, including North America, Greenland, Island, Russia and Scandinavia. ArcGP grid is constantly refined and is considered as the most complete Arctic gravity dataset open to regional geological interpretations. The gravity information for the Russian part of the Arctic ocean was digitized from the open "State map of Free air gravity map gravity map at the scale 1:1 000 000". Gravity data for continental Russia came from the TSNIIGAIK gravity database (Fig. 1.13).

2. Digital 10 × 10 km grid and corresponding gravity map at the sale of 1:5000000 created by Geological Survey of Norway (Gaina et al. 2011), under the Circum Arctic Mapping Project - Geophysical Mapping (CAMP-GM). This map covers not only the oceanic part of the Artic (Free Air gravity), but also its continental part south up to 60°N (Bouguer gravity at 2.67 g/cm^3). Gravity for the continental Russia was digitized from analog contour gravity maps of USSR at the scale 1: 2500000 (Stepanov and Yanushevich 1999), which, in turn, was a compilation of previous Russian gravity surveys of 1: 1: 200000 and 1: 1000000 scales (Litvinova and Glebovsky 2008).

Further refinement of gravimetric information included several steps:

– careful comparative analysis of all available national and international gravity databases;

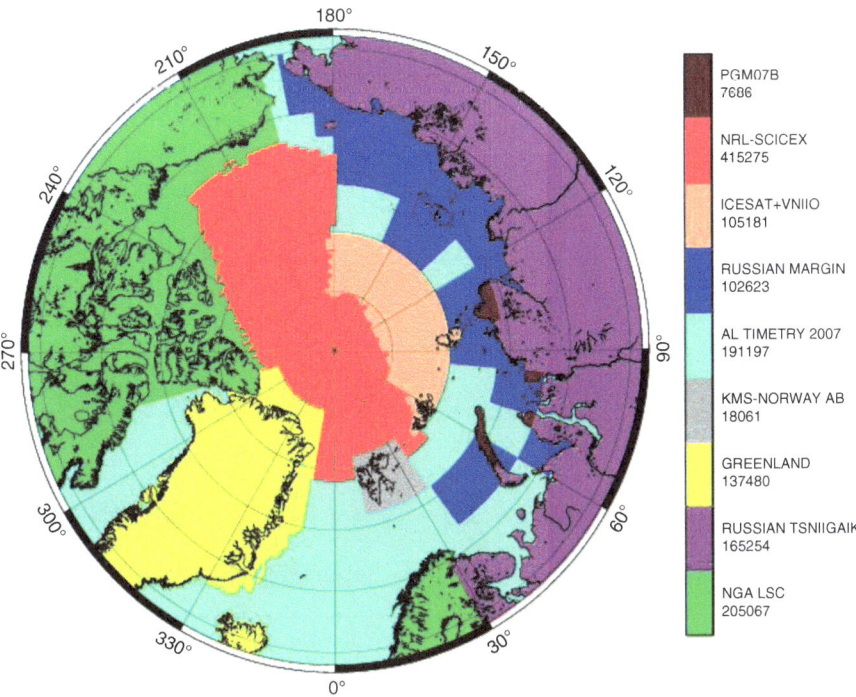

Fig. 1.13 Digital databases used by ArcGP (www.earth-info.nga.mil/GandG/wgs84/gravitymod/)

– selection of new gravity data acquired after ArcGP and CAMP-GM Projects and their reduction to the International Normal Gravity field;
– calculation 3 × 3 km grid of Free Air gravity values based on ArcGP database and later Russian surveys;
– recalculation of refined Bouguer gravity (2.67 g/cm^3) grid with 3D terrain correction using IBCAO v 2.23 grid (Jakobsson et al. 2008);
– compilation of composite 3x3 km digital grid of the Arctic Ocean (Fee Air gravity offshore, Bouguer gravity (2.67 g/cm^3) overland).

The most relevant map of at the scale 1: 5000000 based on the international Projects ArcGP and CAMP-GM and new Russian data (shelf and adjacent territories) is shown on Fig. 1.14.

1.2.3.2   Magnetics

The earliest digital magnetic field map of the Arctic Ocean, North Atlantic and adjacent territories, based on 5x5 km grid, was compiled in 1996 under the international GSC-A (Geological Survey of Canada-Atlantic) Project (Verhoef et al. 1996). VNIIOkeangeologia, from Russian side, and NRL from American, provided the

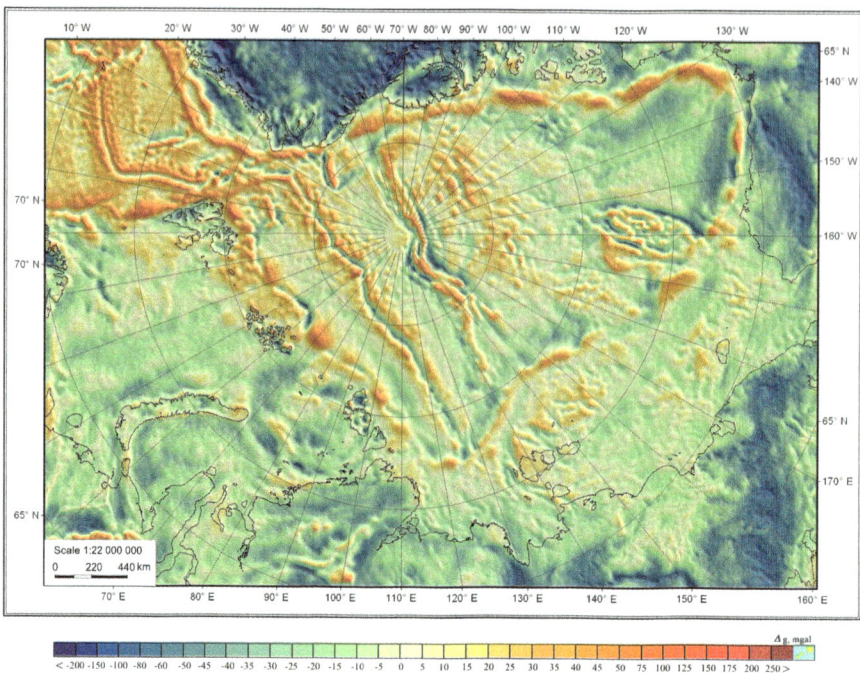

Fig. 1.14 The gravity field of the Arctic Ocean (Fee Air gravity - offshore, Bouguer gravity (2.67 g/cm^3) overland) (Kaminsky et al. 2017)

bulk of information for the central part of the Arctic Ocean, with contribution from Canada (GSC) and other countries. The Russians data were digitized from analog magnetic contour map at 1:6000000 scale (Verba and Volk 1988); the Americans and Canadians - original and digitized data of 1970th vintage.

The sharp differences in the character of the magnetic anomalies depicted by the Russian and American datasets became evident from the very beginning (Macnab et al. 1992). It was also found that, after digitizing and subsequent gridding a small scale map, the Russian magnetic anomalies were shifted to longer wavelength. Nevertheless, this first international grid and map remained the only principal source of digital information for geological interpretation for years to come.

The process of refining the magnetic field model of some parts of the Arctic Ocean continued under the joint Russia-American (VNIIOkeangeologia and NRL). It concentrated on reconciling the numerous aeromagnetic profiles flown by respective organizations. The first results demonstrated that the final product was more reliable and more detail than the original blocks of information (Kovacs and Glebovsky 1993; Glebovsky and Kovacs 1993). It was decided to extend the study to cover the entire and include new American aeromagnetic and aerogravity (1990–1999) surveys covering large portions of the Amerasian, Norwegian-Greenland and western half of the Eurasian Basins, (Brozena et al. 1999), Fig. 1.12.

Next step included the reprocessing of gravity and magnetic data for the Russian shelf run under the research project GRASS (Geology of Russian Arctic Shelf Seas, 1996) jointly operated by VNIIOkeangeologia and Exxon.

Realization of above projects started from digitizing all original analog potential fields data available for the abyssal Arctic Ocean and surrounding territories. Then a composite digital database was created. Being constantly updated, this database currently includes practically all American aeromagnetic data (Brozena et al. 1999, 2003). Missing from the database are aeromagnetic data from Canadian-Norwegian Project LOMGRAV (2009).

After careful analysis, reconciliation and leveling of all blocks of aeromagnetic data, described in (Glebovsky et al. 2002), the contents of the database was interpolated into 5 × 5 km regular grid, a foundation for the new generation of composite gravity and magnetic maps covering the entire Arctic Ocean. This maps were demonstrated in international meetings and published (Glebovsky et al. 1998; Kovacs et al. 2002; Glebovsky et al. 2002).

A considerable part of the above described database was included into compilation of the new digital circumpolar map of magnetic anomalies under Project CAMP-GM by the Geological Survey of Norway (Gaina et al. 2011). This map, based on 2 × 2 km grid, and utilizes all magnetic data available by year 2000, covers not only the oceanic part of the Arctic, but also its continental part south up to 60°N. Continental Russia is represented on this map by 5 × 5 km grid created from digitized contour magnetic map of 1:2500000 scale (Makarova 1978), shelf and transition zone "shelf-ocean"–by the grid of the same size calculated by VNIIOkeangeologia. Currently, the CAMP-GM map is widely used for geological interpretation.

From the beginning of twenty-first century, influx of new digital data and new methods of geophysical data reconciling, substantially refined the Project CAMP-GM map within the Russian domain (Fig. 1.11). The integration of this new aeromagnetic information (Kaminsky et al. 2017), similar to gravity surveys above, included several steps:

- reprocessing of all existing CAMP-GM and new Russian-acquired data using Microleveling technology, developed by GETECH (Leeds, GB) (Chernykh et al. 2015);
- calculation of a new magnetic field digital grid based on normal magnetic field model and geodetic projection used in CAMP-GM model;
- compiling the refined model of Circumpolar Arctic by merging new grid with CAMP-GM model (with preference to the former).

Fig. 1.15 demonstrates the final map of the magnetic anomalies of the Circumpolar region based on 2 × 2 km grid.

1.3   Multi-Channel Seismic Surveys (MCS)

Fig. 1.16 illustrates the major seismic programs undertaken in the Eurasian and Central Amerasian Basins over the last 45 years.

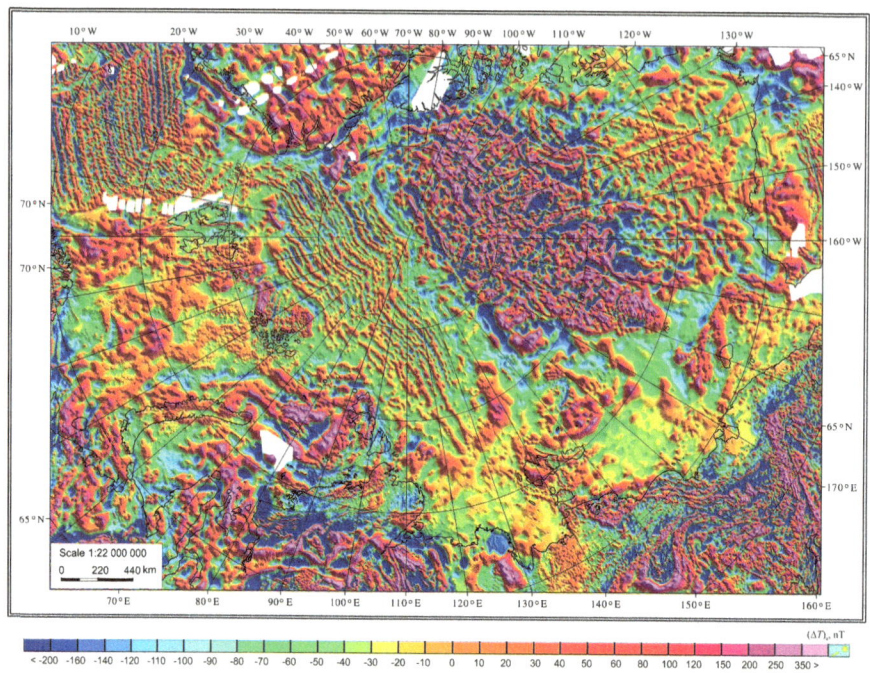

Fig. 1.15 Final map of the magnetic anomalies of the Circumpolar region based on 2 × 2 km grid (Kaminsky et al. 2017)

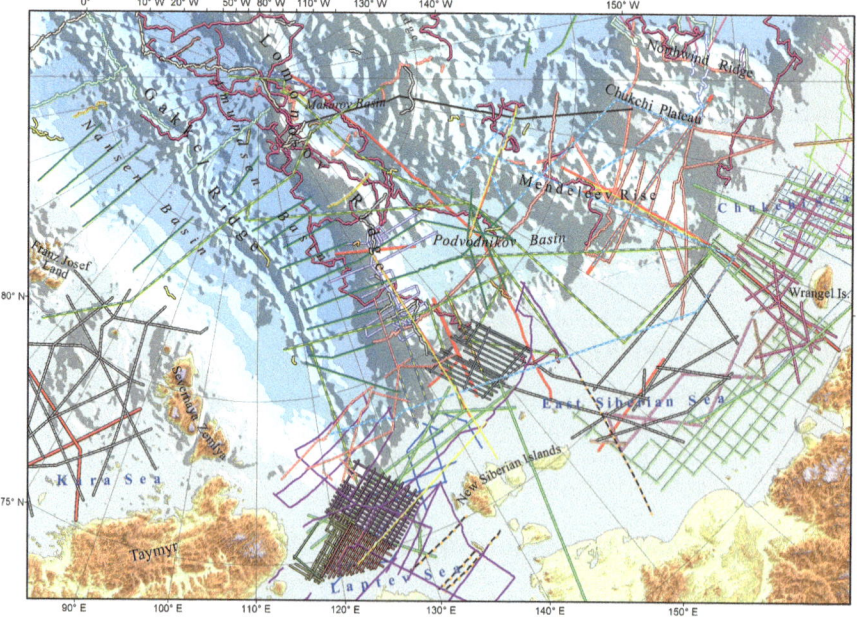

Fig. 1.16 Refraction Deep Seismic Sounding (DSS) and reflection Multi-Channel Seismic (MCS) surveys in the Eurasian and Central Amerasian Basins

Profiles Arktika-2000, 2005, 2007 (Russia, VNIIOkeangeologia, MAGE)

────── Profiles Arktika-2011 (Russia, GNINGI)

· · · Profiles Arktika-2012 (Russia, Sevmorgeo)

═══ Profiles Arktika-2014 (Russia, MAGE)

═══ Drifting stations (Russia, VNIIOkeangeologia, PMGRE)

═══ Expedition "North" (Russia, VNIIOkeangeologia, PMGRE)

═══ Transarctica 89-92 (Russia, VNIIOkeangeologia, PMGRE)

Profiles MAGE (Russia):

────── *1986 -1988, 1990* ═══ *2005* ═══ *2006* ═══ *2007* ═══ *2009* ═══ *2011-2012* ═══ *2014-2015*

- - - LARGE, 1989 (Russia)

Profiles DMNG (Russia):

········ *Project SC-90, 1990*   ═══ *Project ARS-10, 2010-2012*   ═══ *Project ESS-91, 1991*   ──── *Project ESS-10, 2010-2012*

Profiles BGR (Germany) and SMNG (Russia):

────── *1993* ────── *1994* ────── *1997*

Aliance (Russian Academy of Sciences):

═══ *2011* ═══ *2012* ═══ *2013*

──── ODEN-1996, 2001 (Sweden-Norway)

Profiles AWI-1991, 1998, 2001, 2008 (Germany):

═══ *1991*   ═══ *1998*   ═══ *2001*   ═══ *2008*   ═══ *2014*

═══ Profiles ION-GXT, 2006 (USA)

Profiles in the Canada Basin (LSSL, USA, Canada):

──── *2009*   ──── *2010*   ──── *2011*

═══ HEALY-2005 (USA)

Profiles in the Chukchi Sea (USA):

1978   ────── *1981 (WCS)*   ────── *1982*

Profiles Northwind (USA):

═══ *1988* ═══ *1992* ═══ *1993*

═══ LOMROG, 2007-2009 (Denmark-Canada)

━━━ WAR-Refraction profiles, 1989-1991, 1992, 2000, 2005, 2007, 2009, 2012, 2014
(Russia, VNIIOkeangeologia, Sevmorgeo, PMGRE, MAGE)

━━━ WAR-AP (Russia, Sevmorgeo, MAGE)

═══ LOREX-1979 (USA-Canada)

Fig. 1.16   (continued)

The Russian seismic investigations started at the drift-ice camp NP-13 (short for North Pole) in 1965, along its drift track over the East-Siberian Shelf and its continental slope. Later, between 1973 and 1989, the MCS surveys were run from the following ice camps (Figs. 1.17, 1.18):

- NP-21 (Amundsen Basin, Lomonosov Ridge), 1973;
- NP-22 (Canada Basin, Alpha Ridge, Northwind Ridge, Chukchi Plateau, East-Siberian Shelf, Amundsen Basin), 1979–1983;
- NP-23 (Lomonosov Ridge and Amundsen Basins), 1978;
- NP-24 (East Siberian Shelf, Lomonosov Ridge, Amundsen Basin, Gakkel Ridge), 1979–1980;
- NP-26 (Podvodnikov Basin, Mendeleev Ridge, Canada Basin), 1983–1985;

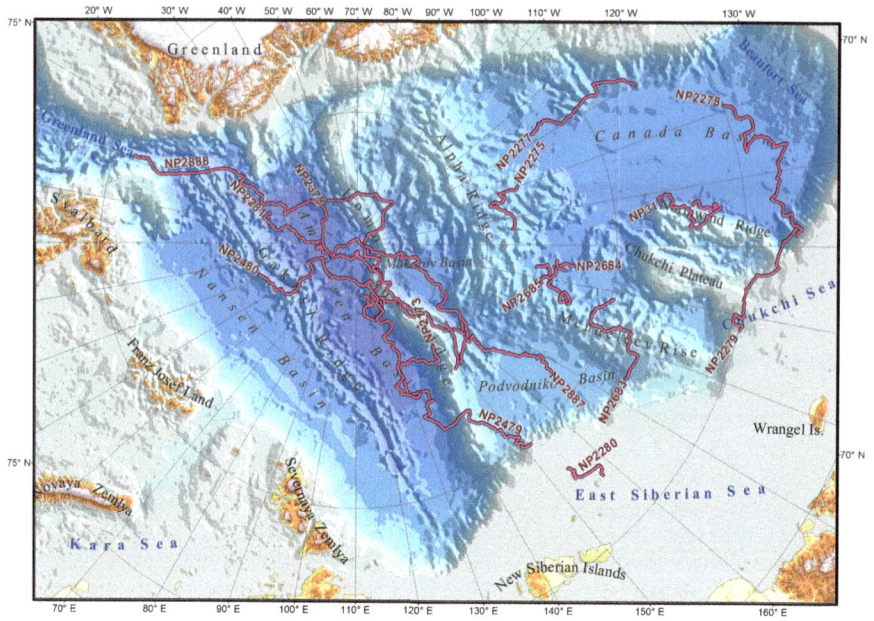

Fig. 1.17 Drifts of the Russian ice stations and tracks of station-based seismic surveys

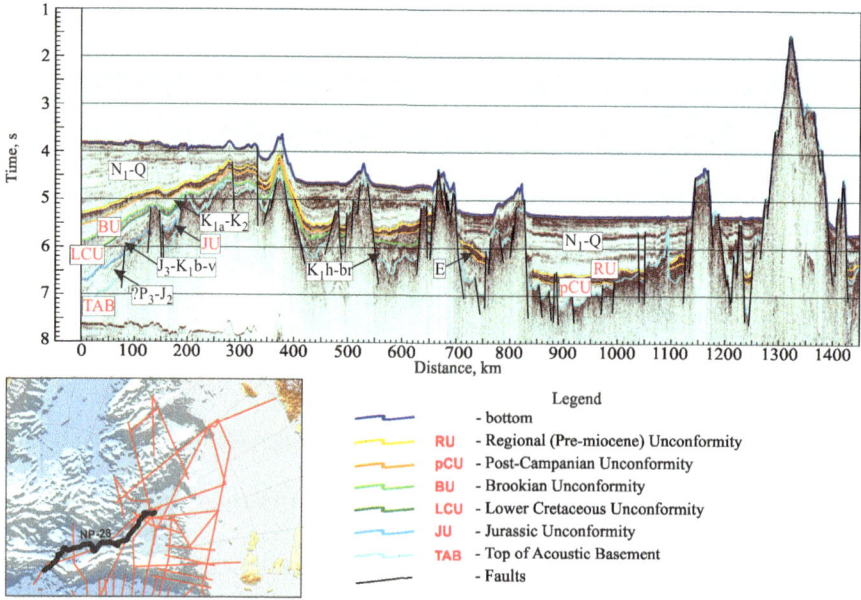

Fig. 1.18 An example of MCS section along SP-28 drift track

- NP-28 (Podvodnikov Basin, Makarov Basin, Lomonosov Ridge, Amundsen Basin), 1987–1989;
- NP31 (Canada Basin, Northwind Ridge), 1988–1989.

Used on majority of the ice stations, the astronomical observations could fix the station coordinates only within ±1.0 km. However, starting in 1983 (stations NP-26, NP-28 and NP-31), the satellite navigation system "Transit" provided the coordinates with accuracy ±0.3 km.

Considerable volume of airborne and ice camp-based reflection seismic was done under the High Latitude Expedition "Sever" ("North") in 1960–1980. Due to classified nature of operations, obtained information was not available for scientific research.

The American and Canadian seismic investigations also started from drift-ice stations in the 1960th.

Station ARLIS II (1961–1962) In this period, the station drifted in Canada Basin between the Lomonosov and Alpha Ridges, collecting bathymetry, seismic and magnetic data (Kutschale 1966).

Station CESAR (1983) About 90 km of single-channel refraction seismic were recorded (Jackson et al. 1986), identifying the high- velocity interface within the lower crust and the Moho discontinuity at 40 km depth. The Maastrichtian and Campanian fossils in the seabed sediments cores collected on Alpha Ridge provided the first direct indication of its age (Mudie et al. 1986). Intraplate type of volcanism within the Alpha Ridge was confirmed by samples of vesicular alkali basalts, dredged from the Ridge' slopes (Van Wagoner et al. 1986).

Station FRAM-I (1979) Multinational scientific project, drifted over Gakkel Ridge.

Stations FRAM-III (1981) and FRAM-IV (1982) Nansen Basin. Seismic observations consisted of single-ray, single fold shooting, with analog recording.

The valuable geological information, collected at the above mentioned drift-ice stations (NP-26, NP-28, project "Transarktika") in Makarov Basin and Greenland-side of the Lomonosov Ridge, became available for interpretation only after 2006–2010 (Langinen et al. 2006, 2009; Lebedeva-Ivanova and Zamansky 2006; Lebedeva-Ivanova 2010). Seismic data were interpreted using data from ACEX drilling project (Backman et al. 2006).

Icebreaker-based MCS surveys began in early 1990th (Kristoffersen and Mikkelsen 2004) (Fig. 1.19). Their technical aspects are highlighted below.

Polarstern – Arktika-91, 1991 During the joint expedition of icebreakers "Polarstern" (Germany) and "Oden" (Sweden), MCS was shot on two profiles (AWI-91091 and AWI-91090) across the Lomonosov Ridge at 88^0 N. (Jokat et al. 1992; Jokat 2005) using 300 m long, 12 channels streamer. Simultaneously, the sonobuoys were deployed to estimate refraction velocities of sediments and bed-

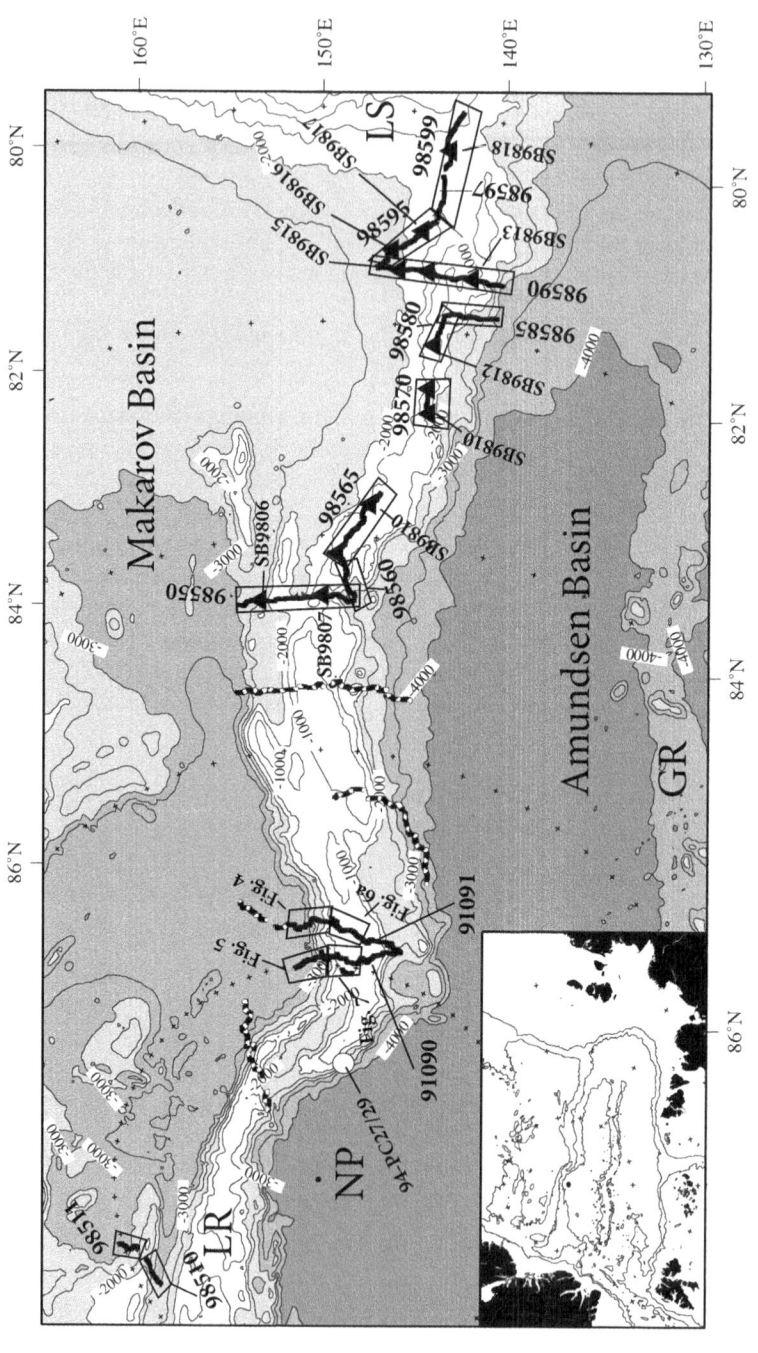

Fig. 1.19 MCS surveys with RV "Polarstern", 1991–1998 (Jokat 2005)

rocks of the Lomonosov Ridge (Jokat et al. 1992; Jokat 2005). Both seismic cross-sections presented similar seismic starigraphy and structure: upper sequence of low-velocity (< 2.2 km/s) nearly horizontal sediments 500 m thick, clearly uncon-formable to the underlying strata with velocities more than 4 km/s. The geometry of seismic reflections below the unconformity points to the Lomonosov Ridge being a part of the continental shelf facing the Makarov Basin (Jokat et al. 1992, Jokat 2005). The seismic information was also used for IODP-ACEX drill sites selection.

Oden – Polarstern, 1996 More than 700 km of MCS (16 channels streamer, 200 m long) and refraction seismic data, combined with sonar scanning, were recorded. The sonar profiling provided valuable information regarding shallow underwater erosion. It was found that the certain parts of the Lomonosov Ridge between 85°24'N and 87°17'N underwent several episodes of glacial erosion (Jakobsson 1999; Jakobsson et al. 2000; Polyak et al. 2001) removing more than 50 m of sedi-ments from the section and creating clearly visible unconformity (Jakobsson 1999). Sonar data from the deeper parts of the basin ($\geq$1km) display well stratified and undisturbed sediments.

Polarstern, 1998 (Jokat 2003, 2005) The "Polarstern" expedition, supported by the Russian icebreaker "Arktika", acquired MCS data along and across the Lomonosov Ridge between 85°N and 80°N and in the central part of the Alpha Ridge.

Based on seismic information, Jokat (2005) classified the Lomonosov Ridge as a continental margin with intermittent sedimentation and complicated internal structure. Two distinct seismic crustal units were identified:

–Low-velocity (<3 km/s) unit associated with Cenozoic and recent sediments;

–High-velocity (>4 km/s) unit representing the acoustic basement with layers deformed by Late Cretaceous or Early Cenozoic rifting events.

Up to 2.2 km of sedimentary sequence (3 s TWT) was accumulated in the abyssal Makarov Basin in proximity to the Lomonosov Ridge foothills. Acoustically strati-fied sediments are undisturbed, except in the areas closer to the ridge.

The central part of the Alpha Ridge was covered by $\approx$900 km of MCS data aug-mented by sonobuoys information (Figs. 1.20, 1.21). Several large amplitude faults were identified at the slopes and regional unconformity related to origin of the Eurasian Basin was mapped and the age of alkali basalts dredged from the ridge were dated at 82 ± 1 Ma ($^{40}Ar/^{39}Ar$) (Mühe and Jokat 1999).

AMORE-2001 The international expedition with RV "Polarstern" and USCG "Healy" investigated the western part of the Eurasian Basin in 2001, acquiring 550 km of MCS and multi-beam bathymetry (Fig. 1.22). Wide range of geological and geophysical investigations shed a new light on the Gakkel Ridge, Amundsen and Nansen Basins structure. The presence of the rift valley with up to 1 km of sedi-ments in it was confirmed by two traverses at 70°E. Up to 4.5 km of sediments was

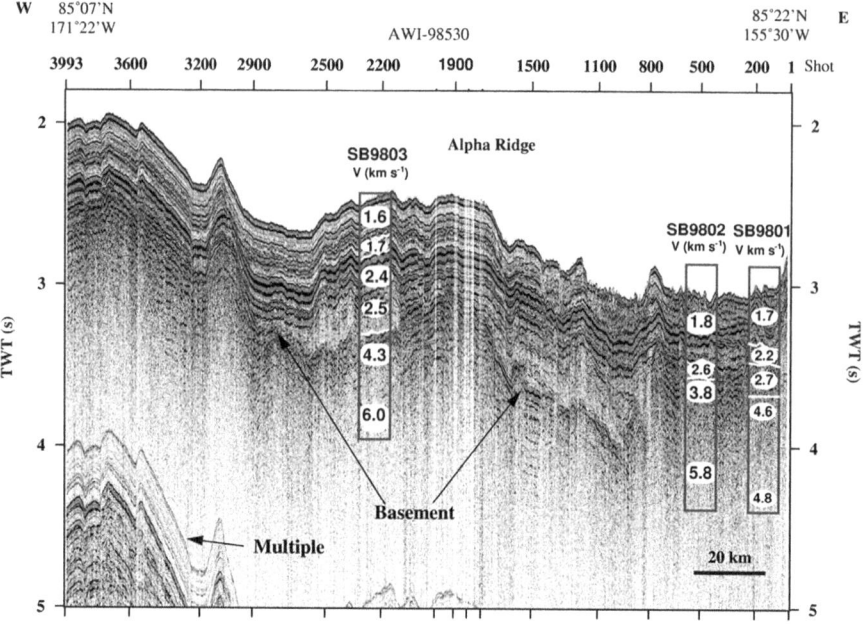

Fig. 1.20 TWT CDP section along traverse AWI-98530 (Alpha Ridge, Polarstern 1998)

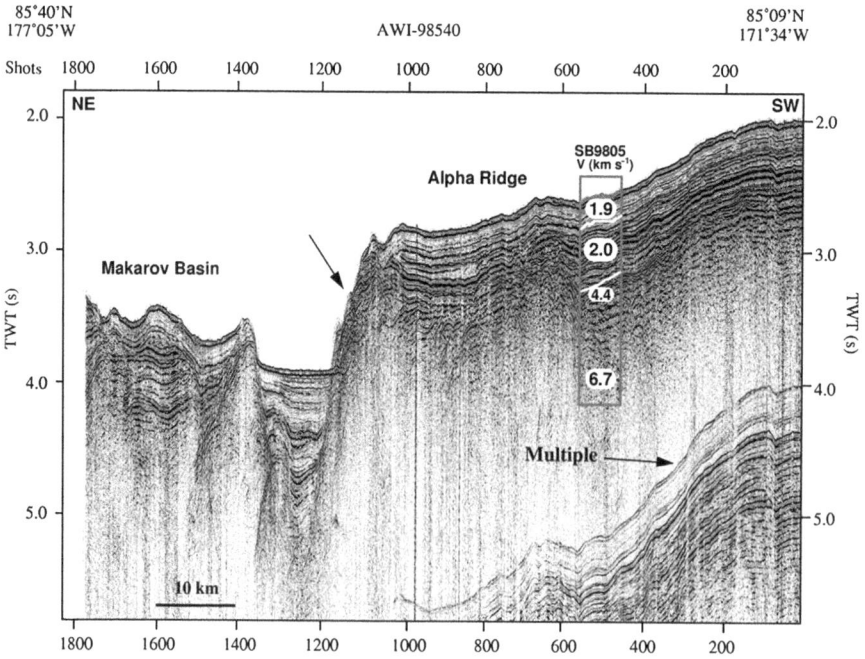

Fig. 1.21 TWT CDP section along traverse AWI-98540 (Alpha Ridge, Polarstern 1998). Arrow indicates dredging site

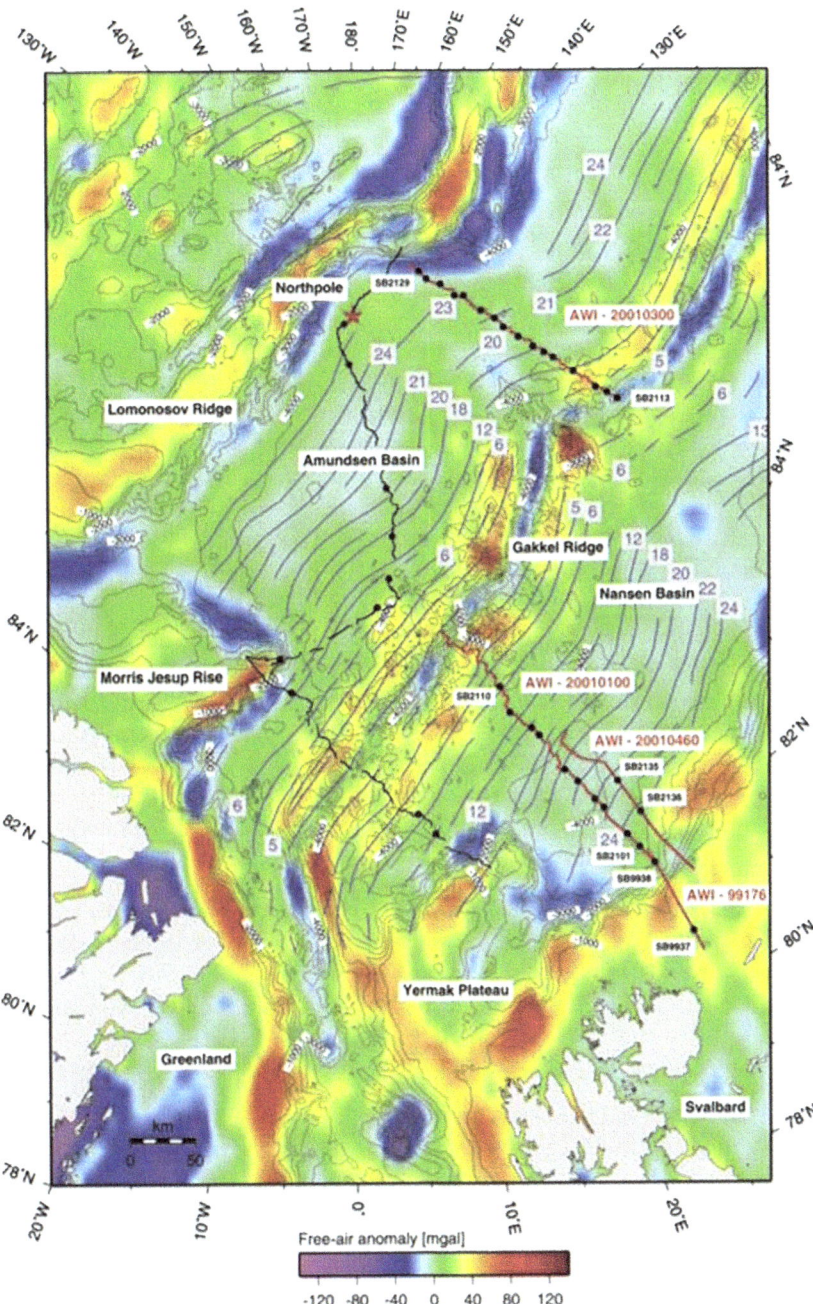

Fig. 1.22 MCS (*red lines*) and DSS (*black dots*) location map from 2001 surveys in the Amundsen and Nansen Basins (Jokat and Micksch 2004)

found in the Nansen Basin and only 1.7–2 km – in the Amundsen Basin, the difference being attributed to the distance from the source of clastic material (Jokat and Micksch 2004).

Healy, 2005 Geophysical surveys on board of icebreaker "Healy" in 2005 (1100 km in total, Fig. 1.23) brought a significant amount of new information about eastern Amerasian Basin, covering junction of Canada Basin with the Alpha Ridge, Mendeleev Ridge and Chukchi Plateau (Dove et al. 2010; Bruvoll et al. 2010, 2012).

Geophysical program included multi-channel profiling, DSS and ship-borne gravimetry (Bruvoll et al. 2012). Two 4 liters-air guns were used as a source, but only half of 24 available channels were active due to adverse ice conditions over the Alpha Ridge.

At the top of the sedimentary section, refraction sounding identified the widely spread low–velocity (Vp 1.5–2.3 km/s) pelagic sediments. Underlying sediments have velocities ranging from 2.3 km/s to 3.8 km/s, occasionally reaching 4.5–4.7 km/s at the base. It was postulated that volcanic "shell", not older than 80 Ma (Campanian), exist in the top of the Mendeleev and Alpha Ridges.

The interface with refraction velocity 2.3–4.0 km/s marks the top of acoustic basement with vertical velocity gradient of $\approx$ 1 km/s per kilometer. The seismic signal could penetrate inside the acoustic basement no deeper than 3 km. The intra-basement reflections forms three type of the seismic facies: 1 – layered continuous strong reflections with frequent abrupt loss of correlation; 2 – weak truncating and dipping reflections; 3 – segmented and often chaotic reflections. These facies may represent volcanogenic formation of interbedded basalt flows, sills, tuffs and, possibly, sediments. It was noted (Bruvoll et al. 2010, 2012) that the basement acoustic and seismic characteristics are closer to these of the Ontong Java or Kerguelen Plateau then to the standard oceanic crust or volcanic margin.

Louis S. St.-Laurent and Healy, 2007–2011 The Canada Basin became an object of geophysical investigation in 2007–2011. In 2007, 300 km of 16-channels reflection seismic were acquired from the Canadian icebreaker "Louis S. St.-Laurent". The operations continued in 2008 with another 1300 km ("Louis S. St.-Laurent"), 1500 km jointly with USCG icebreaker "Healy" and 70 DSS (Fig. 1.24). The project was completed in 2011.

New information led to better understanding of the the structure and history of the Canada Basin itself and the continent-ocean transition zone around the Northwind Ridge. It was suggested that this transition zone is a typical continental margin similar to the West Galicia/West Iberia margin or North-West Atlantic margin around Newfoundland (Grantz et al. 2011). A spreading zone of Cretaceous age was identified in the central part of the basin.

The results of recent MCS surveys (Fig. 1.25) provided the trove of information on morphology, geology and geotectonics of the Eurasian and Amerasian Basins.

Traverse A-7 In 2007, the Russain company "MAGE" run MCS survey along traverse A-7, 832.4 km long, tracing the axial part of the Lomonosov Ridge (ends of

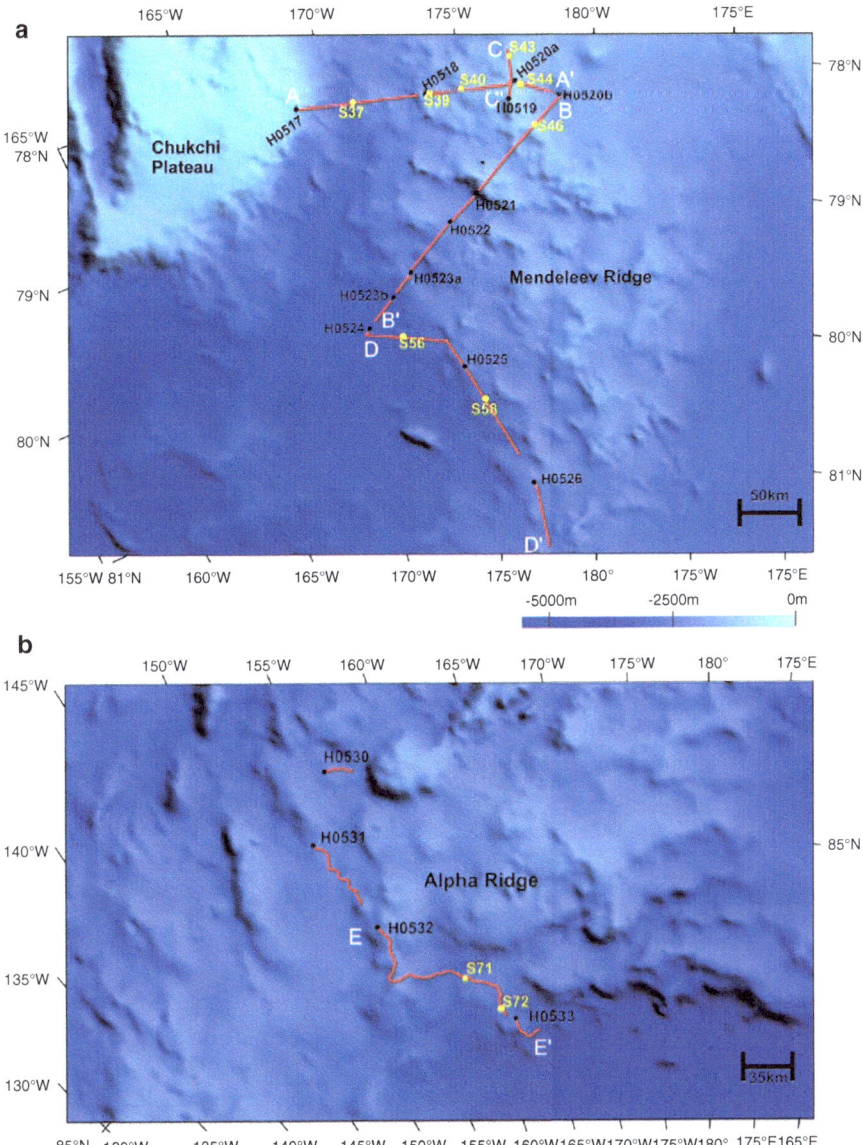

Fig. 1.23 MCS (*red lines*) and DSS (*black dots*) location map at (**a**) - the Mendeleev Ridge; (**b**) – the Alpha Ridge ("Healy", 2005)

line: 83.5688°N, 143.8157°E and 76.1161°N, 143.0602°E; recording station - Sercel SEAL System (France); source - BoltAPG, 1500 in³, 2000 PSI; streamer - SEAL Sentinel Solid 8100 m length, 648 channels; record length - 12 s; fold - 108). Seismic navigation system Spectra was used for positioning. The main objective was the junction the Lomonosov Ridge and East Siberian Shelf.

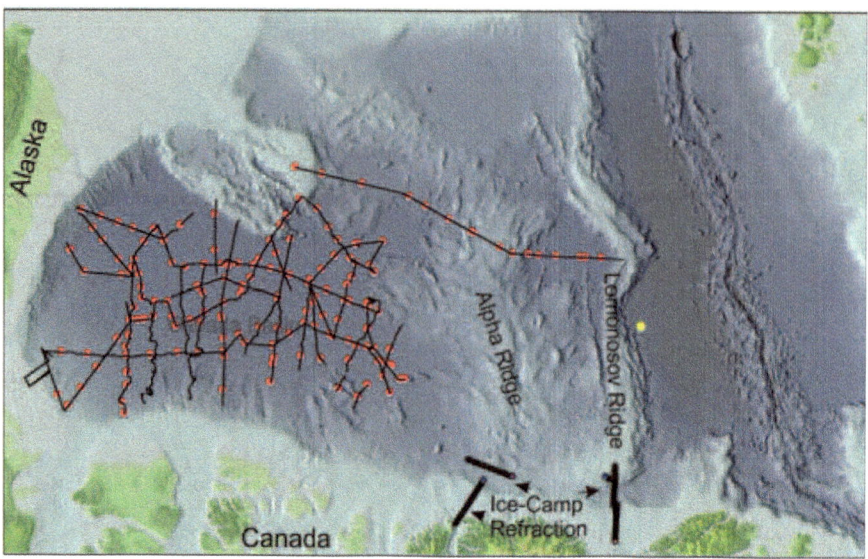

Fig. 1.24 Louis S. St.-Laurent and Healy (2007–2011) seismic tracks and sonobuoy drop locations (*red dots*). The yellow dot - the North Pole (Mosher et al. 2013)

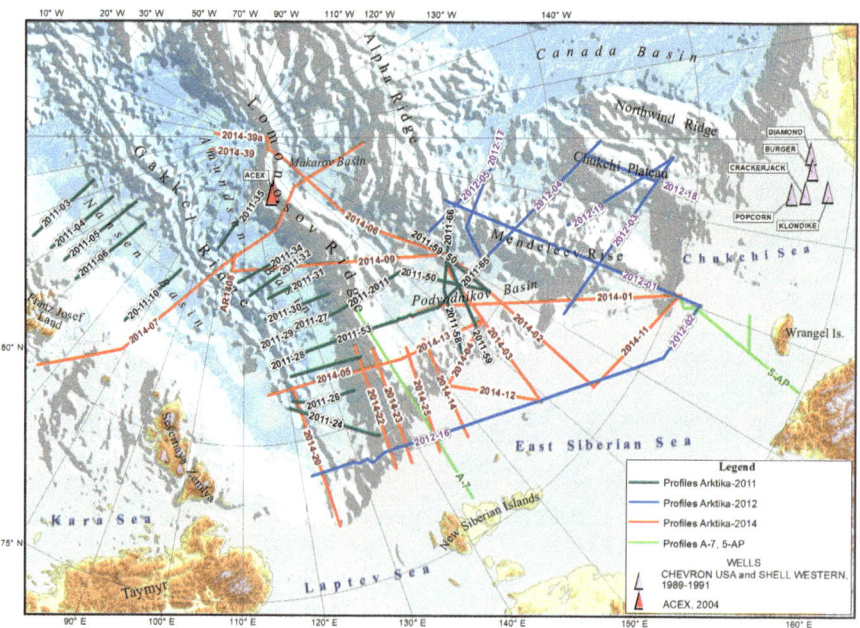

Fig. 1.25 MCS surveys in Eurasian Basin and Central Arctic Uplifts of the Amerasian Basin, 2007–2014

Traverse 5-AR In 2009 "MAGE" completed MCS traverse from Chukchi coast-line to the edge of the Chukchi shelf using RV "Geolog Dmitry Nalivkin" (762.5 km, ends of line: 70.032°N, 175.859°E, and 74.769°N, 179.609°W; recording station - Sercel SEAL System (France); source - I/O Sleeve guns, 4010 in^3, 2000 PSI; streamer - SEAL Sentinel Solid 8100 m length, 648 channels; record length - 15 s, fold - 81). Seismic navigation system Spectra was used for positioning. The survey became a part of the reference grid of the geological-geophysical traverses within the Russian part of the Arctic Shelf.

Project "Arktika-2011" 2011 MCS survey utilized RV "Akademik Fedorov" and icebreaker "Rossia" The operations used solid-filled 48 channels-streamer 600 or 4600 m long, one or two air guns BoltAPG of 1025 or 2050 in^3 and 2050 PSI working pressure. Reflection and refraction sounding using short streamer accompanied CDP shooting on every traverse (3–7 soundings per traverse). Navigation and positioning was provided by the integrated navigation system ORCA built around navigational system SPECTRA and software package SPRINT.

The data were recorded and pre-processed by telemetric system BOX (Fairfield Industries, USA), delivering 14–15 sec long recording, 6 fold for the short streamer and 46-fold – for the long one.

The program's main goal was evaluation of thickness of sedimentary sequences of the Amundsen, Nansen and Podvodnikov Basins which could be used for defining the outer limits of the Continental Shelf using "one percent thickness" criteria.

Project "Arktika – 2012" In 2012 the geological and geophysical activity shifted to the Podvodnikov Basin, Mendelelev Ridge, Chukchi Plain and De-Long High using icebreaker "Kapitan Dranizin" and RV "Dikson" (former icebreaker converted to research vessel equipped to conduct any kind of marine seismic surveys). The principal objective of the project was investigation of the Central Arctic Uplifts and defining the outer limits of the Continental Shelf of the Russian Federation.

The short (600 m) streamer produced 6-fold CDP records, the long one (450 m) – 45-fold records. The longer streamer records provided better definition of seismic velocities and, subsequently, better migration. They also had better signal-to-noise ratio and wider bandwidth allowing for successful suppression of multiples.

Seismic energy was generated by single or double air guns BoltAPG with total volumes 1625/1850/1950/2050 in^3 with working pressure 2050 PSI, navigation and positioning was provided by Trigger Fish system.

The short streamer records demonstrated low sensitivity to variations of stacking velocities. In general, stacking, supported by reliable velocity functions from refraction soundings (50 soundings in total), produced good quality seismic sections, especially in the noiseless environment of abyssal seas.

The data processing used modern technologies and algorithms, such as multiples suppression (SRME), noise reduction (LIFT) and predictive deconvolution. Multiples suppression successfully revealed the underlying primary reflections in the shallow waters. The final product was CDP sections with TWT 15 sec and 6-fold or 45-fold for short and long streamers, accordingly.

Project "Arktika – 2014» In 2014 the MCS programs were ran in the Nansen, Amundsen, Makarov and Podvodnikov Basins, across the Lomonosov Ridge and in the Laptev and East Siberian Seas margins. The main goal – investigation of sedimentary cover structure and startigraphy of major morphological features of the region.

The RV "Akademik Fedorov" and icebreaker "Yamal" served as the base of operation. Technical information: solid-fill streamer 600 m and 4500 m (depending on ice condition), DigiStreamer seimic registration system, air guns BoltAPG with total volumes 1025/1300/2050 in^3 and working pressure 2050 PSI. Positioning - by navigation system QINSy.

Reflection and refraction soundings with short streamer, as a part of the program, defined seismic velocities at major interfaces. The final product was 12 sec - CDP sections with 6-fold or 45-fold for short and long streamers, accordingly.

Project "ALIANCE", Russian Academy of Science, 2011–2013 MCS surveys mainly covered shallow continental shelf, occasionally reaching continental slope (Fig. 1.16). The results were incorporated into regional studies of continental margins of the Russian Arctic.

1.4   Deep Seismic Sounding (DSS)

The investigations of Earth crust of the Central Arctic Ocean began in late 70th of the last century.

Station LOREX, 1979 The station was based on ice floe drifting over central part of the Lomonosov Ridge and Makarov Basin in the vicinity of the North Pole. The continental type of crust 25 km thick was established for the Lomonosov Ridge, and oceanic – for Makarov Basin (10 km thickness) (Mair and Forsyth 1982).

Station CESAR, 1983 Canadian investigations from the drifting station CESAR studied the Alpha Ridge. Refraction seismic discovered high-velocity layer in the lower crust and found Moho discontinuity at depth of 40 km (Jackson et al. 1986).

Project LORITA-2006 In 2006, the Danish-Canadian expedition conducted systematic airborne DSS along 500 km profile over the Lomonosov Ridge–Lincoln Shelf junction zone. The seismic velocity model shows the presence of three sedimentary units and one metasedimentary unit with thickness from first hundred meters to 7 km, traceable from the shelf onto the Lomonosov Ridge. The survey also detected the upper (3–14 km thick), lower (4–10 km thick) crusts and Moho discontinuity subsiding from 20 km at the junction to 22–26 km under the Lomonosov Ridge (Jackson et al. 2010).

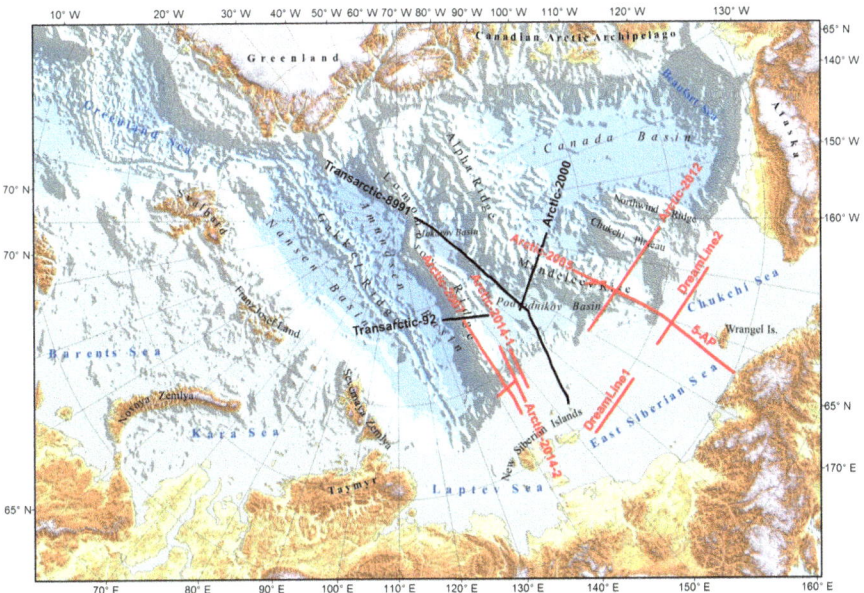

Fig. 1.26 Locations of the Russian DSS profiles. Black lines – profiles completed in 1989–2000; red lines - after 2000th

Project ARTA-2008 In 2008, Canadian geophysicists continued the on-ice DSS program along 350 km long profile crossing the Alpha Ridge – Ellesmere Island Shelf junction zone. Interpretation of the seismic velocity model reveal the similar situation: sedimentary and metasedimentary formations 2–4 km thick continuing from the shelf to the Alpha Ridge, upper and lower continental crusts (4–6 km and 18–20 km thick, accordingly), and Moho dipping from 28 km at the shelf to 31–32 km under the Alpha Ridge (Funck et al. 2011).

The Russians began their own DSS operations in the Central Arctic in 1989 (Fig. 1.26).

Project "Transarktika 1989–1991" The principal objective of the project was comparative analysis of the crust of the continental margins and the deep-water parts of the Arctic Ocean. With this goal in mind, the 1489 km long traverse was laid out from Zhokhov Island on the East Siberian Continental Margin through the Podvodnikov and Makarov Basins to the Lomonosov Ridge in the vicinity of the North Pole. The operation was on-ice with airborne support; planes fly from ice-based base camps.

The scientific part of the program included DSS, seismic reflection, gravity measurement on-ice and aeromagnetic survey.

Completed in three spring field seasons, the project "Transarktika 1989–1991", the first operation of that type, had rather low density of observations. This led to poor determination of low-velocity refractions in sedimentary formations. It was augmented by seismic reflection soundings. More informative were high-velocity refractions at the deep seismic interfaces within the crystalline crust. It was demonstrated that the Moho discontinuity depth changes rather sharply from ≈43 km under the De-Long High to 20–26 km under the Podvodnikov Basin and 13–14 km – under the Makarov Basin.

Project "Transarktika 1992" As in the previous project, it was the on-ice operation with planes flying from ice-based base camps.

The 280 km long nearly latitudinal traverse crossed the Lomonosov Ridge at the latitude 84°N from the Amundsen Basin to the Podvodnikov Basin. The investigations established the continental crust about 20 km thick with approximately equal thickness of upper (5–6 km) and lower (6–7 km) consolidated crust.

Project "Arktika-2000" DSS line (485 km) was shot across northern part of the Mendeleev Ridge and the adjacent Mendeleev and Podvodnikov Basins at 82°N. This time the airborne operations were supported by helicopters from the research vessel "Academic Fedorov' accompanied by the nuclear icebreaker "Rossia". Reflection seismic sounding, gravity measuremets and bottom sampling were done simultaneously with DSS. The thickness of the continental crust under the Mendeleev Ridge axis was estimated at 28 km with lower crust substantially thicker (up to 20 km) than upper (4–5 km).

The data collected by these program served as a base for compilation of the first crustal thickness maps of the Arctic Ocean and developing the Russian conception of the Central Arctic geology. The following works concentrated on the major Central Arctic submarine uplifts and their links with North-Eastern Eurasian Shelf (Fig. 1.26).

Project "Arktika-2005" The similar, both from organizational and scientific aspects, program was run in 2005 along the axial part of the Mendeleev Ridge and its adjoining with shelf (485 km). Additional investigations included refraction shooting at the cross-line 130 km in length and 1:1000000 scale aerogeophysical survey along the main traverse. The results demonstrated that the stratified sedimentary formations, as well as metasedimentarys and crystalline crustal complexes continue uninterrupted from the shelf to the Mendeleev Ridge, supporting the arguments for its geological interconnection. The upper crust thickness was estimated at 4-7 km, lower – 20–22 km and Moho at depth of 31–34 km.

Project "5-AR" Ship-borne DSS and multi-channel seismic reflection (MCS) survey were completed on line 5-AR stretching from the Cape Billings (Chukotka coast) to the southern end of **"Arktika-2005"** traverse between 2008 and 2010. DSS line was completed using self-emerging 4-component oceanic bottom stations (OBS) (geophones X, Y, Z and hydrophone H) and 7300 in^3 airgun. The data confirmed the previous program's finding and established thick (up to 17 km) sedimentary section under the North-Chukchi Basin. The thickness of the upper crust drastically increases southward from only ≈ 3 km under the North-Chukchi Basin to ≈20 km under the Wrangel-Herald Uplift on shelf. The lower crust here is 9–10 km thick. The Moho was found at depth from 30 to 34 km.

Project "Arktika-2007" Under this project, DSS, seismic reflection, aerogeophysics and bottom sampling were conducted along the regional traverse 650 km in length from the near Siberian part of the Lomonosv Ridge to the Laptev Sea Shelf north from Kotelny Island. Helicopters, based on the nuclear icebreaker "Arktika", provided the logistical support.

Under this project, an additional MCS survey was done along the same traverse. The survey was run using, for the first time, 8 km-long streamer in open water (R/V "Professor Kurentsov", 832 km), from New-Siberian Islands up to the intersection with **Transarktika 1992** profile.

Again, the major Mezo-Cenozoic lithostratigraphic units with associated unconformities, along with metasedimentary complex, were traced from the Laptev Shelf onto the Lomonosov Ridge. Upper crust within the ridge is 6–9 km thick, lower – 9-12 km. The Moho depth changes from 28 at the northern Laptev Shelf to 22–23 km under the Lomonosov Ridge.

Project "Arktika-2012" The 480 km-long traverse, from the Vilkitsky Trough to Chukchi Plateau, crossed south part of the Mendeleev Ridge at approximately 72–73°N. DSS data were collected using self-emerging 4-component OBS (geophones X,Y,Z and hydrophone H). The sedimentary section was studied by MCS survey with 4.5 km-long streamer. Seismic data demonstrated continuity of basic crustal layers between the Mendeleev Ridge and the Chukchi Plateau.

The Mendeleev Ridge is underlain with very thick continental crust - the Moho discontinuity detected here at 34 km depth. Under the Chukchi Basin it raises to 22–25 km, thinning the Earth crust, mostly at the expense of almost disappearing upper crust. In contrast, under the Chukchi Plateau the upper crust is noticeably thicker and the Earth crust here returns to more "conventional" continental margin composition.

Project "Dream Line" DSS traverse 925 km in length using self-emerging 4-component OBS and accompanied MCS survey were completed in 2009. It was found that the sediments thickness almost doubles from 7–8 km in the west to 15–16 km in the east; the upper crust changes from 7–12 km to 6–10 km thick in the same direction; Moho at 27–30 km depth – overall, the typical crustal composition of continental margins.

Project "Arktika-2014" The ship-borne (R/V "Nikolay Trubyatchinsky") DSS survey was done on two traverses (250 km and 350 km), using self-emerging 4-component digital OBS and 7300 in^3 airgun. Located in the De-Long Uplift and linkage area, the survey was aimed at the continuity and structure of the crustal complexes. The sharp increase of sedimentary cover thickness from 0.5 km at the De-Long Uplift to 7 km in the Vilkitsky-Podvodnikov Basin was documented. Thickness of upper crust varies from 20 km at the De-Long High to 7 km in the

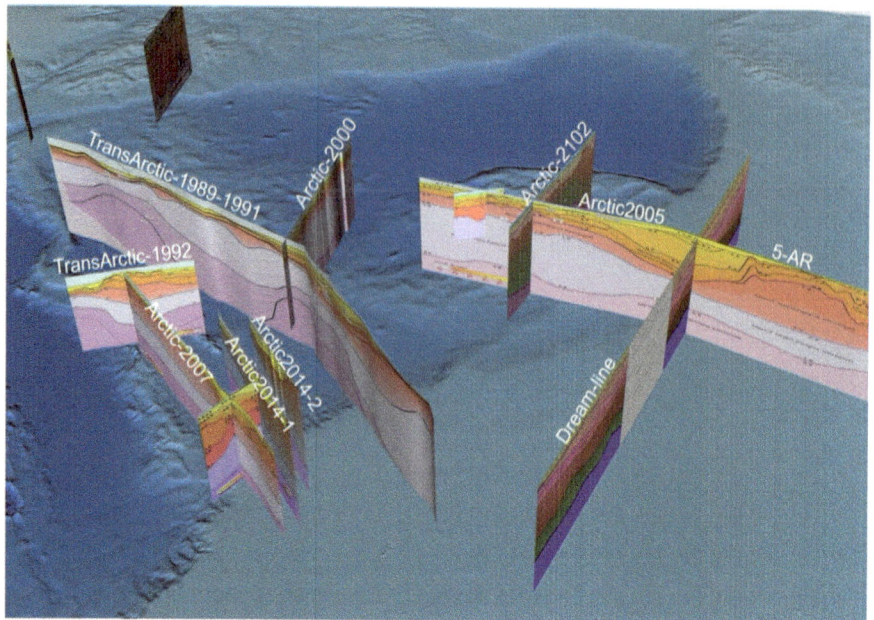

Fig. 1.27 The velocity models of Earth crust along the seismic traverses acquired by the Russian geophysicists in the Central Arctic and adjacent Eurasian Shelf

Vilkitsky-Podvodnikov Basin. Moving from the shelf towards the deep-water region, Moho raises from 32 km to 23–35 km.

The data proved uninterrupted shelf-to-basin continuity of the major layers of the consolidated crust without any significant changes of seismic velocities.

Fig. 1.27 demonstrates the velocity models of Earth crust along the seismic traverses acquired by the Russian geophysicists in the Central Arctic and adjacent Eurasian Shelf.

1.5   Seismicity of the Arctic

The first earthquakes in the Arctic Region (the territories inside the Arctic Circle) were recorded by seismic stations installed in 1910–1912. In following 30–40 years, few widely scattered stations recorded only the strongest events. The network grew rather sharply in 1956–1957 after Soviet stations "Tiksi", "Hayes", "Apatit", "Yakutsk", and later – "Iultin" and "Norilsk" were activated under the International Geophysical Year. Currently, there are close to 50 stations in the Arctic and vicinity sending information to the International Seismological Centre (ISC). The

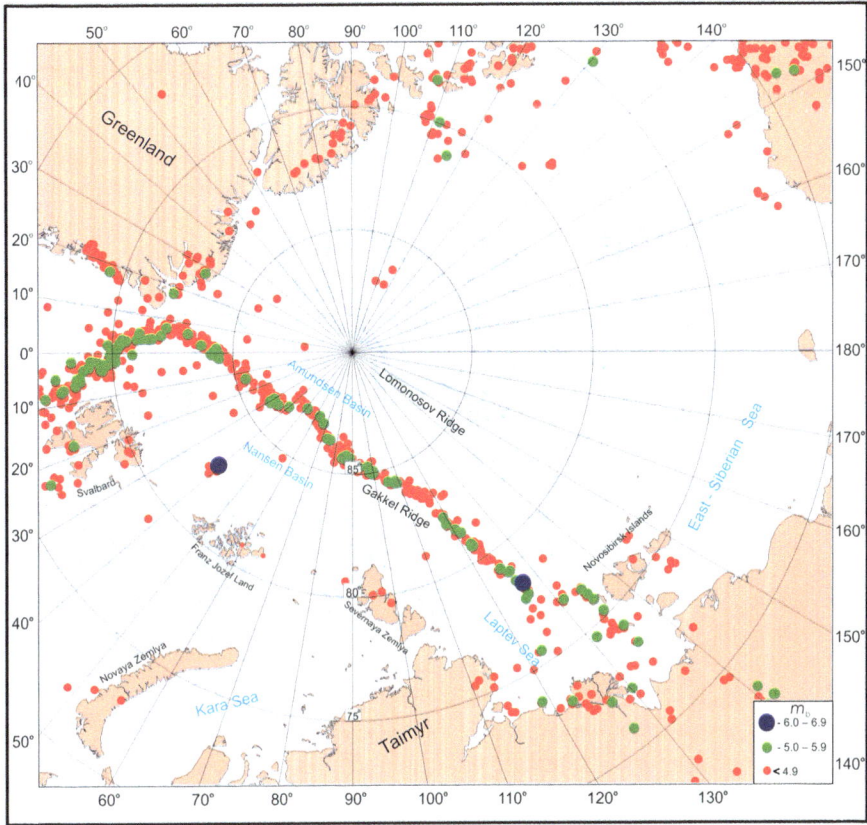

Fig. 1.28   Earthquake epicenters of the Arctic Ocean and its periphery

information is also coming from stations in Yakutia, Kola Peninsula, Archangelsk region, Scandinavia, Spitsbergen and Alaska. This network allows for representations of 4.5 m_b events, or better.

Among many seismically active zones of various intensity, sizes, and shapes presently known in the Arctic, the principal is the Mid-Arctic Seismic Belt, marking the divergent boundary between the Eurasian and North American Plates. Several intraplate zones of the lower order are also present in the peripheral Arctic seas and Canada Arctic Archipelago (Fig. 1.28).

This map was compiled using the database APC created by VNIIOkeangeologia. The database has hypocenter locations of all earthquakes recorded from the start of instrumental observations. It also contains all published information on fault-plane solutions (often by several authors providing conflicting information). Today only the most reliable and objective methods of focal mechanisms solution, e.g. centroid tensor moment, are utilized.

1.5.1   Mid-Arctic Seismic Belt

Mid-Arctic Seismic Belt delineate the abyssal part of modern-day divergent boundary between the Eurasian and North American Plates, marked by the system of
underwater spreading ridges from Iceland trough Norwegian-Greenland Eurasian
Basins to the Laptev Sea continental margin. Its lateral and vertical characteristics
are determined by combination of two major factors: a) orientation of tectonic
forces and b) physical properties of the pre-rift lithosphere. The latter factor mostly
responsible for shaping parameters of each specific segments of the boundary.
Previously was found (Avetisov 1993a, b; Avetisov 1996) that three scenarios are
possible.

In case of nearly homogenous and isotropic medium, the orientation of possible
fracture zone will be mainly determined by the laws of mechanics and orientation
of applied forces. If medium contains some pre-existing zone of weakness (fracture
zones), a forming fracture will deviate away from theoretically possible orientation
towards the pre-existing fracture zones. Finally, if a forming fracture meets some
rigid block head-on, it may bypass it either along single orthogonal fault or by
multiple en-echelon fractures. Each of these scenarios will create its own pattern of
distribution of epicenters and their own type of focal mechanism.

The examples of all three scenarios can be found along the Mid-Arctic Seismic
Belt and Gakkel Ridge is clearly corresponds to the first (Fig. 1.29).

Several features - linearity of the ridge closely following the median line of seismic belt toward the opening centre in Northern Yakutia; clearly dominant normal
faulting mechanism of earthquakes; orthogonal relation between tension axis and
ridge strike; epicenters' proximity to the central parts of the ridge – lead to the conclusion that rifting was initiated in relatively homogenous lithosphere and followed
a "classical" scheme. Ridge geomorphology – its narrow (less than 200–250 km)
width, rugged terrain sharply contrasting (500–1500 m) with adjoining abyssal
plains, presence of long (almost entire length of the ridge), narrow (1–5 to 10–30 km)
and deep central rift valley – support this statement. An exception to this pattern
exists between 35°E and 80°E where 300 km long segment of Gakkel Ridge is offset northward by 100–120 km and then gradually returns to the principal trend further east. Some strike-slip fault-plane solutions typical for transform faults zones
exist in the offset area. Bathymetry data show that the central rift valley becomes
less conspicuous and, at places, disappear completely, while the ridge itself becomes
morphologically less and less prominent. Incidently, the boundaries of the offset
segment are aligned with north-south trending Frants-Victoria (80°N, 41°E) and
Voronin (80°N, 87°E) Troughs on Eurasian Continental Shelf, with several 5.5 m_b
epicenters recorded at their intersection with the continental slope. Quite possibly,
these structures served as suspected zones of weakness, distorting the otherwise
"classical" rifting process in this particular area.

Similar situation exists at the Mid-Arctic Seismic Belt junction with Laptev
Shelf where the system of superimposed Mesozoic troughs is genetically related to
the oceanic rifting of the Eurasian Basin (Gramberg et al. 1990).

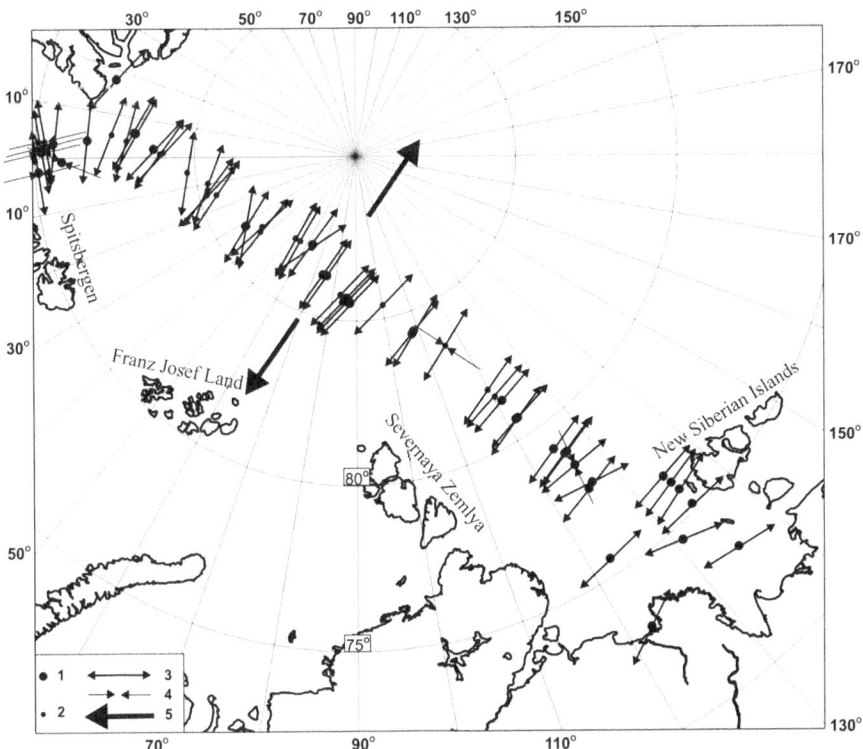

Fig. 1.29 Focal mechanisms of the Eurasian Basin earthquakes. Magnitude (m_b): **1** – ≥ 5; **2** – 4 ↔ 4,9; **3** – tension axes; **4** – compression axes; **5** – plate movement direction

Earlier, (Avetisov 1993a, b, 1996, 1999, 2000, 2002) we suggested that rift propagation into the Laptev Shelf might lead to formation of either microplate or system of transform faults similar to Spitsbergen system. Fig. 1.30 illustrates that, after exclusion of weak seismic events and incorporating the latest seismic data, the presence of the Laptev Microplate at the center of Laptev shelf is confirmed.

The Laptev Microplate is best defined from the north, east and south, less confidently from the west, with two triple junctions - northern at around 126°E and southern at the vicinity of Bour-Khaya Gulf. From the northern triple junction the boundary continues south-east at acute angle to the existing structural elements. It turns south at about 134–135°E following Bel'kovsky–Svyaty Nos Trough and Bel'kovsky Uplift (Vinogradov et al. 1974; Geology of the USSR and Distribution of Mineral Deposits 1984) down to 75°N. Further south, at 73–73.5°N, the boundary obliquely crosses several existing structures, then turns south and, nearly orthogonally crossing some of the above, reaches Bour-Khaya Gulf.

The southern boundary of the Laptev Microplate follows Lena-Taimyr system of peripheral uplifts and reaches western shores of the Olenek Gulf. The western boundary can be inferred with some degree of confidence only at its northern half.

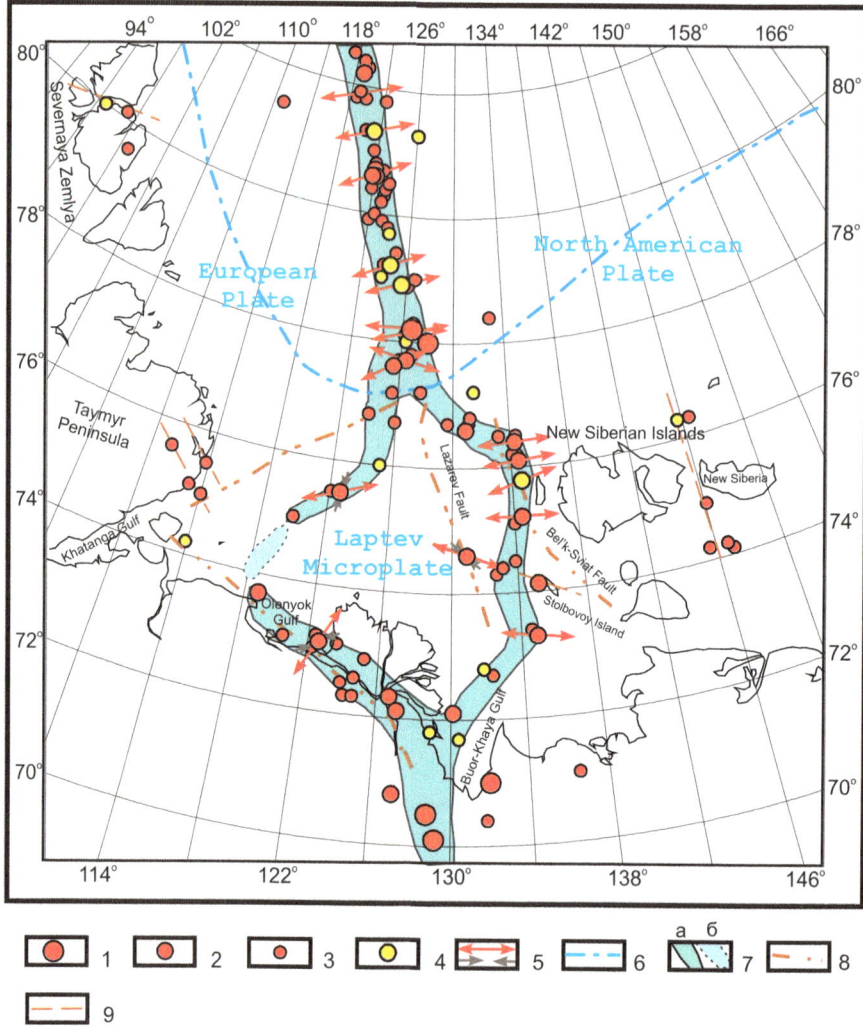

Fig. 1.30 The Laptev Shelf earthquake epicenters and their focal mechanisms: **1–4** – magnitudes: ≥6; 5–5,9; 4–4,9; <4, accordingly. **5** – principal tension and compression axes (arrow length proportional to cosine of dip angle). **6** – continental shelf boundary. **7** – Plate boundaries, (**a**) – proven, (**b**) – inferred. **8** – major fault zones. **9** – local faults

This striking north-east segment is in sharp discord with existing structural plan, cutting across Trofimov dome, Ust'-Lena depression and Minin swell.

The majority of focal mechanism information came from the eastern boundary of the Laptev Microplate with five locations providing consistent normal fault-plane solutions with horizontal tension axis orthogonal to the boundary strike (Fig. 1.29). Both nodal planes, dipping at 40–50° are striking north, close to the boundary trend and orientation of known structures.

The only solution available for the southern border also indicates the same focal mechanism and tension axis perpendicular to the boundary and to the orientation of the Lena-Taimyr system of peripheral uplifts. Nodal planes strike 114° and 315° pitching 36° and 56°, accordingly. Slight left-hand strike-slip component, present there, hints on clockwise plate rotation. According to macro-seismic information available for this event, the 5 m_b isoseis has an oval shape with long axis striking NW-SE parallel to the tectonic trend.

Another event with resolved focal mechanism was recorded at the western border. It also show similar normal faulting environment with subordinate strike-slip component and sub-horizontal tension axis trending at acute angle to the inferred border direction. One nodal plane strikes 8° along the border) dipping at 40° under the Microplate; another - strikes 142°, parallel to Ust'-Lena depression, dipping 59° to SW. If the fault plane coincides with former nodal plane, then the resulting motion will be again left-handed, confirming the clock-wise rotation of the microplate.

Manifestly, the Laptev Microplate, being surrounded by substantially active seismic zones, must be under constant tectonic stress which could de relieved through the fault zones inside the plane acting as the lines of weakness. This might be illustrated by the chain of rather strong earthquakes tracing W-N-W from Stolbovoy Island across the microplate' eastern border to intersection with the Lazarev Fault. At this intersection, the 5.1 m_b earthquake was resolved to a normal-fault mechanism with minor strike-slip component. Nodal planes strike 31° and 175° and pitch at 65° to SE and 29° to W, accordingly. The latter plane attitude coincide well with that of the Laptev Fault. Apart from the described chain, there are numerous weak earthquakes recorded inside the microplate, which support the notion of tectonic forces acting upon the microplate. We may assume that the entire Laptev Shelf currently is under the latitudinal horizontal tension pulling the lithosphere apart.

Fig. 1.30 demonstrates that the Lazarev Fault, considered to be the major spreading line of at the inception of rifting of the Laptev Sea (Hinz et al. 1998), presently is, practically speaking, aseismic. A sporadic epicenters tend to concentrate in places of maximal weakness, e.g. intersections with other faults. The same can be said regarding striking to NE Severny' Fault which was thought to act as a transform fault at the early rifting stages (Hinz et al. 1998).

1.5.2 Active Seismic Zones of Eurasian and Norwegian- Greenland Continental Margins

Generally speaking, intraplate seismic activity of the Arctic is controlled by combination of three major factors: partial strain relaxation in intraplate zones, crust subsidence caused by accumulation of sedimentary sequences and post-glacial rebound of lithosphere.

Evidently, the first factor is the leading one determining the geographical distribution and strength of the intraplate earthquakes. The strong correlation (c.c. 0.7–0.8, with time lag from 0 to 3–4 years) of interplate seismicity of the North Atlantic

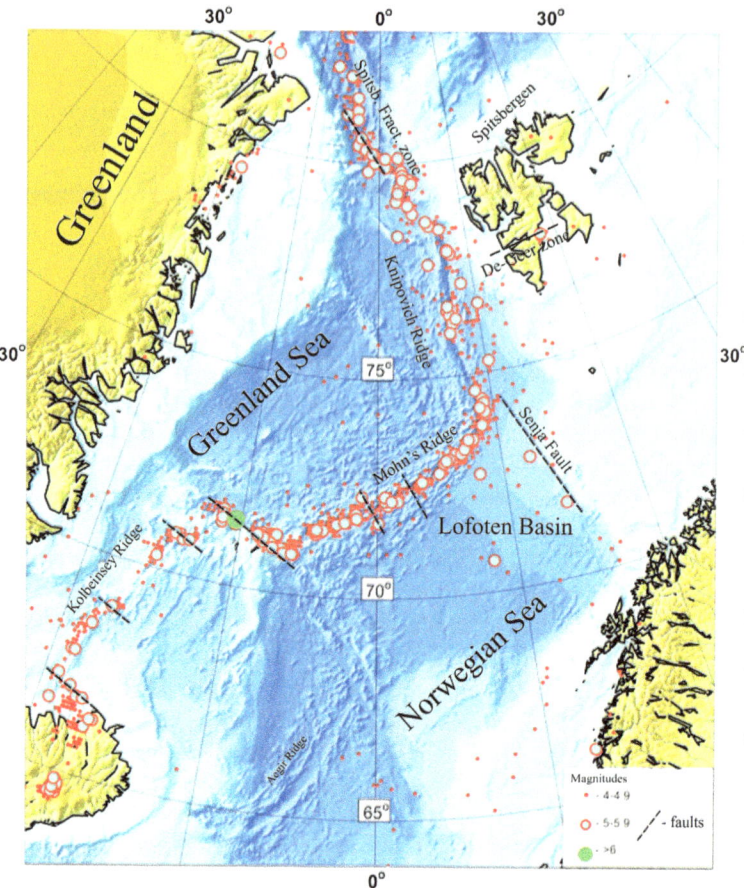

Fig. 1.31 Earthquakes epicenters in Norwegian-Greenland Basin of the Arctic Ocean and its margins

including the Norwegian-Greenland Basin and intraplate seismicity of Fennoscandia was demonstrated in Scordas et al. (1991).

It is expected that spreading activity in the basins axial zones will be transformed into shear or compression forces inside the continental margins. This is clearly manifested along the Eurasian margin of the Norwegian-Greenland (Fig. 1.31) and Eurasian (Fig. 1.28) Basins where the earthquakes tend to concentrate along transcurrent faults and troughs – note the clusters of epicenters in Lofoten Basin close to Senja Fault, in Frants-Victoria and Voronin Troughs, some seismic activity in St. Anna Trough (Avetisov and Golubkov 1971; Avetisov 1971).

Focal plane solutions available there demonstrate strike-slip and reverse fault – strike-slip deformations with attitude of sub-vertical nodal plane close to that of a fault plane. Preponderance of horizontal compression is visible in the posterior parts of continental margins (Kola Peninsula, Novaya Zemlya (Assinovskaya 1994)).

Regions with superimposed compression regime on margins of extension zones exist in the Laptev Sea (Avetisov 1975).

Spitsbergen proximity to the Mid-Arctic Seismic Belt, with no known seismically active lineaments directly connecting the two, may explain Spitsbergen seismicity as a process of relaxation of stress generated within spreading zone (Sykes and Sbar 1973; Bungum et al. 1982; Chan and Mitchell 1985). Epicenters of Western Spitsbergen Island tends to group along De Geer zone which is orthogonal to the Knipovitch Ridge (Fig. 1.31), the closest segment of mid-ocean ridge, and not along the intersecting N-NW). Similarly, the earthquakes within Nordaustlandet (Severo-Vostochny) Island are lining up along faults orthogonal to the Gakkel Ridge. The focal solutions of the earthquakes in this region gave a shear mechanism. At the southern part of this island, a single fault-plate solution indicates reverse faulting mechanism caused by subhorizontal northward oriented stress vector.

A peculiar tendency of earthquake epicenters of Barents-White Seas region to concentrate within areas of reduced crustal heat flows was noted by the Russian geophysicists from Kola Peninsula (Tsibulya et al. 1993). It can be an indirect confirmation of superimposed, or, as we called it earlier (Avetisov 1975), "passive", seismicity triggered by relaxation of external tectonic stress generated elsewhere, with cooler, and, therefore, more brittle blocks of lithosphere being more susceptible to faulting.

Two others causes of seismicity mentioned above are secondary and act only on a local scale. For instance, seismic surveys in Lofoten Basin discovered thick sedimentary section at the base of the continental slope where the thinner continental crust cannot support the enormous load of those formations without cracking (Husebye et al. 1975). Similar situation, in combination with regional tectonic stress, exists in the Lincoln Sea (Basham et al. 1977).

The post–glacial crustal rebound is cited as a cause of increased seismicity in coastal Fennoscandia and eastern Greenland (Husebye et al. 1975).

1.5.3 Canada Arctic Margin

Uneven geographical distribution of epicenters, lack of lineal seismically active zones and connections with existing global seismic belts, inconsistency of focal mechanisms - all this leads to the most likely conclusion that the seismicity of the region is the result of reactivation of pre-existing fracture zones under the same forces which affects another parts of the Arctic outside the Mid-Arctic Seismic Belt.

The subsidence of the Canada Basin oceanic crust under the load of the extraordinary thick (up to 9 km) sediments may cause stress in the adjacent continental crust, mechanically linked to the oceanic crust (Grantz et al. 1990). The hypocenters are only a few kilometers deep and, evidently, are concentrated within the sedimentary cover. The distribution of alternating zones with strong and weak seismic activity is, most probably, dictated by thickness of sediments and by presence or absence of fracture zones. Generally stressed condition of the Canada Margin due to the sediments load reduces the stress threshold required for triggering faulting and earthquakes.

The glacial factor is more significant in the margin interior parts. The correlation between the differentiated vertical movements and levels of seismicity is confirmed by higher concentration of epicenters along stronger gradients of isostatic anomalies. Moreover, the Canada Margin lithosphere is under the constant compressive stress coming from the active Mid-Arctic Ridge.

Generalizing all available information regarding the distribution of epicenters and their focal mechanisms, we may conclude that:

- the divergent boundary between the North-American and Eurasian lithospheric plates marked by the Mid-Arctic Seismic Belt, is the major factor determining the present geodynamics of the Arctic;
- subhorizontal and nearly orthogonal to the boundary tensional forces, and subvertical compressive forces responsible for normal faulting, are constantly active in the ridges abyssal parts. The sheer stress with resulting strike-slip environment exists in proximity to transform faults displacing ridge fragments laterally;
- a stable compressive stress exists at the continental part of the Mid-Arctic Seismic Belt (Northern Yakutia) signifying the transit of the boundary through the opening center;
- regional seismicity outside of the Mid-Arctic Seismic Belt is of superimposed, or "passive" nature, and related to:
- a) partial relaxation of rifting generated stress at intraplate fractured zones;
- b) weight load of anomalously thick sediments on the lithosphere;
- c) isostatic crustal post-glacial rebound of the crust.

1.6   Geological Sampling and Drilling

Geological investigations of the abyssal Arctic Ocean were run from ice stations and specially equipped research vessels (RV's) and, as a rule, in conjunction with geophysical surveys. Initially, only tubular corers were used, bringing up unconsolidated clastic material, with occasional rocky debris.

In 1958 the US ice station "Alpha" commenced systematic sampling of sediments at the Alpha Ridge and continued next year at the Chukchi Plateau from the ice station "Charley". The considerable volume of information was collected in 1951, 1957 and 1958 by icebreakers expeditions in the Beaufort Sea. In 1983, the ice stations LDGO (USA, Lamont-Dougherty Geological Observatory) and CESAR (Canada) worked in the Alpha Ridge and Canada Basin.

Large amount of bottom sediment core and coarse material was collected from the Alpha and Northwind Ridges and in Canada and Chukchi Basins during 1984, 1985, 1989, 1992—1994 expeditions, concentrating their attention on stratigraphy and dating of Late Pliocene –Holocene formations.

In 1995 the icebreaker "Polarstern" obtained 16 cores (mostly aleuro-pelites with occasional thin sandy interlayers) from the Lomonosov Ridge crest between

81°N and 82°N. Geotraverse "Arktika-2000" crossing the Mendeleev Ridge, Podvovdnikov and Mendeleev Basins along 82^0 N provided important geological data from 41 dredging stations (Kaban'kov et al. 2004).

Samples of alkali basalts lifted from the Gakkel Rridge by RV "Polarstern" (AMORE-2001) pointed to its volcanic origin (Michael et al. 2003). It was also noted that basalts are prevalent on the east slope of the Gakkel Ridge while the western one is enriched by the mantle material (peridotites, pyroxenites and gabbroids).

The Upper Cretaceous sediments were established at the central part of the Alpha Ridge (Mudie et al. 1986). The continental aleurolites of Lower Jurassic and Lower Creataceous age were found in the same core taken by US icebreaker "PolarSea" core 1520 m-deep location at the Eurasian flank of the Lomonosov Ridge nerar North Pole in 1994 (Grantz et al. 2001). The only existing oriented samples of Paleozoic and Mesozoic rocks came from the Nortwind Ridge.

USCG icebreaker "Healey" undertook the tubular coring and dredging in 2005 (Adler et al. 2009).

The same year, the RV "Akademik Fedorov", under the geological mapping program, cored 15 locations with 6 m-long tubular and box corers. Later, in 2007 the same RV cored the eastern slope of the Mendeleev Ridge (2 stations, 4 m–long core) (Gusev et al. 2013).

In 2008 the icebreaker "Polarstern" visited the same region with box corer, mega corer and tubular corer (Stein et al. 2010).

In the last decade, dredging, capable of delivering large samples of coarse bedrock materials became more and more poular. However, the major breakthrough came with deep drilling at the Lomonosov Ridge. Before that, all concepts of geology of the Arctic Ocean were based on interpretation of geophysical surveys, geology of the adjacent lands, some Norwegian-Greenland Basin and four short cores (CESAR and ice island T-3), containing Eocene, Campanian and Maastrichtian sediments (long after this well was drilled, the Upper Miocene deposits were identified in core PS87/106 from the Lomonosov Ridge (Stein et al. 2016)).

Project IODP-ACEX The first deep-water drilling operations took place under the **IODP-ACEX** (**I**ntegrated **O**cean **D**rilling **P**rogram - **A**rctic **C**oring **Ex**pedition) program in 2004. The ideas of the Cenozoic evolution of the central part of the Arctic Basin were substantially refined The well sites were chosen at the crest of the Lomonosov Ridge along the seismic traverse "AWI-91090" (Jokat et al. 1992) (Fig. 1.32).

In total five wells, each with different TD and core yield, were drilled (Backman et al. 2006). Two (M0002 and M0004) were merged to produce composite stratigraphic cross-section with four lithological units and total thickness of 428 m. (Figs. 1.33, 1.34).

The Russian expedition "Arktika-2005" on the Mendeleev Ridge (2005) collected numerous samples using tubular corers, mega-box corers and dredging (Fig. 1.35).

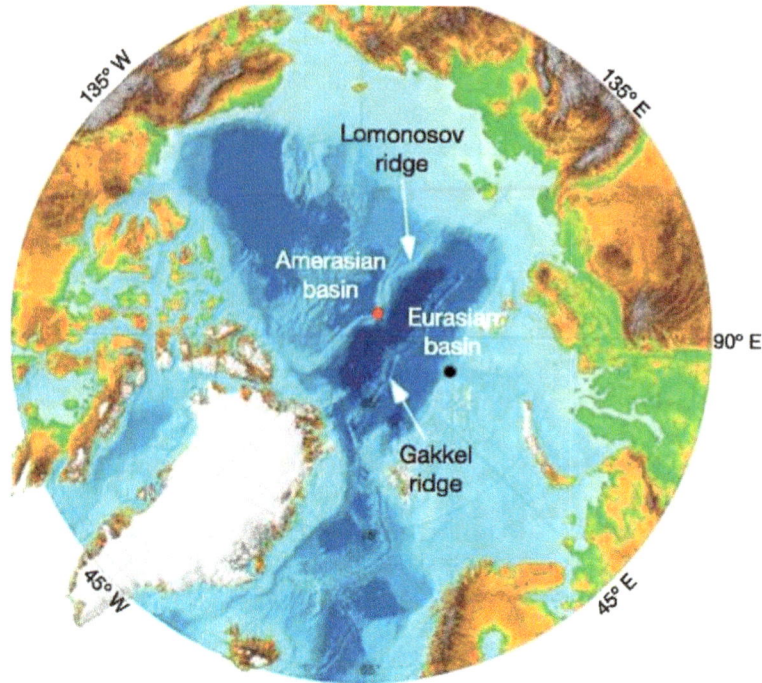

Fig. 1.32 IODP-ACEX drilling location (*red dot*)

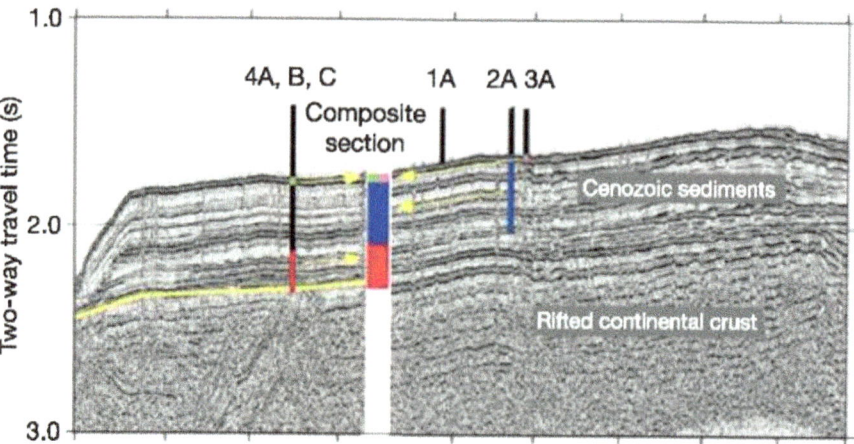

Fig. 1.33 TWT seismic section (line AWI9090 across the Lomonosov Ridge) with ACEX well locations

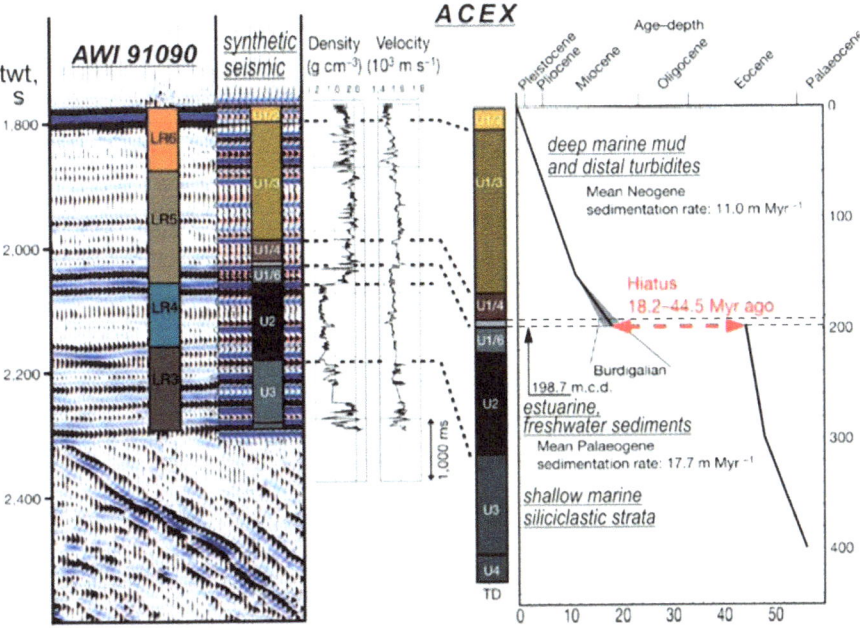

Fig. 1.34 Left – Seismic section (with seismic units LR6-LR3) synthetic seismogram-well stratigraphy correlation (Jokat et al. 1995); Center - density and velocity logs ACEX; Right - depth, age, lihology (Jakobsson et al. 2007)

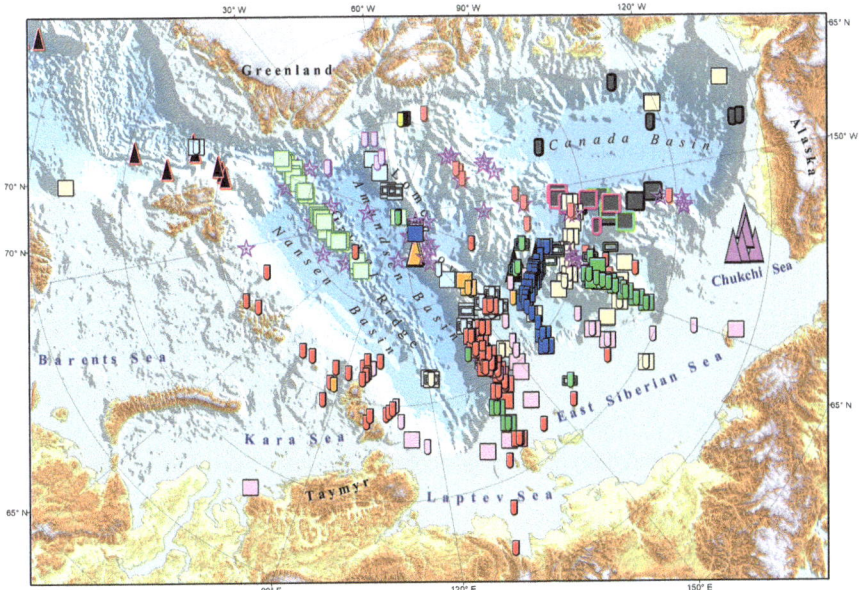

Fig. 1.35 Distribution of sampling sites in the Arctic Ocean (as of 2014).
Original scale – 1:5000000

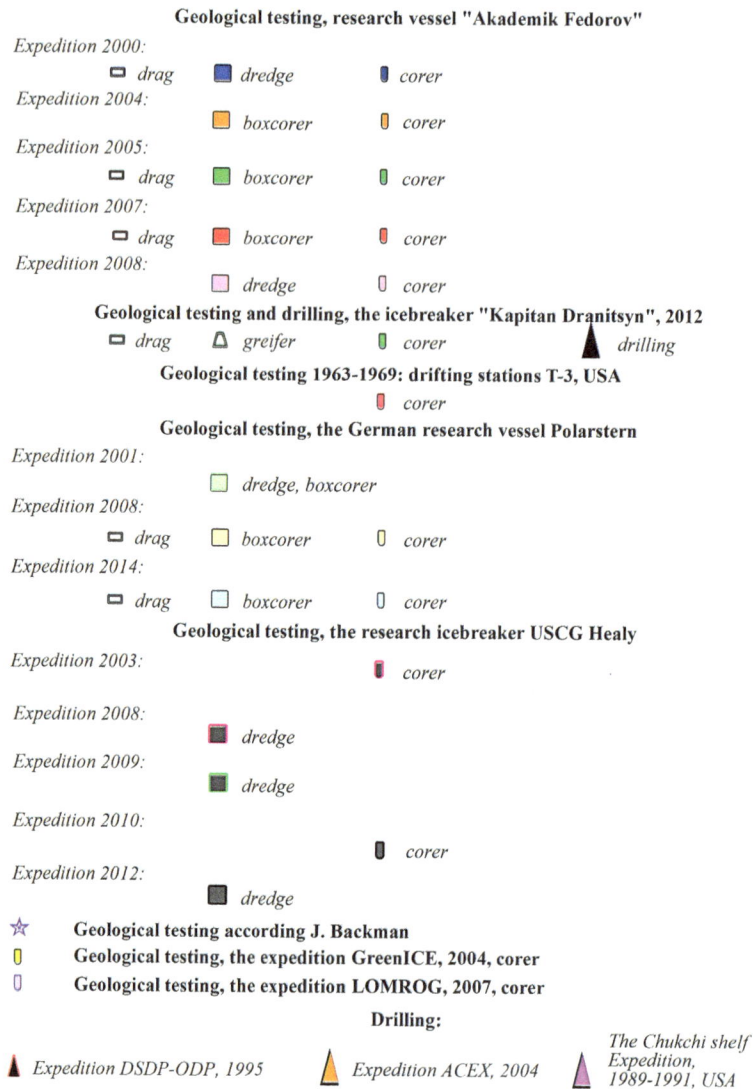

Fig. 1.35 (continued)

Additional information on upper part of the Lomonosov Ridge was obtained in 2007 by Danish-Swedish expedition of icebreaker "Oden" assisted by the Russian nuclear icebreaker from several wells drilled on the southern tip of the Lomonosov Ridge and Morris Jessup Plateau. No more than 6 m of sandy clay was penetrated. The same year, the Russian nuclear icebreaker "Rossia" was sampling the Lomonoisov Ridge - Continental Margin junction using tubular corers, box corers and dredging.

A new approach to underwater exploration was used in 2012 during the expedition "Arktika-2012" on Mendeleev Ridge when sampling and well site selection was accomplished from piloted submersible equipped with high resolution profiler, side scan sonar and TV camera. Uniformity of samples composition, poor fragments roundness and sandy matrix - all indicated "in-situ" origin of most of the samples (few rounded fragments may have glacial origin). Video recordings documented great deal of bedrock outcrops and supported the local origin of clastic material.

The core extracted from southern part of the Mensdeleeev Ridge contained mostly trachybasalt, less - trachyandesite and brecciated trachybasalt lava, dated as Permian-Triassic and Cretaceous (Morozov et al. 2013).

The expedition run a special sampling program using grabbers, tubular cores, seafloor drilling rig with core length up to 1.15 m and arm manipulator operated from submarine. The area of operations roughly coincided with that of 2000 and 2005 (Fig. 1.36). The grabber, drilling and arm manipulator procedures were videotaped and visually guided. The sampling sites were preferably concentrated near the base of escarpments located beforehand by seismic surveys (Gusev et al. 2014, 2017).

The petrographic composition of coarse material dredged from the Mendeleev Ridge is rather consistent: 90% of it represented by sedimentary rocks (crystalline dolomites, quartz sands and limestones, occasionally with Mid-Upper Paleozoic fossils; the remaining 10% - by igneous (Archean or Early Proterozoic granite-gneisses, Late Proterozoic gabbro-dolerites) and metamorphic (green schists, metabasites, quartzites and gneisses). All of these rocks may belong to the acoustic basement underlying the unconsolidated Meso-Cenozoic sediments.

Rock samples collected in the Central Arctic Basin by expeditions "Arktika-2000, 2005, 2007", RV "Akademik Fedorov" and icebreakers "Kapitan Dranicyn" and "Polastern" (expedition ARK-XXIII/3, 2008), from the Mendeleev and Lomonosov Ridges and Podvodnikov Basin, were subjected to variety of isotopic, geochemical and age dating analyses.

Detrital zircons (found in pelagic silt and clastic rocks) and accessory zircons (extracted from fragments of felsic and mafic igneous rocks) were dated at the Center for Isotopic Research, VSEGEI. St. Petersburg. So far, cross-bedded quartz sandstone with illite cement collected at the northern Mendeleev Ridge (site 06, sample SS12–06-5 m) was dated at 550–350 Ma (Silurian-Devonian); detrital zircon from massive quartz sandstone with metacarbonate cement (site 01) gave Triassic- Jurassic age at 250–200 Ma. Out of 100 detrital zircon crystals from both sandstones subjected to testing, some turned out to be as old as Archean.

Chemical composition of basalts is important indicator of oceanic (Jackson et al. 1986; Tarduno et al. 1998; Weber and Sweeney 1990) or continental (Coles et al. 1978; Ivanova et al. 2006; Johnson et al. 2013; King et al. 1966; Miller et al. 2006) origin of the Mendeleev and Alpha Ridges. By 2012, the samples from ten stations (Fig. 1.37). unambiguously proved presence of intraplate alkali, or alkali-calcic, basalts all over the Central Arctic Uplifts. Unaltered basalts from the Mendeleev Ridge (cored by "Arktika-2012" wells at the top of acoustic basement) are chemically identical to the Late Mesozoic-Early Cenozoic basalts at the Chukchi Plateau

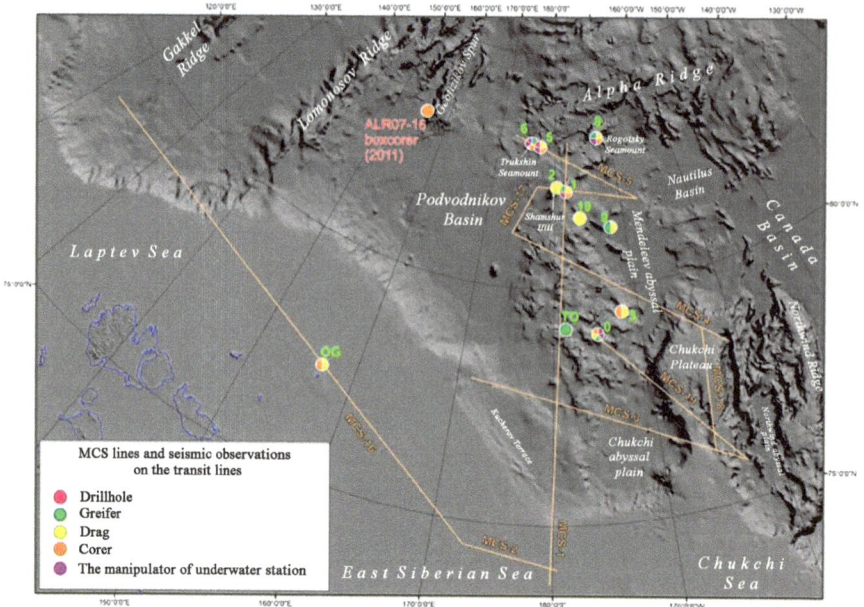

Fig. 1.36 "Arktika-2012" sampling locations on the Mendeleev Ridge

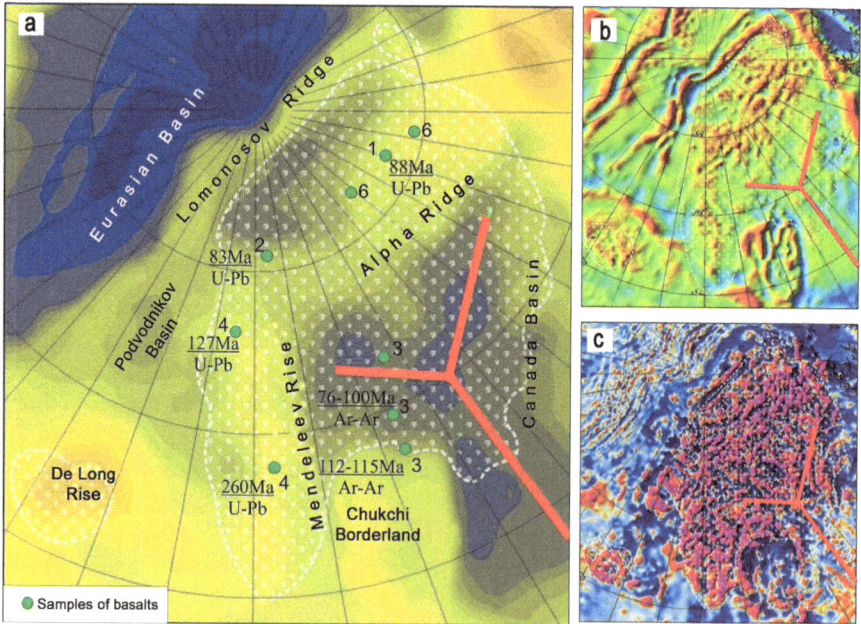

Fig. 1.37 The Central Arctic Basalt Areal of the HALIP on: (**a**) – Moho depth map; – Free Air gravity map; (**b**) – Magnetic anomalies mapNumbered circles – sampling stations location: 1 – CESAR (1983); 2 – Polarstern (1998); 3 – Healy (2008 and 2009); 4 – Arktika-2012; 5 – US ice station T3 (1963). Red lines – implied triple junction

seamounts, Alpha Ridge, Frantz –Joseph Land, De-Long Islands and other parts of the High Arctic Large Igneous Province (HALIP).

Vesicular trachybasalt from well KD12–00-33b (lower slope of the T-3 plateau) yielded 25 zircon crystals. The two youngest marked the lava as Permian (~ 260 Ma). The age of remaining crystals formed five age clusters: Late Archean (2.7 Ga, 3 crystals); Early Proterozoic (1.9 Ga, 6 crystals); Early Riphean (1.65 Ga,7 crystals); Late Riphean (0.8–1.1 Ga,5 crystals); Early Vedian (668 Ma, 1 crystals) (Morozov et al. 2013). $^{40}Ar/^{39}Ar$ dating indicated Early Paleozoic age of Mendeleev Ridge basalts (Vernikovsky et al. 2014).

More recent $^{40}Ar/^{39}Ar$ data from T-3 plateau well hinted on three distinct magmatic episodes: 112 Ma (Albian), 100 Ma (Albian-Cenomanian) and 85–73 Ma (Campanian). Additional dating came from shallow well KD12–06-21b (Mendeleev Ridge, Trukshin Mount) where single well formed multi-zonal zircon crystal was separated from brecciated trachybasalt (U concentration - 600 g/t, Th/U – 0.6, age - 127.5 ± 2.7 Ma) (Morozov et al. 2013). Therefore, Cretaceous age of basalts overlaying the acoustic basement at the Mendeleev and Alpha Ridges can be considered as proven.

Research demonstrated profound difference between the Mendeleev-Alpha basalts and tholeitic MORB of the Gakkel Ridge and Lena Trough. Nauret (Nauret et al. 2011) demonstrated that basalts of active spreading zones, by chemical composition and Sr—Nd—Pb—Hf isotopic ratios, are identical to typical MORB. Along with the basalts, gabbros and dolerites were dredged, and serpentinized and deconsolidated peridotites of the upper mantle were raised on individual amagmatic sections of the ultra-slow spreading Gakkel Ridge. At the same time, in the central part of the Lena Trough, alkaline differences were discovered with relatively high SiO_2 (51.0—51.6%), Al_2O_3 (18.1—18.4%), Na_2O (4.0—4.2%), K_2O (1.0—1.6%), Mg-number (60—65), relation K_2O/TiO_2 (0.6—0.9), (La/Sm)PM (1.4—1.8) and low FeO content (6.5—6.8%). Variations in the isotopic ratios in these lavas: $^{87}Sr/^{86}Sr$ from 0.70361 to 0.70390, $^{143}Nd/^{144}Nd$ from 0.51283 to 0.51290 (εNd from +3.7 to +5.2), $^{176}Hf/^{177}Hf$ from 0.28313 to 0.28322 (εHf from +11.6 to +14.9), and $^{206}Pb/^{204}Pb$ from 17.752 to 17.884, $^{207}Pb/^{204}Pb$ from 15.410 to 15.423, $^{208}Pb/^{204}Pb$ from 37.544 to 37.670 — manifest contamination of the basaltic magma inside the magma chamber by crustal material containing phlogopite, amphibole, and garnet, which allows the authors to classify this segment of the Lena Trough as the continent-ocean transition.

The plateau basalts were dredged on the Healy seamount (see Fig. 1.8) and on the northern spur of the Northwind Ridge (O'Brien et al., 2016). A preliminary study of the samples showed that the rocks from both stations belong to alkaline and moderately alkaline basalts, with the samples of seamount Healy (79° 44′31"N, 155° 06′43"W) representing tholeiitic underwater pillow - lavas, and the rocks of the Nordwind Ridge (78° 32′02"N, 156° 42′00"W) are erupted in a sub-aerial environment and are close to the continental plateau basalts by geochemical characteristics (Mayer et al. 2008; Andronikov et al. 2008).

In short, the rocks chemistry (Fig. 1.38) suggests that both lavas, being currently at depth of more than 3500 m, were effused through a thick continental crust into shallow watery environment.

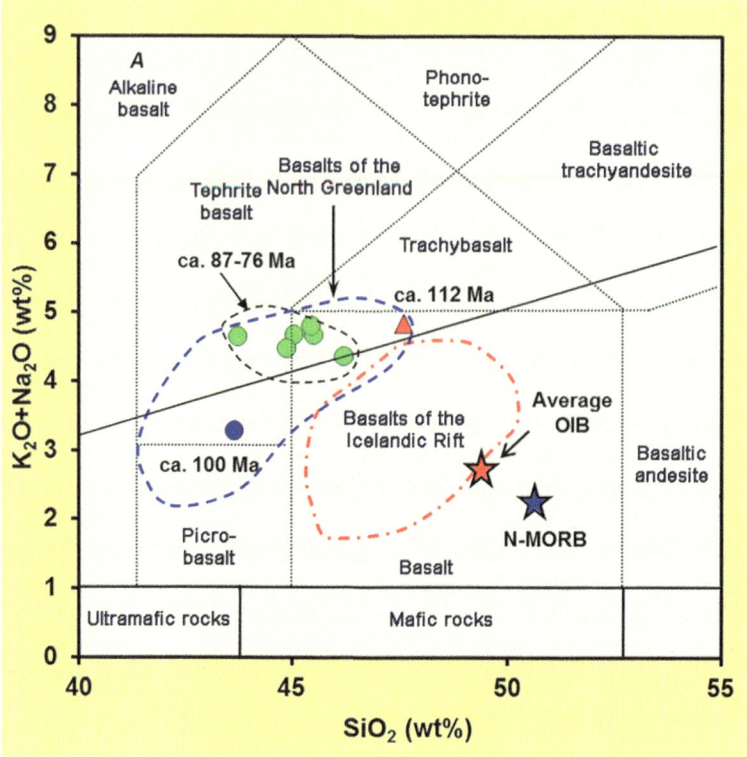

Fig. 1.38 The chemical composition of HALIP Cretaceous basalts dredged from the underwater mountain Healy - the samples of moderately alkaline tholeiitic basalts with 40Ar/39Ar 100 Ma age (*blue dot*) and alkaline basalts with 40Ar/39Ar, 83–76 Ma age (*green dots*); and also dredged from the northern extremity of the Northwind Ridge - trachybasalt with a 40Ar/39Ar age of 112 Ma (*red triangle*) (Mukasa et al. 2012)

These data point to a strong extension, rapid deep sinking of the continental crust of the submarine continuation of the Chukchi Plateau and, in the opinion of the authors, do not support the notion of oceanic crust present in the periphery of the Canadian Basin.

References

Adler RA, Polyak L, Ortiz JD et al (2009) Sediments record from the western Arctic Ocean with an improvement late quaternary age resolution: HOTRAX core HLY0503-8JPC, Mendeleev ridge. Glob Planet Change J 68:18–29

Alexeev SP, Kostenich AV, Starov KG et al (2012) Continental foot of the Arctic Basin. Arktika: Ekologiya i Ekonomika [The Arctic: Ecology and Economy] 1(5):82–91. in Russian

Andersen OB, Knudsen P, Kenyon S et al (2014) Global and Arctic marine gravity field from recent satellite altimetry. 76th EAGE conference extended abstracts

Andronikov A, Mukasa S, Mayer LA et al (2008) First recovery of submarine basalts from the Chukchi borderland and Alpha. Eos Trans AGU 89(53.) Abstract V41D-2124

Arctic Gravity Project http://earth-info.nga.mil/GandG/wgs84/gravitymod/

Assinovskaya BA (1994) Barents Sea seismicity. RAN, Moscow

Atlas of the Oceans (1980) The Arctic Ocean. GUGK, Moscow

Avetisov GP (1971) Seismic regionalization of Frants-Joseph Archipelago. Geol Expr Arctic J 6:28–134

Avetisov GP (1975) The Laptev Sea seismicity in relation to the seismicity of the Eurasian Basin. Tectonics Arctic J 1:31–36

Avetisov GP (1993a) Geodynamics of seismic zones of the Arctic region. J Geology Homel 10:52–62

Avetisov GP (1993b) Some aspects of lithospheric dynamics of Laptev Sea (English translation). J Phys Solid Earth 29(5):402–412

Avetisov GP (1996) Seismic zones of the Arctic. VNIIOkeangeologia, Saint Petersburg

Avetisov GP (1999) Geodynamics of the zone of continental continuation of Mid-Arctic earthquakes belt (Laptev Sea). Phys Earth and Planet Inter J 114:59–70

Avetisov GP (2000) More about the Laptev Sea earthquakes. Geol Geophys Arctic Region Lithos J 3:104–114

Avetisov GP (2002) Lithospheric plates boundaries of the Laptev Sea shelf. Doklady RAN J 385(6):793–796. (in Russian)

Avetisov GP, Golubkov VS (1971) Tectonic-seismic regionalization of Eurasian Basin of the Arctic Ocean and adjacent seas. Geology and mineral deposits of the Northern Siberia, NIIGA Leningrad, p 66–73

Backman J, Moran K et al (2006) Sites M0001-M0004. Expedition 302 Scientists. Proceedings of the Integrated Ocean Drilling Program 302, p 169

Basham PW, Forsyth DA, Wetmiller RJ (1977) The seismicity of Northern Canaga. Can J Earth Sci 14:1646–1667

Bowin C, Warsi W, Milligan J (1982) Free-air gravity anomaly atlas of the world. Woods Hole Oceanographic Institution Geological Society of America Map and Chart Series MC-46

Brozena JM, Childers VA, Daniel ED (1999) New compilation of potential field and bathymetry maps on the basis of joint digital processing of US and RF data sets in the high seas Arctic. Eos Trans American Geophysical Union Boston, S193

Brozena JM, Childers VA, Lawver LA et al (2003) New aerogeophysical study of the Eurasia Basin and Lomonosov ridge: implications for basin development. Geol J 31(9):825–828

Bruvoll V, Kristoffersen Y, Coakley B et al (2010) Hemipelagic deposits on the Mendeleev and northwestern alpha submarine ridges an the Arctic Ocean: acoustic stratigraphy, depositional environment and an inter-ridge correlation calibrated by ACEX results. Mar Geophys Res J 31:149–171

Bruvoll V, Kristoffersen Y, Coakley B et al (2012) The nature of the acoustic basement on Mendeleev and northwestern alpha ridges, Arctic Ocean. Tectonophysics J 514:123–145

Bungum H, Mitchell BJ, Kristoffersen Y (1982) Concentrated earthquake zones in Svalbard. Tectonophysics J 82:175–188

Canadian Hydrographic Service (1979) General Bathymetric Chart of the Oceans (GEBCO). Sheet 5.17: scale 1 : 6 000 000. Canadian Hydrographic Service, Ottawa

Central Arctic Basin, the Map (2002) Scale 1:2500000 (at 75°N). GUNiO MO RF Saint Petersburg, №91115

Chan WW, Mitchell BJ (1985) Intraplate earthquakes in Northern Svalbard. Tectonophysics J 114:181–191

Chernykh AA, Glebovsky VY, Korneva MS et al (2015) "Microleveling" – modern technology for balancing traverses of areal geophysical surveys. Geophys J 4:40–46

Christensen AN, Andersen OB (2015) Comparison of satellite altimeter-derived gravity data and marine gravity data. 77th EAGE Conference and Exhibition

Christensen AN, Andersen OB (2016) Comparison of satellite altimetric gravity and ship-borne gravity offshore Western Australia. 25th ASEG Conference and Exhibition

Coakley BJ, Cochran JR (1998) Gravity evidence of very thin crust at the Gakkel Ridge (Arctic Ocean). Earth Planet Sci Lett J 162:81–95

Coakley BJ, Cochran J, Edwards M (1999) Internal structure of the Lomonosov ridge and location of the continent-ocean boundary from SCICEX data. Eos, Trans. AGU. San Francisco 80(48):F998

Coles RL, Hannaford W, Haines GV (1978) Magnetic anomalies and the evolution of the Arctic. Arctic geophysical review. Earth Phys 45:51–66

Dibner VD, Gakkel YaYa, Litvin VM et al (1965) Geomorphological map of the Arctic Ocean. NIIGA Trans 143

Diets RS, Shumway G (1961) Arctic basin geomorphology. Geol Soc Am Bull J 72(9):1319–1330

Døssing A, Jackson HR, Matzka J et al (2013) On the origin of the Amerasia Basin and the high Arctic large Igneous Province (results of new aeromagnetic data). Earth Planet Sci Lett J 363:219–230

Døssing A, Hansen TM, Olesen AV et al (2014) Gravity inversion predicts the nature of the Amundsen basin and its continental borderlands near Greenland. Earth Planet Sci Lett J 408:132–145

Dove D, Coakley B, Hopper J et al (2010) HLY0503 geophysics team. Bathymetry, controlled source seismic and gravity observations of the Mendeleev ridge: implications for ridge structure, origin, and regional tectonics. Geophys J Int 183(2):481–502

Explanatory notes to the Arctic Basin maps: Orographic map of the Arctic Basin (1999) The Arctic ocean seafloor topography (Text) Saint Petersburg, p 39

Forsberg R, Olesen AV, Keller K (2001) Airborne gravity survey of the North Greenland continental shelf. IAG proceedings volume of Gravity Geoid and Geodynamics conference Banff 2000, 123:235–240

Fridman BS (2007) Hydrographic and topographic investigations related to delineation of the exterior boundary of the Russian continental shelf in the Arctic Basin. NAUKA, Saint Petersburg

Funck T, Jackson H et al (2011) The crustal structure of the alpha ridge at the transition to the Canadian polar margin: results from a seismic refraction experiment. J Geophys Res Solid Earth 116(B12):26

Gaina C, Werner S, Saltus R et al (2011) The CAMP-GM grouP. Circum-Arctic mapping project: new magnetic and gravity anomaly maps of the Arctic. In: Spencer AM (ed) Arctic petroleum geology, vol 35. Geological Society, London, pp 39–48

Gakkel YY (1957) Science and development of the Arctic. Morskoi Transport, Leningrad

Gakkel YaYa (1959) IGY in the Arctic. Nauka I Zhizn' J 1:23–26. (in Russian)

Gakkel YY (1960) Research and development of the Arctic regions. Collected papers «Sov. Geografia». Geografgiz, Moscow, pp 420–437

Gakkel YaYa, Belov NA, Dibner VD et al (1968a) Morphostructure and seafloor sediments of the Arctic Basin. Trans AANII 285:15–27

Gakkel YaYa, Dibner VD, Litvin MV (1968b) Principle features of endogenic geomorphology and tectonics of the Atlantic-Arctic province of the Arctic Ocean. Trans AANII 285:28–36

Gardner JV, Mayer LA, Larry A (2008) From the Arctic to the tropics: the U. S. UNCLOS bathymetric mapping program. Center for Coastal and Ocean Mapping, p 422

Geology of the USSR and Distribution of Mineral Deposits. The Arctic seas (1984) (red Gramberg IS, Pogrebitsky UE) NEDRA, Leningrad, 9:p 270

Geomorphological Aspects of the Russian Continental Shelf Exterior Boundary in the Arctic (2005) In: Naryshkin GD (ed) p 58. GUNIO MO RF, Saint Petersburg

Glebovsky VY, Kovacs LC (1993) The adjustment of aeromagnetic data in the deep Amerasian Basin for mapping and geological interpretation. Zonenshain Memorial Conf. on Plate Tectonics, Moscow, p 62

Glebovsky VY, Kovacs LC, Maschenkov SP et al (1998) Joint compilation of Russian and US navy aeromagnetic data in the Central Arctic Seas. J Polarforshung 68:35–40. erschienen 2000

Glebovsky VY, Zai'onchek AV, Kaminsky VD et al (2002) Digital databases and the Arctic Ocean potential fields maps. In: The Russian Arctic: geological history, mineralogy, geoecology. VNIIOkeangeologia, Saint Petersburg, pp 134–141

Glebovsky VY, Verba VV, Kaminsky VD (2008) Potential fields of the Arctic Basin: history of investigations, older analog and modern digital compilations. In: 60 years in the Arctic, Antarctic and World Ocean. VNIIOkeangeologia, Saint Petersburg, pp 93–110

Glebovsky VY, Chernykh AA, Kaminsky VD et al (2012) Structural-tectonic regionalization of potential fields in the Arctic Ocean for the latest compilation of circumpolar tectonic map of the Arctic. In: Collected articles. Geology and geophysics of the Arctic region lithosphere, vol 8. VNIIOkeangeologia, Saint Petersburg, pp 20–29

Gramberg IS, Demenitskaya RM, Sekretov SB (1990) The rift grabens of the Laptec Sea Sheld as amoissing link of the Gakkel–Moma Rifts. ReP.AN. SSSR 311(3):689–694

Grantz A, May SD, Taylor PT et al (1990) Canada Basin. In: Grantz A, Johnson L, Sweeney J (eds) The geology of North America. The Arctic Ocean region. Geological Society of America, Boulder, pp 379–402

Grantz A, Pease VL, Willard DA et al (2001) Bedrock cores from 89° North: implications for the geologic framework and Neogene paleooceanology of the Lomonosov ridge and a tie to the Barents shelf. Geol Soc Am Bull 113(10):1272–1284

Grantz A, Hart PE, Childers VA (2011) Geology and tectonic development of the Amerasia and Canada basins, Arctic Ocean. Geol Soc Lond Mem 35:771–800

Gusev EA, Maksimov FE, Kuznetsov VY et al (2013) Stratigraphy of bottom sediments in the Mendeleev Ridge area (Arctic Ocean). Dokl Earth Sci 450(2):602–606

Gusev EA, Lukashenko RV, Popko AO et al (2014) New information on slope structure of the Mendeleev ridge seamounts (the Arctic Ocean). Doklady RAN J 455(2):184–188

Gusev E, Rekant P, Kaminsky V et al (2017) Morphology of seamounts at the Mendeleev rise, Arctic Ocean. Polar Res J 36:2–16

Heezen BC, Ewing M (1961) The mid-oceanic ridge and its extension through the Arctic Basin. In: Raaschm G (ed) Geology of the Arctic. University of Toronto Press, Toronto, pp 622–642

Heezen BC, Tharp M (1971) (Bathymetric Compilers). Arctic Ocean Floor. Scale 1 : 9 757 000. National Geographic Society

Heezen BC, Tharp M (1975) Map of the Arctic Region. World 1 : 5 000 000. American Geographical Society

Hinz K, Block M et al (1998) Deformation of continental lithosphere on the Laptev Sea shelf, Russian Arctic. III International Conference on Arctic Margins. Celle (Germany) 12–16 October 1998

Husebye E, Gjoystdal H, Bungum H et al (1975) The seismicity of the Norvegian and Greenland seas and adjacent continental shelf areas. Tectonophysics J 26:55–70

Imaging of Underwater Terrain on Naval Maps (1973) Leningrad, GUNiO MO RF, p 162

International Bathymetric Chart of the Arctic Ocean (IBCAO) http://ibcao.org/

Ivanova NM, Sakoulina TS, Roslov YV (2006) Deep seismic investigation across the Barents–Kara region and Novozemelskiy Fold Belt (Arctic shelf). Tectonophysics J 420:123–140

Jackson HR, Forsyth DA, Johnson GL (1986) Oceanic affinities of the alpha ridge, Arctic Ocean. Mar Geophys Res J 73:237–261

Jackson HR, Dahl-Jensen T, The LORITA working group (2010) Sedimentary and crustal structure from the Ellesmere Island and Greenland continental shelves onto the Lomonosov ridge, Arctic Ocean. Geophys J Int 182(1):11–35

Jakobsson M (1999) First high-resolution chirp sonar profiles from the Central Arctic Ocean reveal erosion of Lomonosov ridge sediments. Mar Geol J 158:111–123

Jakobsson M, Lølie R, Al-Hanbali H et al (2000) Manganese and color cycles in Arctic Ocean sediments constrain Pleistocene chronology. Geol J 28:23–26

Jakobsson M, Backman J, Rudels B et al (2007) The early Miocene onset of a ventilated circulation regime in the Arctic Ocean. Nature J 447:986–990

Jakobsson M, Macnab R, Mayer L et al (2008) An improved bathymetric portrayal of the Arctic Ocean: implications for ocean modeling and geological, geophysical and oceanographic analyses. Geophys Res Lett J:35. https://doi.org/10.1029/2008GL033520

Jakobsson M, Mayer L, Coakley B et al (2012) The International Bathymetric Chart of the Arctic Ocean (IBCAO) Version 3.0. Geophy Res Lett 39(12):n/a–n/a. https://doi.org/10.1029/2012GL052219

Johnson GL, Pogrebitsky J, Macnab R (2013) Arctic structural evolution: relationship to paleoceanography. The polar oceans and their role in shaping the global environment: American Geophysical Union Geophysical Monograph 85, pp 285–294 (Copyright 1994 by the American Geophysical Union)

Jokat W (2003) Seismic investigations along the western sector of alpha ridge, Central Arctic Ocean. Geophys J Int 152:185–201

Jokat W (2005) The sedimentary structure of the Lomonosov ridge between 88°N and 80°N. Geophys J Int 163:698–726

Jokat W, Micksch U (2004) Sedimentary structure of the Nansen and Amundsen basins, Arctic Ocean. Geophys Res Lett J 31:1–4

Jokat W, Kristoffersen Y, Rasmussen TM et al (1992) ARCTIC 91:Lomonosov ridge – a double-sided continental margin. J Geol 20:887–890

Jokat W, Weigelt E, Kristoffersen Y et al (1995) New insights into evolution of the Lomonosov ridge and the Eurasian Basin. Geophys J Int 122:378–392

Kaban'kov VY, Andreeva IA, Ivanov VN et al (2004) The geotectonic nature of the Central Arctic morphostructures and geological implications of bottom sediments for its interpretation. Geotectonics J 38:430–442

Kaminsky VD, Glebovsky VY, Kiselev YG et al (2000) History of geological and geophysical investigations in the Arctic Ocean and its continental margins in relation to the shelf boundary delinitaion problem. In: Geology and geomorphology of the Arctic Ocean and exterior boundary of the Russian Federation continental shelf in the Arctic Basin VNIIOkeangeologia, Saint Petersburg, pp 17–30

Kaminsky VD, Poselov VA, Avetisov GP et al (2014) Russian Arctic Geotransects (results of geological and geophysical studies). VNIIOkeangeologia, Saint Petersburg, p 164

Kaminsky VD, Poselov VA, Avetisov GP et al (2017) Russian Arctic Geotransects (results of geological and geophysical studies). VNIIOkeangeologia, Saint Petersburg, p 180

Kenyon S, Forsberg R, Coakley B (2008) New gravity field for the Arctic. Eos Trans AGU 89(32):289–290

Khain VE, Lobkovsky LI (2003) Continental slopes: morphology, tectonics, deep structure and resources. In: Actual problems of oceanology. Nauka, Moscow, pp 63–81

King ER, Zietz I, Alldredge R (1966) Magnetic data on the structure of the central Arctic region. Geol Soc Am Bull J 77:619–646

Kiselev YG (1979) Seafloor structure and geological evolution of the abyssal Arctic Ocean. In: Geophysical investigations of the world ocean. NIIGA, Leningrad, pp 5–13

Kiselev YG (1986) Deep-laid geology of the Arctic Basin. NEDRA, Leningrad. (in Russian)

Kovacs LC, Glebovsky VY (1993) Adjusting and Combining Russian and US Navy Aeromagnetic Data in the Deep Amerasian Basin. Eos Trans AGU

Kovacs LC, Bernero C, Johnson GL et al (1984) Residual magnetic anomaly chart of the Arctic Ocean region. Scale 1 : 6 000 000, Map and chart series MC-53. Geological Society of America, Boulder

Kovacs LC, Johnson GL, Srivastava SP et al (1987) Residual magnetic anomaly chart of the Arctic Ocean region. Scale 1 : 6 000 000, vol L. Geological Society of America, Geology of North America

Kovacs LC, Glebovsky VY, Maschenkov SP et al (2002) New map and grid of compiled magnetic anomalies from the Arctic Ocean. Eos Trans AGU 83(47):F 1330

Kristoffersen Y, Mikkelsen N (2004) Scientific drilling in the Arctic Ocean and the site survey challenge: Tectonic, paleoceanographic and climatic evolution of the Polar Basin. JEODI Workshop, Copenhagen., 2003, p 85

Kulakov YN, Dibner VD, Egiazarov BX et al (1986) Morphostructure of the Arctic Basin of the Arctic Ocean. In: Structure and history of development of the Arctic Ocean. Sevmorgeologiya, Leningrad, pp 40–53

Kutschale H (1966) Arctic Ocean geophysical studies: the southern half of the Siberian Basin. Geophys J 31(4):683–710

Langinen AE, Gee DG, Lebedeva-Ivanova NN et al (2006) Velocity structure and correlation of the sedimentary cover on the Lomonosov Ridge and in the Amerasian Basin, Arctic Ocean. IV International Conference on Arctic Margins (ICAM IV), Canada, 30 September – 3 October 2003

Langinen A, Lebedeva-Ivanova N, Gee D et al (2009) Correlations between the Lomonosov ridge, Marvin spur and adjacent basins of the Arctic Ocean based on seismic data. Tectonophysics J 472:309–322

Lastochkin AN, Naryshkin GD (1989) Orography of the Arctic Ocean. Vestnik LGY 7/2(14):45–54

Laxon S, McAdoo D (1994) Arctic Ocean gravity field derived from ERS-I satellite altimetry. Science J 265:621–624

Laxon S, McAdoo D (1998) Satellites provide new insights into polar geophysics. Eos Trans AGU 79(6):69–73

Lebedeva-Ivanova N (2010) Geophysical studies bearing on the origin of the Arctic Basin. Dissertation, Acta Universitatis Upsaliensis

Lebedeva-Ivanova NN, Zamansky YY (2006) Seismic profiling across the Mendeleev ridge at 82°N: evidence of continental crust. Geophys J Int 165(2):527–544

Litvinova T, Glebovsky V (2008) New map and grid of compiled magnetic anomalies from the Arctic Ocean and adjacent continental part of the Russian Federation. Geophisical Research Abstracts, vol 10

Macnab R, Verhoef J, Srivastava SP (1992) Magnetic observations from the Arctic and North Atlantic oceans. Eos TransAGU 73:123–124

Mair JA, Forsyth DA (1982) Crustal structures of the Canada Basin near Alaska, the Lomonosov ridge and adjoining basins near the north pole. Tectonophysics J 89:239–253

Makarova ZA (1978) Map of magnetic anomalies of the continental USSR (scale 1:2 500 000, earth normal magnetic field). VSEGEI, Leningrad

Maschenkov SP, Glebovsky VY, Zayonchek AV (2001) New digital compilation of Russian aero-magnetic and gravity data over the north Eurasian shelf. Polarforschung J 69:35–39

Matzka J, Rasmussen T, Olesen A et al (2010) A new aeromagnetic survey of the north pole and the Arctic Ocean north of Greenland and Ellesmere Island. Earth Planets Space J 62(10):829–832

Mayer LA, Brumley K, Andronikov A et al (2008) Resent mapping and sampling on the Chukchi Borderland and Alpha/Medeleev Ridge Complex. Eos Trans AGU 89(53.) Abstract C11C-0516

Michael PJ, Langmuir CH, Dick HJB et al (2003) Magmatic and amagmatic seafloor generation at the ultraslowspreading Gakkel ridge, Arctic Ocean. J Nature 423(26):956–961

Miller EL, Toro J, Gehrels G et al (2006) New insights into Arctic paleogeography and tectonics from U–Pb detrital zircon geochronology. Tectonics J 25(3)

Morozov AF, Petrov OV, Shokalsky SP et al (2013) New geological evidence justifying the nature of the continental area of the Central Arctic elevations. Reg Geo Metallog J 53:34–55

Mosher DC, Chapman C, Shimeld J et al (2013) High Arctic marine geophysical data acquisition. Lead Edge J 32(5):524–536

Mudie PJ, Stoffyn-Egli P, Van Wagoner NA (1986) Geologic constraints for tectonic models of the alpha ridge. J Geodyn 6:215–236

Mühe R, Jokat W (1999) Recovery of volcanic rocks from the Alpha Ridge, Arctic Ocean: prelimi-nary results. AGU 1999 Fall Meeting. Transaction AGU 80(№46):F1000

Mukasa SB, Mayer LA, Brumley K et al (2012) New geochemical and 40Ar/39Ar data for the first Intraplate lavas recovered from the Arctic Ocean: bearing on the origin of the Amerasian Basin. International scientific workshop on extended continental shelf in the Arctic, New Hampshire, USA 7–8 November 2012

Nauret F, Snow JE, Hellebrand E et al (2011) Non-peridotitic source in mid-ocean ridge melts from Lena trough, Arctic Ocean. Petrol J 52:1185–1206

O'Brien TM, Miller EI, Benowitz JP et al (2016) Dredge samples from the Chukchi borderland: implications for paleogeographic reconstruction and tectonic evolution of the Amerasia Basin of the Arctic. Am J Sci 316:873–924

Olesen OG, Gellein J, Habrekke H et al (1997) Magnetic Anomaly Map of Norway and adjacent ocean areas. Scale 1 : 3 000 000. Geological Survey of Norway, Oslo

Orographic Map of the Arctic Basin (1995) [Maps] Naryshkin GD (ed) Scale 1 : 5 000 000. Helsinki, Karttakeskus

Ostenso NA (1962) Geophysical investigations of the Arctic Ocean basin. Research Report University of Wisconsin Geophysical and Polar Research Center, N4, p 124

Pogrebitsky YE (ed) (1984) The geological structure of the USSR and regularities of minerals seas of the Russian Arctic. NEDRA, Leningrad. (in Russian)

Pogrebitsky YE, Goriachev YV, Trukhalev AI (2005) Tectonic zoning of the Central Arctic Basin. Razvedka i okhrana nedr J 6:24–27

Polyak L, Edwards MH, Coakley BJ et al (2001) Ice shelves in the Pleistocene Arctic Ocean inferred from glaciogenic deep-sea bedforms. Nature J 410:453–457

Riddihough RP, Haines GV, Hannaford W (1973) Regional magnetic anomalies of the Canadian Arctic. Can J Earth Sci 10(2):157–163

Schenke HW, Zarayskaya Y, Accettella D, Armstrong A, Anderson RM, Bienhoff P, Camerlenghi A, Church I, Edwards M, Gardner JV, Hall JK, Hell B, Hestvik O, Kristoffersen Y, Marcussen C, Mohammad R, Mosher D, Nghiem SV, Pedrosa MT, Travaglini PG, Weatherall P (2012) The International Bathymetric Chart of the Arctic Ocean (IBCAO) Version 3.0. Geophys Res Lett 39(12):XXX–XXX

Scordas E, Meyer K, Olsson R et al (1991) Causality between interplate (North Atlantic) and intra-plate (Fennoscandia) seismicities. Tectonophysics J 185:295–307

Seafloor Topography of the Arctic Ocean (1998) [Map] scale 1 : 5 000 000, stereographic projection. GUNiO MO RF. VNIIOkeangeologia, Saint Petersburg

Sobczak LW (1977) Bathymetry of the Arctic Ocean north of 85°N latitude. Tectonophysics J 42:T27–T33

Sobczak LW, Sweeney JF (1978) Bathymetry of the Arctic Ocean. Arctic Geophys Rev J 45:7–14

Sobczak LW, Hearty DB, Forsberg R et al (1990) Gravity from 64°N to the north pole. The Geological Society of America: The Arctic Ocean Region, vol LP:101–108

Stein R, Mattheissen J, Niessen F et al (2010) Towards a better (Litho-) stratigraphy and reconstruction of quaternary Paleoenvironment in the Amerasian Basin (Arctic Ocean). Polarforschung J 79:97–121

Stein R, Fahl K, Schreck M et al (2016) Evidence for ice-free summers in the late Miocene Central Arctic Ocean. Nat Commun 7:1–13. https://doi.org/10.1038/ncomms11148

Stepanov PP, Yanushevich MA (1999) Bouguer (2.67 g/cm3) Gravity Map of the USSR, scale 1 : 2 500 000. VNIIGeofizika, Moscow

Sweeney JF, Weber IR, Blasco SM (1989) Continental ridges in the Arctic Ocean: Lorex contraints. Tectonophysics J 89:217–238

Sykes LR, Sbar ML (1973) Intraplate earthquakes, lithospheric stresses and the driving mechanism of plate tectonics. Nature J 245:298–302

Symonds PA et al (2000) Characteristics of continental margins. In: Cook PJ, Carleton CM (eds) Continental shelf limits: the scientific and legal Interface, pp 25–63

Tarduno JA, Brinkman DB, Renne PR et al (1998) Evidence for extreme climatic warmth from late cretaceous Arctic vertebrates. Science J 282:2241–2244

Taylor PT, Kovacs LC, Vogt PR et al (1981) Detailed aeromagnetic investigation of the Arctic Basin. J Geophys Res 86:6323–6333

The Arctic and Pacific Oceans (1985) Nauka, Leningrad (in Russian)

The General Bathymetric Chart of the Oceans (GEBCO) http://www.gebco.net

The Geological Map (2011) Scale 1:1 000 000. Series Ocean. In: U-53-56 Lomonosov Ridg. VSEGEI, Saint Petersburg

Treshnikov AF (ed) (1985) Atlas of the arctic. GUGK, Moscow, p 204. (in Russian)

Tsibulya LA, Levashkevich VG, Kremenetskaya EO (1993) Thermal flow and seismicity of the Barents-White Sea region. Geotermia seismichnykh i aseismichnykh zon. Nauka, Moscow, pp 27–32. (in Russian)

Van Wagoner NA, Williamson MC, Robinson PT et al (1986) First samples of acoustic basement recovered from the alpha ridge, Arctic Ocean: new constraints for the origin of the ridge. Geodynamics J 6:117–136

Verhoef J, Roest WR, Macnab R et al (1996) Magnetic anomalies of the Arctic and North Atlantic oceans and adjacent land areas. Geological survey of Canada. Open File 3125a:577

Vernikovsky VA, Morozov AF, Petrov OV et al (2014) New data on the age of dolerites and basalts of Mendeleev rise (Arctic Ocean). Dokl Earth Sci 454(2):97–101

Vinogradov VA, Gaponenko GI, Rusakov IM et al (1974) In: Papers L (ed) Tectonics of the East-Arctic shelf of the USSR, vol 171. NEDRA, Leningrad, p 144

Vogt PR, Taylor PT, Kovacs LC et al (1979) Detailed aeromagnetic investigation of the Arctic Basin. J Geophys Res 84:1071–1089

Weber JR (1980) Exploring the Arctic seafloor. GEOS J 9(3):2–7

Weber JR, Jackson HR (1985) CESAR bathymetry. In: Jackson H (ed) Initial geological report on CESAR. The Canadian expedition to study the alpha ridge. Arctic Ocezn. Geological survey of Canada. Paper, vol 8422, pp 15–17

Weber JR, Roots EF (1990) Geophysical and geological explorations in the Arctic Ocean region: Summory of the most important expeditions and surveys that have contributed to geophysical and geological knowledge of the Arctic Ocean. Mar Geol J 93:1–13

Weber JR, Sweeney JF (1990) Ridges and basins in the Central Arctic Ocean. In: The geological society of America: the Arctic ocean region, vol L, pp 305–336

Zinchenko AG (2004) Orographic zoning and general characterization of seafloor topography. In: Geology and mineral deposits of Russia. VSEGEI, Saint Petersburg 5(1)15–25

Chapter 2
Seismic Stratigraphy of Sedimentary Cover

Victor A. Poselov, Victor V. Butsenko, Artem A. Kireev, Oleg E. Smirnov, and Sergey M. Zholondz

Abstract Three major unconformities (Regional pre-Miocene, Early Eocene, post-Campanian) are regionally traced in the depressions of the Eurasian Basin. They are dated either by well data (ACEX IODP 302 on the Lomonosov Ridge), or by correlation of their onlaps on acoustic basement with the nearest identifiable linear magnetic anomaly.

The interpretation of pre-Cenozoic stratigraphy in the Central Arctic Uplifts Complex (CAUC) was based on tracing several regional seismic reflectors from seismic lines tied to wells drilled onshore and offshore Alaska into the North Chukchi Trough, with identification of Brookian, Lower Cretaceous and Upper Jurassic unconformities in the latter. This procedure was facilitated by previously established continuity of these unconformities from North Chukchi Trough into Vilkitsky Trough and Podvodnikov Basin. The established reflectors through the system of the Russian MCS profiles were further traced onto the entire Central Arctic Uplifts Complex.

Keywords Arctic Basin · Stratigraphy · Magnetic anomalies · MCS · Sampling and drilling

2.1 Eurasian Basin

Seismostratigraphic model of Eurasian Basin sedimentary section was based mainly on well ACEX IODP 302 drilled on the Lomonosov Ridge (Backman et al. 2006). Moreover, termination of the major sedimentary units against acoustic basement (TAB) were correlated with linear magnetic anomalies (LMA) (Glebovsky et al. 2006), defining age of this particular location (Fig. 2.1).

V. A. Poselov (✉) · V. V. Butsenko · A. A. Kireev · O. E. Smirnov · S. M. Zholondz
All-Russian Research Institute of Geology and Mineral Resources of the World Ocean
(VNIIOkeangeologia), Saint Petersburg, Russia
e-mail: szh@vniio.nw.ru; vicb@vniio.nw.ru

© Springer International Publishing AG, part of Springer Nature 2019　　　　71
A. Piskarev et al. (eds.), *Geologic Structures of the Arctic Basin*,
https://doi.org/10.1007/978-3-319-77742-9_2

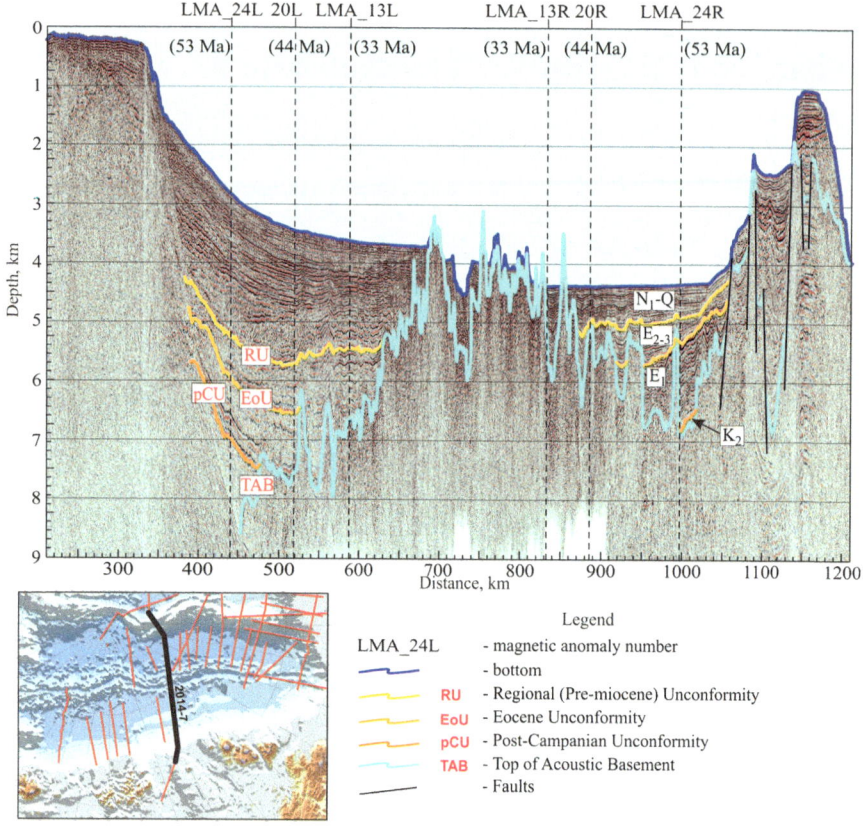

Fig. 2.1 MCS line 2014–7, Barents Shelf - Lomonosov Ridge

Amundsen Basin Seismic sections 2011–27, 2011–30, 2011–34 and 2011–35 demonstrate that three major unconformities are traced in the Amundsen Basin (Fig. 2.2–2.5).

The uppermost unconformity was identified as regional pre-Miocene (RU) by its onlap on acoustic basement (TAB) close to Linear Magnetic Anomaly 13 (LMA 13, 33 Ma, Oligocene). ACEX well data also show the similarly dated, and much more pronounced, unconformity on the Lomonosov Ridge. The difference in definition in two locations may indicate the difference in sedimentation regime in the Lomonosov Ridge and the Amundsen Basin. At the former, this unconformity reflects a complete hiatus of 18–44 Ma (Backman et al. 2006) (or drastic, up to an order of magnitude, drop of sedimentation rate between 12–36 Ma (Poirier and Hillaire-Marcel 2011)), while at the latter, the sedimentation continued uninterrupted. Evidently, RU reflects changing kinematic regime of terrigenous material transportation from the Lomonosov Ridge. This change, in turn, might be triggered by opening of the Fram Strait (dated by LMA 13) with subsequent influx of North Atlantic waters into the Eurasian Basin.

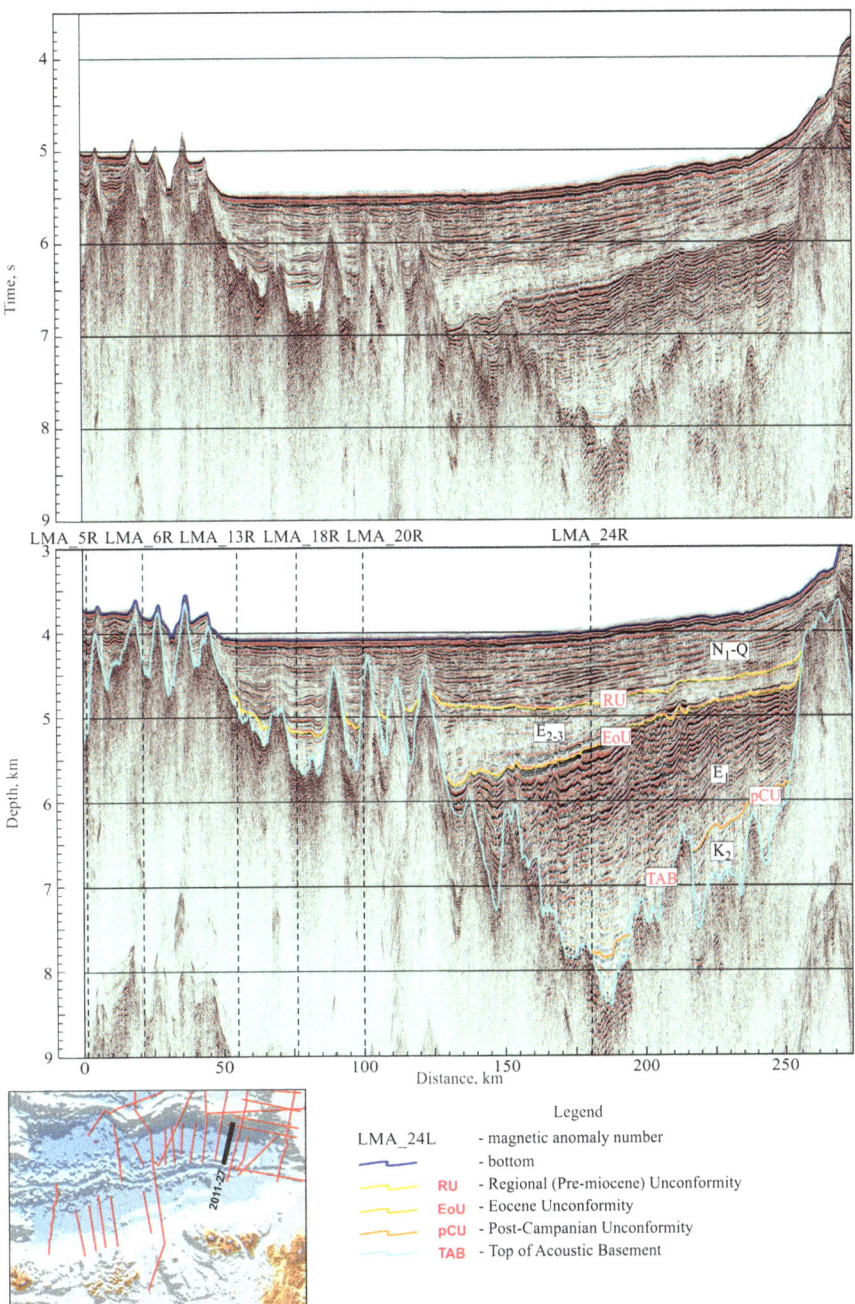

Fig. 2.2 MCS line 2011–27, Amundsen Basin – Lomonosov Ridge

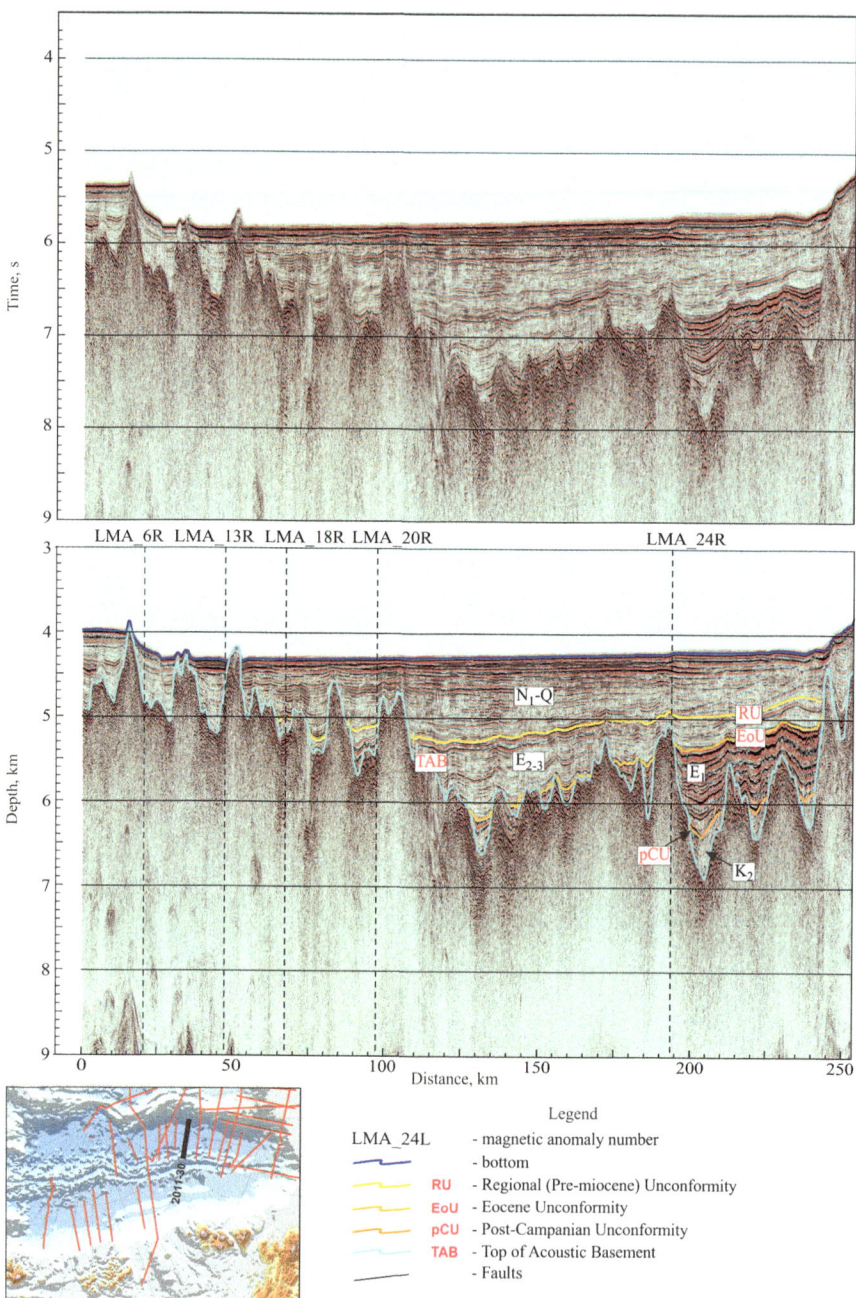

Fig. 2.3 MCS line 2011–30, Amundsen Basin – Lomonosov Ridge

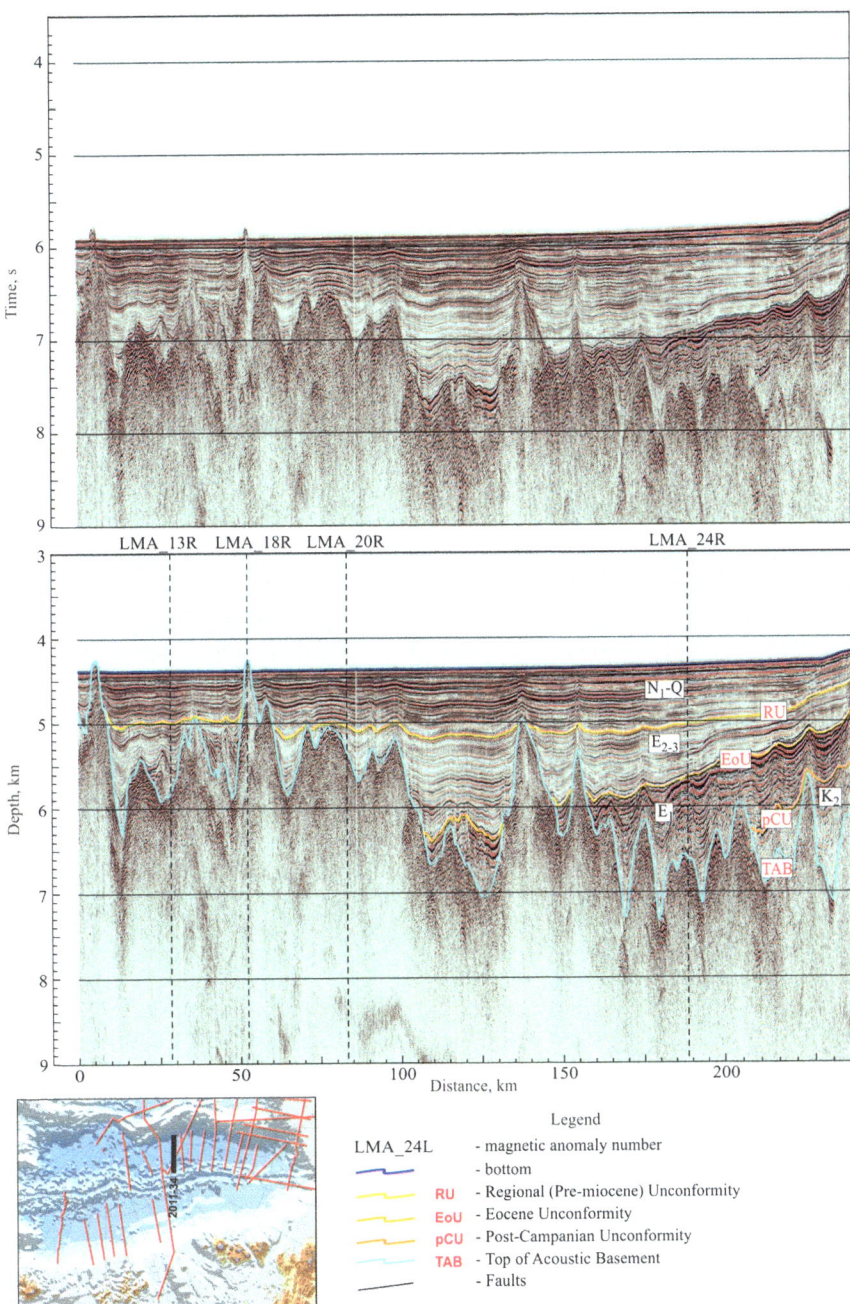

Fig. 2.4 MCS line 2011–34, Amundsen Basin – Lomonosov Ridge

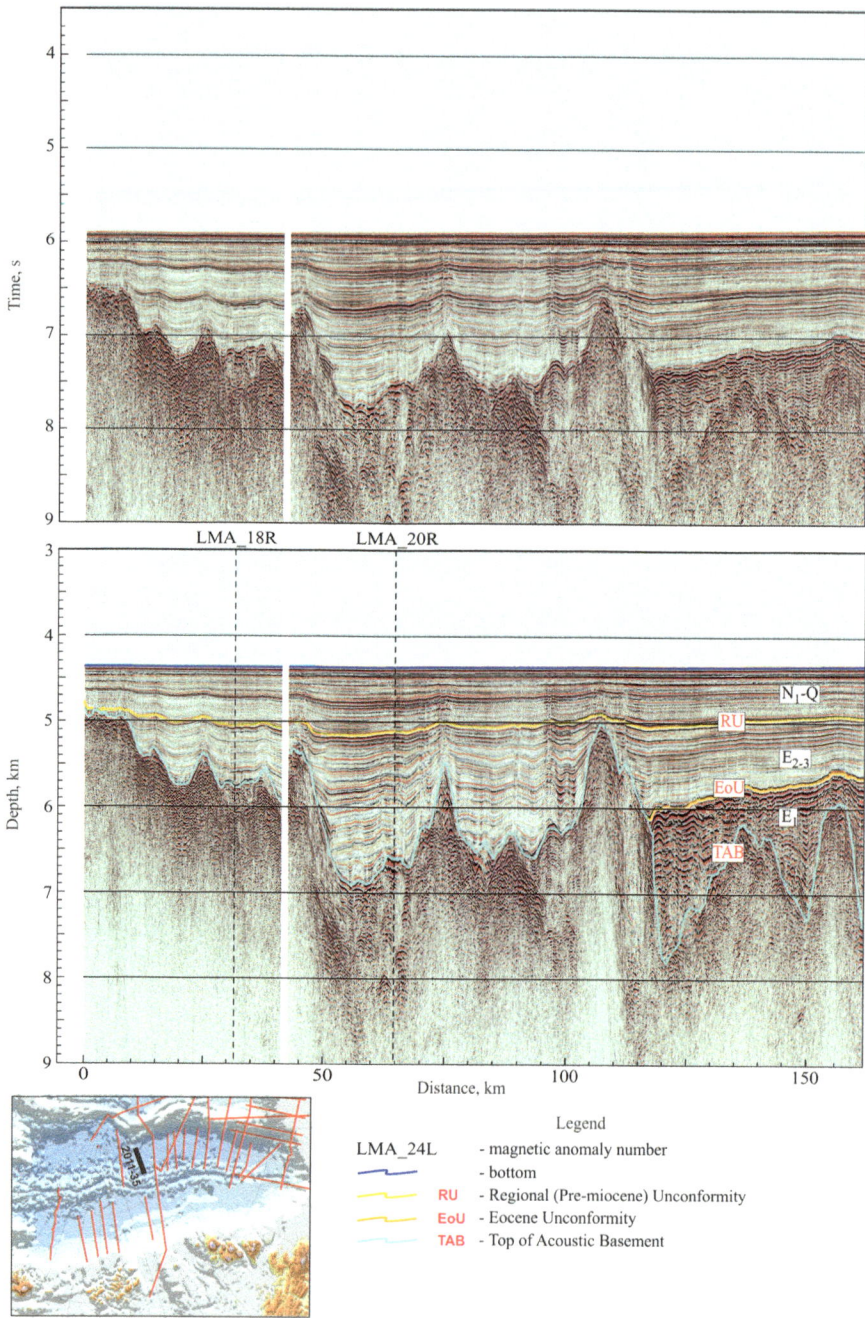

Fig. 2.5 MCS line 2011–35, Amundsen Basin – Lomonosov Ridge

The lowermost pCU unconformity is truncated by TAB at LMA 24 (54 Ma, Late Paleocene). Again, the nearly synchronous cessation of sedimentation was recorded in ACEX well at 57–80 Ma and interpreted as post-Campanian. As it accepted by the majority of researchers, this unconformity marks the top of upper Cretaceous syn-rift complex formed in rifting episodes preceding the breakup of Laurasia into North-American and Eurasian plates.

The intermediate (and the most distinguish in the Amundsen Basin) unconformity was identified between RU and pCU. This interface truncates against TAB between LMA 24 (53 Ma, Late Paleocene) and LMA 20 (44 Ma, Middle Eocene) and, accordingly, was identified as Early Eocene (EoU). In sharp contrast to sub-horizontal overlaying formations and seafloor, the EoU unconformity noticeably dips toward the Gakkel Ridge.

The above described unconformities divide the sedimentary cover into four depositional sequences. Integrated interpretation of reflection (MCS) and refraction (sonobuoy) seismic surveys provided enough information to create several detail 2D interval velocity models which support such subdivision by assigning the specific ranges of P-waves interval velocities to each sequence: 1.8–2.3 km/s for Miocene-Pleistocene (seafloor – RU); 2.4–2.7 km/s - Eocene-Oligocene (EoU-RU); 2.8–3.5 km/s – Paleocene-Eocene (pCU-EoU); and 3.6–4.1 km/s – Upper Cretaceous (TAB-pCU).

Nansen Basin Three major unconformities are also traced in the Nansen Basin (Fig. 2.6, 2.7). Dated, as above, by the age of LMA nearest to the location of unconformity termination, they are designated as Post-Campanian (pCU), Early Eocene (EoU) and regional Pre-Miocene (RU).

Interpretation of seismic horizons in the Nansen Basin is complicated by weaker reflections amplitudes and multiples interference, especially near Barents-Kara shelf slope. The acoustic basement is also untraceable at some locations. Interval P-wave velocities of the Nansen Basin major depositional sequences are: 1.9–2.3 km/s for Miocene-Pleistocene (seafloor – RU); 2.7–3.3 km/s – Eocene-Oligocene (EoU-RU); 3.5–4.2 km/s km/s – Paleocene-Eocene (pCU-EoU).

The comparative analysis of interval velocities in synchronous (as synchronous as identified LMAs are, of course) sequences demonstrates that while Miocene-Pleistocene velocities in the Amundsen and Nansen Basin are practically identical, they are noticeably higher in older formations of the Nansen Basin: 2.7–3.3 km/s against 2.4–2.7 km/s for Eocene-Oligocene sequence; 3.5–4.2 km/s against 2.8–3.5 km for Paleocene-Eocene.

The difference in interval velocities between the two basins may lie in the ways the sedimentation process proceeded within these depression. The Nansen Basin sediments came mainly from the classic passive continental margin – Barents-Kara shelf - from which they were removed by active erosional processes creating high volume of coarse and poorly sorted material. At the same time the Amundsen Basin was fed not from the passive margins but from the underwater morphological

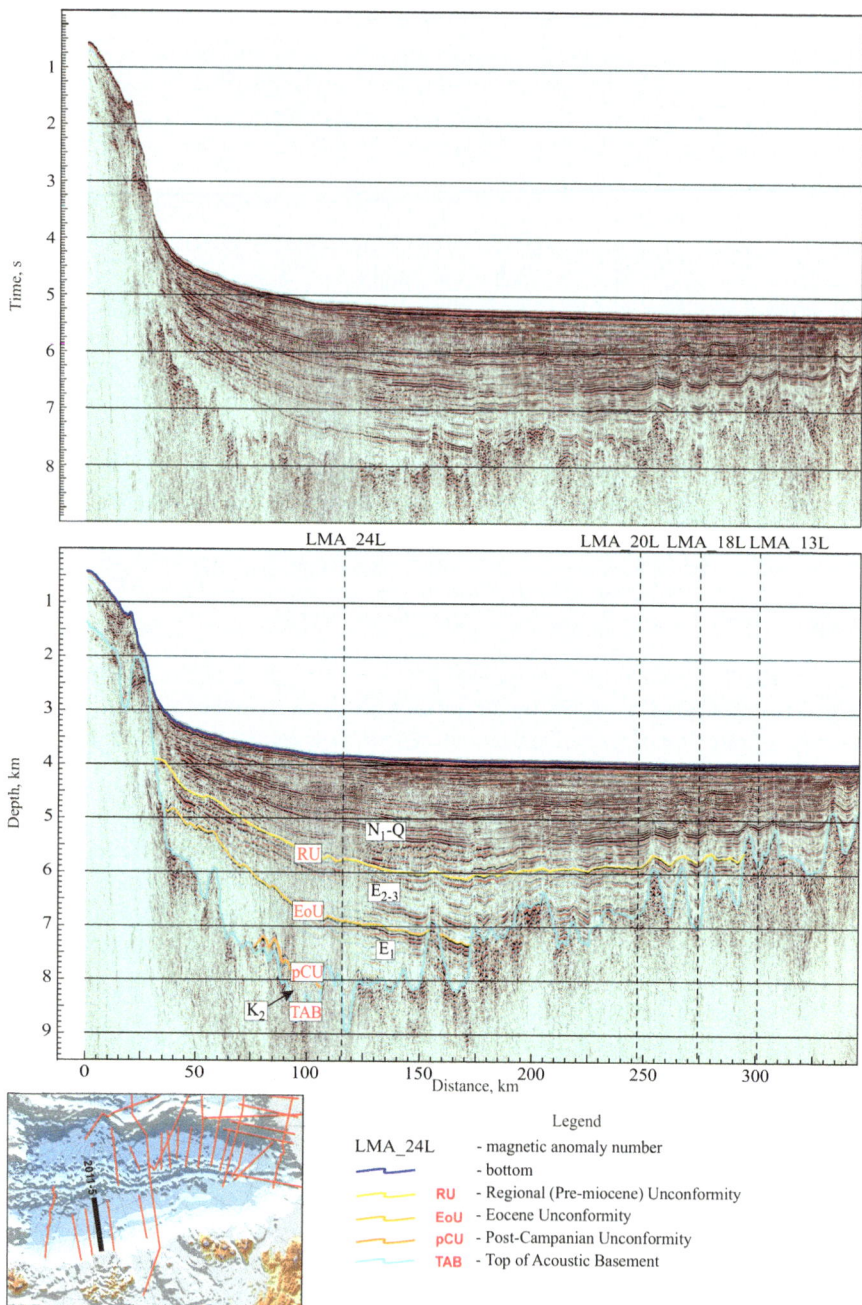

Fig. 2.6 MCS line 2011–5 (Barents-Kara shelf slope – Gakkel Ridge)

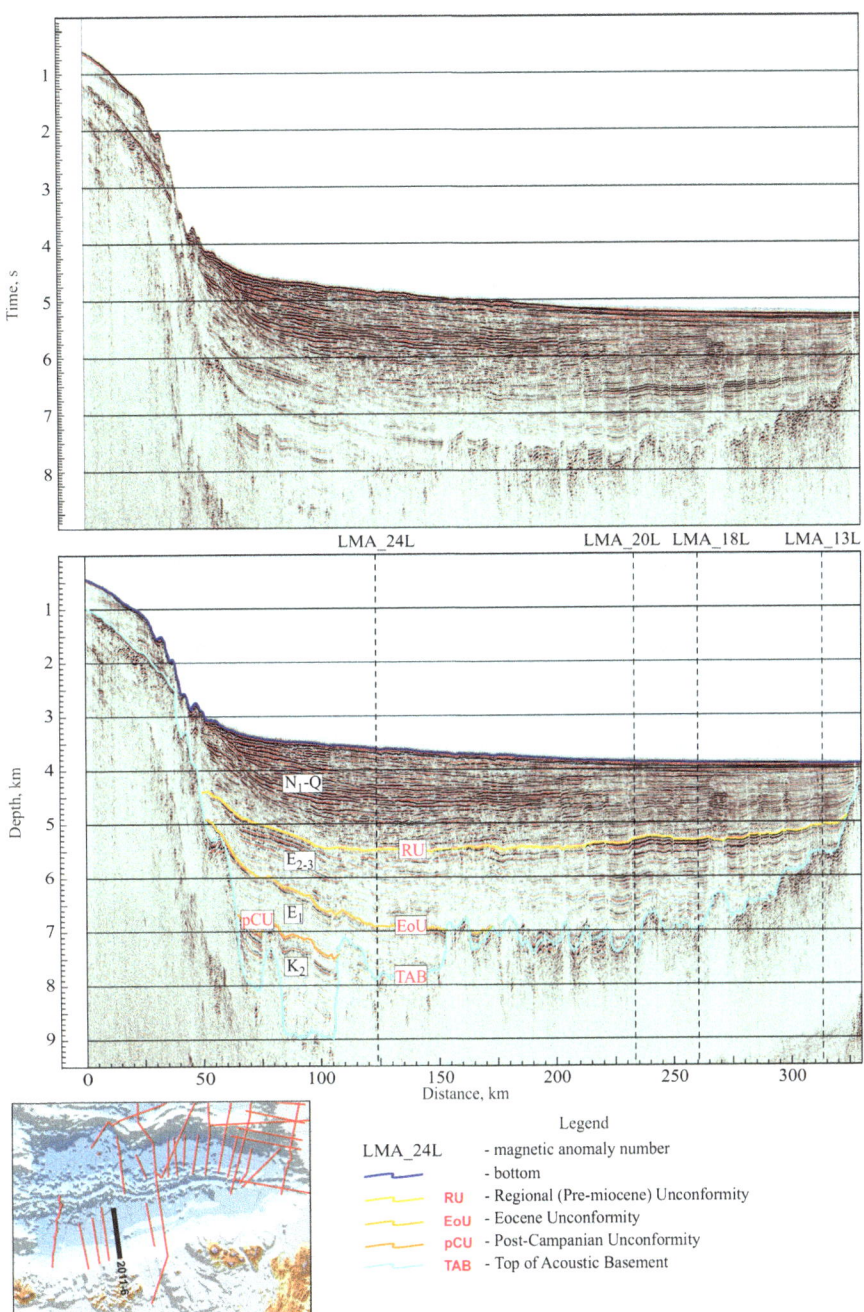

Fig. 2.7 MCS line 2011–6 (Barents shelf slope – Gakkel Ridge)

structures known as the Central Arctic Uplifts Complex inside the peripheral block of the North American lithospheric plate. Being already under more than kilometer of water, slower erosional processes produced lesser volume (≈ 1.5 times less than in the Nansen Basin) of fine-grained, well sorted and stratified sediments. That may explain not only higher interval velocities of the Nansen Basin, but also their relatively poor reflectivity.

Any ambiguity in identification of LMAs and, subsequently, in dating and correlation of the stratigraphic sequences, may also be the reason of observed differences. This problem will be discussed later.

The following sums up the seismostratigraphic features of the Eurasian Basin sedimentary cover:

1. Three major unconformities are regionally traced in the depressions of the Eurasian Basin. They are dated either by well data (ACEX IODP 302 on the Lomonosov Ridge), or by correlation of their onlaps on acoustic basement with the nearest identifiable linear magnetic anomaly.
2. The upper regional pre-Miocene unconformity (RU) terminates against acoustic basement in the vicinity of LMA 13 (33 Ma, Oligocene). The ACEX well on the Lomonosov Ridge relates this unconformity to the major depositional hiatus (or depositional slowdown). The unconformity may reflect the change of deposition rate due to opening the Fram Strait.
3. The middle (and the most pronounced) Early Eocene (EoU) unconformity contacts acoustic basement between LMA 24 (53 Ma, Late Paleocene) and LMA 2o (44 Ma, Mid Eocene).
4. The lower unconformity truncation is located close to LMA 24 (53 Ma, Late Paleocene) and is identified as post-Campanian (pCU). This unconformity is synchronous to the second depositional hiatus (57–80 Ma) in the ACEX well on the Lomonosov Ridge. It marks the top of Upper Cretaceous synrift sedimentary sequence and may precede the onset of Eurasian Basin spreading.
5. The above described unconformities mark the regional boundaries of four depositional sequences of the Eurasian Basin: Miocene-Pleistocene (seafloor – RU); Eocene-Oligocene (EoU-RU); Paleocene-Eocene (pCU-EoU) and Upper Cretaceous (TAB-pCU), each one of them with its own pattern of seismic reflectivity and range of interval velocities.
6. It was suggested that noticeably higher interval velocities of the lower three sequences (and thickness of sediments) in the Nansen Basin compare to the Amundsen Basin may be caused by different depositional environment, namely, by different source of terrigenous material: Barents-Kara shelf, classic continental margin of Eurasian continental plate for the Nansen Basin, and submerged Central Arctic Uplifts Complex for the Amundsen Basin.

2.2   Central Arctic Uplifts Complex

The Central Arctic Uplifts Complex includes the Lomonosov, Mendeleev, Alpha Ridges, Chukchi Plateau and adjacent Podvodnikov, Makarov, Mendeleev, Chukchi Basins.

The seismic stratigraphy of the Province sedimentary cover is based on:

– correlation of major Cenozoic unconformities and lithostratigraphy of the wells ACEX IODP 302 on the Lomonosov Ridge (Backman et al. 2006) and POPCORN in Chuklchi Sea of Alaska (Sherwood et al. 2002) (the closest one to the North Chukchi Trough);
– on the age of the pre-Cenozoic Chukchi Sea sediments from the US seismic surveys and wells drilled onshore (INIGOK) (Bird 1994) and offshore (Sherwood et al. 2002) Alaska;
– seafloor sampling and shallow drilling on Mendeleev Ridge.

Cenozoic Sediments The major Cenozoic unconformities identified in the well ACEX IODP 302 on the Lomonosov Ridge were projected onto the Russian MCS A-7 seismic line running along the Ridge axis from 83.5°N south to Novosibirsk Islands. Using this seismic section as a reference, all major Cenozoic unconformities were traced from the Lomonosov Ridge to the Podvodnikov Basin, Vilkitsky Trough and further through the Chukchi Shelf to the Mendeleev Ridge, Chukchi Basin and Chukchi Plateau (Fig. 2.8).

The distinct and strong reflection is visible through whole length of the line A-7 at the bottom of the Lomonosov Ridge uppermost semi-pelagic unit (Fig. 2.9). It corresponds to the interface between organically rich neritic material of mid-Eocene and organically poor semi-pelagic Early-Mid-Miocene sediments in the ACEX well (Backman et al. 2006). This interregional pre-Miocene unconformity (RU), traceable throughout the entire Central Arctic Province (Butsenko and Poselov 2006), may manifest the cardinal change in sedimentation regime related to opening of the Fram Strait (Late Paleogene-Early Neogene) and massive influx of the North Atlantic waters into the Arctic Ocean (Poselov et al. 2014).

Another prominent seismic event below RU unconformity is associated with the second depositional hiatus (57–80 Ma). The hiatus was identified in the ACEX core, dated by Campanian (and, possibly, Maastrichtian) fossils present below the erosional interface as post-Campanian (Backman et al. 2006) and correlated with strong seismic reflection on line AWI-91090 passing through the well location (Fig. 2.10), (Jokat et al. 1995). Formation of this unconformity (pCU) is thought to be related to cardinal changes of the depositional regime at the end of Cretaceous period triggered by opening of the Eurasian Basin (Poselov et al. 2014), initial stages of event.

As line A-7 (Fig. 2.9) clearly demonstrates, the interval between RU and pCU unconformities decreases northwards and at the Lomonosov Ridge they practically merge together. We continue to interpret the merged (RU + pCU) unconformity as RU unconformity, because its age is dictated by the age of overlying horizons and

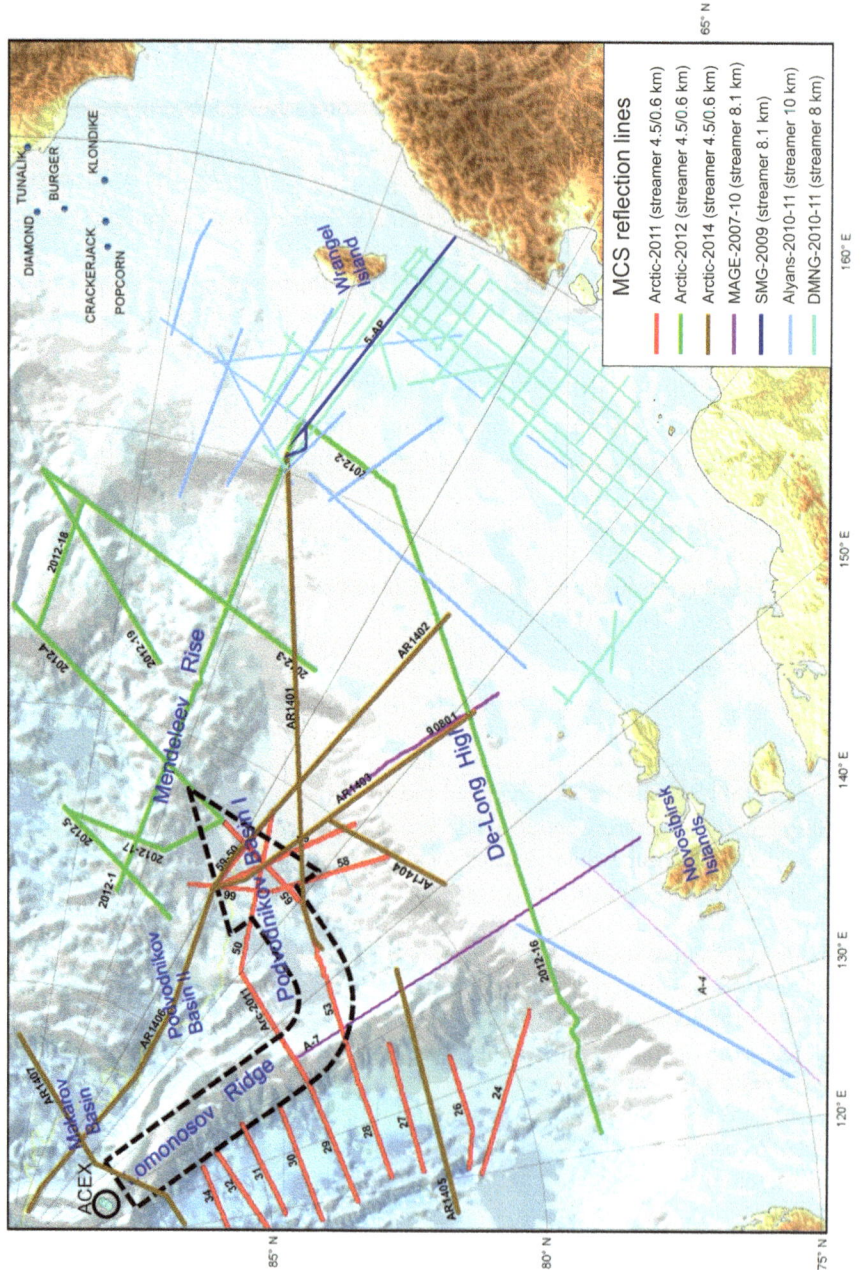

Fig. 2.8 MCS seismic surveys in the Central Arctic Uplifts

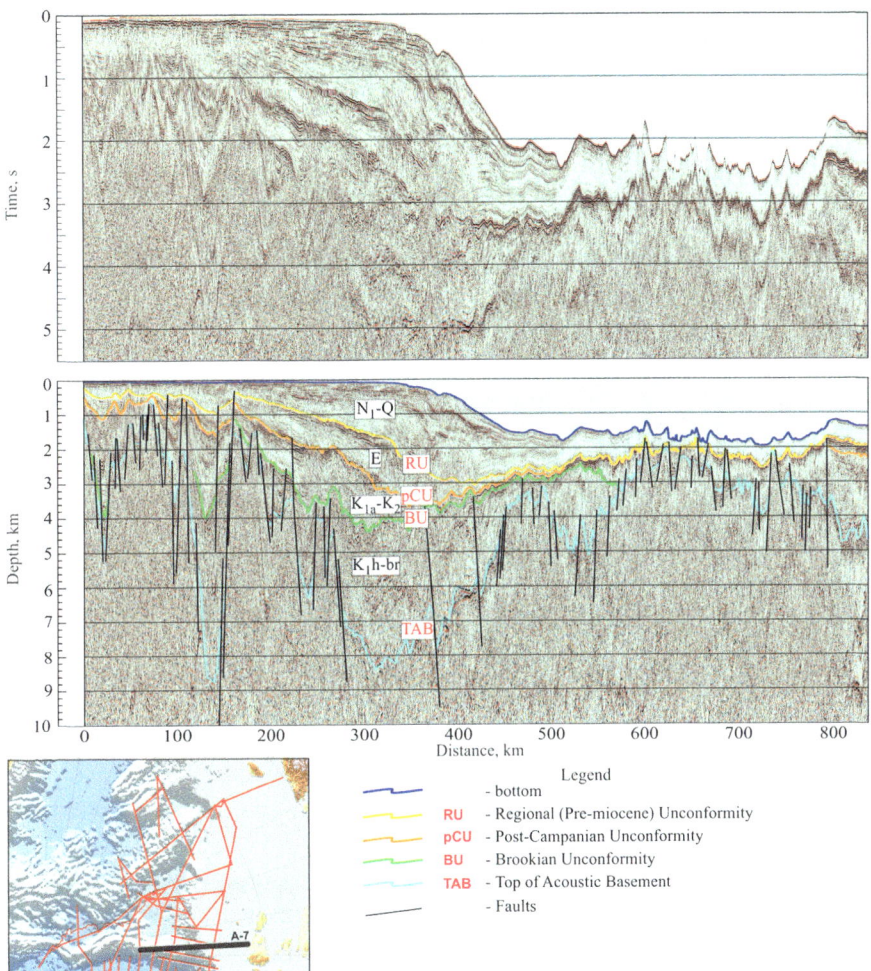

Fig. 2.9 MCS Seismic line A-7

the same horizon of semi-pelagic sediments continues to be present above it even at the crest of the Lomonosov Ridge. It seems that this regional unconformity is the result of drastic decrease of sedimentation rate at the crestal parts of the Ridge and on its slopes - by erosion. In the transitional zone between the crest of the Lomonosov Ridge and its slope, as well as in the shelf itself, the merged unconformities separate again and continue to be marked as pre-Miocene and post-Campanian.

As the ages of major unconformities define the ages of depositional sequences bounded by them, overall four sequences – Lower Cretaceous, Upper Cretaceous, Paleogene and Lower Neogene-Pleistocene - are identified on seismic section A-7 (Fig. 2.9). All four are present in the depositional fore-deep outlined by deepening of the acoustic basin under the transitional zone between outer shelf and the Lomonosov Ridge. At the Ridge crest the Paleogene sequence is either completely

Fig. 2.10 Seismic Line
AWI-91090 (fragment) and
unconformities penetrated
by ACEX 302 well

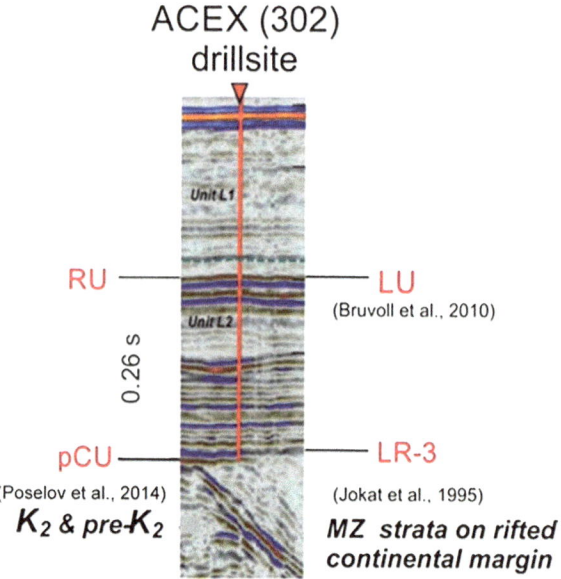

eroded or reduced to first hundred meters (~200 m in ACEX well). Therefore, up to
Early Miocene, the Siberian segment of the Lomonosov Ridge was above sea level,
or slightly below it (Poselov et al. 2014).

Seismic line AWI-91090, shot through the well site ACEX IODP 302, clearly
displays two unconformities – LU (Bruvoll et al. 2012) and LR-3 (Jokat et al. 1992,
1995). Angular unconformity LR-3 separates upper undisturbed sub-horizontal
units from underlying dipping and, at places, chaotic reflections. The LR-3 unconformity was originally thought to represent the eroded surface of moderately dislocated Early-Middle Cretaceous formations overlaid by undisturbed Cenozoic cover
(Jokat et al. 1995).

The LU and LR-3 unconformities coincide with RU (pre-Miocene) and pCU
(post-Campanian) unconformities (Butsenko and Poselov 2006; Poselov et al. 2014)
(Fig. 2.9). The present stratigraphic subdivision is in agreement with previously
proposed stratigraphy of the Laptev Sea shelf (Franke et al. 2001).

Evidently, the Cenozoic formations continue from the Eurasian shelf trough transitional zone to the Lomonosov Ridge uninterrupted, preserving their seismostratigraphic and seismofacial characteristics. That would not be the case if there was
substantial strike-slip movement between the ridge and the Laptev Sea continental
margin.

Fig. 2.11 show latitudinal seismic section from the crest of the Lomonosov Ridge
with minimal thickness of the Paleogene sequence (merged RU + pCU) towards the
Podvodnikov Basin. The meridional section (Vilkitsky Trough – Podvodnikov
Basin) is presented on Fig. 2.12.

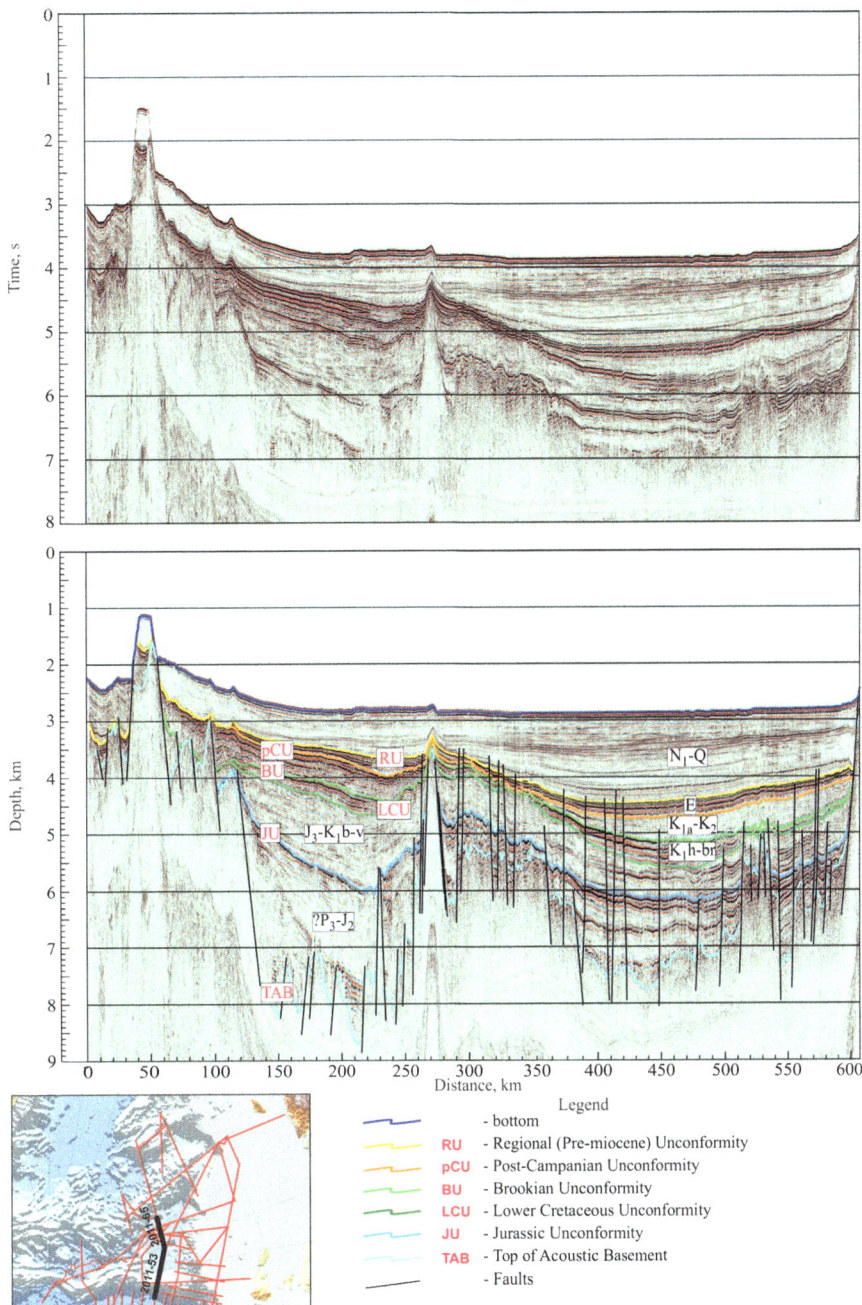

Fig. 2.11 MCS traverse 2011–53 + 2011–56. The Podvodnikov Basin seismic stratigraphy

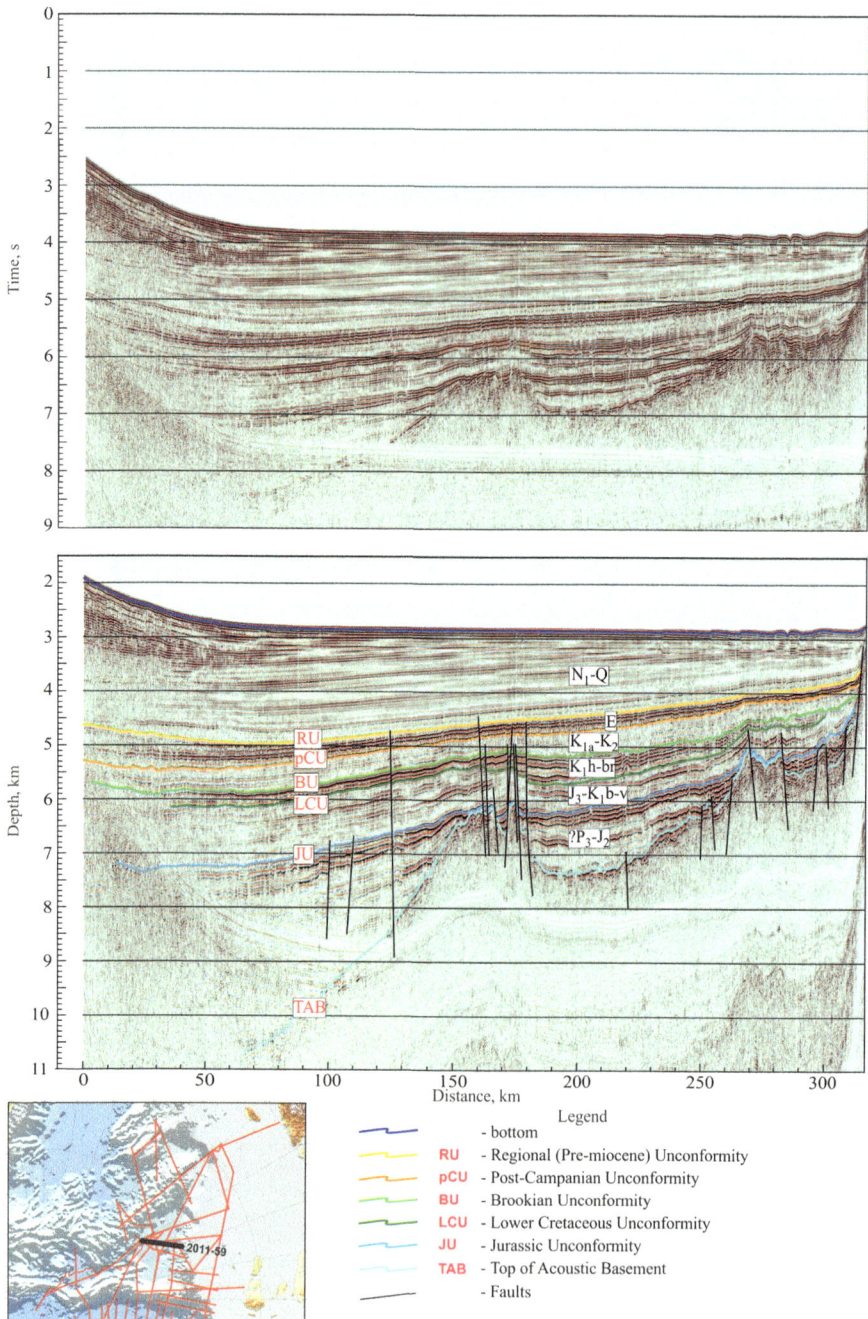

Fig. 2.12 MCS seismic line 2011–59 (Vilkitsky Trough – the Podvodnikov Basin)

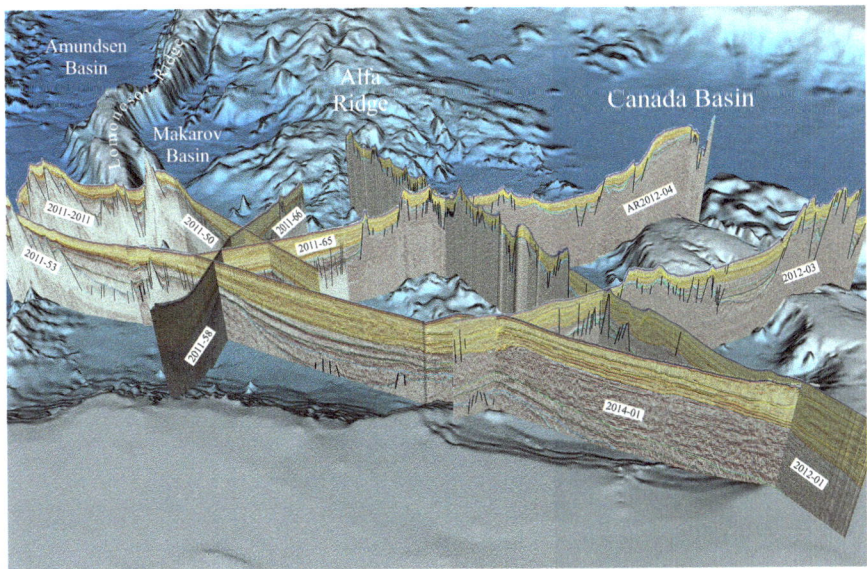

Fig. 2.13 3D rendition of Cenozoic seismic stratigraphy of the Podvodnikov Basin and adjacent uplifts

These two seismic section illustrate the recurring pattern of merging unconformities RU and pCU into RU + pCU at the Lomonosov Ridge, their separation in the Vilkitsky Trough and tendency to merge again approaching the Mendeleev Ridge.

Fig. 2.13 shows the fence-diagram of seismic sections in the Podvodnikov Basin and adjacent uplifts.

Total thickness of Cenozoic sediments in the Podvodnikov Basin reaches 1500–2600 m, dropping down to 900 m or less at the Mendeleev Ridge. The ACEX well on the Lomonosov Ridge penetrated 430 m of Cenozoic sediments. The largest volume Cenozoic sediments (2660–2900 m) was found in the Vilkitsky Trough bordering the Podvodnikov Basin from the shelf side. A relatively thin formation of Paleocene neritic sediments, widely present in the western and north-western part of the Podvodnikov Basin suggests that this segment of the basin is a submerged slope of the Lomonosov Ridge. This conclusion was supported by W. Jokat after R/V Polarstern expedition (Jokat et al. 2013).

There is one more important fact confirming our interpretation of seismic stratigraphy of Cenozoic sequences. Using system of intersecting MCS lines it is possible to trace pCU (or mBU, as in (Sherwood et al. 2002)) from POPCORN well (Chukchi Sea, Alaska shelf), through the North Chukchi Trough into the Vilkitsky Trough and Podvodnikov Basin with only minor gap (Fig. 2.14). Incidentally, the seismic reflection marking the base of Cenozoic section, traced on the system of profiles from the ACEX well on the Lomonosov Ridge, and the same reflection, traced from Popcorn well, met each other in the Podvodnikov Basin at the same phase.

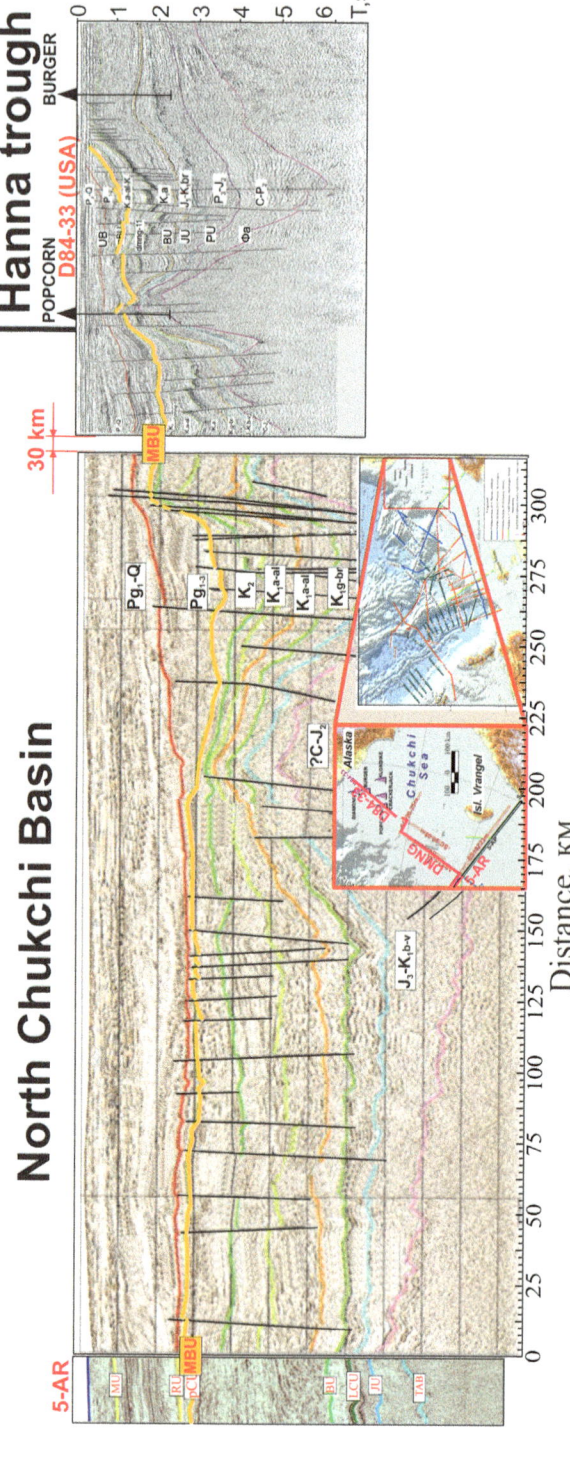

Fig. 2.14 MCS seismic sections (from right to left): D84-33 through Burger-Popcorn wells; DMNG composite traverse(North Chukchi Trough); 5-AR (SEVMORGEO)

Based on the above information, the Cenozoic sediments in the Central Arctic Uplifts Complex are divided in two seismostratigraphic complexes, from top to down, – SSC-1 and SSC-2.

SSC-1 - (N$_1$-Q, Lower Miocene-Pleistocene), bound by sea floor and pre Miocene regional unconformity RU, consist of semi-pelagic alevrolites with sandy interlayers (as per ACEX core) deposited after the Fram Strait opening. Interval velocity (by both refraction and reflection surveys) - 1.8–2.8 km/s.

SSC-2 - (E$_1$-E$_2$, Upper Paleocene – Middle Eocene, possible Oligocene in the Podvodnikov Basin), confined between unconformities RU and pCU. ACEX core from this interval displays mostly alevro-pelites and bi-silicate clays. The complex was formed in neritic environment after Laurasia separation onto Eurasian and North American plates. Post-Campanian unconformity at the base of the complex separates its neritic sediments from underlying Campanian prograding/deltaic littoral sandstones and argillites. Interval velocity varies from 2.9 to 3.3 km/s.

The regional pre-Miocene unconformity is well developed in the Makarov Basin underlying a considerable section of semi-pelagic, well stratified Miocene-Pleistocene sediments. In contrast, the Paleogene interval is reduced here, as demonstrated by closely spaced RU and pCU unconformities (Fig. 2.15). The underlying Upper Cretaceous unsubdivided sediments fill numerous grabens and half-grabens clearly visible in the acoustic basement, and often display the features of underwater turbidities seismofacies.

The stratigraphic classification of (Sherwood et al. 2002) relates the Cenozoic deposits to the Upper Brookian Complex.

Pre-Cenozoic Sedimentary Sequences The seismostratigraphic interpretation of pre-Cenozoic formations in the Central Arctic Uplifts Complex, ambiguous as it was, often suffered from subjective and speculative alterations to suit various tectonic models. The breakthrough came after the Russian expedition "Arctic-2014" acquired (with streamer 4.5 km long) the MCS line 1401 (Fig. 2.16). This seismic line is unique, because it unquestionably demonstrates that all principal unconformities of the North Chukchi Trough sedimentary cover continue uninterrupted into the Podvodnikov Basin. So, in order to unlock the stratigraphy of the Podvodnikov Basin, the North Chukchi Trough sediments should be dated correctly. That, in turn, could only be done by tying the North Chukchi Trough unconformities to numerous wells drilled on Alaska shelf (Sherwood et al. 2002), which, having a dense system of seismic lines in the vicinity of the wells, seemed possible. However, those wells were all drilled on the uplift between North Chukchi Trough and Hanna Trough with complicated structure distorting the reflectivity and making the horizons correlations with the wells all but impossible (except for tying the Cenozoic unconformities to the POPCORN well). Therefore, it was imperative to find some regional seismostratigraphic reference points, or benchmarks, inside the pre-Cenozoic formations, which could be traced into the North Chukchi Trough.

Such benchmarks were established on MCS lines R13, R8, R14 (Fig. 2.16, 2.17) shot on the Coastal Alaska Plain (USA) and verified by INIGOK well (Bird 1994).

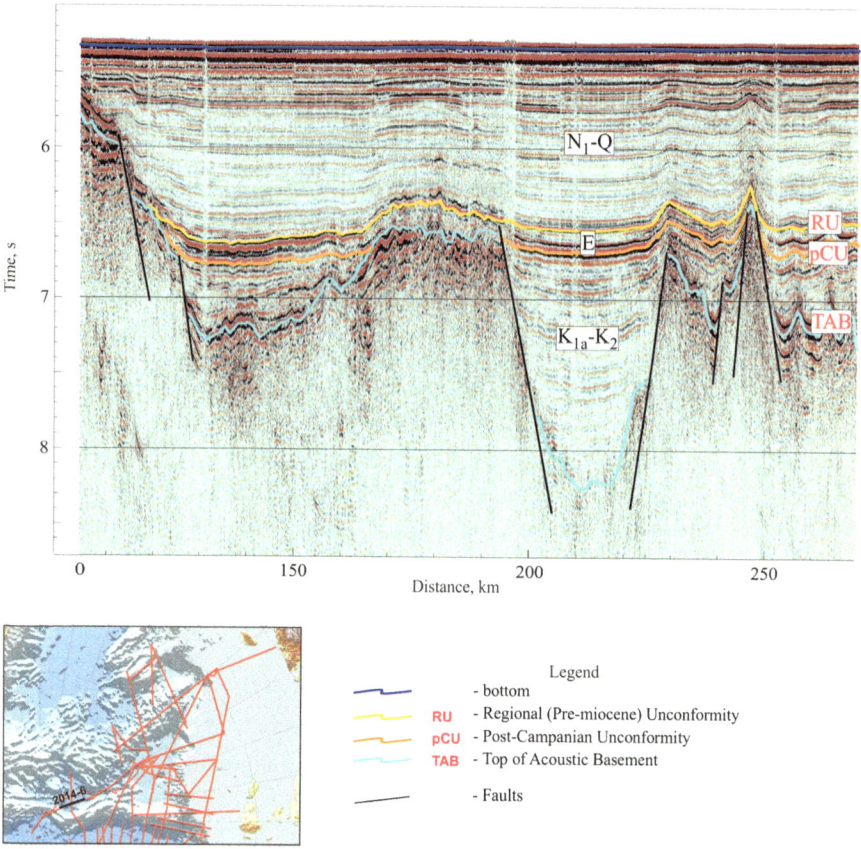

Fig. 2.15 The TWT MCS seismic section 2014–6 (fragment)

The seismic lines SEG-Y digital files are available at https://www.usgs.gov/ (U.S.Geological Survey).

The first benchmark represents clinoformal downlapping sequence (Fig. 2.17), dated by INIGOK well as Aptian-Albian. The top of the complex, with characteristic upward flattening of clinoforms in proximity to pCU unconformity, is defined only approximately and corresponds with Lower-Upper Cretaceous boundary. Its base is marked by Brookian unconformity (BU) which often merges with Lower Cretaceous Unconformity LCU. The shelf progradation and paleoslopes formation continued through either transgressions or regressions, therefore this benchmark was named transgressive-regressive complex (TRC).

The second benchmark is confined between LCU (or BU + LCU) at the top and Upper Jurassic unconformity JU (INIGOK well data), visible on seismic section as a strong double-phase reflector. This interval is presented by chaotic low-amplitude seismofacies and interpreted (Klemperer et al. 2002; Sherwood et al. 2002) as

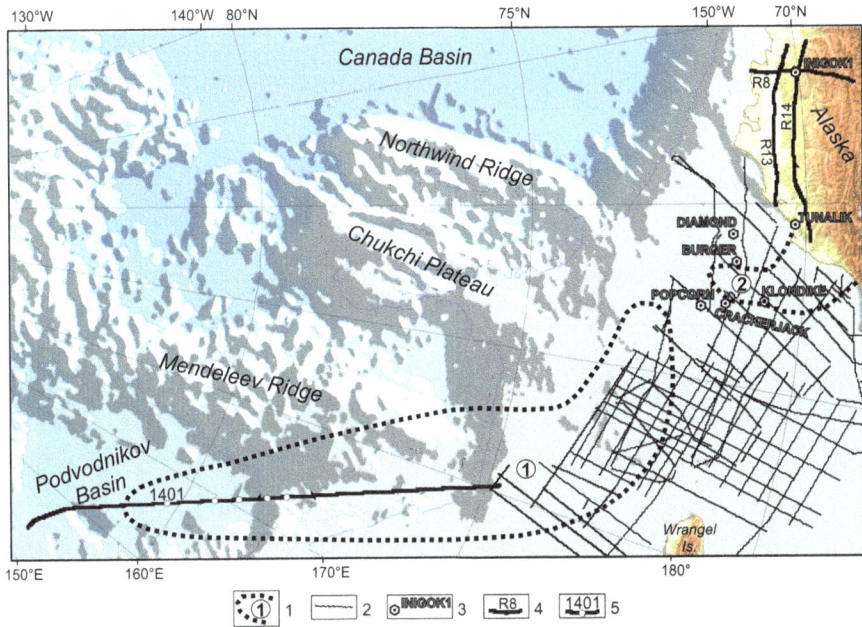

Fig. 2.16 The database used for seismostratigraphic interpretation: 1– Outline of North Chukchi Trough (1) and Hanna Trough (2); 2 – existing seismic MCS surveys; 3 – wells (USA); 4 – overland Alaska seismic lines (in text); 5 – seismic line 1401("Arctic-2014", in text)

synchronous to the Neocomian opening of the Canada Basin rift complex, separating Brookian and Ellesmerian structural stages (Fig. 2.17).

The regional character of these benchmarks, in our opinion, is related to a critically important regional event – opening of the Canada Basin. If the second (rift) benchmark is synchronous to the this event, then the first (TRC) was formed during subsequent high-and low stands of one large body of water creating series of downlapping clinoforms as its prominent seismostratigraphic feature.

Fig. 2.17 convincingly displays the presence of thick clinoformal downlapping sequence at the shelf end of line 1401, which, considering the above comments, can be identified as the first, Aptian-Albian TRC. Consequently, the boundary of on which all clinoforms downlap, is identified as BU, with LCU short TWT down. The second (rift) benchmark, with its typical chaotic low-amplitude reflections and strong double-phased basal reflector (JU), is also easily identified (Poselov et al. 2017).

Incidentally, this interval of chaotic low-amplitude reflections (with no fully developed TRC clinoforms above it) was identified on many seismic lines in North Chukchi Trough without clear understanding of its nature. But now we may say that the full picture with the two benchmarks present will be visible only on lines oriented along the direction of progradation, from North Chukchi Trough depositional center towards the Podvodnikov Basin depocenter – exactly the orientation of

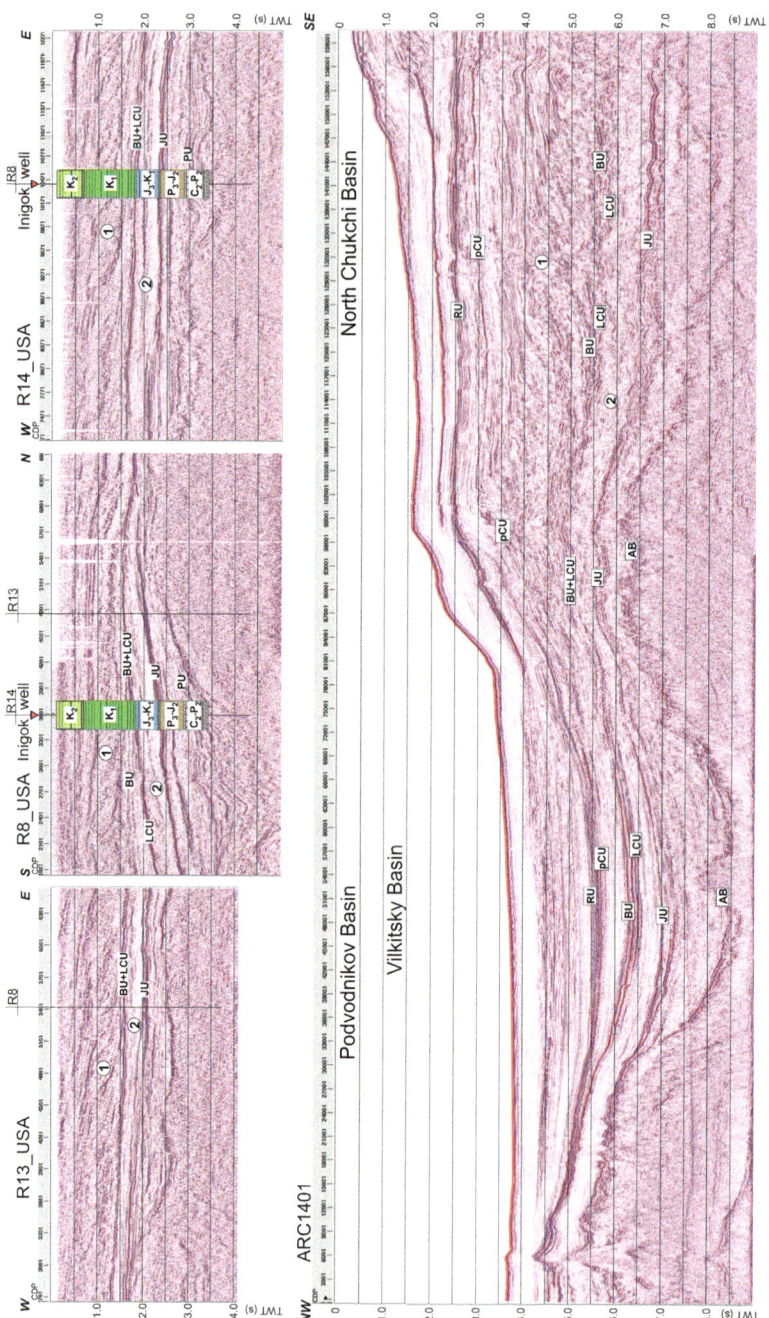

Fig. 2.17 Correlation of Alaska MCS lines (R13, R8, R14) and MCS line 1401 ("Arctic-2014", Podvodnikov Basin- North Chukchi Trough). Major uncon-formities: RU – regional pre-Miocene; pCU – post-Campanian; BU – Brookian; LCU – Lower Cretaceous; JU – Upper Jurassic; AB – acoustic basement. Seismostratigraphic benchmarks: transgressive-regressive complex (1), chaotic low amplitude reflections (rift) complex (2)

seismic line 1401. Seismic lines oriented along the paleo-shelf edge, would clearly display only the second benchmark and only partially the first one (Fig. 2.16).

The stratigraphic analysis of POCORN well provides some clues regarding the source of terrigenous material forming TRC prograding clinoforms in the North Chukchi Trough. POPCORN was drilled on the high between the North Chukchi Trough and Hannah Trough (Fig. 2.16) and demonstrated drastic reduction of the Aptian-Albian sequence (Sherwood et al. 2002). All other wells in the area also show complete absence of Upper Cretaceous horizons. Therefore, it would be logical to place the source of TRC clastic material at the structural uplifts surrounding the North Chukchi Trough from the East and South-East. Due to complete erosion of the Upper Cretaceous horizons, the age brackets of TRC benchmark within the North Chukchi Trough should be extended to include the Upper Cretaceous and the share of these sediments in TRC is also expected to increase along the prograding (and rejuvenation) direction towards the Vilkitsky Trough (Fig. 2.17).

Another, younger prograding complex is visible close to the south-east end of the line 1401 at 0.5–2. 5 s TWT (Fig. 2.17). The same complex at the same time interval was recorded by R/V Polarstern expedition in 2008 at the northern slope of North Chukchi Trough (Hegewald and Jokat 2013). The German scientists identified the Middle-Brookian Unconformity (mBU, analog of pCU)) at the base of the Cenozoic complex and proposed that the Cenozoic clinoforms were formed by sea level changes, getting stronger after Early Miocene (post-RU) opening of the Fram Strait and inrush of North Atlantic to previously isolated Arctic Ocean. (Hegewald and Jokat 2013).

Judging by considerable thickness of Cenozoic sediments in wells drilled on the Alaska shelf including POPCORN, their source was located south from North Chukchi Trough, possibly Wrangel-Gerald High. Therefore, direction of Chukchi paleo-shelf propagation changed from north-western for Aptian-Albian TRC to northern for Cenozoic clinoforms. This change may be related to the one of the most important event in the Arctic Region history, namely the breakup of Laurasia into Eurasian and North American Plates.

This "switch" of Chukchi paleo-shelf propagation at the Mesozoic-Cenozoic boundary can be by comparing the seismic signatures of clinoformal sequences on lines 1401(Fig. 2.17, SE-NW) and 5-AR (Fig. 2.18, SW-NE). At the NE end of the line 5-AR North Chukchi Trough clinoforms are clearly visible above 3.0 s TWT above pCU, while at TWT > 3.0 s they became more chaotic (line 1401 display the opposite pattern). There is also no indication of the Wrangel-Gerald High being a source of terrigenous material to form the lower clinoform complex dumping of clastics eroded from it.

It is important to note, that using the POPCORN well as a starting point, the unconformity pCU (mBU) could be almost continuously traced on the entire system of seismic lines on the North Chukchi Trough (Fig. 2.14). This procedure proved that everywhere in the Trough this interface always could be found at 3.0 s TWT, and the base of lower clinoform complex - at 5.5–6.0 s TWT (Fig. 2.17).

Therefore, interpretation of the base of the upper clinoform complex as the base of Cenozoic section confidently establishes Aptian-Albian-Late Cretaceous age of

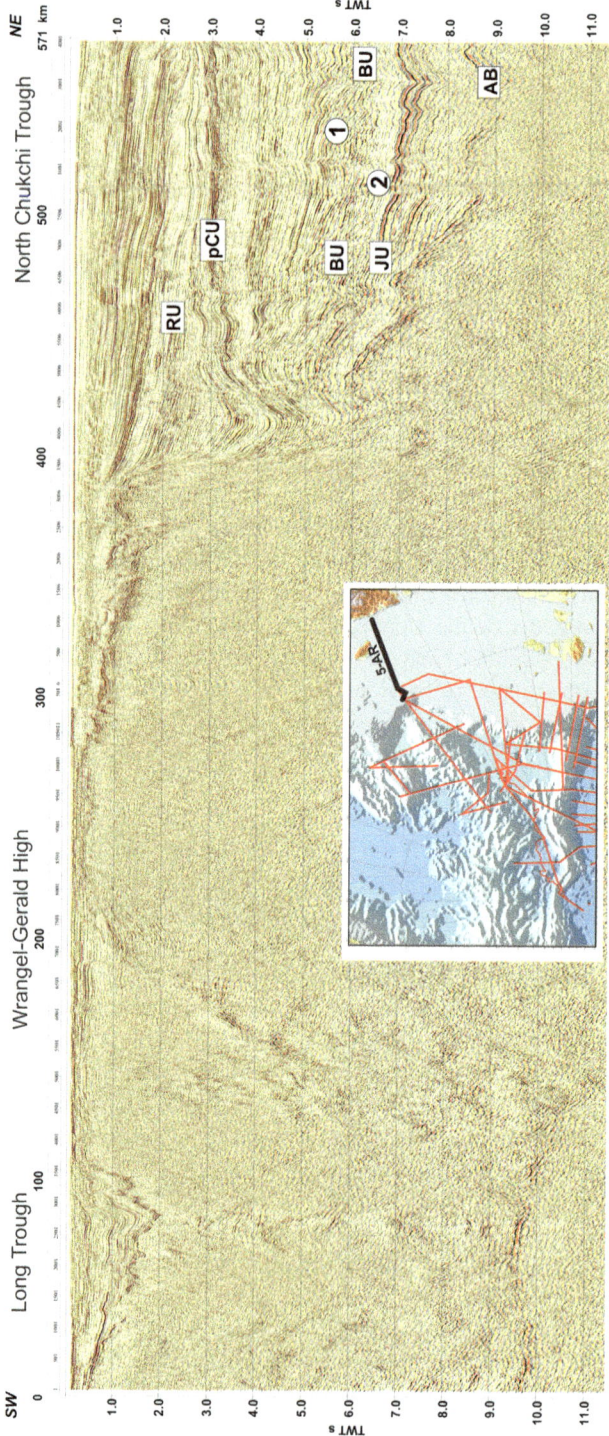

Fig. 2.18 MCS line 5-AR (North Chukchi Trough). Unconformities: RU – regional pre-Miocene; pCU – post-Campanian; BU – Brookian; JU – Upper Jurassic; TAB – acoustic basement. Seismostratigraphic benchmarks: (1) - transgressive-regressive complex; (2) - chaotic low-amplitude complex

the lower TRC complex (with increasing volume of Late Cretaceous towards the Podvodnikov Basin).

Consequently, uninterrupted continuation of BU, LCU and JU unconformities from North Chuckchi Trough shelf to the Podvodnikov Basin unambiguously proves the presence of both Cretaceous and Jurassic sediments there. The major unconformities were traced all over the Central Arctic Uplifts Complex using the existing seismic lines system and line 1401 as a reference. In the process it was found that the Jurassic is present only in the most depressed parts of the Podvodnikov (Vilkitsky Trough) and Chukchi Basins.

Therefore, the Pre-Cenozoic sediments are subdivided in four seismostrati-graphic complexes: SSC-3, SSC-4, SSC-5, and SSC-6 (continuing numbering down from two Cenozoic ones).

SSC-3 – bound by pCU and BU unconformities (Lower Brookian complex by (Sherwood et al. 2002)) and associated to the last HALIP magmatic event (80–90 Ma) – includes Aptian-Albian-Upper Cretaceous terrigenous sediments. The complex seismofacies include paleo-shelf progradational series in the North Chukchi Trough and turbidital littoral series –in Vilkitsky Trough and Podvodnikov Basin. The SSC-3 continues from the shelf North Chukchi Trough into Podvodnikov Basin with its thickness reducing from 3000–5000 m on shelf to ≈1000 m in Vilkitsky Trough down to 100–300 (or even completely pinching out) in the Podvodnikov Basin. At the transition from the Vilkitsky Trough to the Mendeleev Ridge the complex is 200–300 m thick. Concordant (reflection and refraction P-wave combined) interval velocities in North Chukchi Trough are 3.8–3.9 km/s, Podvodnikov Basin - 3.5–3.7 km/s and in Mendeleev Ridge - 2.6–3.1 km/s (Fig. 2.20).

SSC-3 (perhaps, only its Upper Cretaceous part) is also present in Makarov Basin where its average thickness of 800–1000 m increases to ≈2500 m in the central graben (Fig. 2.21).

SSC-4 – bound by BU and LCU (Lower Brookian complex by (Sherwood et al. 2002)) and associated with the first HALIP magmatic event (130–120 Ma) – includes Lower Cretaceous (K_1h-br) shales, alevrolites and sandstones (Alaska shelf wells data). In the North Chukchi Trough and Chukchi Plateau its thickness does not exceeds 300–400 m, pinching out in the northern and western parts of Podvodnikov Basin and completely absent in Makarov Basin. Within limits of the Mendeleev Ridge its thickness varies from first hundred meters to values below the resolution of MCS survey. The concordant interval velocities are: the North Chukchi Trough - 3.9-4.0 km/s; the Podvodnikov Basin – 3.8 km/s; Chukchi Basin – 3.6 km/s; Chukchi Plateau – 3.5 km/s. (Fig. 2.19–2.21, 2.23).

SSC-5 – bound by LCU (or BU + LCU) and JU unconformities, includes upper Jurassic –Lower Cretaceous (J_3-K_1b-v) terrigenous turbidities (mainly shales and sandstones), synchronal with rift complex of the Alaska shelf which separates the Brookian and Ellesmerian structural stages, i.e. regional structural re-alignment. In the North Chukchi Trough up to 2000 m of high velocity (≈ 4.5 km/s) are present. In transitional zone between the North Chukchi Trough to the Podvodnikov Basin the complex thins to 800–1000 m and stay that way throughout the whole basin. In the Podvodnikov Basin the SSC-5 interval velocities stay in the 3.8–4.0 km/s range

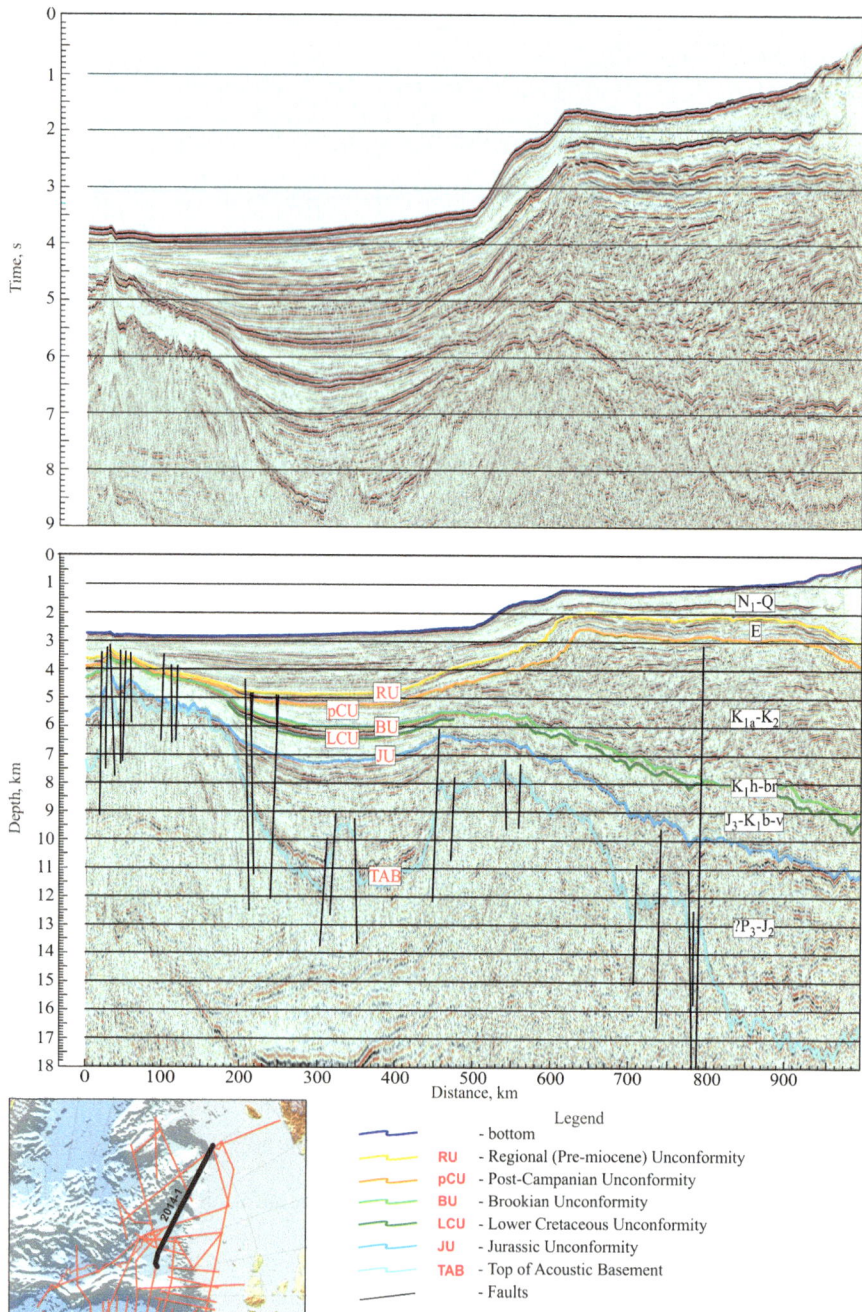

Fig. 2.19 Interpretation of MCS line 2014–01, North Chukchi Trough – Podvodnikov Basin

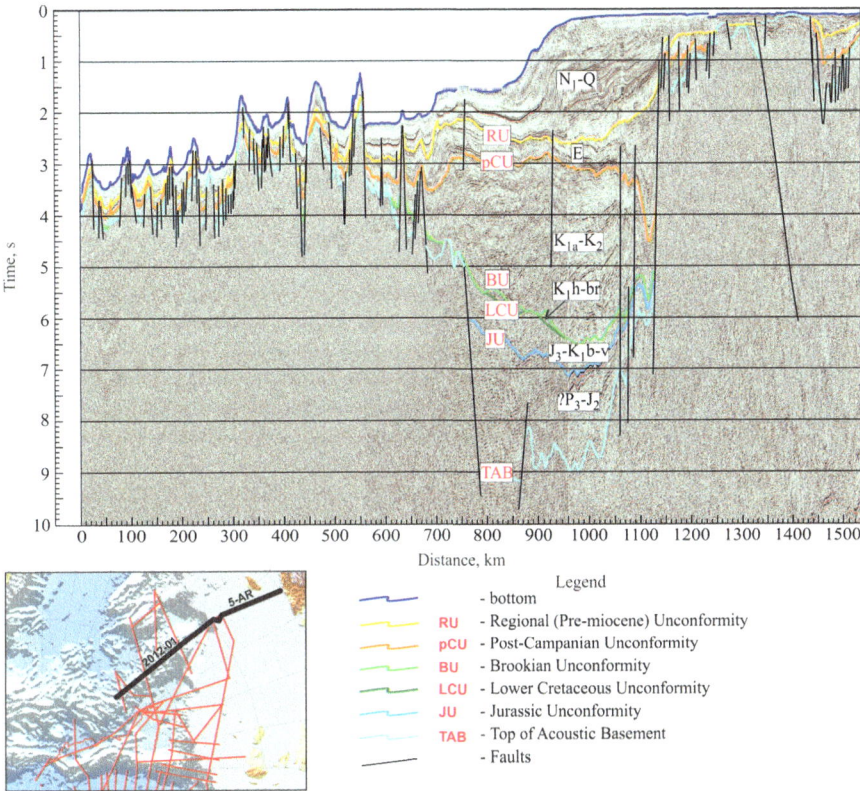

Fig. 2.20 Interpretation of MCS line 2012–01, North Chukchi Trough – Mendeleev Ridge

(Fig. 2.19). The complex is not present at the Mendeleev Ridge and the Makarov Basin. In the Chukchi Basin and the central graben of the Chukchi Plateau SSC-5 complex is ≈600 m thick with interval velocities around 4.0 km/s (Fig. 2.23).

The 3D rendition of the pre-Cenozoic complexes SSC-3 to SSC-6 in the Podvodnikov Basin, Mendeleev Ridge, Chukchi Plateau and East Siberian sea shelf is presented on the fence-diagram of appropriate seismic sections (Fig. 2.22).

SSC-6 is enclosed between JU unconformity and acoustic basement (TAB). In (Sherwood et al. 2002) classification the complex is defined as Upper Ellesmerian and included Early and Middle Jurassic, possibly even older, sediments. As Fig. 2.17, 2.19–2.20 illustrate, it displays implicit disrupted reflectivity typical for poorly stratified terrigenous deposits. The complex can be as 6000 m thick in the depositional center of the North Chukchi Trough, with interval velocity 4.8–5.8 km/s. The complex continues to the Vilkitsky Trough where its thickness decreases to ≈4000 m and interval velocity drops to 4.1–4.3 km/s (Fig. 2.21). Much thinner SSC-6 was identified in the Chukchi Basin, and it is not traced beyond Vilkitsky Trough and Chukchi Basin (Fig. 2.23).

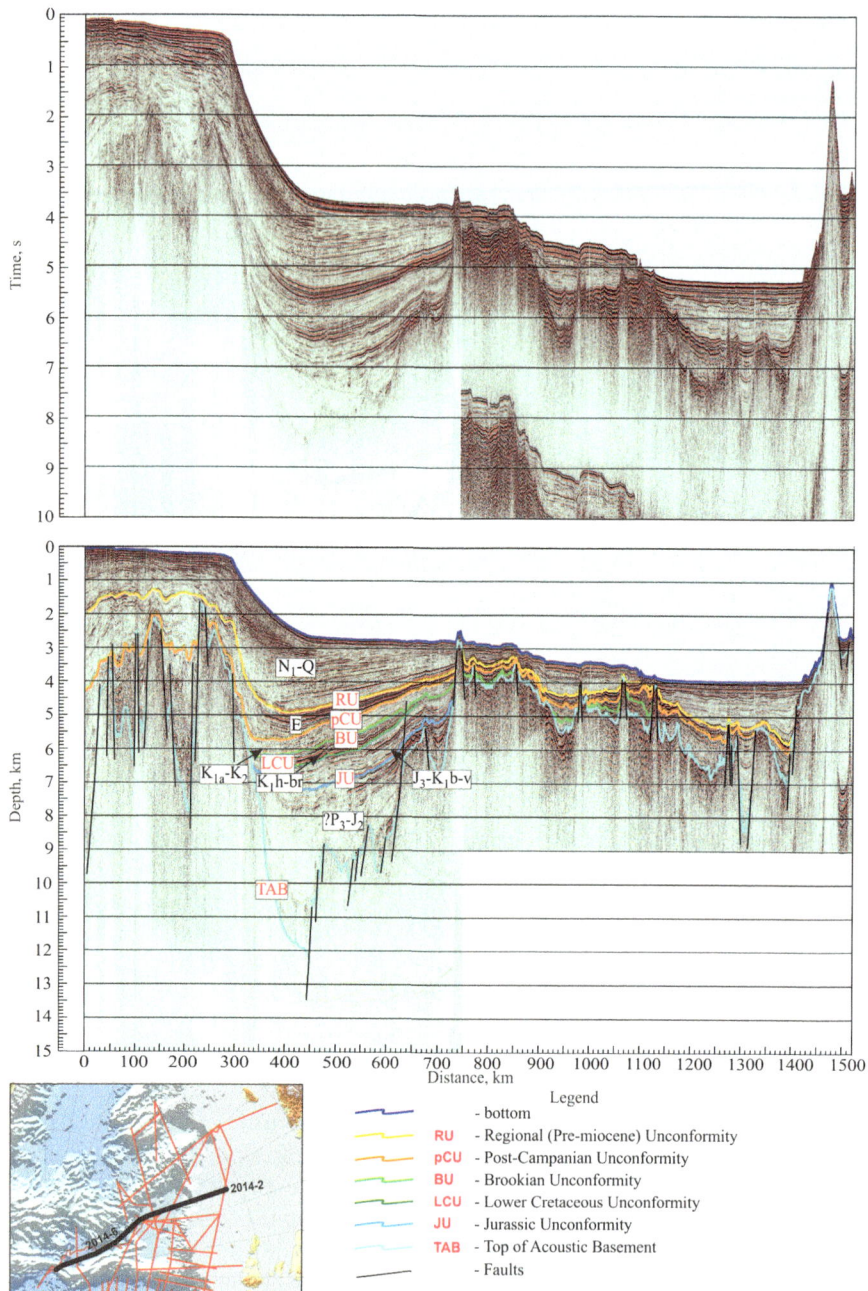

Fig. 2.21 Interpretation of MCS line 2014–02, De-Long Shelf –Vilkitsky Trough - Podvodnikov Basin-Makarov Basin

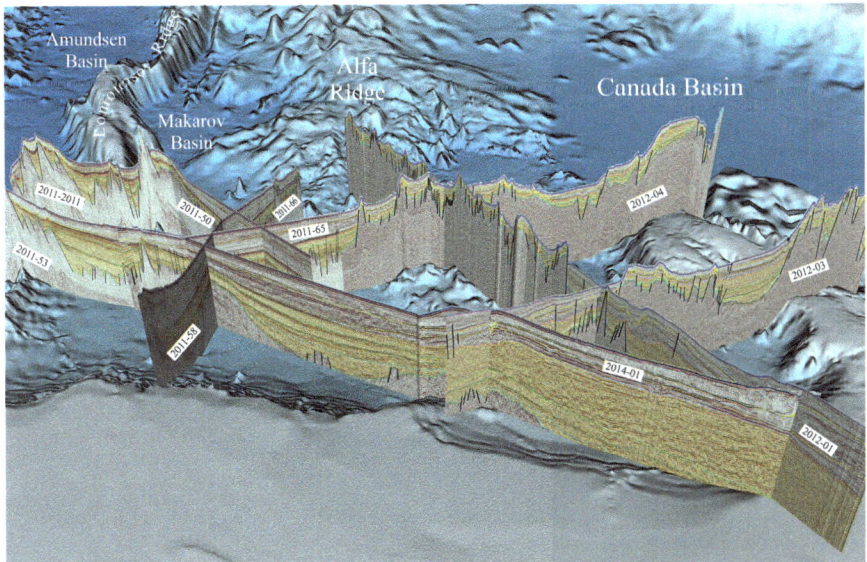

Fig. 2.22 3D rendition of pre-Cenozoic formations in Podvodnikov Basin and the adjacent uplifts

Finally, as far as the sedimentary cover of the Central Arctic region is concerned, it can be stated that:

1. The presented seismostratigraphic model of Central Arctic Uplifts Complex sedimentary cover (Poselov et al. 2014, 2017), (Fig. 2.24), agrees with previously developed stratigraphic schemes (Hagewald and Jokat 2013; Bruvoll et al. 2012; Jokat et al. 2013, 2016).
2. The Cenozoic unconformities interpreted from the recently acquired within the Central Arctic Uplifts Complex seismic data correlate with that of earlier seismic data (line AWI-91090 tied to the well ACEX IODP 302 drilled in the Lomonosov Ridge).
3. The Cenozoic sediments in the Central Arctic Uplifts Complex are divided in two seismostratigraphic complexes, from top down: SSC-1 - (N_1-Q, Lower Miocene-Pleistocene), bound by sea floor and pre-Miocene regional unconformity RU, and SSC-2 - (Upper Paleocene –Middle Eocene, possible Oligocene in the Podvodnikov Basin, E_1-E_2/E_3), confined between unconformities RU and pCU. Both complexes continue uninterrupted from the Eurasian shelf through the transit zone to the Podvodnikov Basin, Vilkitsky Trough, Mendeleev and Lomonosov Ridges.
4. The complexes are traced throughout the Province preserving their seismostratigraphic and seismofacial characteristics. It proves that there is no major strike-slip movement between the Central Arctic Uplifts Complex and the Laptev Sea continental margin. (whatever normal faulting is detected, it does not affect the continuity of correlation).

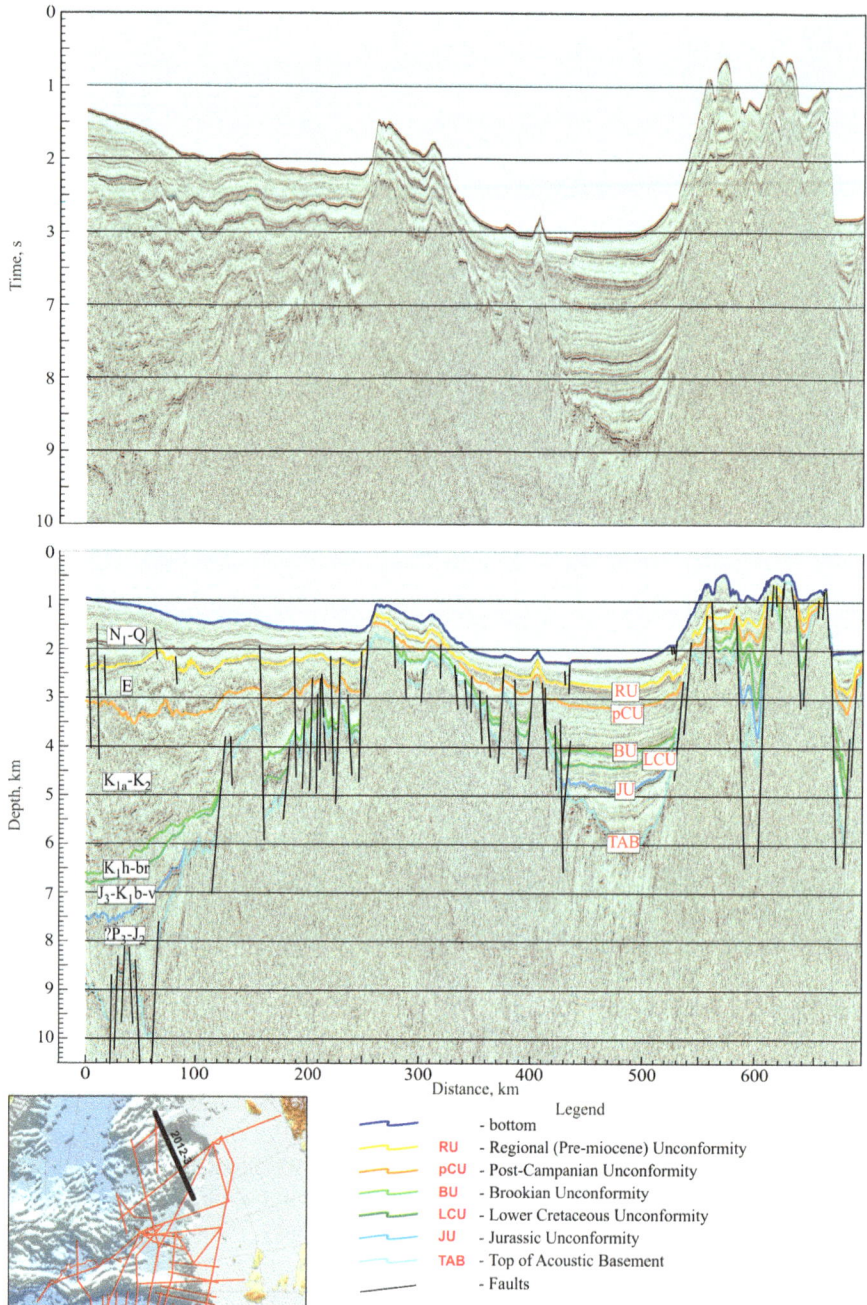

Fig. 2.23 Interpretation of MCS line 2012–03, North Chukchi Trough - Mendeleev Ridge – Chukchi Basin –Chukchi Plateau

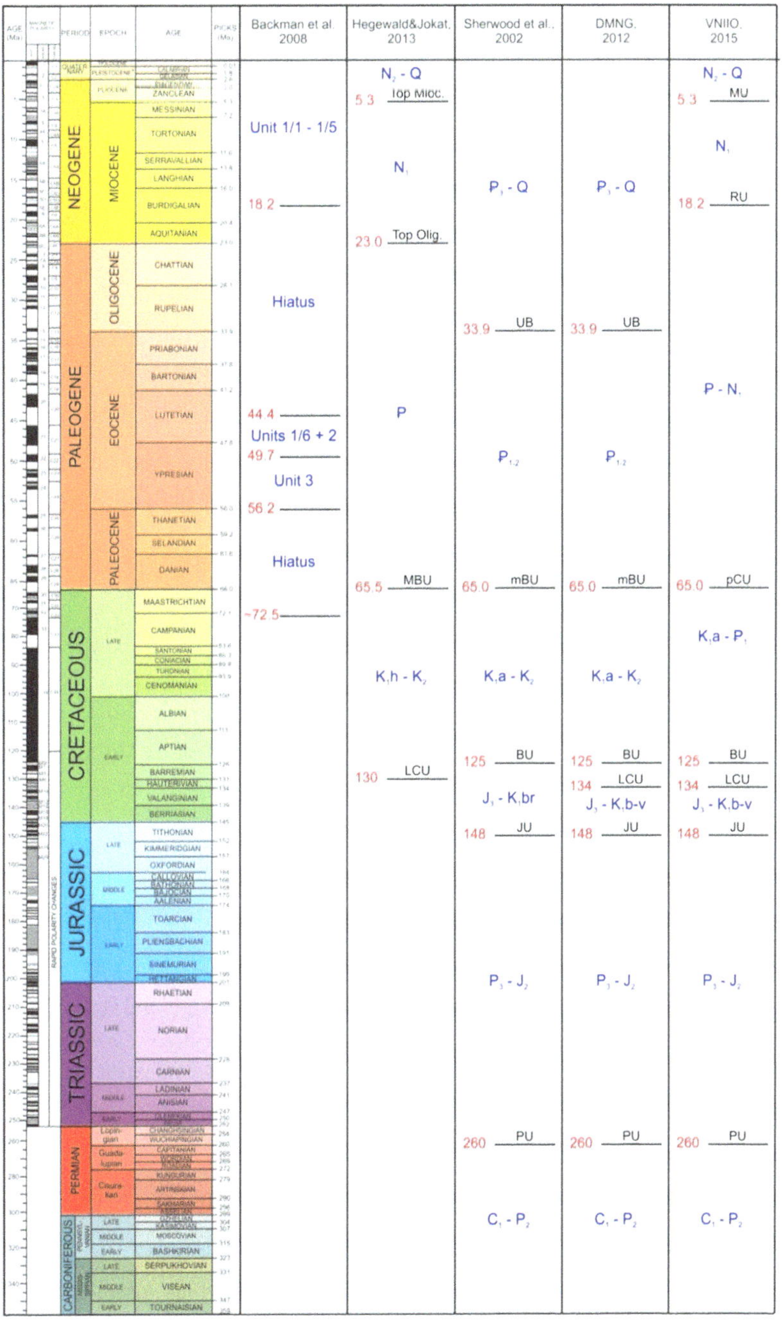

Fig. 2.24 Seismostratigraphic constructs of Chukchi Sea shelf and Central Arctic Uplifts Complex sedimentary cover

5. SSC-1 consists of semi-pelagic alevrolites with sandy interlayers (as per ACEX core) deposited after the Fram Strait opening. Interval velocity - 1.8-2.8 km/s.

6. SSC-2 in ACEX core consists of neritic alevro-pelites and bi-silicate clays ($\geq$ 200 m) deposited after Laurasia breakup into Eurasian and North America plates. Interval velocity varies from 2.9 to 3.3 km/s.

7. The lower terrace of the Podvodnikov Basin (Arlis Gap) and most part of the Podvodnikov Basin itself, being a morphological uplift till the end of Paleogene, can be considered as subsided flank of the Lomonosov Ridge. It also possible for the Arlis Gap to act as structural saddle (basement sag) between the Lomonosov and Mendeleev ridges.

8. In the Makarov Basin, the regional pre-Miocene unconformity underlying a considerable section of semi-pelagic Miocene-Pleistocene well stratified sediments, is well developed. Structurally, the Makarov Basin differs from the Podvodnikov Basin - in the Makarov Basin the underlying Upper Cretaceous un-subdivided sediments fill numerous grabens and half-grabens clearly visible in the acoustic basement, and often display the features of underwater turbidities seismofacies.

9. The interpretation of pre-Cenozoic stratigraphy was based on tracing several regional seismostratigraphic benchmarks from well-proven seismic lines onshore and offshore Alaska into the North Chukchi Trough, with identification of BU (Late Brookian), LCU (Lower Cretaceous) and JU (upper Jurassic) unconformities in the latter. This procedure was facilitated by previously established continuity of BU, LCU and JU from North Chukchi Trough into Vilkitsky Trough and Podvodnikov Basin. The established benchmarks through the system of the Russian MCS profiles, were further traced onto the entire Central Arctic Uplifts Complex.

10. Pre-Cenozoic section was subdivided in four seismostratigraphic complexes SSC-3, SSC-4, SSC-5, and SSC-6.

11. SSC-3, bound by pCU and BU unconformities (Lower Brookian) and associated to the last HALIP magmatic event (80–90 Ma), includes Aptian-Albian-Upper Cretaceous terrigenous sediments. The complex seismofacies include paleo-shelf prograding series in the North Chukchi Trough and turbidital littoral – in Vilkitsky Trough and Podvodnikov Basin. The SSC-3 continues from the shelf North Chukchi Trough into Podvodnikov Basin with its thickness reducing from 3000–5000 m on shelf to $\approx$1000 m in Vilkitsky Trough down to 100–300 (or even completely pinching out) in the Podvodnikov Basin. SSC-3 (possibly only its Upper Cretaceous portion) continues into Makarov Basin were it is 800–1000 m thick (up to $\approx$2500 m in the central graben). At the transition from the Vilkitsky Trough to the Mendeleev Ridge the Complex is 200–300 m thick. Interval velocities in the North Chukchi Trough - 3.8-3.9 km/s, Podvodnikov Basin - 3.5-3.7 km/s and in Mendeleev Ridge - 2.6-3.1 km/s.

12. SSC-4, bound by BU and LCU, associated with the first HALIP magmatic event (130–120 Ma), includes Lower Cretaceous (K_1h-br) shales, alevrolites and sandstones (Alaska shelf wells data). In the North Chukchi Trough and Chukchi Plateau its thickness does not exceeds 300–400 m, pinching out in the

northern and western parts of Podvodnikov Basin and completely absent in Makarov Basin. Within limits of the Mendeleev Ridge its thickness varies from first hundred meters to values below the resolution of MCS survey. The interval velocities are: the North Chukchi Trough - 3.9-4.0 km/s; the Podvodnikov Basin – 3.8 km/s; Chukchi Basin – 3.6 km/s; Chukchi Plateau – 3.5 km/s.

13. SSC-5, bound by LCU (or BU + LCU) and JU unconformities, includes upper Jurassic –Lower Cretaceous (J_3-K_1b-v) terrigenous turbidities (mainly shales and sandstones), synchronous with rift complex of the Alaska shelf which separates the Brookian and Ellesmerian structural stages. In the North Chukchi Trough up to 2000 m of high velocity ($\approx$ 4.5 km/s) are present. In transitional zone between the North Chukchi Trough to the Podvodnikov Basin the complex thins to 800–1000 m and stay that way throughout the whole basin. In the Podvodnikov Basin the SSC-5 interval velocities stay in the 3.8–4.0 km/s range. The complex is not present at the Mendeleev Ridge and the Makarov Basin. In the Chukchi Basin and the central graben of the Chukchi Plateau SSC-5 complex is $\approx$600 m thick with interval velocities around 4.0 km/s.

14. SSC-6 (Ellesmerian) is enclosed between JU unconformity and acoustic basement and included Early and Middle Jurassic, possibly even older, sediments with implicit disrupted reflectivity typical for poorly stratified terrigenous deposits. The complex can be as 6000 m thick in the depositional center of the North Chukchi Trough, with interval velocity 4.8–5.8 km/s. The complex continues to the Vilkitsky Trough where its thickness decreases to $\approx$4000 m and interval velocity drops to 4.1–4.3 km/s (Fig. 2.21). Much thinner SSC-6 was identified in the Chukchi Basin, and it is not traced beyond Vilkitsky Trough and Chukchi Basin.

15. All described seismostratigraphic complexes are continually traced, without any noticeable interruption, from shelf-located North Chukchi Trough to Vilkitsky Trough, Podvodnikov and Chukchi Basin, and upper ones – to Makarov Basin, Lomonosov and Mendeleev Ridges and Chukchi Plateau. The continuity of sedimentary units from shelf to deep-water regions points to the common history of geologic evolution of the East Arctic continental margin and the Central Arctic Uplifts Complex.

References

Backman J, Moran K et al (2006) Sites M0001-M0004. Expedition 302 Scientists. Proceedings of the Integrated Ocean Drilling Program 302, p 169

Bird KJ (1994) Ellesmerian petroleum system, North Slope of Alaska, USA AAPG special volumes. Chapter 21(V): pp 339–358

Bruvoll V, Kristoffersen Y, Coakley B et al (2012) The nature of the acoustic basement on Mendeleev and northwestern Alpha ridges, Arctic Ocean. Tectonophysics J 514:123–145

Butsenko VV, Poselov VA (2006) Regional paleotectonic interpretation of seismic data from the deep-water central Arctic. IV International Conference on Arctic Margins (ICAM IV), Canada, 30 September – 3 October 2003

Franke D, Hinz K, Oncken O (2001) The Laptev Sea rift. Mar Pet Geol J 18:1083–1127

Glebovsky VY, Kaminsky VD, Minakov AN et al (2006) Formation of the Eurasia Basin in the Arctic Ocean as inferred from geohistorical analysis of the anomalous magnetic field. Geotectonics J 4:21–42

Hegewald A, Jokat W (2013) Tectonic and sedimentary structures in the northern Chukchi region, Arctic Ocean. J Geophys Res 118(7):3285–3296

Jokat W, Kristoffersen Y, Rasmussen TM et al (1992) ARCTIC 91:Lomonosov ridge – a double-sided continental margin. Geology J 20:887–890

Jokat W, Weigelt E, Kristoffersen Y et al (1995) New insights into evolution of the Lomonosov ridge and the Eurasian Basin. Geophys J Int 122:378–392

Jokat W, Ickrath M, O'Connor J (2013) Seismic transect across the Lomonosov and Mendeleev ridges: constraints on the geological evolution of the Amerasia Basin, Arctic Ocean. Geophys Res Lett 40(19):5047–5051

Jokat W, Lehmann P, Damaske D et al (2016) Magnetic signature of North-East Greenland, the Morris Jesup Rise, the Yermak Plateau, the central Fram Strait: constraints for the rift/drift history between Greenland and Svalbard since the Eocene. Tectonophysics J 691:98–109

Klemperer SL, Miller EL, Grantz A et al (2002) Crustal structure of the Bering and Chukchi shelves: deep seismic reflection profiles across the North American continent between Alaska and Russia. GSA Special Paper 360:1–24

Poirier A, Hillaire-Marcel C (2011) Improved Os-isotope stratigraphy of the Arctic Ocean. Geophys Res Lett 38(14):L14607

Poselov VA, Butsenko VV, Chernykh AA et al (2014) The structural integrity of the Lomonosov Ridge with the North American and Siberian continental margins. VI International Conference on Arctic Margins (ICAM VI), Fairbanks, Alaska, 30 May – 2 June 2011

Poselov VA, Butsenko VV, Zholondz SM et al (2017) Seismic stratigraphy of sedimentary cover in the Podvodnikov Basin and North Chukchi trough. J Dokl Earth Sci 474(2):688–691

Sherwood KW, Johnson PP, Craig JD et al (2002) Structure and stratigraphy of the Hanna Trough, U. S. Chukchi Shelf, Alaska. Geological Society of America Special Paper 360:39–66

U.S. Geological Survey. https://www.usgs.gov (USGS)

Chapter 3
Eurasian Basin

Vasily A. Savin, Georgy P. Avetisov, Daria E. Artem'eva, Dmitry V. Bezumov, Andrey A. Chernykh, Vladimir Yu. Glebovsky, Gennady S. Kazanin, and Alexey L. Piskarev

Abstract The Eurasian Basin, ~2000 km long and ~900 km wide, consists of abyssal plains of Nansen and Amundsen Basins separated by mid-oceanic Gakkel Ridge. The magnetic field of the Eurasian Basin consists mostly of linear magnetic anomalies (LMA). According to classic hypothesis, anomalies of Cenozoic sequence, from LMA 24 (chron 24, ~53 Ma) to present, were identified and dated. Steady decline of spreading velocity from 22–27 mm/yr. to 5–9 mm/yr. was established for 53–20 Ma interval; later it grew slightly to 7–12 mm/yr.

At the same time pronounced asymmetry of geological, morphological and bathymetry features of the abyssal depressions, as well as demonstrated by seismic data asymmetry of distribution and thickness variations of sedimentary formations inside them, lead us to conclusion that the considerable part of the Eurasian Basin was probably formed in pre-Cenozoic time.

Keywords Eurasian Basin · Nansen Basin · Amundsen Basin · Gakkel Ridge · Spreading velocity · Cenozoic oceanic basement · Pre-Cenozoic oceanic basement

V. A. Savin (✉) · G. P. Avetisov · D. E. Artem'eva · D. V. Bezumov · V. Y. Glebovsky
All-Russian Research Institute of Geology and Mineral Resources of the World Ocean (VNIIOkeangeologia), Saint Petersburg, Russia
e-mail: savinvasily@rambler.ru; v.yu.glebovsky@vniio.ru

A. A. Chernykh · A. L. Piskarev
All-Russian Research Institute of Geology and Mineral Resources of the World Ocean (VNIIOkeangeologia), Saint Petersburg, Russia

Saint Petersburg University, Saint Petersburg, Russia
e-mail: a.a.chernykh@vniio.ru

G. S. Kazanin
Marine Arctic Geological Expedition, Murmansk, Russia
e-mail: info@mage.ru

© Springer International Publishing AG, part of Springer Nature 2019
A. Piskarev et al. (eds.), *Geologic Structures of the Arctic Basin*,
https://doi.org/10.1007/978-3-319-77742-9_3

3.1   Origin and Evolution of the Eurasian Basin According the Traditional Interpretation

The Eurasian Basin, ~2000 km long and ~900 km wide, consists of abyssal plains of Nansen and Amundsen Basins separated by mid-oceanic Gakkel Ridge. From the west and south the Basin is bound by passive Barents-Kara and rift Laptev Sea (Jokat and Micksch 2004) continental margins, accordingly; from the east- by the Lomonosov Ridge (Fig. 3.1). Abyssal plains of the Eurasian Basin are the deepest ones in the Arctic Ocean, sloping by less than 0,1°. Spitsbergen transform fault, also known as De Geer fault, acts as the Basin western boundary.

The majority of Russian and Western scientists attribute the origin of the Basin to the ocean floor grows through Cenozoic Period caused by constant introducing of new portions of oceanic crust at the axis of the Gakkel Ridge (Brozena et al. 2003; Glebovsky et al. 2006). The oceanic spreading was predated by continental rifting initiated in Late Cretaceous and continued in Paleocene. In the process, the Lomonosov Ridge, initially being part of the Eurasian Plate, was split from it and drifted eastward until reaching current position. The Gakkel Ridge, marking the Basin axial line, has a distinction to be one of the slowest segment of mid-oceanic ridges global system and exhibit a certain features of ultra-slow spreading. Khain (2001) postulated several stages of the Eurasian Basin evolution. During the first stage, from ~55–56 Ma to ~33–34 Ma, averaged spreading velocity was ~12 mm/yr. At the Early Oligocene, spreading started at the Fram Strait with simultaneous reducing of Gakkel Ridge spreading velocity up to ultra-slow ~5 mm/yr. In Late

Fig. 3.1 Eurasian Basin and its major structures

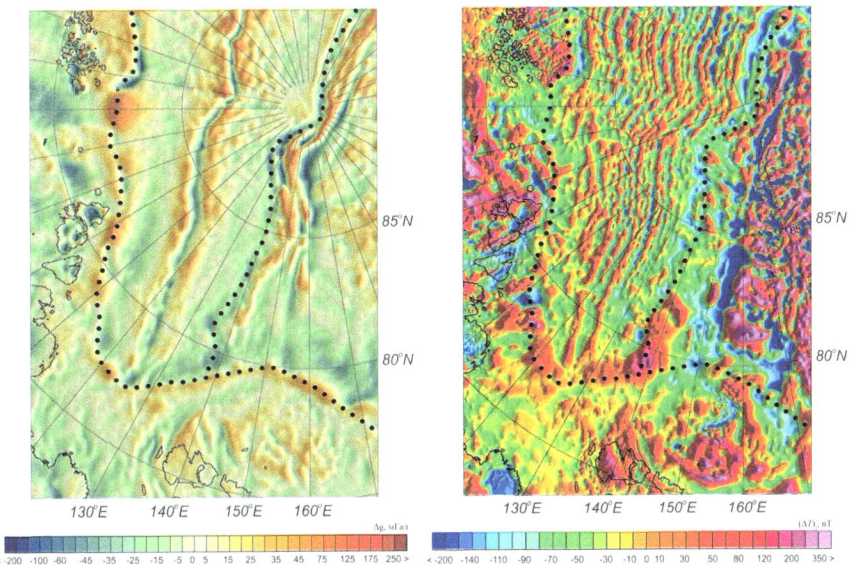

Fig. 3.2 Free Air gravity (**a**) and magnetic anomalies (**b**) of the Eurasian Basin. Dotted line – outlines of abyssal regions

Paleocene, the intraplate volcanic plateau was formed on the continental crust at the Basin western closure, later split by spreading into Morris –Jessup Plateau, west from the Gakkel Ridge, and Ermak Plateau, east from it. Both uplifts display intensive normal faulting and steep escarpments facing the abyssal oceanic depressions.

Many morphological elements visible at the ocean floor are associated with normal or transform faults. The largest normal faults form not only the axial rift valley of Gakkel Ridge but also sculpture the ridge outer slopes, creating the first-order horst-like structure in the center of the Eurasian Basin.

Diverse composition and structure of the Eurasian Basin basement is vividly reflected in anomalies of potential fields, qualitative and quantitative interpretation of which helps to delineate basement structural blocks and estimate their probable composition. Magnetic data were used by A.M. Karasik (1974, 1980, 1981) in his general outline of structure and evolution of the Eurasian basement, laterly specified by Brozena et al. (2003) and Glebovsky et al. (2006).

The magnetic field of the Eurasian Basin (Fig. 3.2b) consists of linear anomalies (LMA) of alternating polarity, roughly parallel and symmetrical to the central anomaly located at the axis of the Gakkel Ridge. Alternating polarity of magnetic anomalies reflects the magnetization of basement formations according to established periods of inversions of the Earth magnetic field. Using Russian and Western magnetometric investigations, linear magnetic anomalies (LMA) of Cenozoic sequence from 24 (chron A24, ~53 Ma) to present-day anomaly (Brozena et al. 2003; Grantz et al. 2001, Karasik 1980, 1981; etc.).

Free Air gravity (Fig. 3.2a) also display a certain asymmetry of gravity anomalies relative to the Gakkel Ridge axis.

A new page in history of geo-tectonic analysis of the Eurasian Basin was turned after compilation of a new updated digital map of magnetic profiles (Glebovsky et al. 1998, 2006). The multi-layered analysis of this map utilizing original processing software (Merkuriev 1991) included iterative modeling of bodies and layers, identification of magnetic isochrones, calculation of linear and angular spreading velocities, rotation poles and drift lines of the North-American and Eurasian tectonic plates.

The process of LMA age identification was initiated from the western part of the Eurasian Basin adjacent to Spitsbergen and Greenland, mostly because this segment was covered by relatively dense net of the US Navy magnetic profiles (Brozena et al. 2002, 2003). In the latter, the comparison of isochrones maps by (Brozena et al. 2003) and maps mentioned in the previous paragraph found them to be identical, thus confirming the reliability of the Russian identification.

Consequently, all key LMA of the Cenozoic sequence, starting from LMA 24, were identified and traced in the western part of the Eurasian Basin, with LMA 2a, 5, 6, 13, 18, 20 and 24, as the most reliable (Fig. 3.3). Problematic LMA 21, 22 and 23 were identified using more sophisticated models taking into account jumps of spreading axis.

It was shown (Glebovsky et al. 1998, 2002) that veracity of magnetic information in the east of the Basin was inferior to that in the west due to low density of observations, and poor positioning accuracy. Furthermore, the LMA identification process here was handicapped by "crowding" of LMA of different age in proximity to rotation pole and ultraslow and irregular spreading velocity combined with frequent inversions of the Earth magnetic field. All these factors often make the standard simplified models of magnetic anomalies unusable. However, the LMA identification in the basin eastern was completed using more sophisticated models and turned our to be in general agreement with actual isochrones (Fig. 3.3) based on profile-to-profile correlation.

Rotation poles positions and spreading velocities were determined by iterative approximations, selecting as a final the position which minimizes the difference between computed spreading velocity with its empirically obtained value. Linear spreading velocity (Fig. 3.4) for different stages of the basin evolution was estimated as a ratio of distance between two conjugated points on synchronous anomalies and their age.

General spreading asymmetry (in km) was defined by (Hayes 1976, Müller et al. 1997) as difference in distance between two conjugated points on synchronous anomalies from the spreading center measured along drift lines. The distance between two neighboring LMA on opposite sides of spreading axis was defines as staged spreading asymmetry as an indicator of difference in crustal growth on opposite sides of the spreading axis (Nansen and Amundsen Basins in our case) during various stages of their evolution. By joining the couples of synchronous LMA starting at chron 2 (3.5 Ma) and continent-ocean boundary (COB), on opposite sides of the mid-ocean ridge (Cande and Kent 1995), the position and configuration of the oceanic basins and continents in distant past could also be reconstructed.

Eventually, the complete picture of ocean floor growth and evolutional history of the Eurasian Basin were developed (Glebovsky et al. 2006). It was postulated that

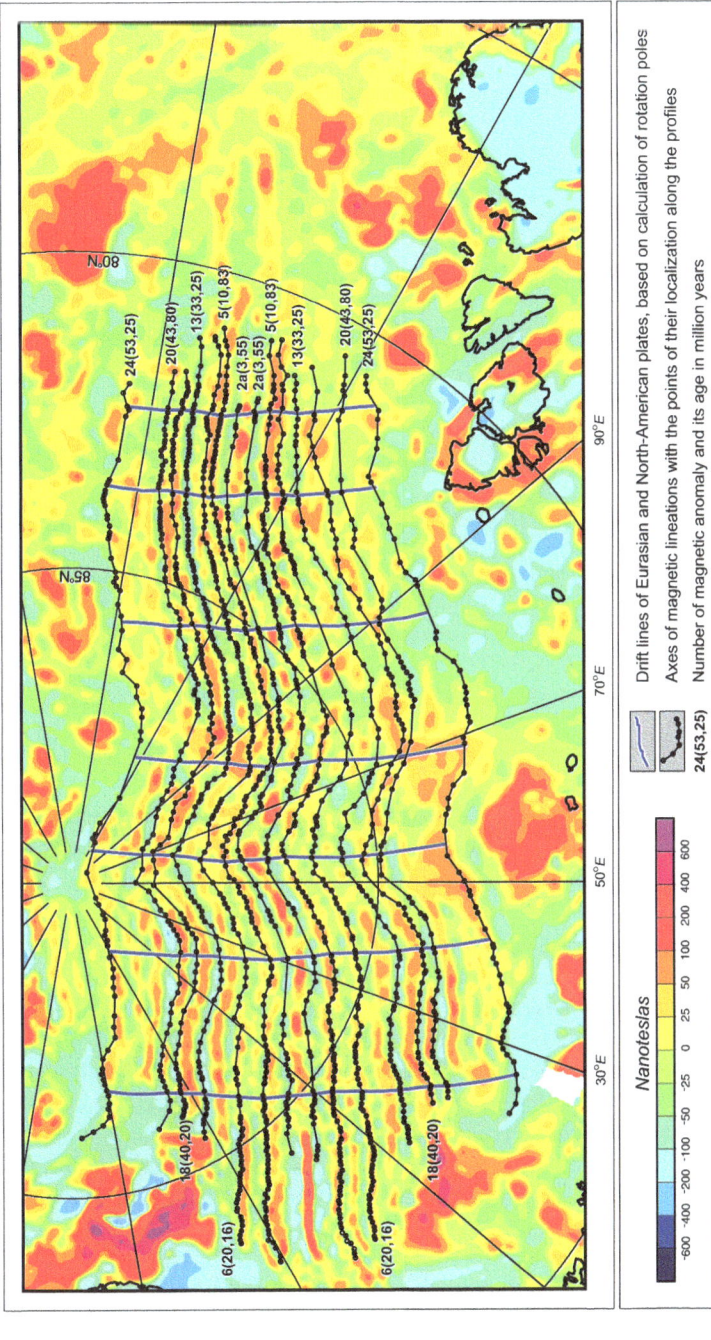

Fig. 3.3 Key LMA (magnetic chrons) in the Eurasian Basin (Glebovsky et al. 2006)

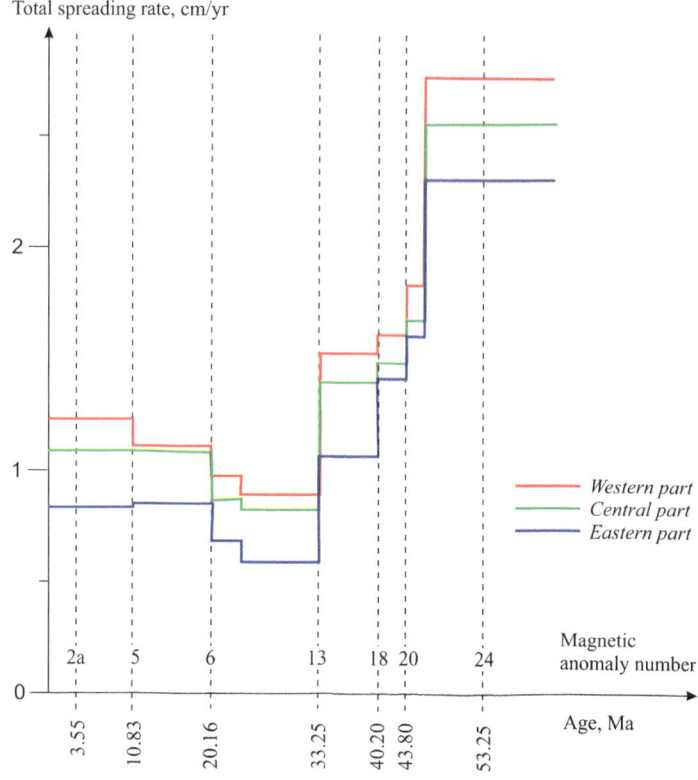

Fig. 3.4 Eurasian Basin – average spreading velocity vs time (Glebovsky et al. 2006)

general pattern of spreading velocity changes in the Eurasian Basin, on background of overall de-acceleration, remains constant for the entire length of the Gakkel Ridge (Fig. 3.4). Initially (Early -Mid Eocene, 53–44 Ma, chrons 24–20), the velocity was relatively high – 22–27 mm/yr. Then it started to decline and in Oligocene –Early Miocene (crones 13–6) dropped to 5–9 mm/yr. At about 20 Ma (chron 6), the spreading velocity grew slightly to 7–12 mm/yr. and presently remains the same. Incidently, the similar pattern was found at Reykjanes Ridge and in the Labrador Sea (Glebovsky et al. 1990; Glebovsky 1995).

Spreading asymmetry was established in (Glebovsky et al. 2006). It was noted that not only presumed difference in plates speed, but also periodic spreading axis jumps (particularly at the early stages) might cause the asymmetrical position of continental margin base and older LMA on the opposite sides of the Gakkel Ridge.

Summarizing the results of the Eurasian Basin LMA analysis, we me say that:

• Prior to onset of rifting, the Lomonosov Ridge was a part of an immense platform and was attached the present days Barents-Kara shelf. Later it was split from Eurasia and migrated to the present position.

- Eurasian seafloor spreading started about 58 Ma ago and proceeded during numerous inversion of the Earth magnetic field. It was preceded by Norway-Greenland Basin spreading and initially was connected to the opening Labrador Sea through the Nares Strait, until termination of the Baffin Bay spreading.
- Simultaneous initial opening of the North Atlatntic (north from Charlie-Gibbs fracture zone) and Labrador Sea triggered the northern shift of Greenland towards North America 53 Ma ago
- In their movement, the Geenland acted as a dull wedge penetrating into space created by spreading of the Eurasian Basin. Judging by changing LMA pattern, the Eurasian Basin seafloor shrunk by at least 100 km due to lateral compression caused by this move. At this stage, the jointed Ermak and Morris-Jessup Plateaus were part of the Eurasian platform.
- The northward drift of Greenland ceased at about 33–35 Ma and connection between the Labrador Sea and the Eurasian basin was terminated. Greenland became a part of the North American Plate; a newly born divergent plate boundary north form the Norway-Greenland Basin split apart Eurasian continental margin into present Ermak and Morris-Jessup Plateaus.
- From this time on, the Eurasian Basin spreading was related to North Atlantic and took place along an entire length (until it meets the Laptev shelf) of single axial line, which is now the central part of the Gakkel Ridge.

The information, critically important for further understanding of evolution of the Eurasian Basin was collected along profile 2014–07 by RV" Academic Fedorov "(JSC MAGE) in 2014, the first profile crossing the entire Eurasian Basin from Barents-Kara shelf to the Lomonosov Ridge (and beyond). In addition to reflection and refraction seismic, gravity was recorded using on-board gravity meters "Chekan-AM" and "Shelf-E". Figure 3.5 shows the profile location and seismic velocity model derived from seismic ray modeling:

In the Eurasian Basin (500–1070 km marks) the following units were identified (from top down).

Miocene-Pleistocene complex between seafloor and regional pre-Miocene unconformity RU (Vp 1.8–1.9 km/s, 0.6–0.7 km –thick in the Amundsen Basin, and twice as much – in Nansen).

Mid-Late Eocene-Oligocene complex between the RU and EoU unconformities with Vp 2.3–2.5 km/s and thickness growing toward the Gakkel Ridge from several hundred meters to 1.7 km in the Amundsen Basin, higher velocities, up to 2.7–3.1 km/s and thickness also growing toward the Gakkel Ridge from 1.5 to 2 km - in the Nansen Basin (in both Basins the complex completely pinches out against acoustic basement at location close to chron 13, 33 Ma, Oligocene).

Paleocene-Early Eocene complex between EoU unconformity and acoustic basement (AB) with Vp 2.8–3.1 km/s and thickness from several hundred meters to 1.5 km in the Amundsen Basin and 3.4–3.5 km/s and 1–2 km - in the Nansen Basin, accordingly (again, in both Basins the complex pinches out against acoustic basement, only there - between chrons 24, 53 Ma, Late Paleocene, and 20, 44 Ma, Mid Eocene).

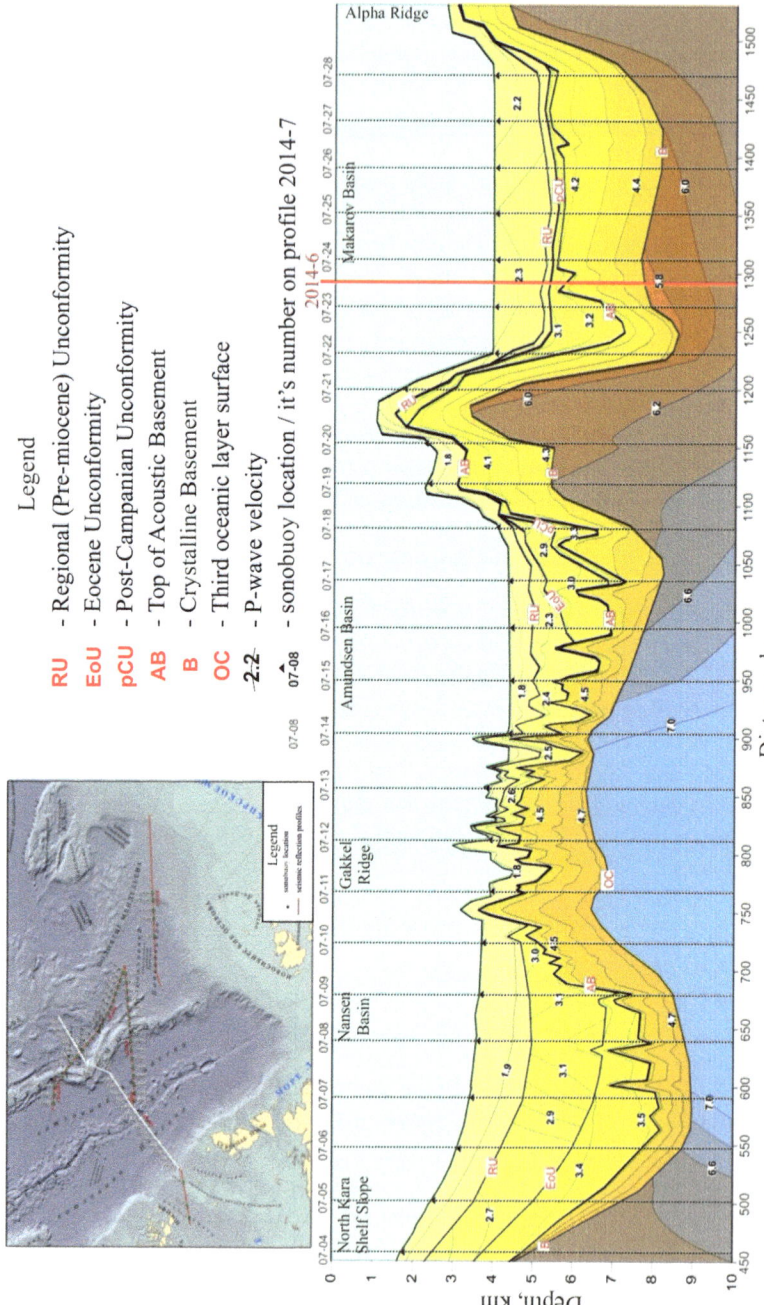

Fig. 3.5 Seismic velocity model along line 2014–07

Cretaceous, possibly Early Paleocene in its upper part, complex under post-Campanian unconformity, with Vp 3.3–3.4 km sec and about 1 km thick found only in the Lomonosov Ridge-Amundsen Basin transitional zone where it pinches out against acoustic basement at chron 24, 563 Ma).

Acoustic Basement (AB), identifies as the IInd oceanic (volcano-sedimentary) layer with Vp 4504.7 km/s, with thickness from several hundred meters in local depressions to 2–2.5 km under the Gakkel Ridge.

Crystalline basement (B) in Lomonosov Ridge-Amundsen Basin and Kara shelf – Nansen Basin transitional zones, with Vp 6.2–6.6 km/s.

Top of IIIrd oceanic layer (OC) with Vp ≥ 7 km/s, 6.5 km-deep under the Gakkel ridge and plunging to ≈7.5 km under the Amundsen Basin and to ≈9 km – under the Nansen.

The seismic-gravity modeling was used to define the velocity-density crustal models along many seismic profiles crossing all principal structures of the abyssal parts of the Artcic Ocean. The standard procedure solves the inverse problem of geophysics by iteratively solving the direct problem, altering both geometry (structure) and density (composition) of the causative bodies until their computed gravity response is acceptably close to observed Δg values.

The original software (GM-SYS, Gravity/magnetic Modeling Software, v. 4.7) which runs the procedure, requires input of gravity, subsurface geometry and density. Gravity values, usually in form of Free Air anomalies, for MCS seismic lines come from the ship-borne surveys run simultaneously with seismic operations (e.g. RV "Academic Fedorov", 2014); for DSS profiles - were digitized from general digital gravity database (Glebovsky et al. 2012) with 1 km increment. The subsurface geometry, or structure, was entered into the system as a copy of time-converted interpreted seismic; density was assigned to each structural element using empirical velocity - density function by (Brocher 2005) reflecting lateral change of velocity, if any exist within it. the initially assigned densities of Fig. 3.6. presents the results of seismic - gravity modeling of MCS line 2014–07, so far, the only one crossing the entire Eurasian Basin. The initially assigned densities of sedimentary formations varied from 1.8 g/cm³ (Vp 1.7 ÷ 1.8 km/s) at seafloor to 2.65 g/cm³ (Vp 5.6 ÷ 6.3 km/s) in the deep depressions; for modeling, seismic complexes SSC-5 and SSC-6 (JU-PU) (see Chap. 2) were merged together.

The computed density of sedimentary section was fixed at, upper crust - 2.70 g/cm³, lower crust- 2.90 g/cm³, and mantle- 3.30 g/cm³. The Lithosphere-asthenosphere density contrast was accepted after (Deep Structure and Evolution of the Lithosphere in the Central Atlantic Ocean 1998) as 0.11 g/cm³.

In Kara Sea, the crust is 29 km thick, with 2–3 km of sediments upper consolidated crust- 11–17 km and lower-7–11 km. In the Nansen Basin there is two sedimentary complexes 3–4 km thick with density 2.00–2.35 g/cm³, the crust thins from ≈14 km in the Kara shelf-Basin transitional zone to ≈7 km in the central part and further to ≈ 5 km towards the Gakkel Ridge.

Inside the Gakkel Ridge, sedimentary package in the central rift valley may contain up to 1.5 km of sediments and in peripheral depressions – less than 1 km.; the upper crust thickness grows from 1 km under the central rift valley to 4 km under the slopes; lower crust – 2-3 km.

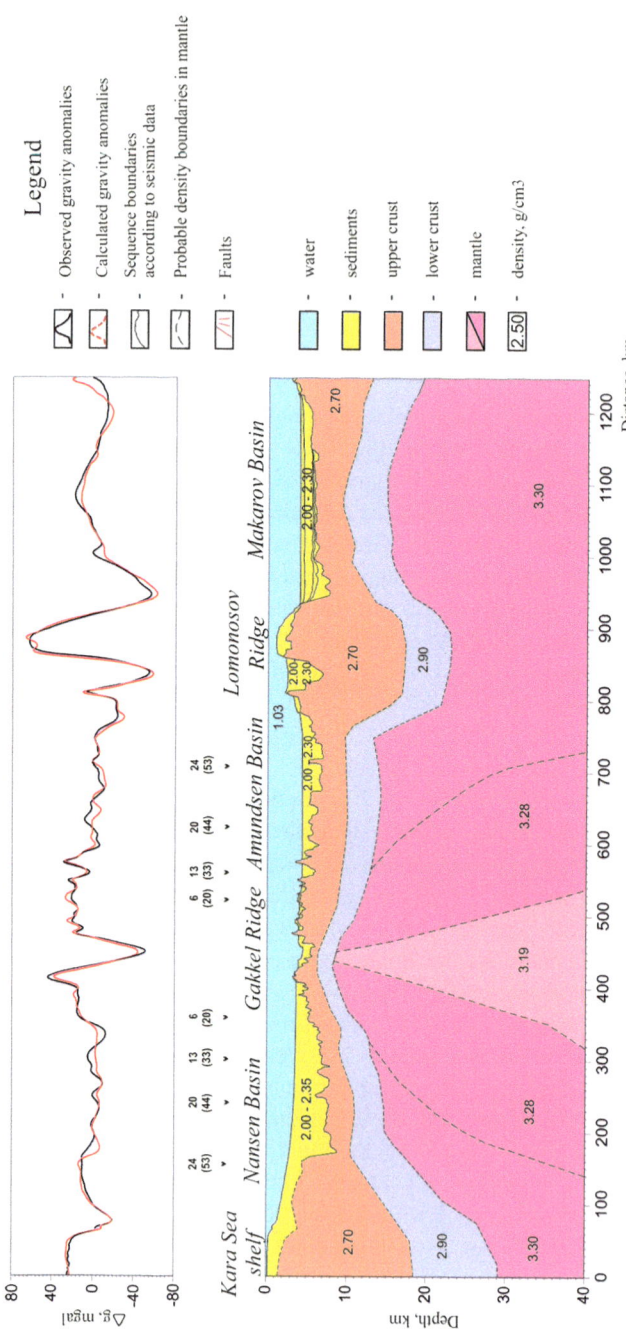

Fig. 3.6 Crustal seismic-gravity model along line 2014–07

The Amunddsen Basin also contains two sedimentary complexes with density 2.00–2.30 g/cm³ and maximum thickness 2.5 km in the central part. The crust thickness here was estimated at 9 km.

Strong,~70 mGal, negative gravity anomaly is associated with axial zone of the Gakkel Ridge, the crust generating center of the Eurasian Basin. This anomaly is thought to be caused by negative density contrast of about 0.11 g/cm³ between hotter upper mantle under the Gakkel Ridge and cooler masses on the periphery of the spreading zone (Breivik et al. 1999; Chernych and Golmstock 2009; Deep Structure and Evolution of the Lithosphere in the Central Atlantic Ocean 1998). Accordingly, the two mantle blocks with density 3.28 g/sm³ were introduced on both flanks of the central block with density 3.19 g/cm³.

3.2 The Gakkel Ridge

The Gakkel Ridge stands out as 1800 km- long and 200 km-wide linear uplift with complex and rugged morphology, surrounded from both sides by abyssal plains of the Nansen and Amundsen Basin, only at the near-Laptev Sea region contacting with the continental rise (Fig. 3.7). East from 70°E the ridge is visibly asymmetric, with Nansen Basin abyssal plain almost in contact with the central rift valley, while from the Amundsen Basin side there is wide plateau elevated above abyssal plain by 200–400 m. Morphology of the axial rift valley changes along ridge length forming four distinct segments. Depth of the rift valley floor varies from 5000–5200 m in the

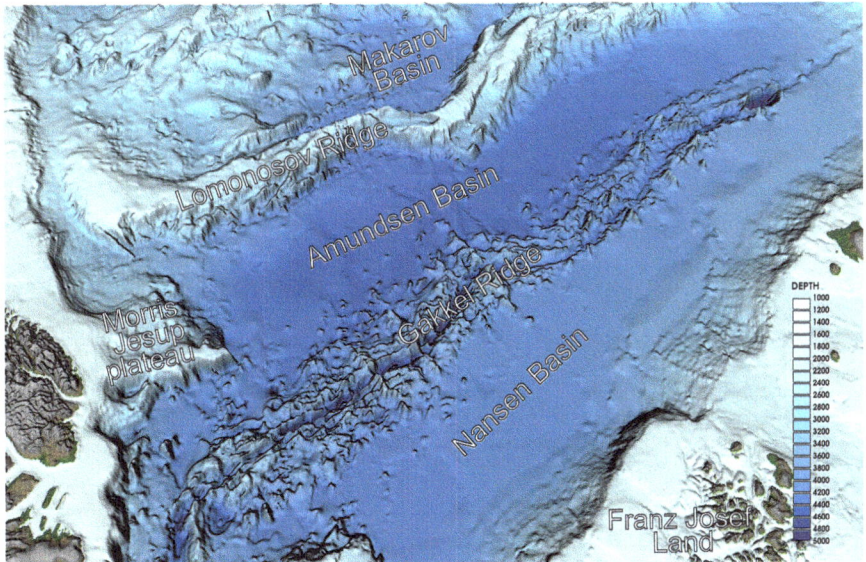

Fig. 3.7 3D image of the Gakkel Ridge (IBCAO v.3.0)

near-Laptev Sea segment to 4300 m in central segment and 4500–5000 m - in near-Greenland (Naryshkin 1987; Orographic Map of the Arctic Basin 1995).

The Gakkel Ridge has the lowest spreading velocity in the global system of mid-ocean ridges varying from 12.8 mm/yr at the near-Greenland part to 6.5 mm/yr near Siberian continental margin (Cochran 2008).

Detail bathymetry and gravity data were collected by US Navy submarine Hawkbill (Cochran et al. 2003) covering 850 km-wide strip from 5°E to 97°E demonstrating some similarities of the Gakkel Ridge segment with the highest spreading velocity of 1–12 mm/yr (5°E -63°E) with Mid-Atlantic Ridge with rate 2–4 times higher. Rift valley in this interval is 15–20 km wide and1–2 km deep, with slopes formed by fault planes. According to interpretation of nearby seismic, the crust under the Gakkel Ridge is only 2–3 km thick.

However, volcanic activity here is much lower, than at the axial zone of the Mid-Atlantic Ridge. In 32°E–63°E interval, and east from 80°E, isolated volcanoes are 25–95 km apart; large volcano at 69°E is the only one for 175 km between 63°E and 80°E. Significant lava flows are identified in sediment-filled auxiliary rift troughs in this segment of the Gakkel Ridge. Evidently, slow spreading reduces the amount of rising magmatic material, thereby "starving" volcanic activity. In this region the Gakkel Ridge axis is at 25–40° angle from orientation of the normal to general spreading vector which in this area runs along 140°E meridian.

International expeditions AMORE 2001 based on icebreakers "Polarstern" and "Healy" brought a trove of bathymetry, magnetism and seafloor sampling information (Michael et al. 2003; Jokat et al. 2003), delineating three large segments of the Gakkel Ridge: western volcanic, central amagmatic and eastern volcanic zones. Western and Eastern zones, with spreading velocity reaching 14 mm/yr, contain unaltered vitreous basaltic pillow-lavas. In the central zone, with very slow rate of amagmatic spreading, basalts are absent, but sampling showed presence of peridotite and diabase as derivatives of the prevailing here third oceanic level of the standard oceanic crust.

Combined interpretation of seismology information, bathymetry and potential fields anomalies in (Engen et al. 2003) demonstrated that the Gakkel Ridge can divided into four regional segments. Spitsbergen transform system includes several short stretches and transform segments. Further east, an important, first-order changes in morphology and pattern of geophysical anomalies happened at longitude 63°E .West from this point, the Gakkel Ridge has a very pronounced terrain and strong magnetic anomalies, while east from it the terrain become less contrasting and magnetic anomalies – much weaker. Finally, at the continental slope of the Laptev Sea the Gakkel Ridge disappears under the cover of sediments and, as seismology data indicate, ultraslow oceanic spreading gives way to active continental rifting in 60-km wide zone inside the continent-ocean transition zone.

As it was mention before, the seismic line 2014–07 was, in 2014, the first line to cross the entire span of the Gakkel Ridge. The detail fragment of this line is shown oh Fig. 3.8.

Strikingly asymmetric position of the rift valley attracts immediate attention – the ridge north slope is more than 100 km wide measuring from the center of the rift valley, and southern - only 40 km. Sizeable amount of sediments accumulated in the

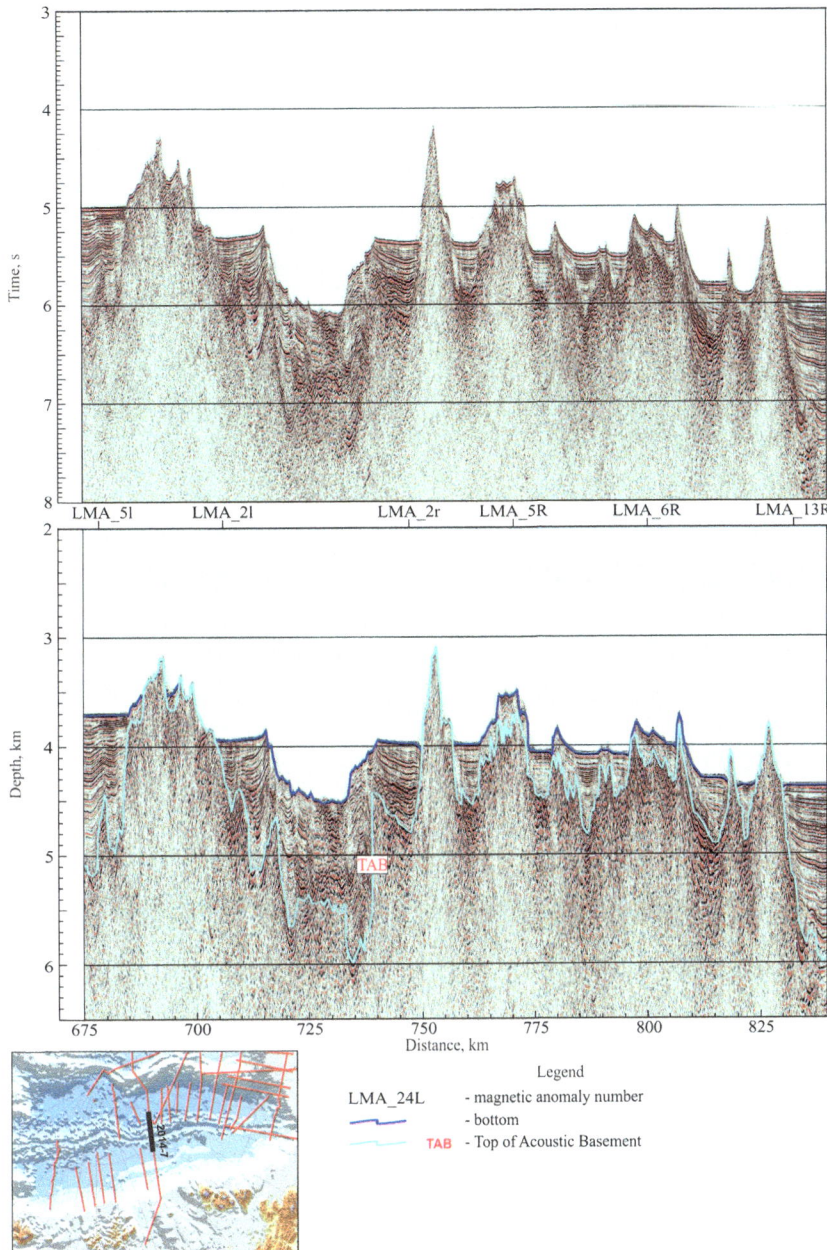

Fig. 3.8 MCS line 2014–07 (Gakkel Ridge)

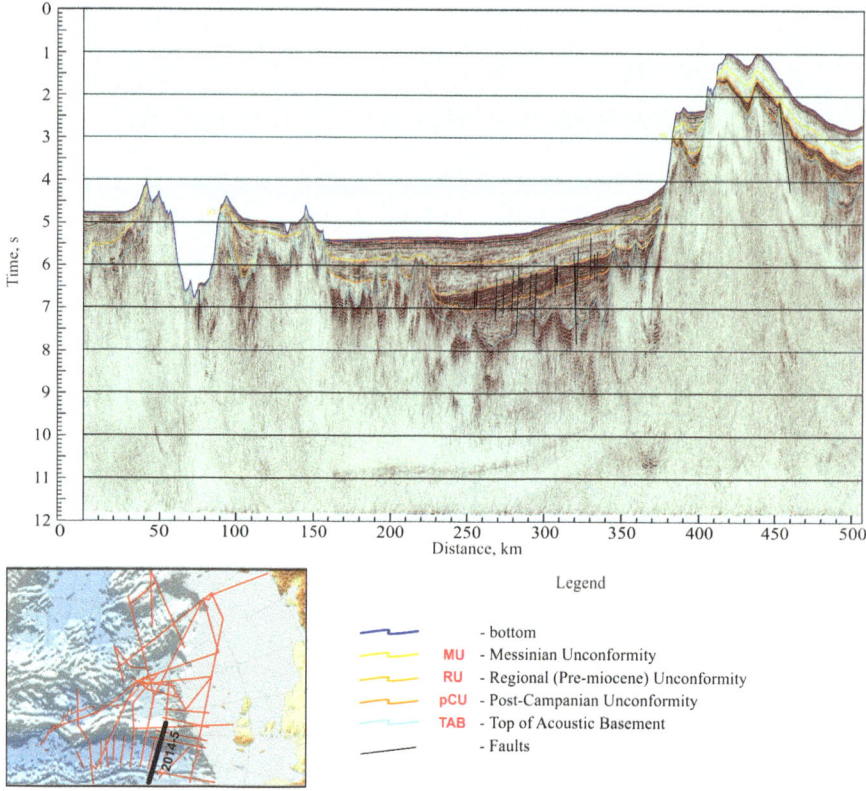

Fig. 3.9 MCS line 2014–05 (Gakkel Ridge-Amundsen Basin)

rift valley also makes the Gakkel Ridge unique among other oceanic spreading ridges.

Similar asymmetrical position of the rift valley is also visible on another seismic line (2014–05) crossing the Gakkel Ridge at around 81° 40'N and 120°E (Fig. 3.9). Incidentally, the line also cuts the unique volcanic structure - large caldera 35 km wide. The time of origin and consequences of its formation will be discussed in the following chapters.

Further south, closer to the Gakkel Ridge junction with the Laptev Sea continental slope at around 79° 17'N and 125° E, the seismic (MCS DOP1501V, Fig. 3.10) shows different Gakkel Ridge structure. Here the rift valley contains all seismic-stratigraphic complexes, from Holocene down to pre-Cenozoic, disturbed by multitude of normal and reverse faults forming the differentially displaced fault blocks and fringe escarpments several hundred meters high.

Evidently, the forming of the rift valley at this part of the continental slope must be attributed to the last stage of tectonic restructuring at the Late Pleistocene - Early Holocene, around 2 Ma.

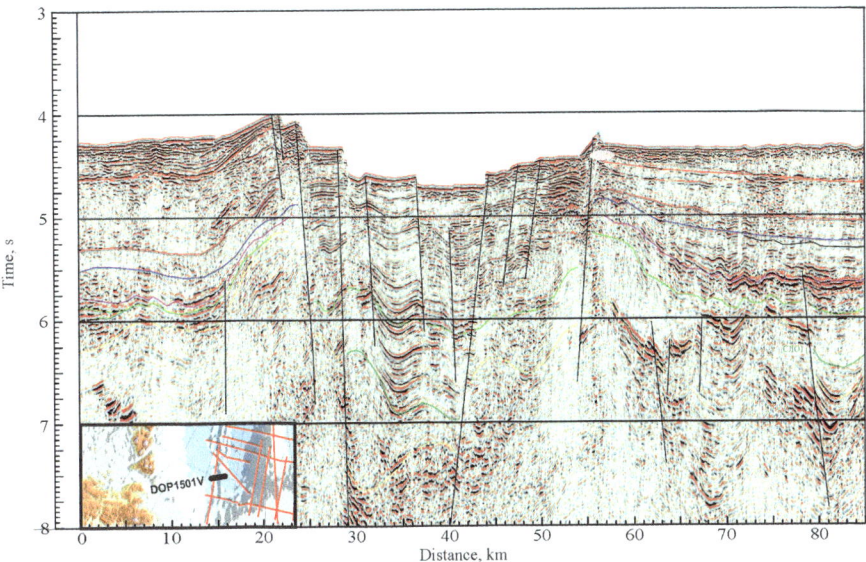

Fig. 3.10 Fragment of MSC line DOP1501V (Courtesy of G.S. Kazanin and T.A Kirilova, MAGE)

3.2.1 Nansen Basin

The seafloor inside both Nansen and Amundsen Basin consist of subhorizontal abyssal plains with maximum depth 4000 m and 4500 m, accordingly (Figs. 3.11, 3.12) The deepest parts in the Nansen Basin gravitate to its western parts, while in the Amundsen Basin they concentrate along its axial line, the difference presumably dictated by specifics of sedimentation process. (Orographic Map of the Arctic Basin 1995). At the near-Greenland parts of the Arctic Ocean where both Nansen and Amundsen abyssal plains are in contact with complex slopes of the Ermak and Morris-Jessup Plateaus, the plains are getting narrower and shallower (to 3500 m) and tilt to the east. The contacts of Nansen and Amundsen abyssal plains with the Gakkel Ridge usually have very complex configuration due to complexity of the morphology of the ridge slopes.

Barents-Kara board of the Nansen Basin can be described as a simple slope with gradual transition to the continental rise at about 3000 m b.s.l., exception being only the stretch around Ermak Plateau which be classified as a complex slope. The W-NW orientation of the board gently changes to EW (sub-latitudinal) NW just east from Kara Sea-Laptev Sea. Mostly concave in cross-section (becoming convex only at the trains of sediments delivered by underwater canyons) the slope usually is 4–8°, seldom 10°, steep. The steepest parts are found north from Frantz-Joseph Land and its vertical drop, as a rule, is more the 2000 m.

The rise is considerably wider in the eastern part of the Nansen Basin than in the western. North from the Severnaya Zemlya archipelago the rise plain formed by

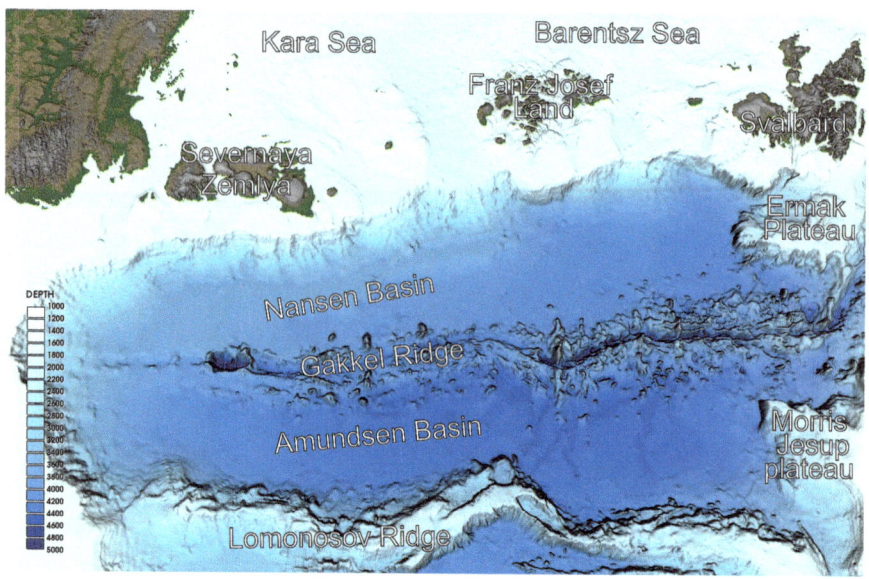

Fig. 3.11 Nansen Basin and Barents-Kara continental slope (IBCAO v.3.0)

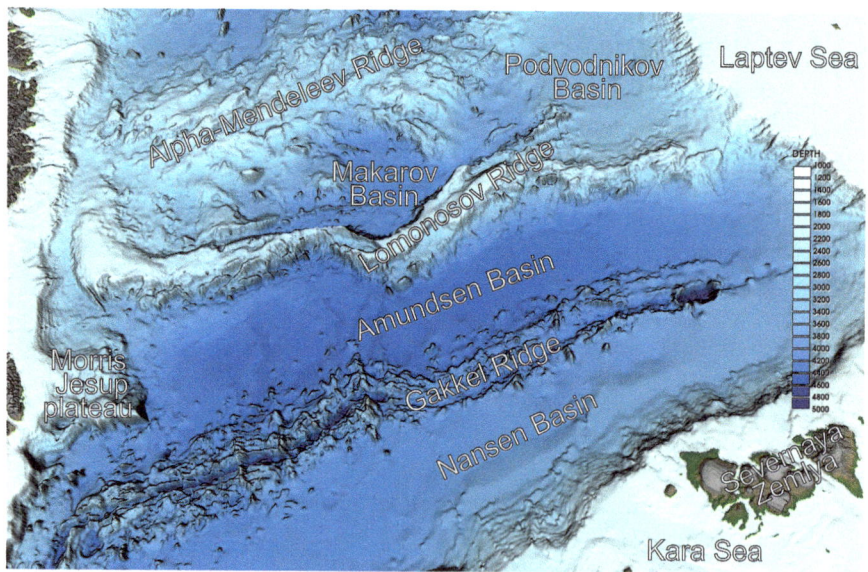

Fig. 3.12 Nansen and Amundsen Basins and Laptev Sea shelf continental slope (IBCAO v.3.0)

Kara Sea wash-down merges with that of the Laptev Sea and becomes so wide that it almost touches the base the Gakkel Ridge western flank. Close to the Laptev Sea, the rise becomes steeper and its upper boundary moves to shallower depths (2600–2800 m), Figs. 3.11, 3.12.

Reliable information on structure and composition of the basins sedimentary cover came from systematic international arctic expeditions in 1999 and 2001 (Jokat and Micksch 2004), Russian - in 2011–2014, as well from compilations and generalizations of the first decade of the XXIst century (Glebovsky et al. 2006; Verba 2008; Moore and Pitman 2011, etc).

German interpretation of profile AWI-100 (Jokat and Micksch 2004) with addition of Norwegian data from 2001 expedition in western part of the Nansen Basin (Engen et al. 2009) shown on Fig. 3.13.

The authors noted apparent thickening of sedimentary section towards the Frantz-Victoria Trough deposition center and identified four seismic-stratigraphic complexes (from top down):

(1) NB-4 (~2.6 Ma) with two glacial phases doubling thickness from Ermak Plateau to Frantz-Victoria Trough;
(2) NB-3 (~10 Ma) with parallel bedding and strong reflector at the base (possibly related to Fram Strait opening);
(3) NB-2 (~23 Ma) - low frequency, variable amplitude reflectivity, constant thickness;
(4) NB-1 (~55 Ma) - semi-transparent continuous reflectivity in the Nansen Basin Paleogene sequence (NB-1A- Eocene, possibly Paleocene, sub-complex with strong reflectivity).

Profile AW1 In the Amundsen Basin (Fig. 3.14) provided the first proof of asymmetry of basement composition in the Nansen and Amundsen Basin relative to modern spreading axis of the Gakkel Ridge.

MCS line 2014–07 (Fig. 3.15) confirmed the asymmetry of the Eurasian Basin.

Seismic clearly shows not only the southward offset of the Gakkel Ridge rift valley relative to its axis, but also apparent lack of correlation between seismic signatures of the Nansen and Amundsen Basins. Sedimentary section in the Nansen Basin is at least twice of that in the Amundsen Basin.

The latest seismic survey helped to identify the specific segments of the Eurasian Basin with no LMAs which could de confidently correlated. These areas include southern parts of the Nansen Basin and near-Laptev Sea portion of the Eurasian Basin excluding narrow contemporary rift zone. In these regions basement is buried under more than 5 km of sediments and could not be traces on seismic sections, but sharp hinge-like suture line is confidently identified on several seismic lines crossing the Nansen Basin (Figs. 3.16, 3.17, and 3.18).

Similar situation is highlighted by previously mentioned in Chap. 2 seismic sections 2011–05 and 2011–06 (Figs. 2.6 and 2.7).

Heterogeneity of the Eurasian Basin western board is, in many aspects, predetermined by non-uniformity of the Barents-Kara continental margin with its Precambrian basement, at places re-activated by the latest tectonic events. Basement outcrops are studied on Spitsbergen and Severnaya Zemlya Archipelagoes and penetrated by 1-Nagurskaya well on Alexandra Land (Franz-Joseph Archipelago). The continental crust under this part of the Eurasian Basin is 30–35 km thick.

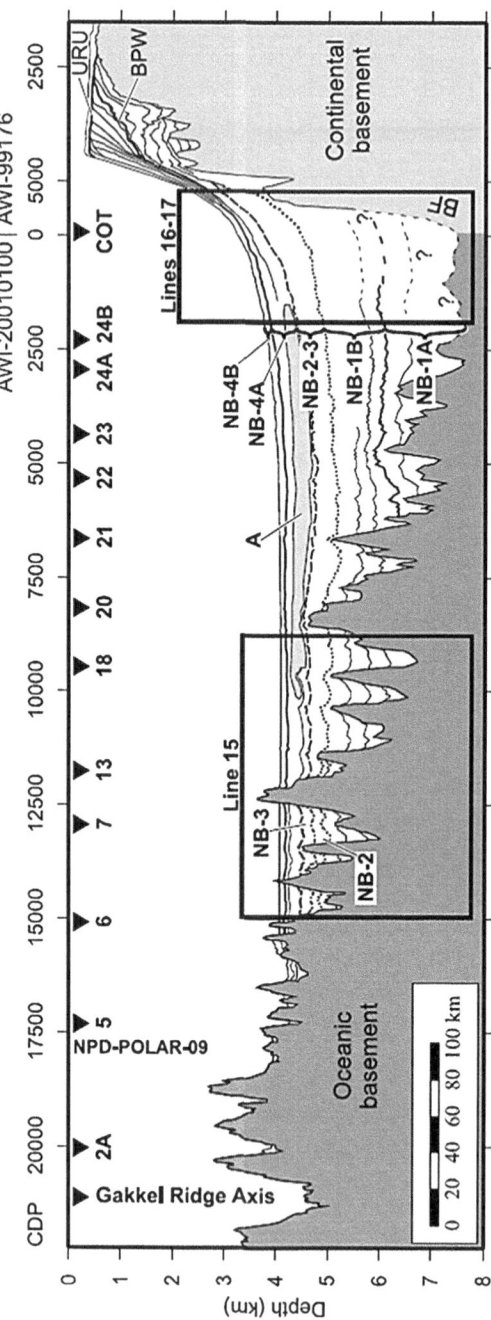

Fig. 3.13 Crustal cross-section of the Nansen Basin (Jokat and Micksch 2004), modified by (Engen et al. 2009) (in blocks)

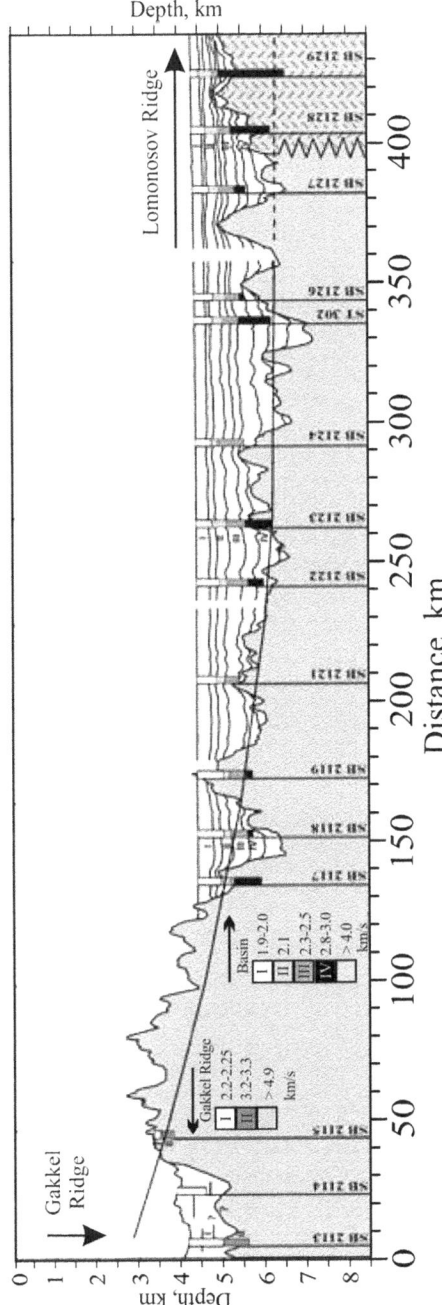

Fig.3.14 Crustal cross-section of the Amundsen Basin (Jokat and Micksch 2004). Solid line – basement surface calculated assuming Cenozoic opening of the Eurasian Basin

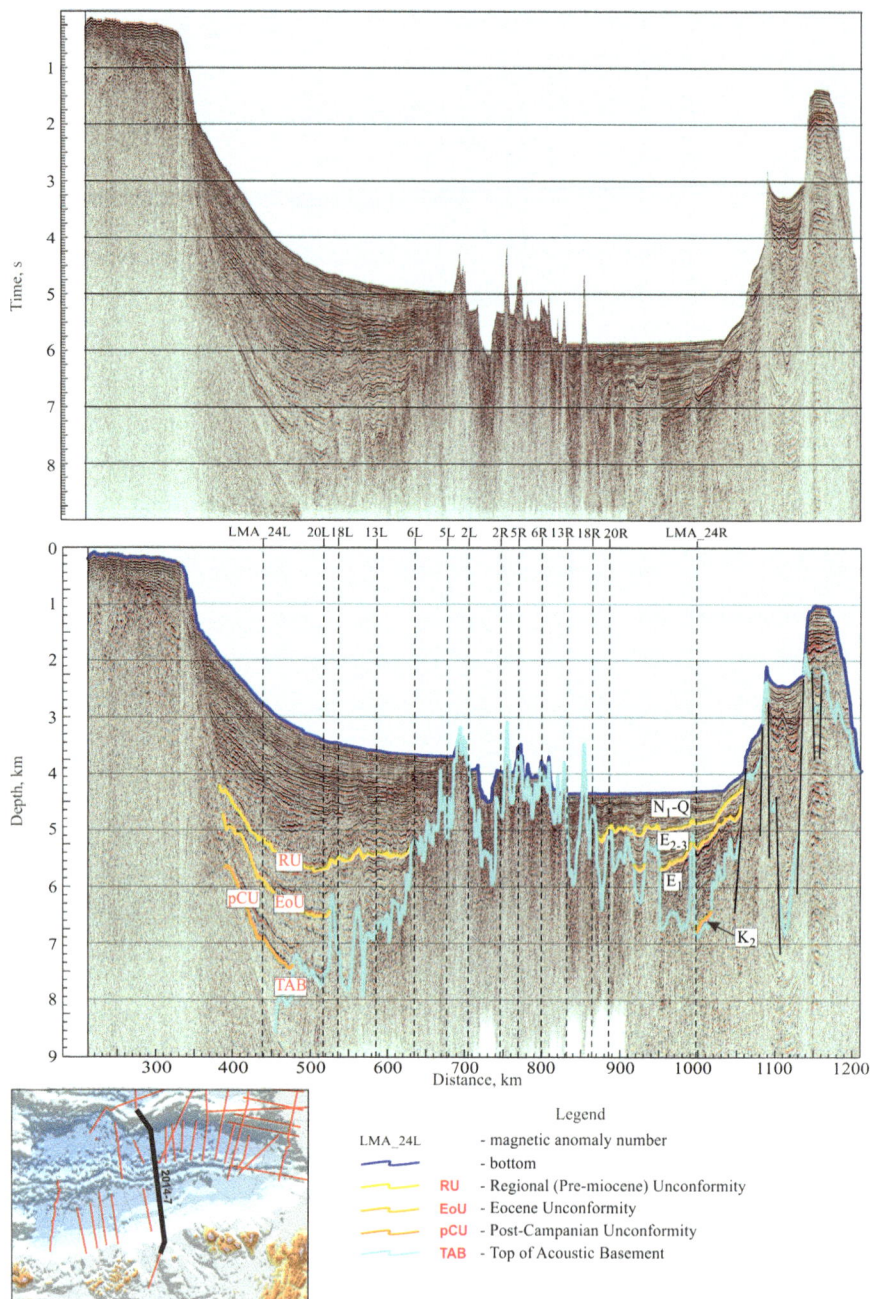

Fig. 3.15 The Eurasian Basin seafloor structure along MSC 2014–07

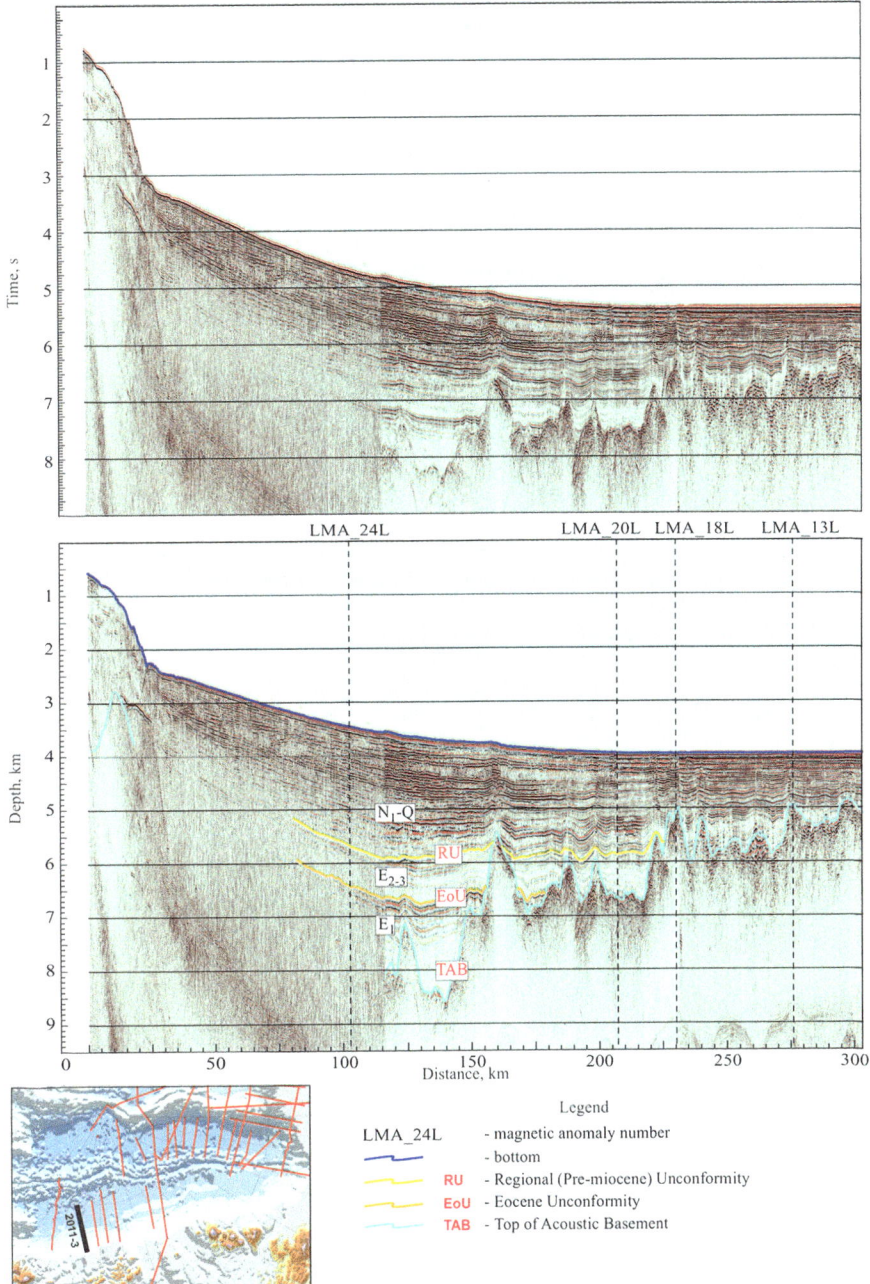

Fig. 3.16   MCS line 2011–03

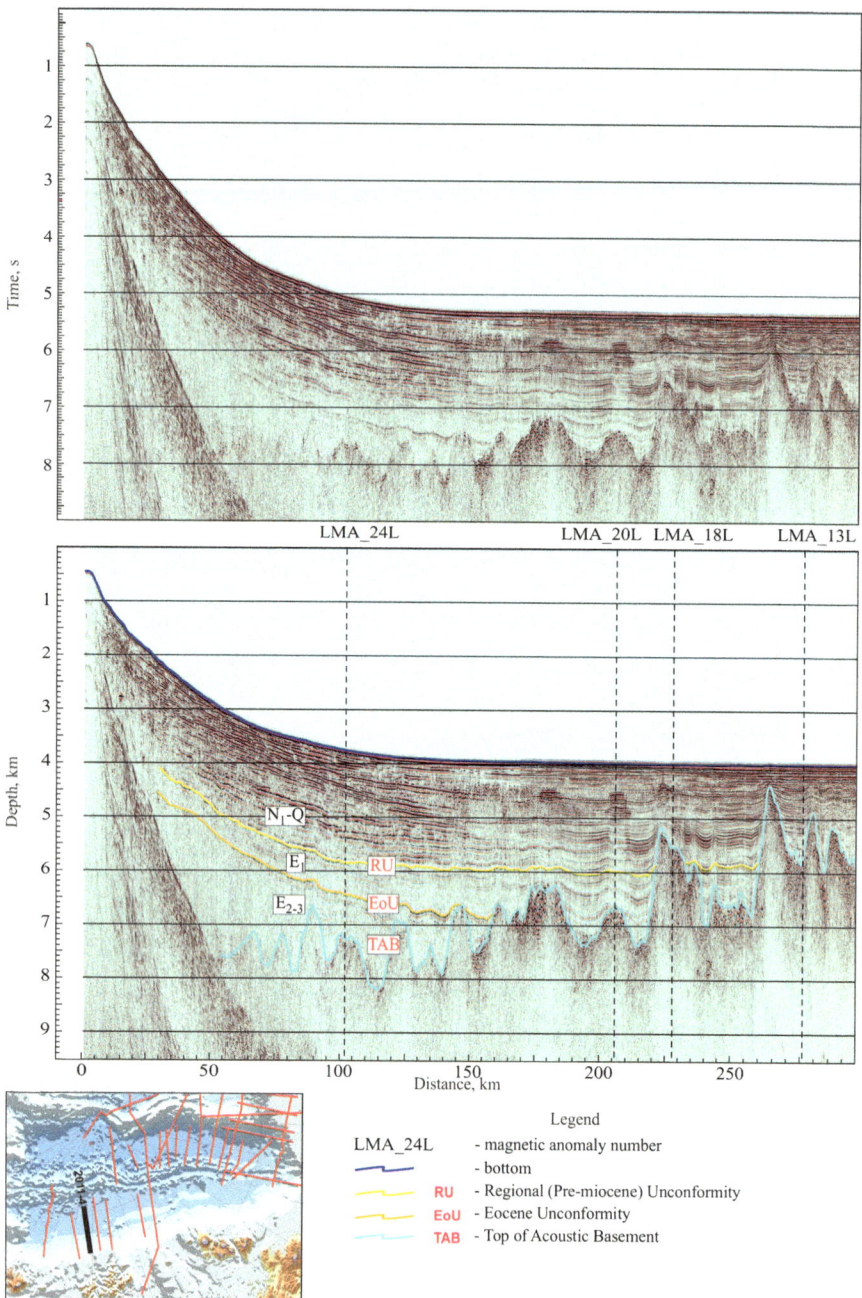

Fig. 3.17 MCS line 2011–04

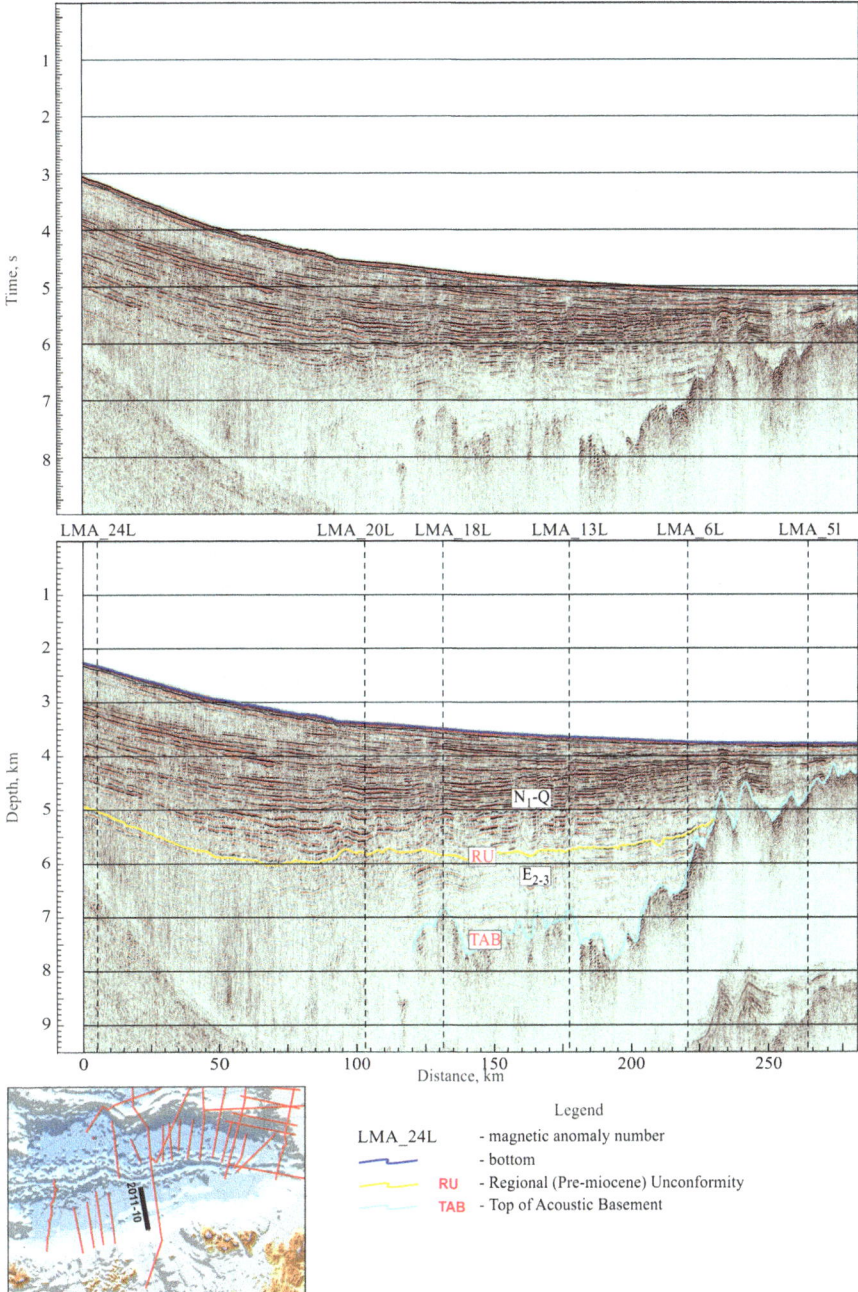

Fig. 3.18 MCS line 2011–10

The sedimentary cover of the peripheral zone consists of Paleozoic and Mesozoic formations with limited content of Cenozoic units (more of them are present at the continental slope).

Magnetic data show that active intraplate magmatism spreads from the western Franz- Joseph Land (where it was well studied) onto adjacent continental slope, maintaining the prevailing NW trend of dolerite dykes swarms.

Enormous dyke swarms of similar petrology, azimuth and age are known in the coastal Greenland, the Ellesmere Island and the Lomonosov Ridge (Døssing et al. 2013). These facts lead to the conclusion that these magmatic formations belonged to the same magmatic province, later separated by spreading.

3.2.2   Amundsen Basin

The boundaries of the area containing correlatable LMAs (possibly anomaly 13) in the Amundsen Basin coincide with easily identifiable seismic features (Figs. 2.3, 2.4 and 3.21), for instance, local, 10 km wide, uplifted basement block. The same uplift, with amplitude 1–2 km, isolate part of the basin, next to the Lomonosov Ridge, filled with large volume of sediments.

In south, the acoustic basemen, as can be seen on lines 2011–24 and 2011–26 (Fig. 3.19, 3.20) broken into numerous fault blocks, each with its own amplitude and overburden pattern. More than 5 km of sediments is present at the lower slopes of the Lomonosov Ridge and Laptev rise, as can be seen on line 2011–24.

There are three major unconformities in the Amundsen Basin, dated by the age of LMA nearest to place where unconformity is truncated by acoustic basement (Figs. 2.2, 3.21, 3.22, 3.23).

Seismic sections show very rugged surface of the Amundsen acoustic acoustic basement with no coherent reflections inside (Figs. 2.4, 2.5, 3.24).

Despite large volume of seismic shot by Russian researchers in the last years, significant portions of the Eurasian Basin remain untouched. They include south-eastern Nansen Basin near Severnaya Zemlya Archipelago and north-western Amundsen Basin which, until 2014, had only one seismic line (Fig 3.14). So, the importance of MCS line 2014-39a (Fig. 3.25) showing the structural complexity of sedimentary cover in the Amundsen Basin – Lomonosov Ridge transition zone cannot be overestimated.

MCS line 2014–09 crosses the Amundsen Basin and previously shot lines 2011–32 and 2011–34. It is well positioned to highlight the specifics of the Amundsen Basin – Lomonosov Ridge junction – Fig. 3.26.

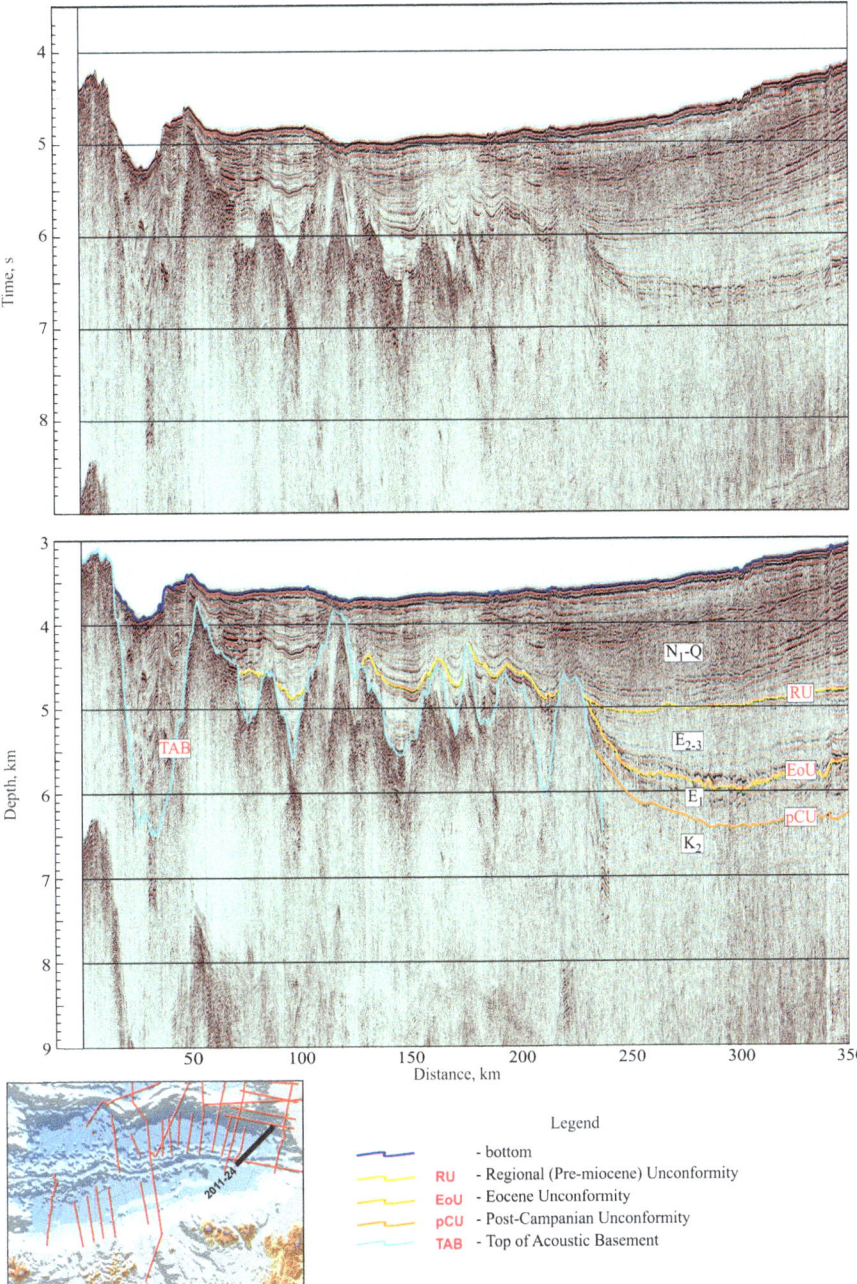

Fig. 3.19 MCS 2011–24. Note deep rift graben on the left, 80% filled by recent sediments; sharp thickening of sedimentary sequences – on the right

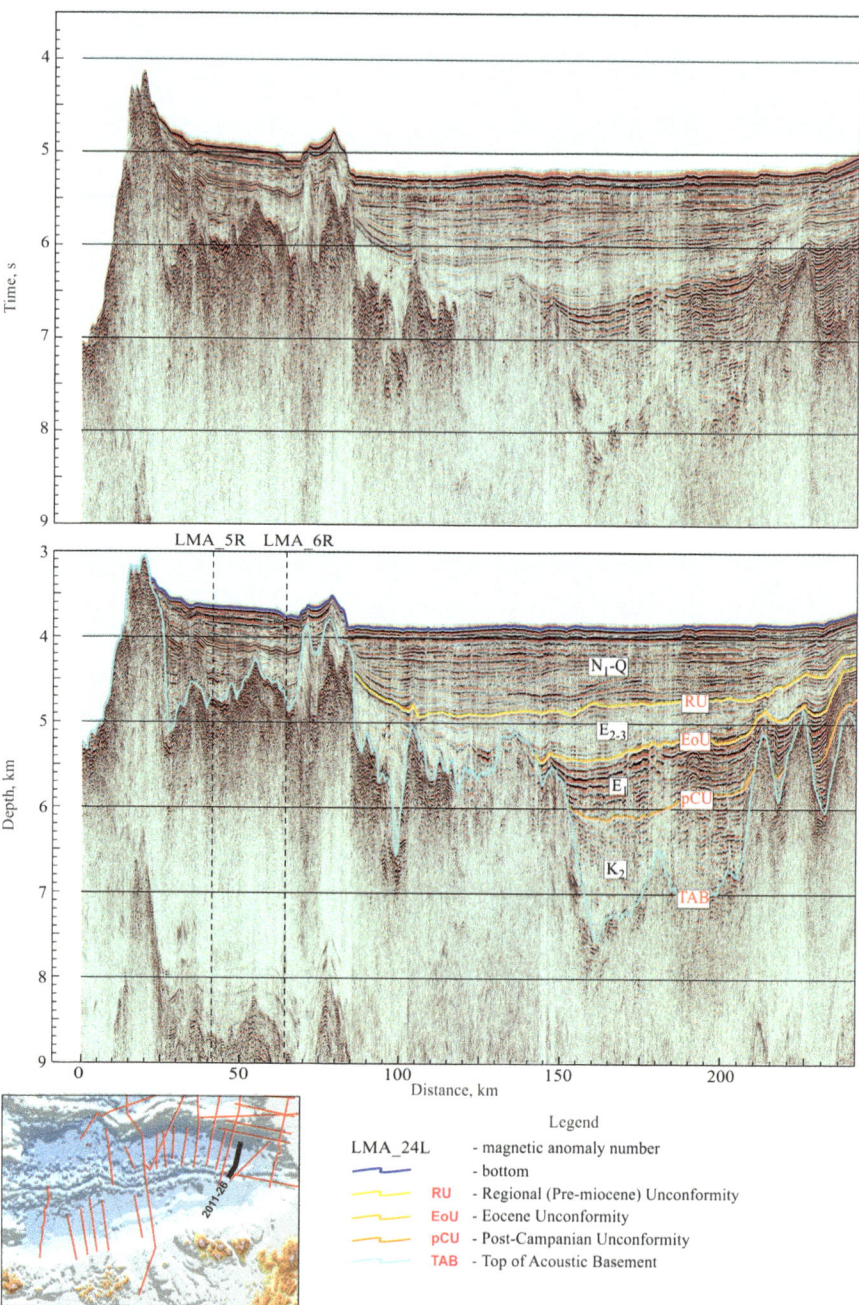

Fig. 3.20 MCS line 2011–26

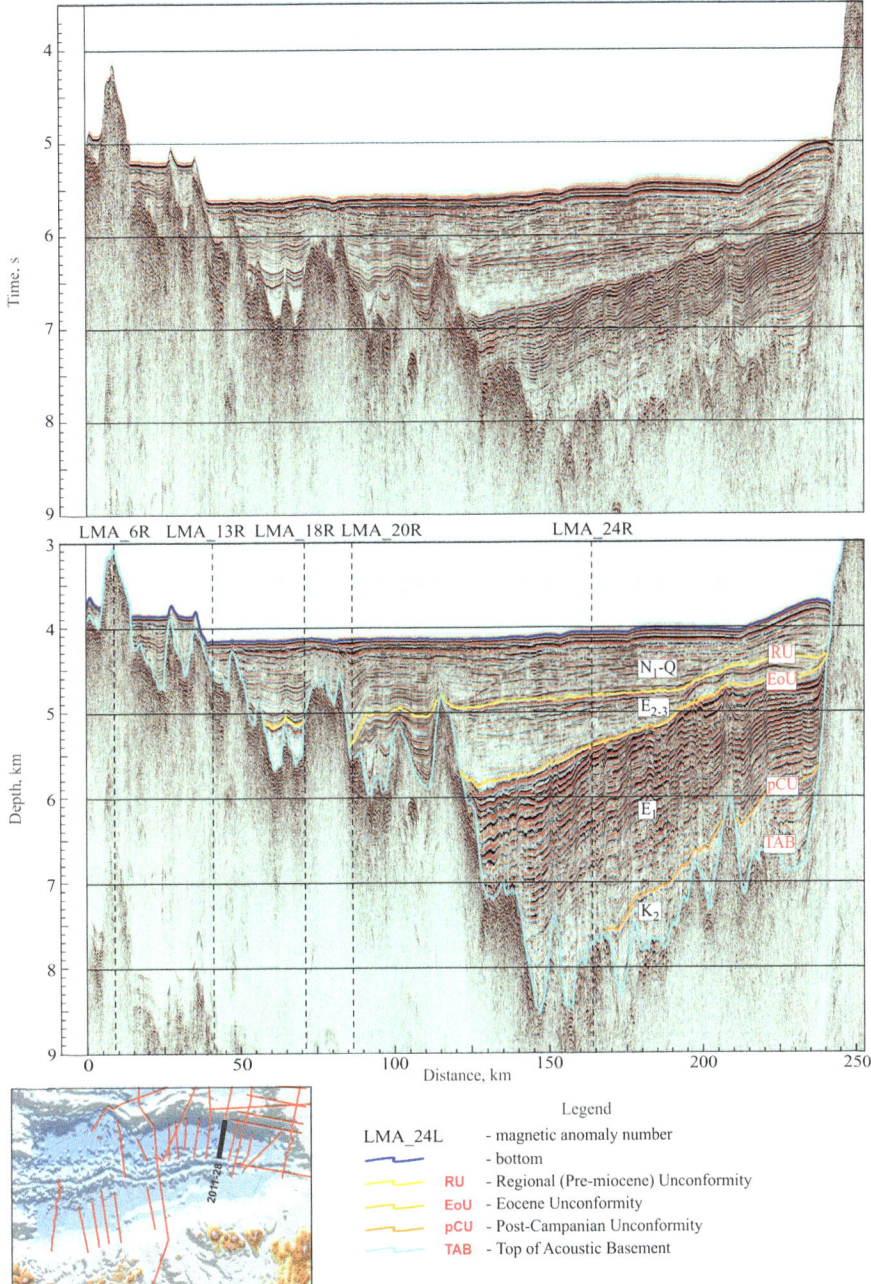

Fig. 3.21 MCS line 2011–28

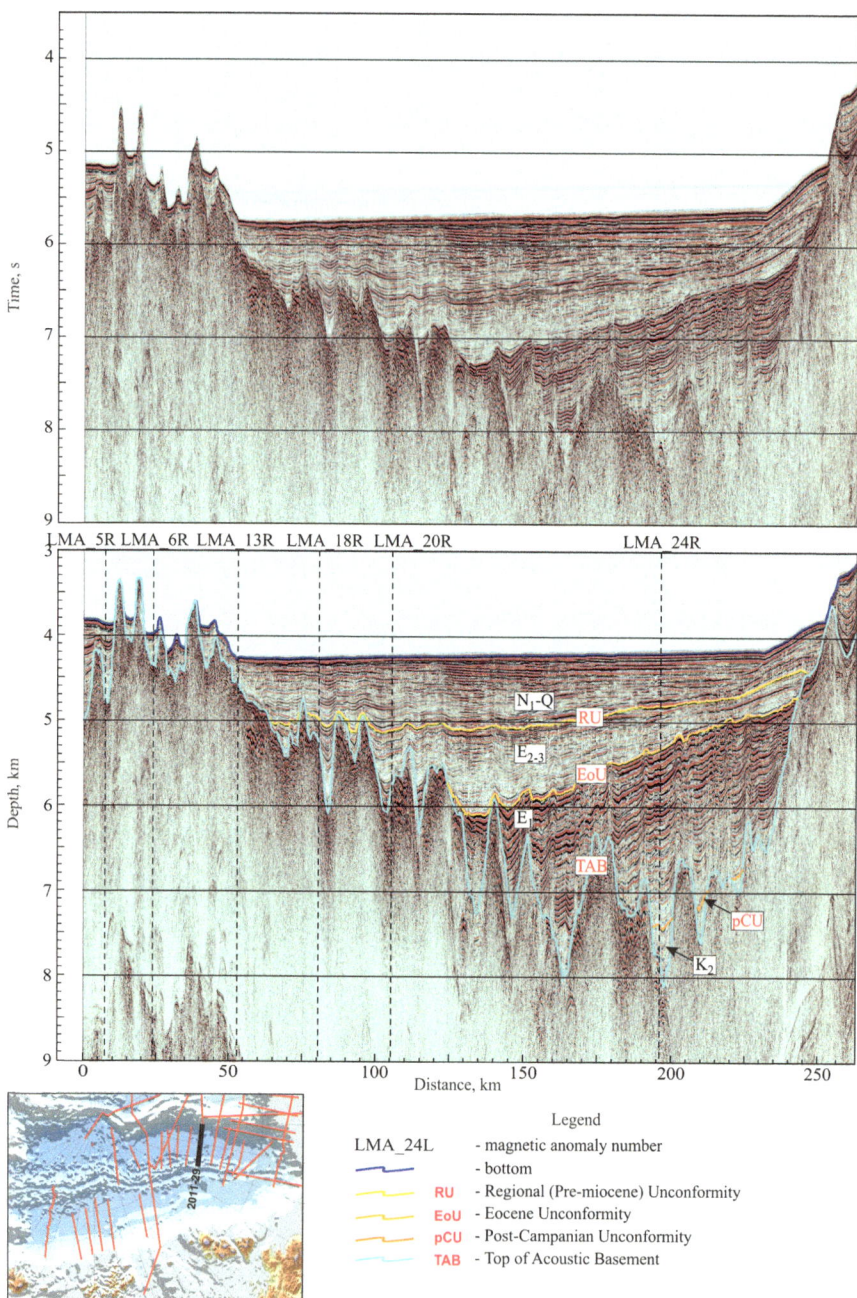

Fig. 3.22 MCS line 2011–29

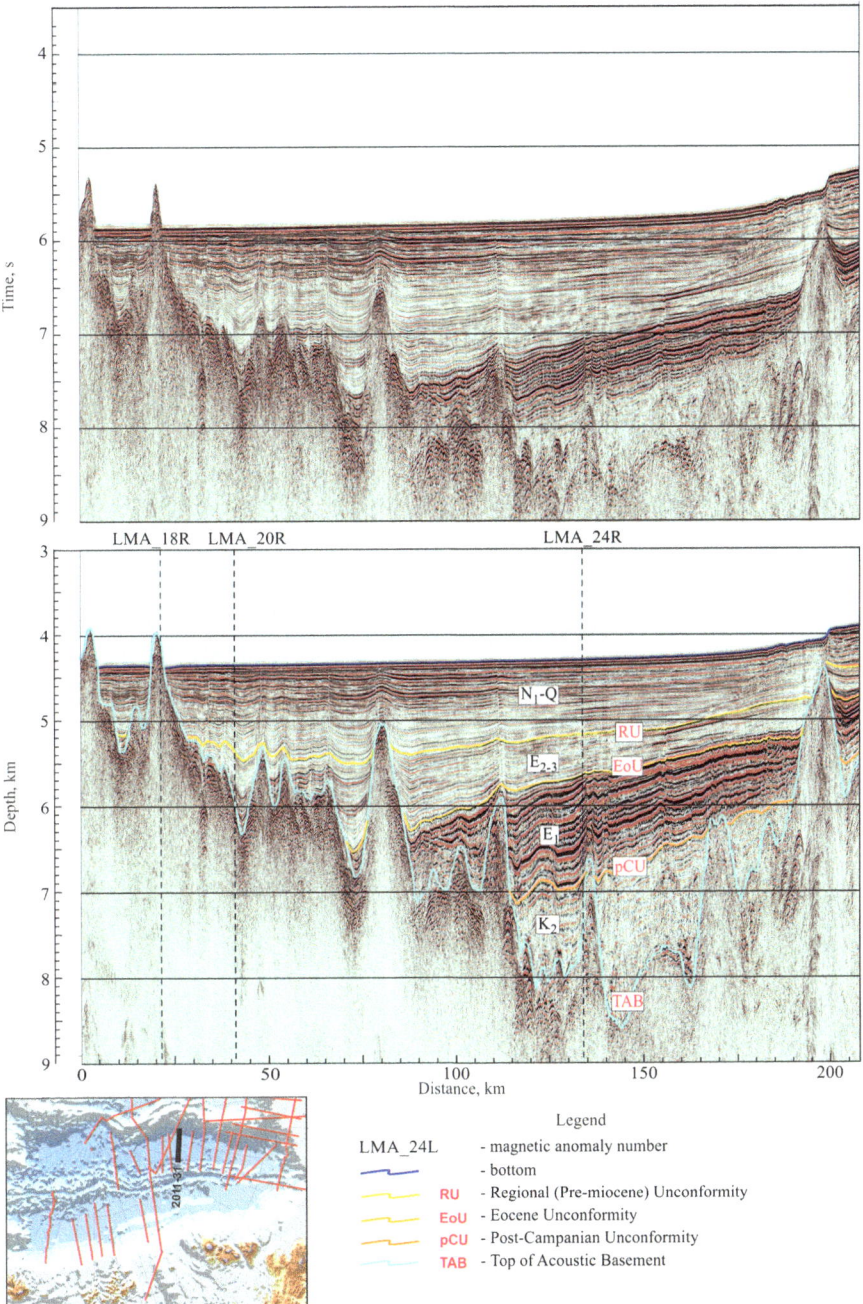

Fig. 3.23 MCS line 2011–31

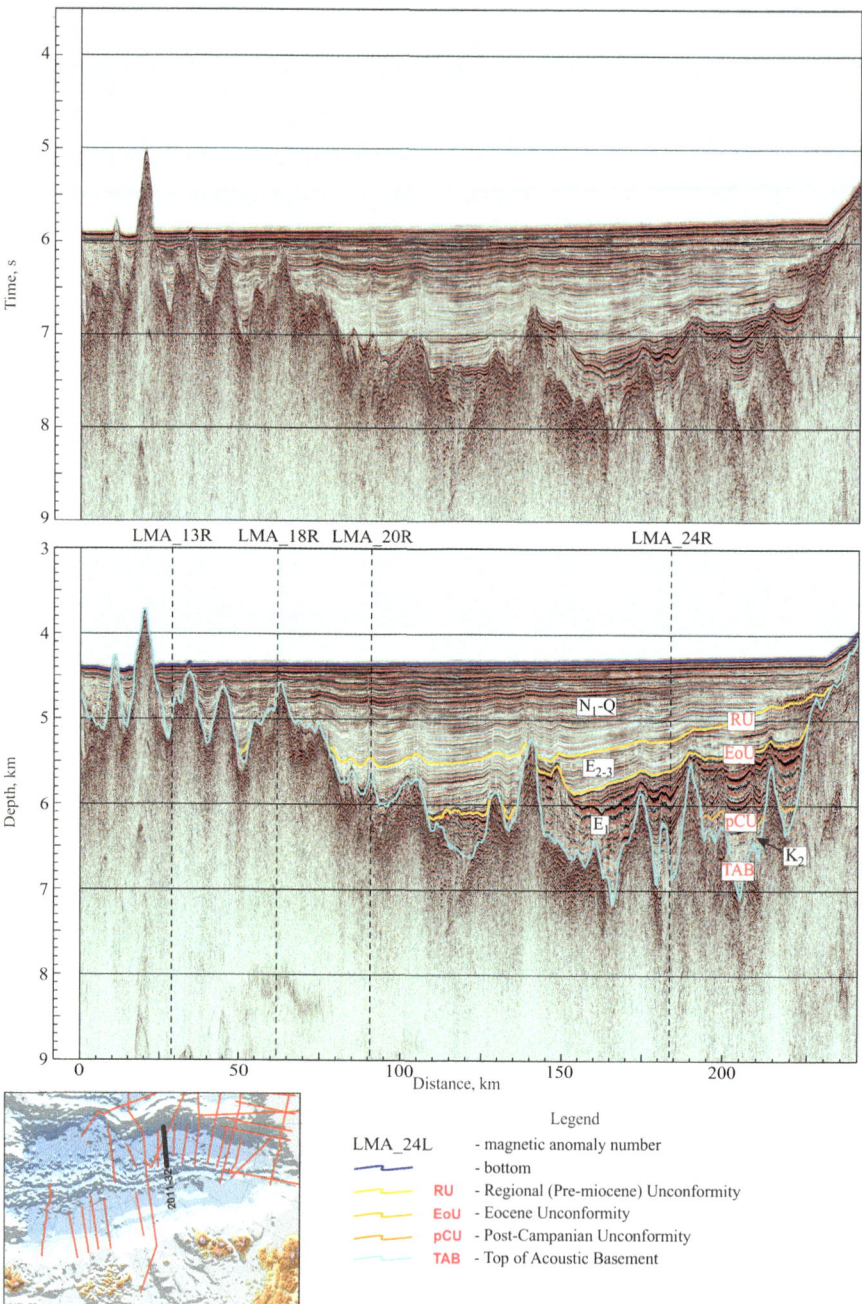

Fig. 3.24 MCS line 2011–32

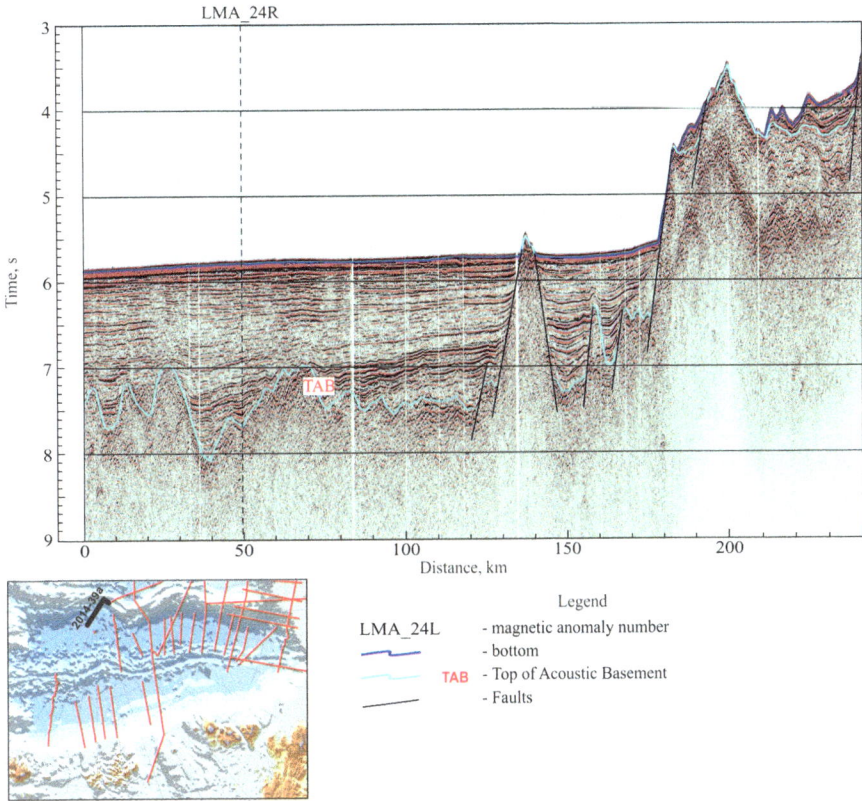

Fig. 3.25 MCS line 2014-39a

3.3 Laptev Sea Continental Margin

The Laptev Sea shelf can be best described as a flat, mostly less as 50 m deep, slightly tilted northward plain with few insular uplifts in the middle part, banks and shallow submarine valleys – a submarine extension of the major continental rivers, sometimes following deeper geological lineaments. Several benches (terraces), few meters high, also present and, depending where the valleys are regarding the benches, the valleys depth may vary from 5–10 m to 20 m. One of the deepest (40–45 m) submarine valleys stretches from the Khatanga estuary around the Taymyr Peninsula. Sharp seafloor drop-off at depth around 100 m marks the shelf edge running NW-SE, E-W and NE-SW in the western, central and eastern parts of the Laptev Sea, accordingly - Fig. 3.27.

The Gakkel Ridge - the northernmost segment of World system of mid-oceanic ridges - joins the continental margin and disappears under the thick blanket of sediments derived from shelf and deposited on the gradually inclined continental slope. The continental slope, shelf, continental rise and large abyssal depression separated

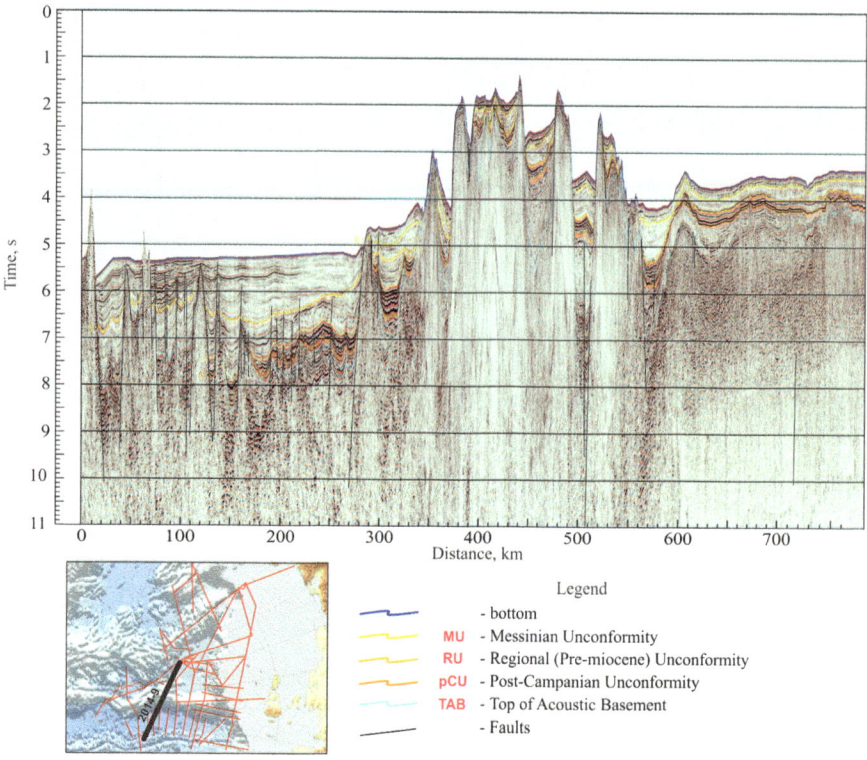

Fig. 3.26 MCS line 2014–09

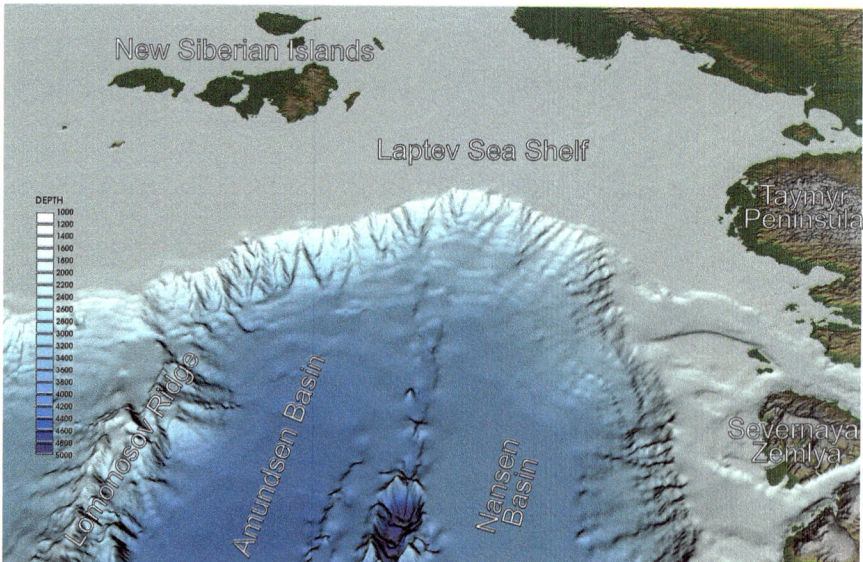

Fig. 3.27 Laptev Sea continental margin

by the Gakkel Ridge into Nansen and Amundsen Basins – are the principal geomorphological elements of the Laptev Sea continental margin.

Along the whole periphery of Laptev Sea shelf, the continental slope is formed
by gently concave surface gradually merging into continental rise at about 2000–
3000 m b.s.l. (GeoCap data), and can be classified as simple slope (Fig. 1.8). The
steepest slopes are found in the upper and middle parts of its north-western and
north-eastern segments, accordingly. Elsewhere, numerous submarine canyons
25–30 km long and up to 10 km wide, dissect both the slope and benches, superimposed on it.

The rise consists of vast train of sediments dumped from shelf. The rise's upper
part merges with the continental slope, lower part - into the Nansen and Amundsen
abyssal plains with no clearly identified boundary. The Gakkel Ridge reaches the
rise in the south-eastern part of the Laptev Sea continental margin. There it loses its
morphological expressiveness - south from 80° N and at depth of 3000 m the ridge
looks like shallow trough between two gentle swells about 100 m high (Naryshkin
1987).

Several distinct zones are identified by gravity and magnetic anomalies within
shelves of Laptev and East Siberian Seas the following zones (Figs. 1.15, 1.16, 3.2):

- **South-Anuy zone** of strong northwest trending gravity and magnetic anomalies
  along almost complete stretch of East Siberian Sea coastline, reflecting the mafic
  and ultra-mafic complexes outcropping on the south coast of Bol. Lyakhovsky
  Island;
- **Laptev Sea zone** with very weak anomalies reflecting thickness variations of
  sedimentary formations (in the abyssal regions following continental slope rather
  the Gakkel Ridge).
- **Kotel'ny - Lyakhov zone** - complicated assembly of anomalies of different
  intensity and sizes caused, as numerical interpretation and modeling show, by
  multitude of sources, including crustal heterogeneity and/or felsic intrusions
  inside basement and sedimentary cover.

Not a single exploration well had been drilled in the Laptev, East Siberian and
Chukchi Seas. Therefore, the composition and age of the sedimentary sequences in
the region remain disputable.

Using the results of seismic interpretation (Franke et al. 2001; Piskarev 2004;
Malyshev et al. 2009; Drachev et al. 2010; Khoroshilova et al. 2014; Shkarubo et al.
2014; Poselov et al. 2016; Nikishin et al. 2017), we identify four reliable regional
seismic marker horizons related to unconformities - F (acoustic basement), LS1,
LS2 and LS3 (Fig. 3.28) and several others, only locally present. Horizon F is identified as pre-Senomanian non-depositional interface (weathering zone) (~130–
125 Ma) related to general pre-rifting uplift which, in turn, might reflect the onset
of formation of Verkhoyanski fold belt in Mesosic. LS1 (66–56 Ma) is correlated to
pre-Late Paleocene hiatus identified in several wells and outcrops in the Laptev Sea
coastal fringes. It marks the start of rifting in the Eurasian Basin. Marker LS2 corresponds to the epoch of major low-stand and re-assembling of the Arctic plates at
the beginning of Eocene (~34 Ma). The longest non-depositional break in Mid-

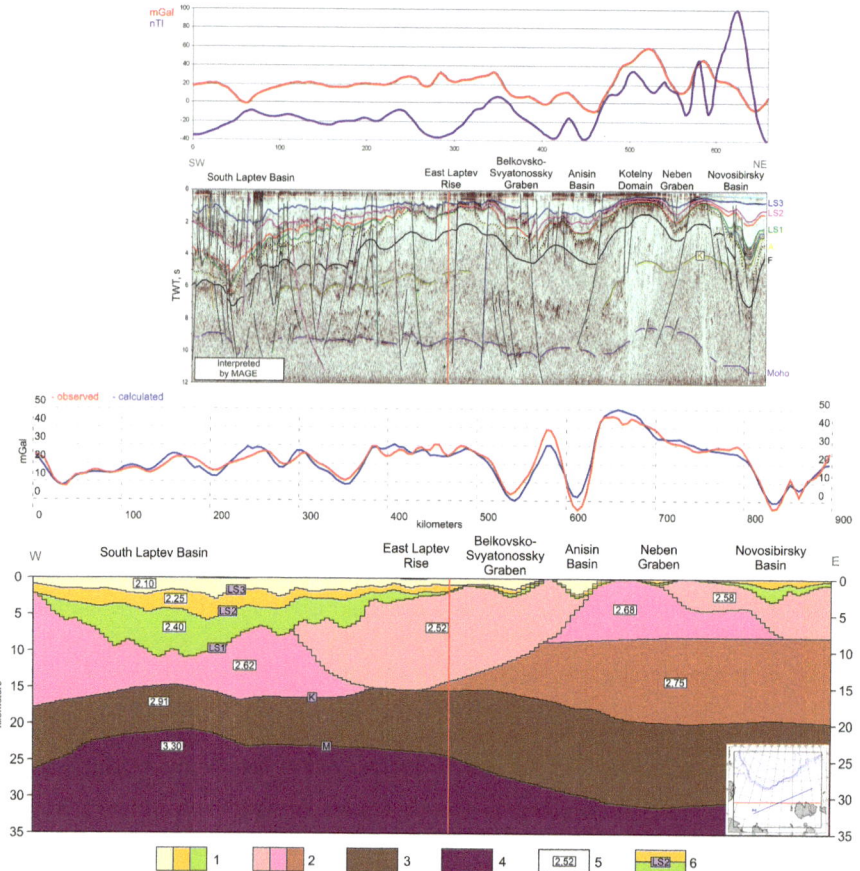

Fig. 3.28 Deep crustal model of Laptev Sea shelf (Bouguer gravity 2.67 g/cm³)
1 –sedimentary complexes; 2 – upper crust blocks of variable density; 3 – lower crust (undifferentiated); 4 – upper mantle; 5 – density (g/cm³); 6 – boundaries of sedimentary complexes and crustal
blocks. Top: MCS line A-4 (TWT section); red line - profiles cross-tie position

Miocene documented in ACEX wells on the Lomonosov Ridge, coincides with
marker LS3.

There was no DSS (Deep Seismic Sounding) surveys in the Laptev Sea, not
counting some limited work in the southwestern part. Therefore, we base our understanding of the regional deep crustal structure mostly on seismic-gravity modeling
utilizing as much of MCS information as possible.

One of such models was created along the line running from East Taymyr
Peninsula coast through central Laptev Sea shelf to point north from Kotel'ny Island
(Fig. 3.28).

Under the western Laptev Sea, the continental crust is about 22 km thick (half of
it - sedimentary formations). In the central part top of consolidated crust is 10–12 km
deep and overall crustal thickness is reduced to 10–15 km. Increasing basement

depth is usually compensated by rising mafic lower crust to 15 km and upper mantle to 22 km, therefore the Basin gravity field does not reflect its internal structure. Keeping in mind rising Moho and extreme depth to which the basement plunges, the ductile state and active metamorphism of deeper crust is expected (Piskarev 2004). (Some signs of plastic creep along 15 km-deep detachments can be glanced on several seismic lines).

In vicinity to Anjou Archipelago the sedimentary section thins down and, at places, completely pinches out, while consolidated crust increases its thickness to 25–30 km,. Under the East Laptev Rise we introduce the substantial (12—13 km thick) crustal bloc with density reduced to 2,52 g/cm^3, typical for terrigenous folded formations, which may represent an offshore continuation of Verkhoyansk-Chukchi fold belt.

Strong, short wave-length gravity anomalies over the East Laptev Rise reflect complexity of shallow basement surface. The eastern part of the East Laptev Rise, along the suture zone of junction with Kotel'ny Massif, display some signs of upper crust weakening – Bel'kovsky-Svyatonos graben and Anisin Basin (the eastern slope of the latter coincides with boundary between two crustal blocks). Contrary to the South Laptev Sea Basin, these structures are isostatically uncompensated and are not accompanied by rising deep boundaries.

The Kotel'ny Massif further east is underlain by standard three-layer crust up to 31.5 km thick. Upper felsic-metamorphic crust is subdivided into upper, with density 2,68 g/cm^3 identified with outcropping there formations, and lower, with density 2.75 g/cm^3, in which at depth 8–16 km, are concentrated sources of magnetic anomalies. At the base of this block the lower mafic crust with density 2.91 g/cm^3 is estimated to be 9–10 km thick.

Interpretation of MCS line A4, the only seismic section displaying deep crustal reflections, confirms validity of this computed model (Fig. 3.28). It also shows thick sedimentary sequence with apparent, but rather chaotic, reflectivity between base of Cenozoic (unconformity LS1) and top of upper crust (horizon F). This sequence may represent well lithified pre-Cenozoic sediments with negligible density contrast between it and upper crust, so it had no influence on model calculate gravity response.

We may postulate with all probability that the basement of Laptev Sea shelf consists of Late Mesozoic (Cimmerian) consolidated fold belts subjected later to intense extension and subsequent differentiated normal faulting and rifting, as a precursors to opening of the Eurasian Basin. Further east, in the East-Siberian Sea shelf (De-Long Massif) older basement (Hercynian or even Baikalian) is possible.

Another model was computed along profile running trough the southern closure of the Eurasian Basin (Fig. 3.29) and highlights the deep structure of the special part of the Gakkel Ridge - Laptev Sea shelf junction. Here the Gakkel Ridge is not visible in the seafloor bathymetrty and, as modeling shows, is not expressed as an active spreading zone in the basement structure, either. Instead, a 100 km-wide depression filled with 10–11 km of sediments, most probably of rifting origin, exists there at the base of the continental slope. It was formed on the upper crust reduced to 3–4 km in the central part and compensated by upper mantle rising to 16.5 km.

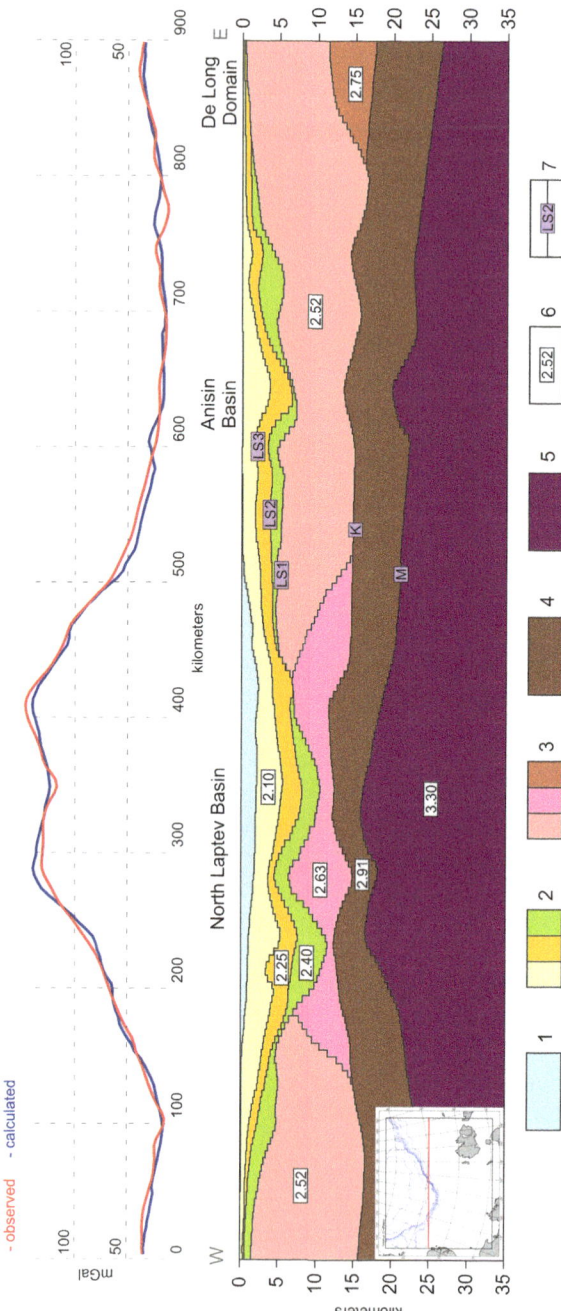

Fig. 3.29 Crustal model of the Laptev Sea continental margin (profile location - see insert), Bouguer gravity 2.67 g/cm³
1 –sea water; 2 –sedimentary complexes; 3 – upper crust blocks of variable density; 4 – lower crust (undifferentiated); 5 – upper mantle; 6 – density (g/cm³);
6 – boundaries of sedimentary complexes and crustal blocks

Avetisov G.P. (Avetisov 2004), found that earthquakes epicenters related to the Gakkel Ridge start to separate on western and eastern branches in this depression.

Reduced continental crust of the continental margin with total thickness of 20–25 km has two-tiered composition with upper crust (2.60—2.75 g/cm^3) twice as thick as the lower crust (2.80–2.90 g/cm^3). The upper crust includes Cimmerian folded basement outcropping on New Siberian Islands and presumed in the southern segment of the Lomonosov Ridge. Older, Caledonian or Baikalian, terranes of the Kotel'ny Massif are imbedded into Cimmerian basement. The discussed model was constructed to reflect exactly this situation. Reduced crustal density in some areas are thought to be caused by local granitization, thermal metamorphism, and migmatization of the crust.

Seismic data for the Laptev Sea continental slope and southern part of the Amundsen Basin (Fig. 3.30.) clearly demonstrate that compacted Cenosoic and partially compressed Mesozoic sedimentary sequences directly overlay high-velocity unidentified formations of the upper crust. It can be basement complexes of the De-Long Massif western margins, folded formations and granitoids of the Kotel'ny Massif - Lomonosov Ridge Cimmerian belt, or HALIP intrusives (as possible sources of strong magnetic anomaly at the southern part of the Lomonosov Ridge).

The igneous intrusions penetrating the entire sedimentary cover can be seen on several MCS lines in the Amundsen Basin southern closure (Figs. 3.31, 3.32, 3.33 and 3.34).

The final interpretation of this information still awaits its turn. Several attempts (Gaina et al. 2015, Rekant et al. 2015) presented alternative interpretations deviating from the classical models of the Eurasian Basin evolution and it seems that only new modern high-resolution aero-geophysical surveys can reconcile the differences.

3.4 On Alternative Hypothesis of Evolution of the Eurasian Basin

There is a multitude of facts contradicting widely accepted hypothesis of the Eurasian Basin and the Lomonosov Ridge evolution described in the previous Chapters. Among them - bathymetric and topographic asymmetry of abyssal depressions (Fig. 3.35); asymmetry of their basement structure and composition; uneven occurrence and thickness of sedimentary formations; asymmetric and discordant distribution of potential fields gradients relative to the Gakkel Ridge (Piskarev 2004).

Moreover, facial analysis of sedimentary sequences suggest the existence of abyssal basin north from Kara Sea in Cretaceous time (Gramberg et al. 1984). Predominantly southward advance of marine transgressions into the Barents-Kara Basin during Late Cretaceous (mainly Santonian) and cooling of shelf seas waters from 10–22 °C in Early Cretaceous to 5–8 °C in Mid-Late Cretaceous

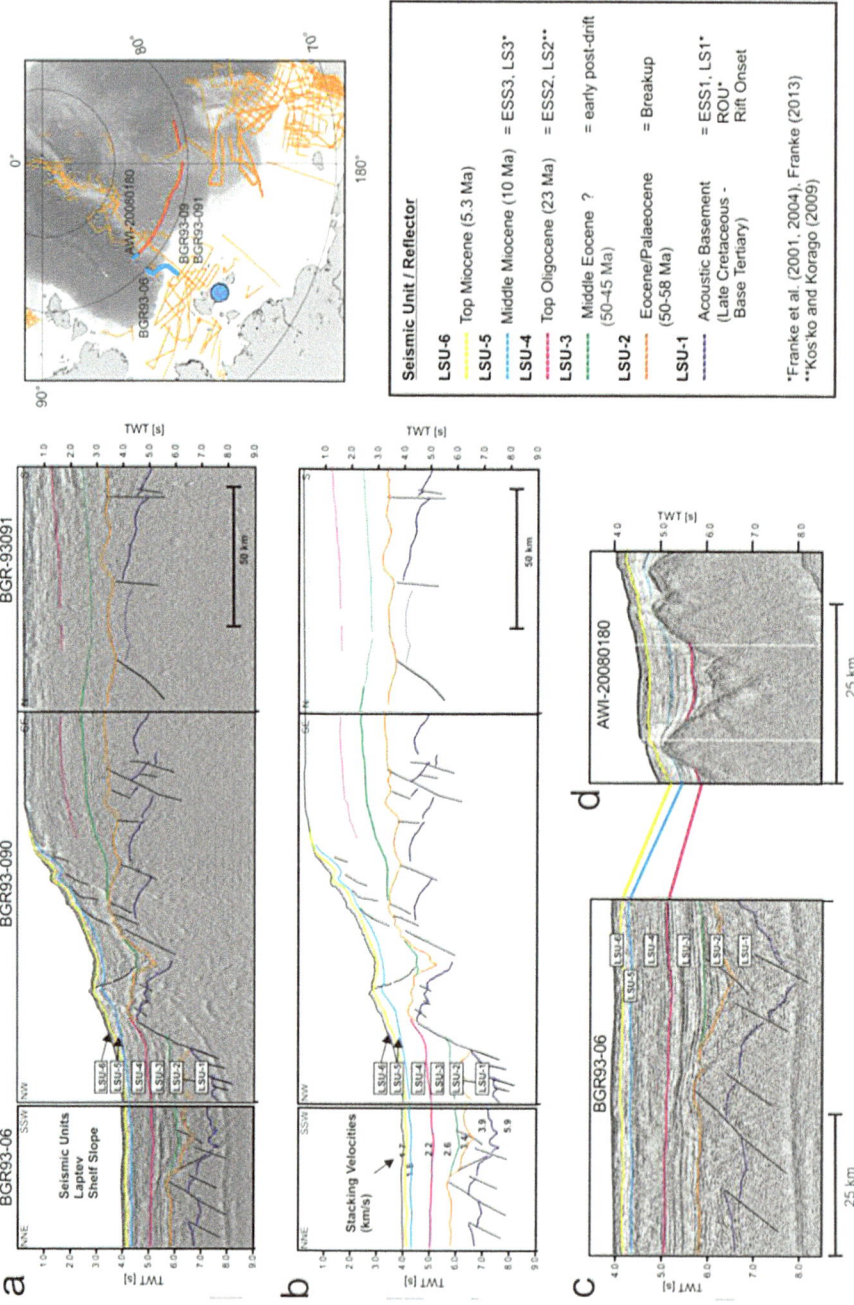

Fig. 3.30 Seismic TWT section across the Laptev Sea continental Slope (locations- see insert) (Weigelt et al. 2014)

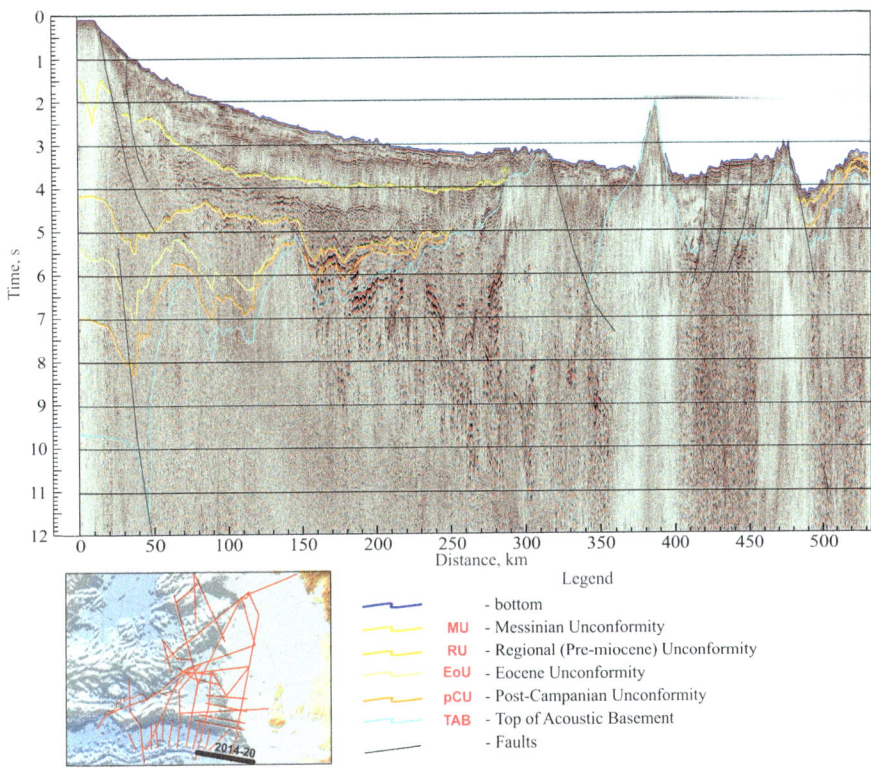

Legend
- bottom
MU - Messinian Unconformity
RU - Regional (Pre-miocene) Unconformity
EoU - Eocene Unconformity
pCU - Post-Campanian Unconformity
TAB - Top of Acoustic Basement
- Faults

Fig. 3.31 MCS line 2014–20

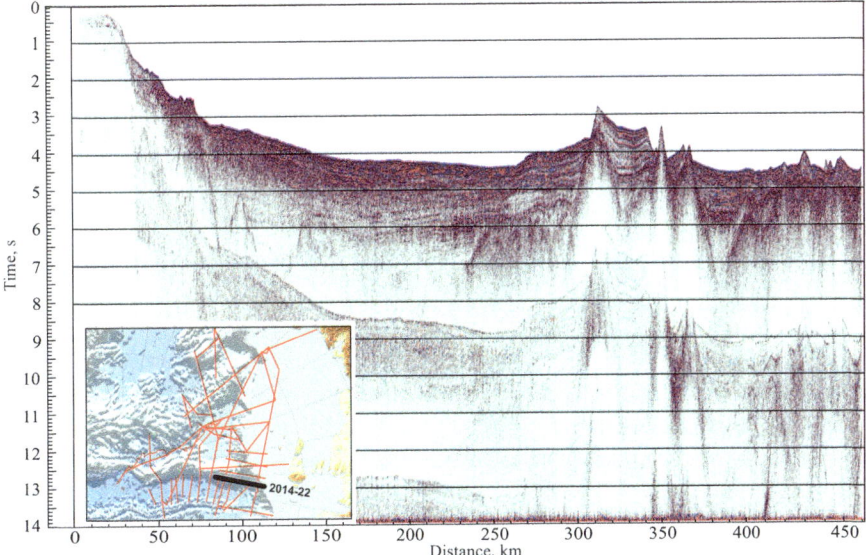

Fig. 3.32 MCS line 2014–22

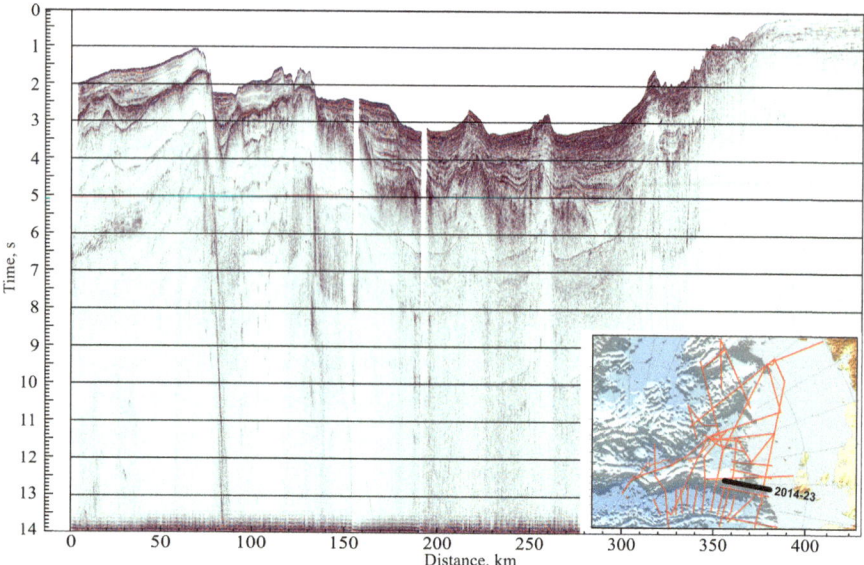

Fig. 3.33 MCS line 2014–23

(Paleogeography of the Northern USSR in the Jurassic Period 1983) – are, perhaps, the first signs of abyssal Arctic Basin coming to being.

Keeping in mind an ambiguity of linear magnetic anomalies (LMA) dating outside the Gakkel Ridge proper, we conclude that the riddle of Eurasian Basin evolution is far from being solved. The latest seismic survey in the Amundsen and Nansen Basins (Figs. 1.17 and 1.26) provided enough new information to undertake a new integrated analysis of geophysical and bathymetry data.

Fig. 3.36 presents the schematic zoning of the Eurasian Basin basement (modified after (Kireev and Piskarev 2014)) superimposed on local (period T <40 km) magnetic anomalies map.

Detail analysis of the aeromagnetic data acquired in the 60th of the last century in the eastern Eurasian Basin and subsequently used for regional correlations and compilations, reveals their very low quality. So low, that some parts of it were completely excluded from consideration and some - degraded as unreliable due to large positioning errors (Fig. 3.37).

After this "cleansing" operation average distance between flight paths reached to 40–60 km and large, up to 10,000 sq. km "white spots" with no data at all, appeared on the map. We made an inevitable conclusion that existing aeromagnetic surveys do not provide reliable information on shape and location of magnetic anomalies in this part of the Eurasian Basin and redirected our attention to analysis of local gravity anomalies recorded in the eastern part of the Eurasian Basin - Fig. 3.38.

The residual gravity anomalies were calculated by filtering out regional (low frequency, T > 200 km)) gravity from observed Free Air anomalies in frequency

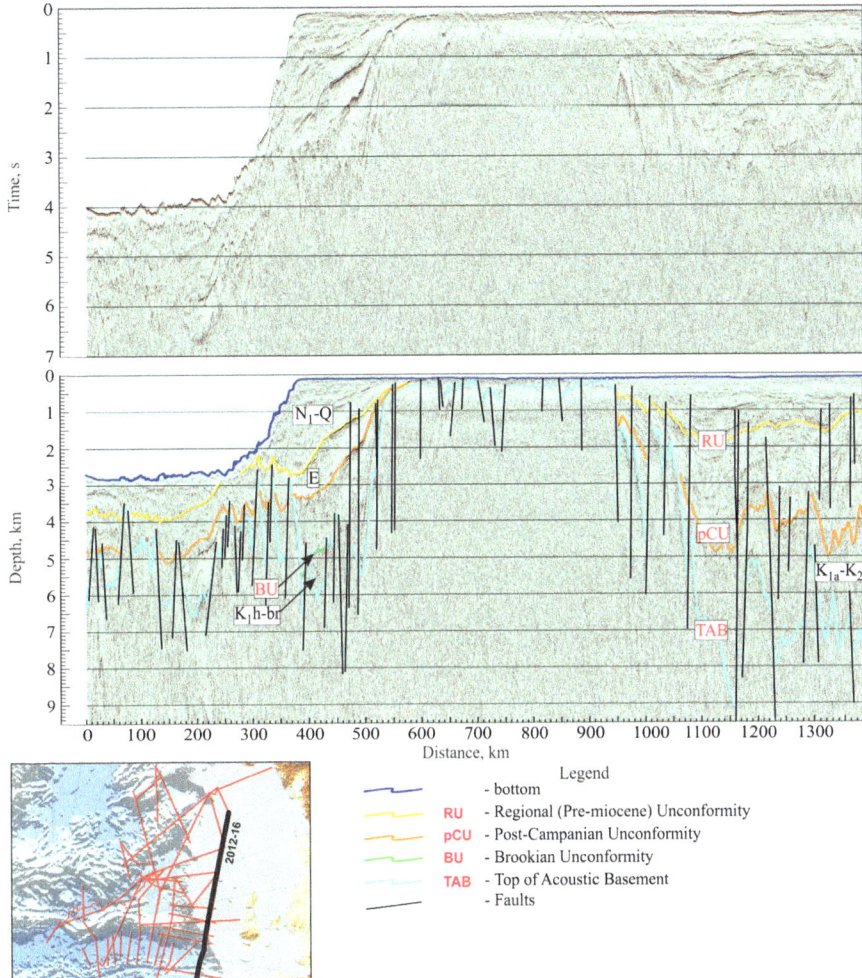

Fig. 3.34 MCS line 2012–16

domain, leaving only medium-short wavelength anomalies. The residual anomalies, being largely free from gravitational influence of regional surficial topography and deep crustal structural elements, better reflect basement geology.

The residual gravity zonality is clearly visible. The distinct linear, 160–200 km wide, belt of strong positive anomalies at the center marks the Gakkel Ridge and equally strong negative - its axial rift valley.

The pattern of residual gravity anomalies noticeably changes along the meridian 75°E from the North Pole to Kara Sea. To the west, linear gradient zones and local anomalies are either parallel with continental margins, or orthogonal to them and to the Gakkel Ridge. To the east of meridian, the picture is sharply different: here lin-

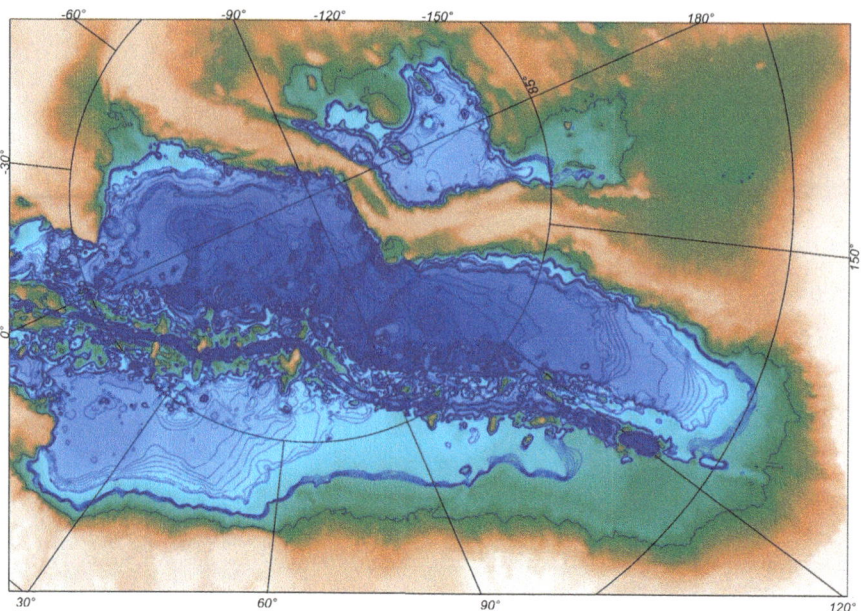

Fig. 3.35 Simplified bathymetry of the Eurasian Basin

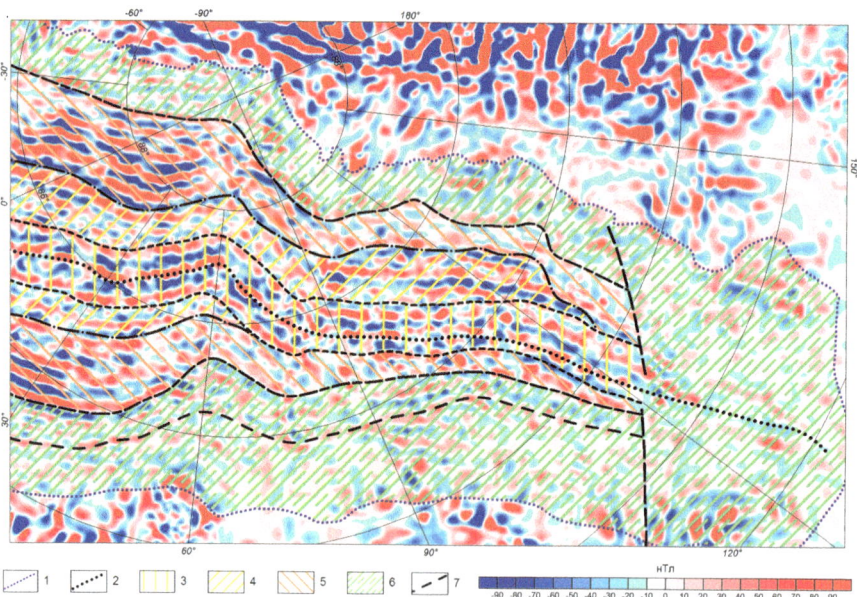

Fig. 3.36 Zoning of the Eurasian Basin basement on the local magnetic anomalies (T<40 km) map. 1 – Eurasian Basin borders; 2 –rift valley axis; 3 – Gakkel Ridge proper (bounded by LMA 5, ~10–12 mA); 4 – post- Early Oligocene crust (close to LMA 13–36 Ma); 5- post- Early Eocene crust (close to LMA 24, 54 Ma); 6 –Mesozoic oceanic crust; 7 –Transverse faults

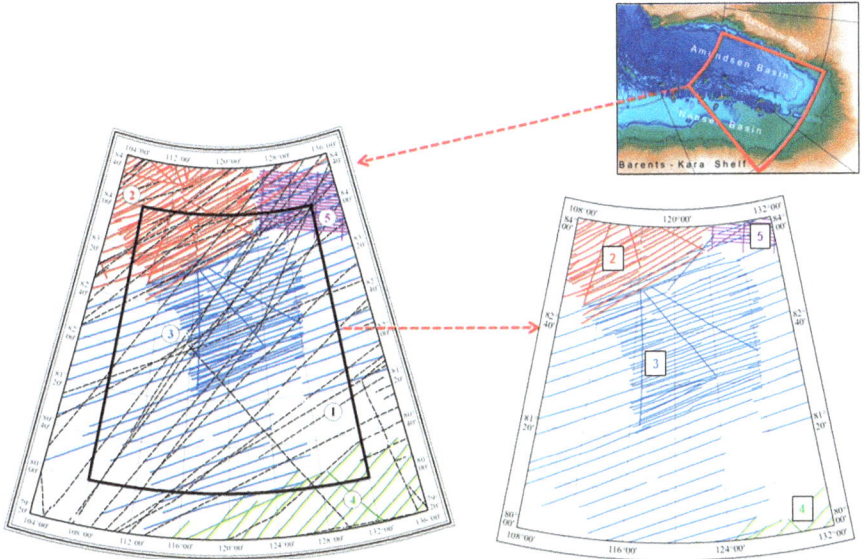

Fig. 3.37 Analysis of the magnetic surveys quality (2017). Survey 1 (1961) is excluded from final map due to absence of correlation with other surveys. Navigation errors of other surveys are from ±35 m (Survey 5, 1992) up to ±8400–30,500 m (Survey 2, 1966)

ear gradient zones and local anomalies are at 30–40°angle to both the Gakkel Ridge (not dissecting or crossing it, as a rule) and margins of the Nansen and Amundsen Basin.

The chains of strong positive anomalies delineate transition zones between shelves and continental slopes along the entire periphery of the Arctic Ocean. The cause of this anomaly, comprehensively analyzed by V.A Litinsky (1972), lies in the nature of Free Air gravity reduction itself which takes into account only the elevation of observation point above sea level, disregarding the mass in-between them. Therefore, the "positive" gravitational pull of gradually rising toward the shelf edge mantle is partially compensated there by "negative" pull of much lighter and rapidly increasing volumes of water and sediments.

Divergent boundary of the rift zone and transition zones between abyssal plains and continental shelves are the most pronounced in the gravity field. At the same time, we can confidently locate the faults parallel to the ridge axis and 80–100 km away from it marking the boundaries between the Gakkel Ridge and adjacent abyssal basins – the situation quite different from the Atlantic where ridge - basins transitions are gradual.

Most of structures in the Amundsen and Nansen Basins are neither parallel to the Gakkel Ridge, nor orthogonal to it (again, contrary to what we observe in the gravity field of abyssal basins near the Atlantic and the Pacific mid-oceanic ridges). Instead, there are linear gradient zones, parallel to the Laptev Sea continental slopes. Complex chain of anomalies and gradient zone traces a suture with left-hand shift

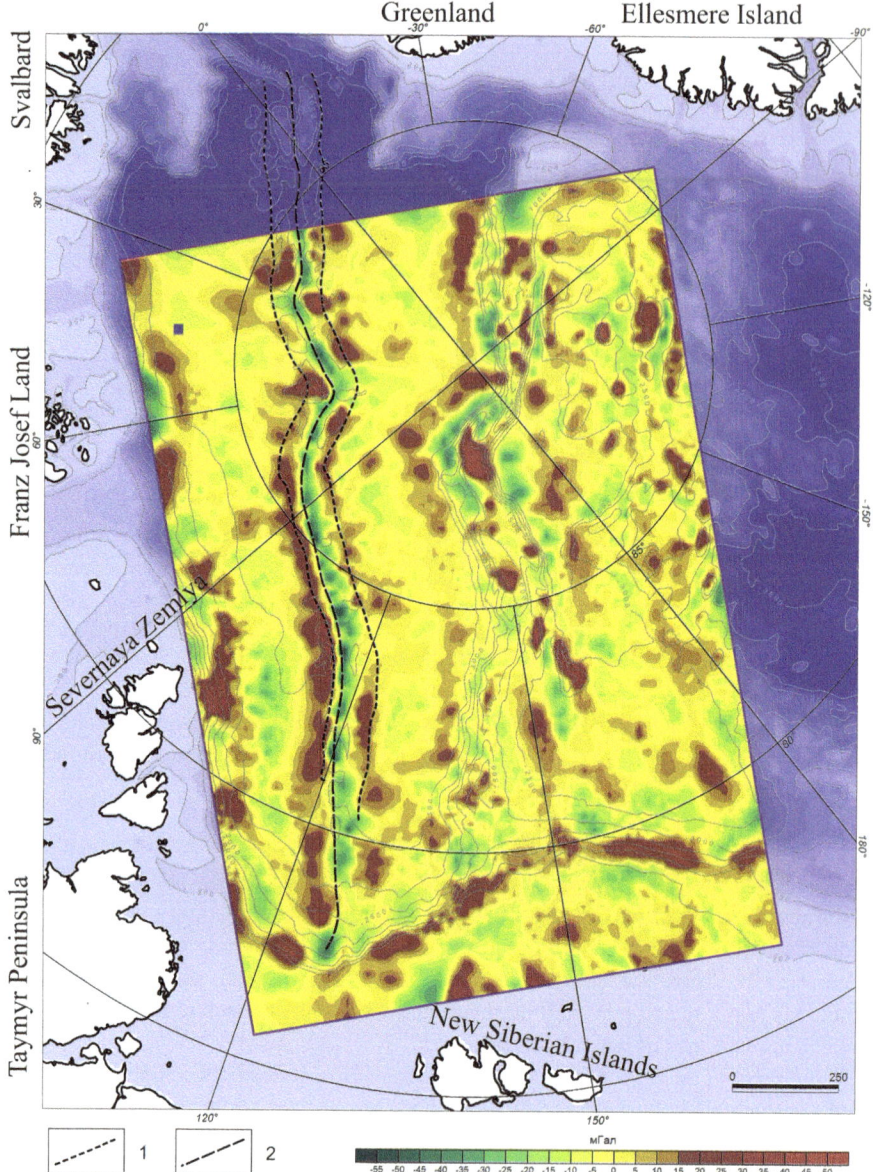

Fig. 3.38 Residual gravity anomalies with T< 200 km where 1 – Gakkel Ridge limits; 2- rift valley

through the Gakkel Ridge between 30°E and 60° E and across the central segment of the Lomonosov Ridge into the Amerasian Basin.

Another important regional suture is visible along ≈ 80° E meridian from Kara shelf through the Gakkel Ridge to the North Pole. West from it, the abyssal basins fracture zones and faults of the Gakkel Ridge, belongs to the same system. On the suture eastern side they do not follow systematic pattern. Importance of this suture for understanding Cenozoic evolution of the Eurasian Basin is illustrated by residual gravity maps (Fig. 3.39). The distinct linear system gravity anomalies accompanying the geomorphological expression of the Gakkel Ridge stands out in the center of map A. Reconstructing the situation which existed 10–12 Ma (estimated age of the Gakkel Ridge) by collapsing this system, two transverse anomalies along 90°E meridian join together, delineating important transform-type fracture zone (Fig. 3.39, map B). It also indicates that the modern Nansen and Amundsen Basin and the Gakkel Ridge were formed in epochs with different orientation of spreading vector.

The gravity anomalies depict Lomonosov Ridge as essentially heterogeneous structure. The general structural trend in the near-Greenland and near-Siberian sectors follows 140°E meridian and accompanied by low-amplitude and long- wavelength geophysical anomalies, while at the central segment it changes to 160°E – 170°E and represented by stronger and isometric short-wavelength anomalies. Inside the crestal areas of these segments of the ridge the anomalous fields are often similar to those in fold belts. Between latitudes 85°N and 87°N the transverse structural belt, not expressed in the seafloor bathymetry, crosses the Lomonosov Ridge into the Amundsen Basin.

We demonstrated that the map of local (T<40 km) magnetic anomalies Fig. 3.36, mostly quite reliable, looses its value in its eastern part, where spacing between half a century old aeromagnetic flight paths may be 10—40 km apart and positioning error may reach tens of kilometers. Nevertheless, while linearity of anomalies was difficult to establish, the amplitude characteristics were found quite consistent. Therefore, a belt of strong magnetic anomalies was established east from 75°E in the Amundsen Basin outside the Gakkel Ridge limits, which is practically absent in the Nansen Basin. In contrast, west from 75°E these strong magnetic anomalies are present in both Amundsen and Nansen Basins symmetrically on both sides of the Gakkel Ridge. This situation might be explained by Miocene spreading axis jump, resulting in its bending around 60–80° E.

The sub-parallel to the present spreading axis LMA are present in the narrow neck of the Nansen Basin and in widest part of the Amundsen Basins. There are no reliable LMA in the near- Laptev Sea part of the Eurasian Basin except along narrow zone of modern spreading axis.

Seismic surveys of 2011 and 2014 provided much needed information on Eurasian Basin sedimentary cover. Among them the MCS line 2014–07, as the only one, to date, to cross the entire basin, was the most representative (Fig. 3.40).

Presence of several hundred meters of sediments in the rift valley tells us that rifting was initiated on the pre-existing ocean floor. The Gakkel Ridge-facing flank of the Amundsen Basin (Oligocene-Miocene) demonstrates highly un-even base-

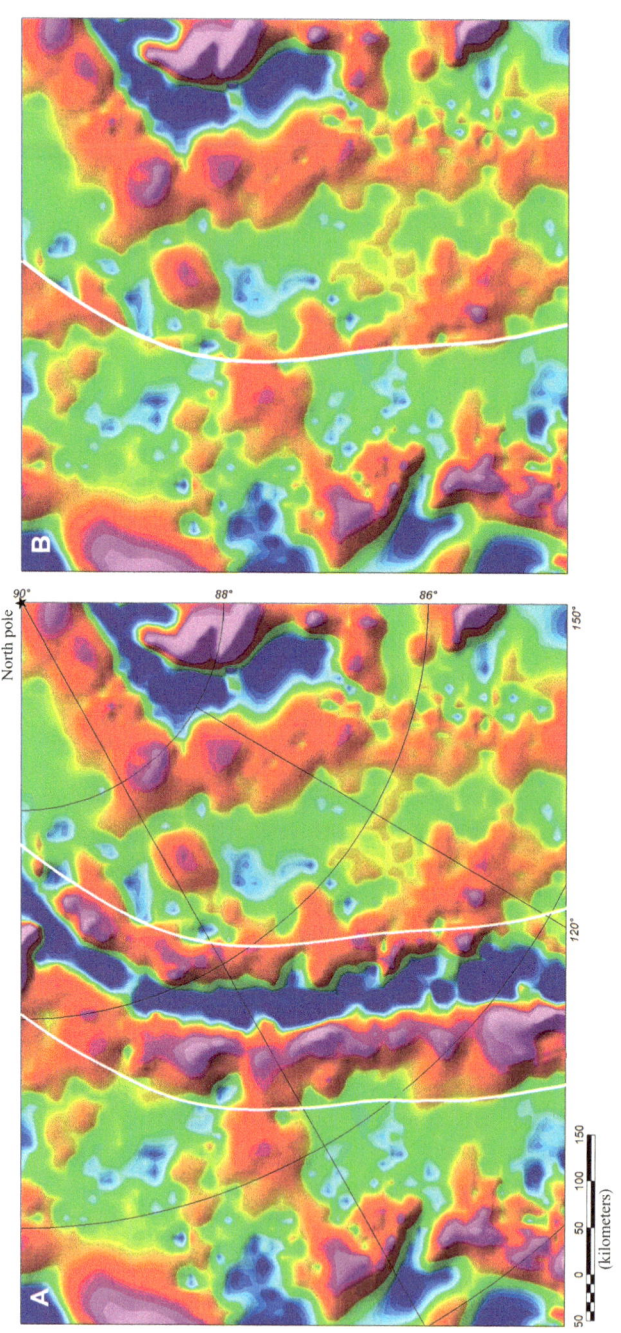

Fig. 3.39 Residual anomalies map. A-gravity anomalies in the near-Pole region. B- Nansen and Amundsen Basins prior to formation of the Gakkel Ridge

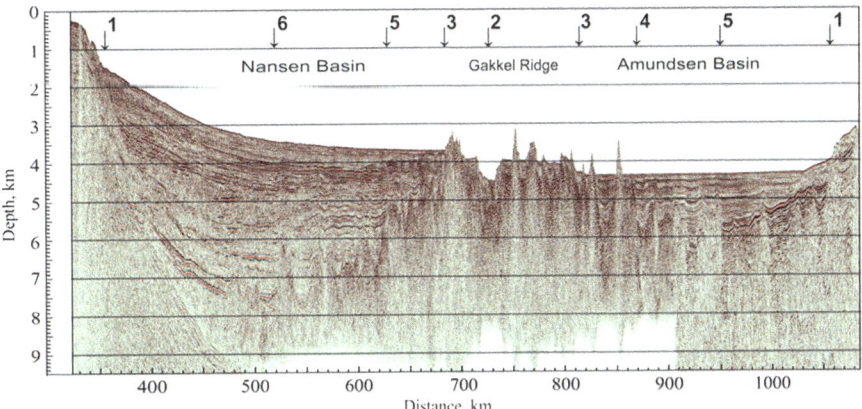

Fig. 3.40 MCS line 2014–07. Markers: 1 – Eurasian Basin boundaries; 2 – axis of the rift valley; 3–3 – Gakkel Ridge area (up to magnetic anomaly 5, ~10–12 Ma); 3–4 – post- Early Oligocene crust (near anomaly 13); 4–5 – post- Early Eocene crust (near anomaly 24); 6 – tectonic suture in the oceanic basement

ment surface and variable thickness of sedimentary cover. The following zones (up to marker 5) on both sides of the Gakkel R have 1.5 km of sediments, on average. Assuming the constant sedimentation rate of 3 cm/kA (which is much higher than recently documented (Levitan 2015)), it would take 50 million years to create this sequence. Therefore, the oceanic basin must have been formed no later the Early Eocene.

Substantial, apparently Late Cretaceous, sedimentary sequence appears beyond marker 5 in both Basins (for Amundsen Basin recently confirmed by (Rekant et al. 2015)). Finally, at the Barents-Kara shelf junction, beyond marker 6, additional sequence (presumably, no older the Early Cretaceous), was identified, but basement could not be interpreted.

Total thickness of sedimentary cover of the Nansen Basin at the Barents –Kara margin exceeds 5 km (Figs. 2.6, 2.7, 3.16–3.18, 3.13). This situation is similar to what we see in the south and south-west of the Canadian Basin were an identified basement is also buried under very thick sedimentary cover.

Seismic data from the near-Laptev part of the Amundsen Basin also show very thick, up to 7 km, Mesozoic sequence, lower part of which may be dated as Cretaceous (Fig. 3.41).

We may conclude that the considerable part of the Eurasian Basin was formed in pre-Cenozoic time. However, the reliable geodynamic model of its evolution as a part of the entire Arctic Basin need more geological and geophysical information.

This model must include the available information regarding extensional structures known on the Laptev and East-Siberian shelves. There are reasons to believe that the modern seismically active Laptev Sea extensional regime in Miocene also existed in the areas east from the New-Siberian Island Archipelago.

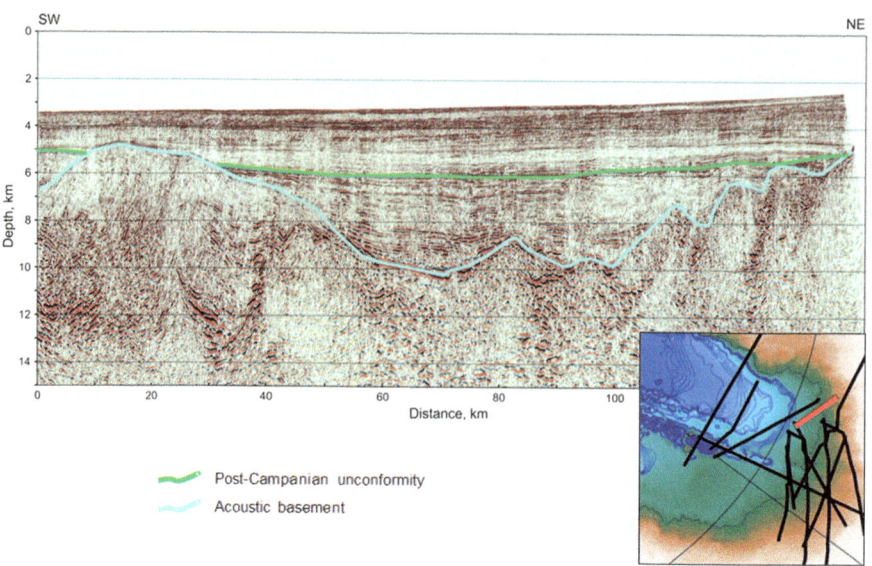

Fig. 3.41 MCS seismic line LAT1403 (MAGE, 2014)

Apparently, the Gakkel Ridge rift valley assumed its present location in the eastern Eurasian Basin only in Pliocene (Fig. 3.36). However, even before that time, the rift valley position was not much different from present, as orientation of LMA (starting from 13th, parallel to the valley) may attest. In Eocene, the spreading axis was at oblique angle to the current orientation.

Formation of narrow elongated extensional structures (grabens) on the Laptev and East-Siberian shelves may be attributed to this stage of evolution (Fig. 3.42).

The residual gravity map was produced by the same method of selective filtering of gravity data as described at the beginning of this Chapter, only in this case applied to the Bouger gravity instead of Free Air. The shaded-relief map is the same map but with gradients, orthogonal to selected direction, artificially exaggerated ("illuminated"). In this case direction of illumination from the west was dictated by predominant north-western and sub-meridional orientation of residual anomalies.

The strongest anomalies are located on the periphery of Kotel'ny and De-Long Massifs. Seismic data confirms that this anomalies coincide with graben-like structures filled with sediments. Incidentally, grabens located in the Laptev Sea are tectonically active and produce earthquakes. To the east, in the East-Siberian sea these structures are tectonically passive due to westward shift of spreading zone which tentatively took place at Oligocene-Eocene interface (chron LMA 13).

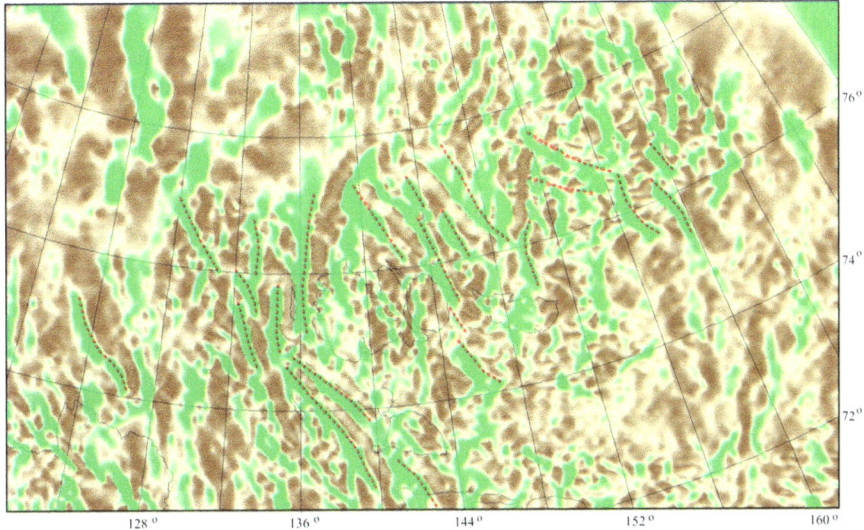

Fig. 3.42 Residual (T<200 km) Bouguer gravity anomalies of Laptev and East Siberian shelves (shaded-relief view, illumination from the west; dotted lines - grabens with in-fill of young low density sediments)

References

Avetisov GP (2004) Geological-geophysical features of the lithosphere of the Arctic region. VNIIOkeangeologia, Saint Petersburg

Breivik AJ, Verhoef J, Faleide JI (1999) Effect of thermal contrasts on gravity modeling at passive margins: results from the western Barents Sea. J Geophys Res 104(B7):15293–15311

Brocher TM (2005) Empirical relations between elastic wavespeeds and density in the Earth's crust. Bull Seismol Soc Am 95(6):2081–2092

Brozena JM, Childers VA, Lawver LA et al (2002) New Aerogeophysical study of the Eurasia Basin and Lomonosov Ridge: implications for basin development. Eos Tran AGU 83(47):F-1266

Brozena JM, Childers VA, Lawver LA et al (2003) New aerogeophysical study of the Eurasia Basin and Lomonosov Ridge: Implications for basin development. Geology J 31(9):825–828

Cande SC, Kent DV (1995) Revised calibration of the geomagnetic polarity time scale for the late Cretaceous and Cenozoic. J Geophys Res 100(B4):6093–6095

Chernykh AA, Gol'mshtock AY (2009) Gravity-thermal model of the Near-Laptev closure of the Eurasian Basin. Voprosy Geofiziki 41:62–79

Cochran JR (2008) Seamount volcanism along the Gakkel Ridge, Arctic Ocean. Geophys J Int 174(3):1153–1173

Cochran JR, Kurras GJ, Edwards MH et al (2003) The Gakkel Ridge: bathymetry, gravity anomalies, and crustal accretion at extremely slow spreading rates. J Geophys 108:2116–2137

Deep Structure and Evolution of the Lithosphere in the Central Atlantic Ocean (1998) Maschenkov SP, Pogrebitsky YuE et al (eds) VNIIOkeangeologia, Saint Petersburg

Døssing A, Jackson HR, Matzka J et al (2013) On the origin of the Amerasia Basin and the high Arctic large igneous province (results of new aeromagnetic data). Earth Planet Sci Lett 363:219–230

Drachev SS, Malyshev NA, Nikishin AM (2010) Tectonic history and petroleum geology of the Russian Arctic shelves: an overview. Petroleum Geology Conference Series, vol 7, pp 591–619

Engen O, Eldhom O, Bungum H (2003) The Arctic plate boundary. J Geophys Res 108(B2):1–17

Engen O, Gjengedal JA, Faleide JI et al (2009) Seismic stratigraphy and sediment thickness of the Nansen Basin, Arctic Ocean. Geophys J Int 176(3):805–821

Franke D, Hinz K, Oncken O (2001) The Laptev Sea rift. Mar Pet Geol 18:1083–1127

Gaina G, Nikishin AM, Petrov EI (2015) Ultraslow spreading, ridge relocation and compressional events in the East Arctic region – A link to the Eurekan orogeny? Arktos J. https://doi.org/10.1007/s41063-015-0006-8

Glebovsky VYu (1995) Magnetic anomalies and the history of the Reykjanes Ridge seafloor spreading. IUGG, Abstracts week A. XXI general assembly, Boulder, Colorado, July 2–14 1995

Glebovsky VYu, Karasik AM, Merkuriev SA et al (1990) The Reykjanes Ridge – Iceland Basin hydromagnetic surveys and spreading of the North Atlantic. Electromagn Induction World Ocean, Part 1:22–34. (in Russian)

Glebovsky VY, Kovacs LC, Maschenkov SP et al (1998) Joint compilation of Russian and US navy aeromagnetic data in the central Arctic seas. Polarforshung J 68:35–40. erschienen 2000

Glebovsky VY, Zai'onchek AV, Kaminsky VD et al (2002) Digital databases and the Arctic Ocean potential fields maps. In: The Russian Arctic: geological history, mineralogy, geoecology. VNIIOkeangeologia, Saint Petersburg, pp 134–141

Glebovsky VY, Kaminsky VD, Minakov AN et al (2006) Formation of the Eurasia Basin in the Arctic Ocean as inferred from geohistorical analysis of the anomalous magnetic field. Geotectonics J 4:21–42

Glebovsky VY, Chernykh AA, Kaminsky VD et al (2012) Structural-tectonic regionalization of potential fields in the Arctic Ocean for the latest compilation of circumpolar tectonic map of the Arctic. In: Collected articles. Geology and geophysics of the Arctic region lithosphere, vol 8. VNIIOkeangeologia, Saint Petersburg, pp 20–29

Gramberg IS, Kos'ko MK, Lazurkin DV (1984) Major stages of the Arctic continental margin Neogeian evolution. J Sovetskaya Geologiya 7:32–40. (in Russian)

Grantz A, Pease VL, Willard DA et al (2001) Bedrock cores from 89° North: implications for the geologic framework and Neogene paleooceanology of the Lomonosov Ridge and a tie to the Barents shelf. Geol Soc Am Bull 113(10):1272–1284

Hayes DE (1976) Nature and implications of asymmetric sea-floor spreading – Different rates for different plates. Geol Soc Am Bull 87:994–1002

Jokat W, Micksch U (2004) Sedimentary structure of the Nansen and Amundsen basins, Arctic Ocean. J Geophys Res Lett 31:1–4

Jokat W, Ritzmann O, Schmidt-Aursch MC et al (2003) Geophysical evidence for reduced melt production on the Arctic ultraslow Gakkel mid-ocean ridge. Nature J 423:962–965

Karasik AM (1974) Plate tectonics of the Eurasian Basin of the Arctic Ocean. In geology of the Earth polar reigons. NIIGA, Leningrad. (in Russian)

Karasik AM (1980) Strtucture and evolution of the of the Arctic Ocean floor in light of aeromangnetic data Marine geology, sedimentology, petrography and ocean geology, pp 178–193. (in Russian)

Karasik AM (1981) Slow ocean floor spreading (Eurasian Basin of the Arctic Ocean) and specifics of geohistorical analysis of anomalous magnetic field. Oceanic magnetic anomalies and global tectonics, pp 162–174. (in Russian)

Khain VE (2001) Tectonics of continents and oceans (year 2000). Scientific world, Moscow

Khoroshilova MA, Franke D, Kirillova T et al (2014) Dating and correlation of reference seismic horizons in the Laptev Sea Basin. Mosc Univ Geol Bull 69(5):271–280

Kireev AA, Piskarev AL (2014) Specifics of the Cenozoic spreading in the eastern Eurasian Basin of the Arctic Ocean. XLVI Tectonic conference (Proceedings II:84–88), Moscow, 28 January – 1 February 2014

Levitan MA (2015) Sedimentation rates in the Arctic Ocean during the last five marine isotope stages. Oceanology J 55(3):425–433

Litinsky VA (1972) Recent and contemporary tectonics and shelf gravity field. Geophys Explor Arctic 7(L):18–34. in Russian

Malyshev NA, Obmetko VV, Borodulin AA et al (2009) The Laptev Sea shelf sedimentary cover according to the newest information. Geology of the Earth Polar Reigons: proceedings of XLII Tectonic conference II:32–37

Merkuriev SA (1991) SPREAD – magnetic field calculation software for Inversion-Spreading model. Doctorat dissertation, Institute of terrestrial magnetism (IZMIRAN ELPO), Saint Petersburg (in Russian)

Michael PJ, Langmuir CH, Dick HJB et al (2003) Magmatic and amagmatic seafloor generation at the ultraslow spreading Gakkel Ridge, Arctic Ocean. Nature J 423(26):956–961

Moore TE, Pitman KJ (2011) Geology and petroleum potential of the Eurasia Basin. J Geol Soc 35(48):731–750

Müller RD, Roest WR et al (1997) Digital isochrons of the world's ocean floor. J Geophys Res 102(B2):3211–3214

Naryshkin GD (1987) Mid-oceanic ridge of the Eurasian Basin (Arctic Ocean). Nauka, Moscow. (in Russian)

Nikishin AM, Gaina C, Petrov EI et al (2017) Eurasia Basin and Gakkel Ridge, Arctic Ocean: crustal asymmetry, ultra-slow spreading and continental rifting revealed by new seismic data. J Tectonophysics. https://doi.org/10.1016/j.tecto.2017.09.006

Orographic map of the Arctic Basin (1995) [Maps] Naryshkin GD (ed) Scale 1 : 5 000 000. Karttakeskus, Helsinki

Paleogeography of the Northern USSR in the Jurassic period (1983) In: Basov VA (ed) Novosibirsk. Nauka

Piskarev AL (2004) Petrophysical models of the Earth crust in the Arctic Ocean. VNIIOkeangeologia, Saint Petersburg

Poselov VA, Kireev AA, Butsenko VV et al (2016) Structure and seismic stratigraphy of sedimentary cover in the Amerasian Basin of the Arctic Ocean. 35th International Geological Congress. Cape Town, 27 August – 4 September 2016

Rekant PV, Petrov OV, Kashubin SN et al (2015) History of formation of the sedimentary cover of Arctic Basin. Multychannel seismic approach. Reg Geol Metallogeny 64:11–27

Shkarubo SI, Zavarzina GA, Zuykova ON (2014) The geological structure of the Eastern Arctic Shelf. Prospect Protection of Mineral Res J 4:23–30

Verba ML (2008) Comparative geodynamics of the Eurasian Basin. Nauka, Saint Petersburg. (in Russian)

Weigelt E, Jokat W, Franke D (2014) Seismostratigraphy of the Siberian sector of the Arctic Ocean and adjacent Laptev Sea shelf. J Geophys Res 119(7):5275–5289

Chapter 4
Lomonosov Ridge

**Alexey L. Piskarev, Victor V. Butsenko, Andrey A. Chernykh,
Mikhail V. Ivanov, Valery D. Kaminsky, Victor A. Poselov,
and Vasily A. Savin**

Abstract The Lomonosov Ridge is situated between oceanic abyssal regions of the
Eurasian Basin, on the one side, and the region of the Central Arctic Uplifts including the abyssal Makarov Basin – on the other. As its morphology and internal structure indicate, it consists of three distinct segments. The shallowest ($\leq$ 400 m)
Canada-Greenland segment is composed mostly by felsic and metamorphic rocks
(basic igneous – at the part facing Alpha-Ridge). In the Near–Polar segment the
ridge has double-crested top with the basin in-between. The structural trend here is
at acute angle to the ridge axis and upper crust consists of basic crystalline rocks.
The 12 km-thick upper crust of the 150 km-wide Siberian segment of the Lomonosov
Ridge consists mainly of felsic and metamorphic rocks.

The ridge's continental crystalline crust is 20–22 km thick, with upper and lower
crusts thickness of about half of the above. Neogene-Pleistocene hemipelagic,
Paleogene bathyal and possible Cretaceous deposits comprise the sedimentary
cover.

Keywords Lomonosov Ridge · Morphology · Sedimentary cover · Continental
crust

Western boundary of the Amerasian Basin (looking north from the Eurasain continent) follows western slope of the Lomonosov Ridge. It is the largest abyssal basin
in the Arctic, and understanding its structure and evolution is imperative for not only
global problems of the planet evolution but also for delineation of the Arctic shelf.

A. L. Piskarev (✉) · A. A. Chernykh
All-Russian Research Institute of Geology and Mineral Resources of the World Ocean
(VNIIOkeangeologia), Saint Petersburg, Russia

Saint Petersburg University, Saint Petersburg, Russia

V. V. Butsenko · M. V. Ivanov · V. D. Kaminsky · V. A. Poselov · V. A. Savin
All-Russian Research Institute of Geology and Mineral Resources of the World Ocean
(VNIIOkeangeologia), Saint Petersburg, Russia
e-mail: vicb@vniio.nw.ru; falcon@vniio.nw.ru; okeangeo@vniio.ru

© Springer International Publishing AG, part of Springer Nature 2019   157
A. Piskarev et al. (eds.), *Geologic Structures of the Arctic Basin*,
https://doi.org/10.1007/978-3-319-77742-9_4

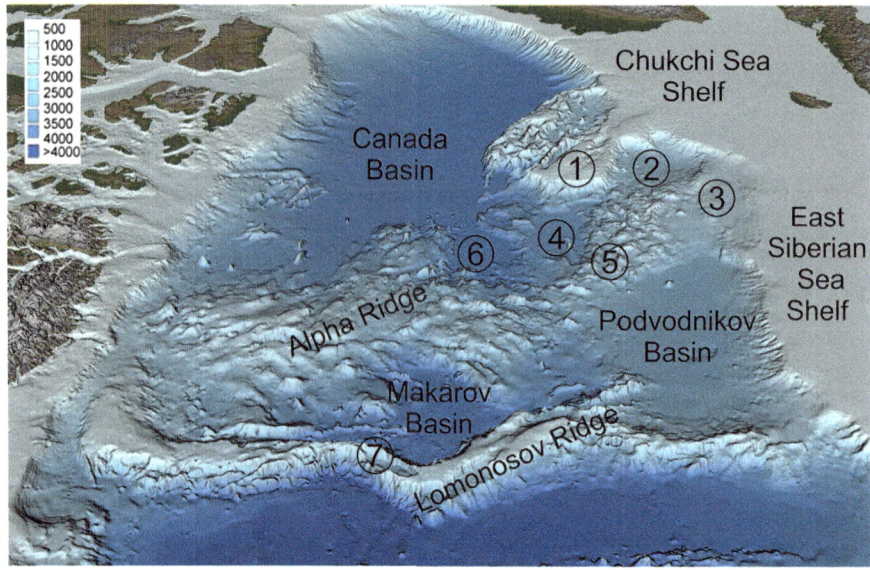

Fig. 4.1 Principal morhological structures of the Amerasian Basin: 1 — Chukchi Plateau; 2 — Chukch Abyssal Plain; 3 — Kucherov Terrace; 4 — Mendeleev Abyssal Plain; 5 — Mendeleev Rise; 6 — Nauilus Gap; 7 — Belov Trough

The largest part of the Amerasian Basin is taken by the Central Arctic Uplifts. In contrast with the Eurasian Basin, no reliable linear spreading-related magnetic anomalies, or valid plate boundaries were found in the Amerasian Basin (Glebovsky et al. 2008; Saltus et al. 2011; Russian Arctic Geotransects 2014).

The Central Arctic Uplift "partions" the central part of the Arctic Ocean enclosed between Greenland-Canada shelf, on the one side, and the East-Asian – on the other. This region includes large first-order positive morphological structures and vast depressions separating them (Podvodnikov, Makaraov and Nautilus Basins, Mendeleev and Chukchi Abyssal Plains), as well as superimposed on the numerous smaller morphological features of the lower order (Fig. 4.1).

In the later years numerous modification of toponymy of the smaller sea floor topographical features took place and were included into the latest version of the IBCAO Arctic bathymetry map. For clarity and convenience, we present the fragment of this map for the Mendeleev Rise-Chukchi Plateau region – Fig. 4.2.

The true diversity of composition and structures of the Amerasian Basin basement vividly depicted by potential field anomalies, as demonstrated by magnetic and gravity (Free Air) anomalies maps for the Central Arctic Uplifts region – Fig. 4.3. These maps reflect physical properties of the geological formations, thus making them instrumental in deciphering the composition and structure of the Amerasian Basin basement.

Several deep crustal regional cross-sections, primarily based on interpretation of seismic data, combined with information from all available sources, were subjected

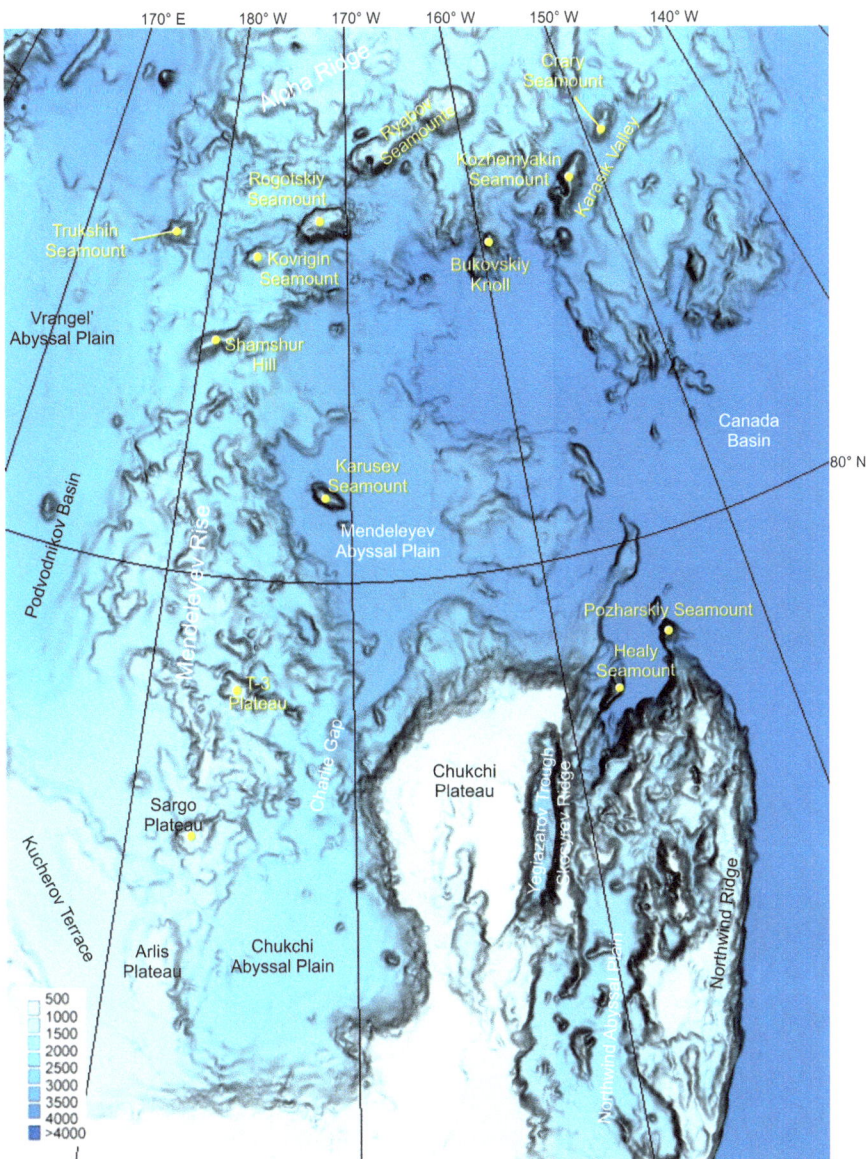

Fig. 4.2 Toponymy of the Mendeleev Rise-Chukchi Plateau region

to specific modeling processes which helped to find plausible interpretation of observed geophysical anomalies.

Correlation of major reference seismic unconformities of the East Arctic Seas shelves and adjacent abyssal parts of the Arctic Ocean was accomplished by referencing the regional seismic lines A-4 and A-7 to the ACEX wells drilled on the

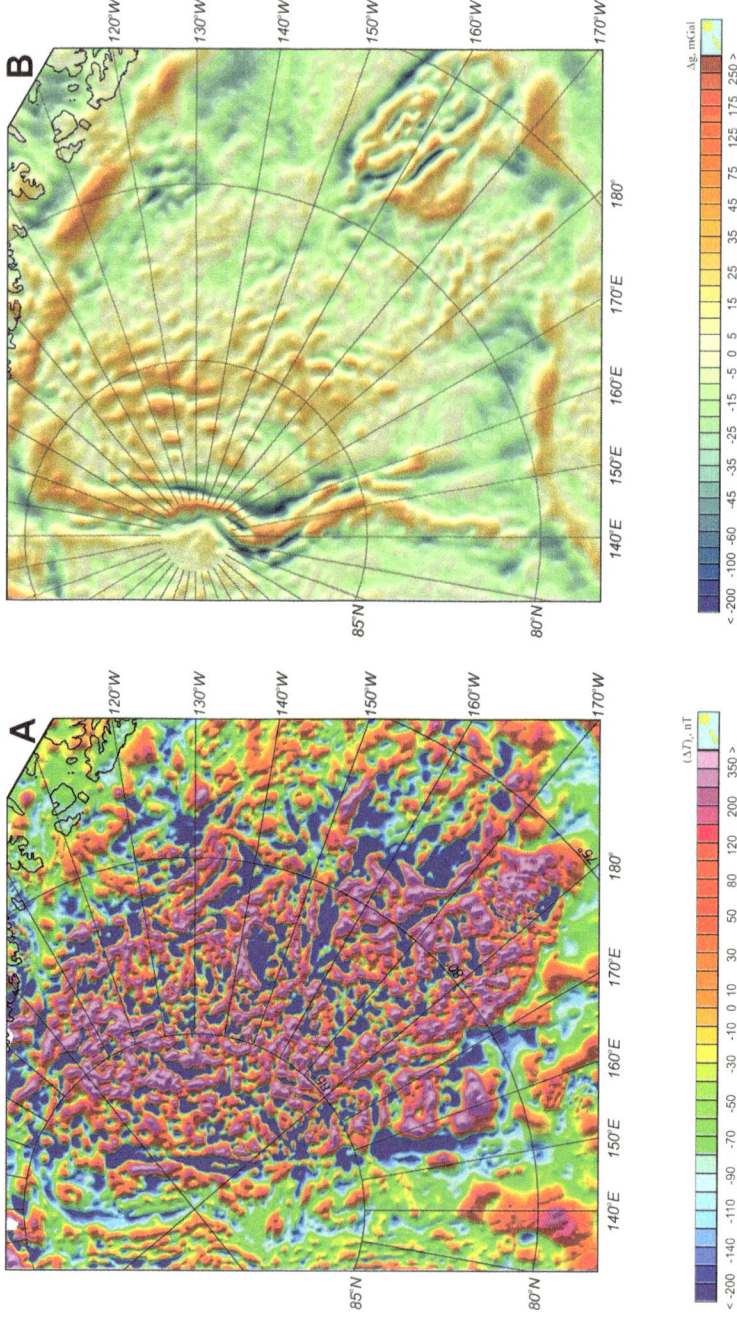

Fig. 4.3 (**a**) Magnetic anomalies of the Central Arctic Uplifts region; (**b**) Gravity anomalies (Free Air) of the Central Arctic Uplifts region

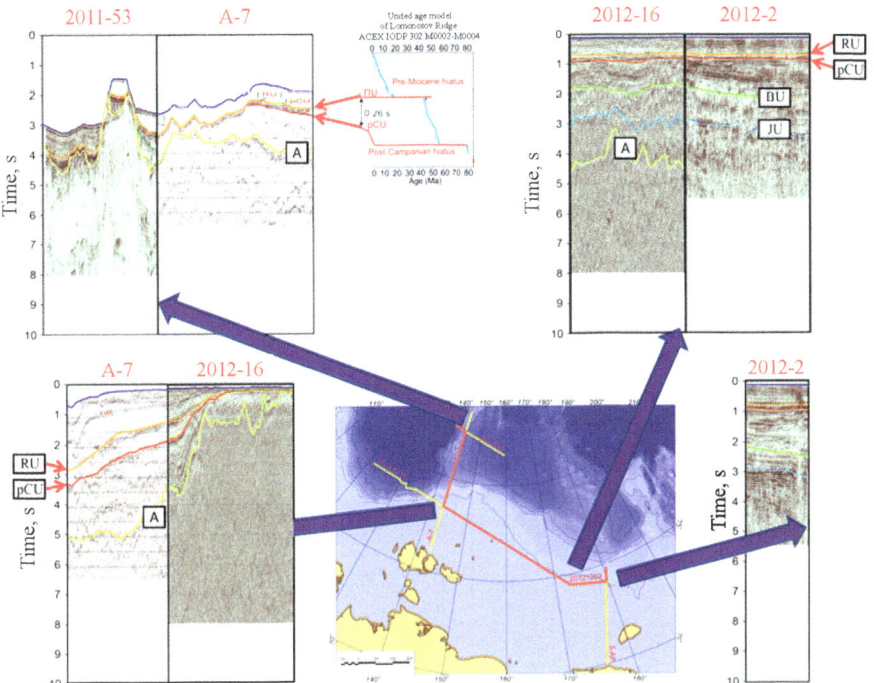

Fig. 4.4 Correlation of the reference seismic unconformites in the Laptev and East-Siberian Seas

Lomonosov Ridge. Additionally, using dense seismic grid (shot by JSC "DMNG") in Chukchi and East-Siberian Seas, and, as a link, the regional MCS line 2012–16, which crosses the Eurasian Basin structures, north-eastern Laptev shelf and East-Siberian continental margin, the seismic data were tied to several wells drilled on Alaska shelf – Fig. 4.4.

4.1 Morphology

The Lomonosov Ridge is ≈ 1700 km long and 50–200 km wide. Its shape mimics the configuration of the Barents-Kara continental margin – the observation traditionally used as a proof of their unity before spreading of the Eurasian Basin. The ridge orientation fluctuates within 20–30° and numerous secondary terrain features - plateaus, saddles, terraces, spurs - follow general direction of the ridge strike (Fig. 4.5). Sea floor depth changes drastically from 1000 m or less at the ridge crest to 3900–4300 at its base. The shallowest part of the Lomonosov Ridge (400 m b.s.l.) is located close to Ellesmere-Greenland shelf junction, where it is separated from the shelf by 2400 m-deep saddle.

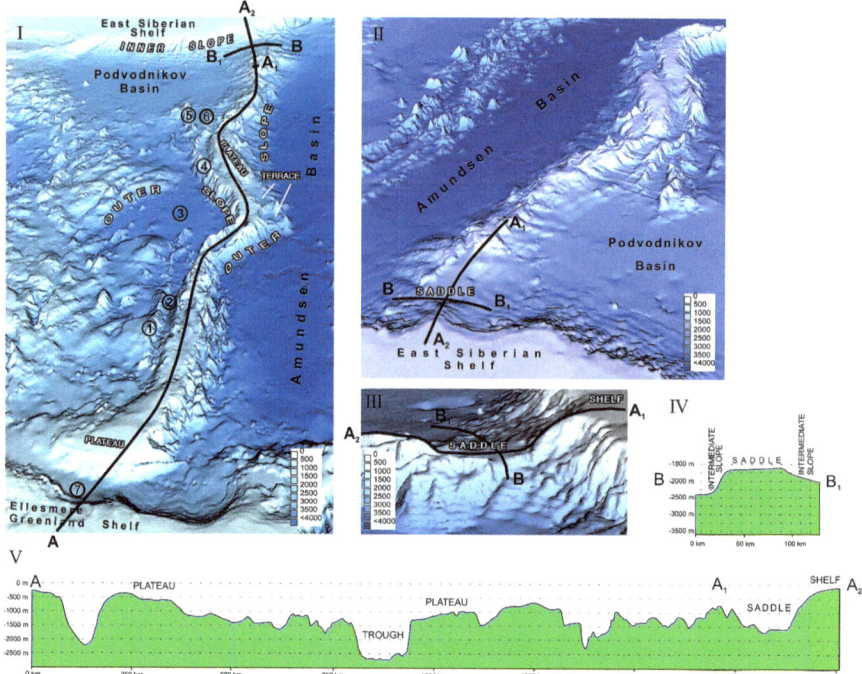

Fig. 4.5 The Lomonosov Ridge: I —3D view looking from the Ellesmere-Greenland shelf; II- 3D view looking from the East-Siberian shelf; III — Saddle at the ridge East-Siberian shelf junction - details; IV — bathymetry profile along B-B1; V — bathymetry profile along A-A1-A2; Numbers in circles: 1 — Marvin Spur; 2 — Marvin Gorge; 3 — Makarov Basin, 4 — Oden Spur; 5 — Geofizikov Spur; 6 – Senchura Spur; 7 – Klenova Ttrough

The Lomonosov Ridge, as a first-order structural element of both Siberian and North American continental margins, acts as an underwater bridge connecting North American and Eurasian continents and, as such, represents a natural extension of the continental Russia on Siberian side and Denmark – on Greenland. The ridge also separate the oceanic abyssal regions of the Eurasian Basin from the Central Arctic Uplifts with abyssal Makarov Basin.

The ridge outer slope runs continuously along the side facing Amundsen Basin. On the opposite side outer slope is clearly visible only when the Lomonosov Ridge verges on the Makarov Basin.

The prominent morphological saddle is present at the Lomonosov Ridge - East Siberian margin junction. The saddle is at 1600 m b.s.l., 1200 m below shelf, but 2000 m above the base of the western slope and 100 m above the floor of the Podvodnikov Basin. The eastern side gently blends into the complex slope of the southern termination of the Podvodnikov Basin; the western side - into the simple southern slope of the Eurasian Basin. Smooth transition of these morphological elements stresses the morphological unity of the Lomonosov Ridge with shelf.

North from the saddle the ridge runs straight northward up to 85°N. In this segment the ridge consists of two parallel ridges, one of which stretches south from 84°40'N along 154°E. South from 83°45'N, the main ridge is also broken in two parallel ridges framing the Lomonosov Ridge from the Podvodnikov Basin side. The top of the main ridge here is 677–1231 m b.s.l. Eurasian flank typically consist of many tilted fault blocks dropping stepwise toward the abyssal Eurasian Basin – the structure typical for uplifts in spreading environment.

In the near–Pole region, the neighboring abyssal plains on both sides of the Lomonosov Ridge are getting closer to each other and, as a result, its outer and intermediate slopes are getting steeper (up to 18°) and fault escarpments – more prominent (Zinchenko 2004).

Large amount of detail bathymetry and gravity data were collect by American SCICEX program (Coakley and Cochran 1998). Some of its profiles and 2011–2014 seismic surveys confirm extensive normal faulting of ridge flanks.

Several sizable spurs (Geofizikov, Oden, Senchu etc., Fig. 4.5) flank the Amerasian segment of the Lomonosov Ridge. Apparent tectonic origin of these morphological units points out to axial tensional tectonic regime. The Geofizikov Spur, crested at 780 m at 81°12' N (line Archerfish3) and 1128 m at 80°54'N (line Cavallal) (Cochran et al. 2006). Accompanied by strong linear gravity anomaly the Geofizikov Spur stretches south along 158°E and, gradually getting less prominent, disappear under the sedimentary cover of the Siberian continental margin.

Crestal depth of the Lomonosov Ridge reaches 1400 m (profiles Cavalla3 at 88° N and Archerfish2, north of 80°N, Fig. 4.6). Strong gravity anomaly detected on profile Pogy 6, crossing the ridge at ≈ 82°45'N, proves that, at this latitude, the Lomonosov Ridge still is a significant structure at the basement level. As a boundary of the Eurasian Basin, and linear shallow terrain feature, the Lomonosov Ridge continues south until disappearing under the Siberian continental slope south from 80°N.

Further north, around of the prominent "bend", the configuration of the Lomonosov Ridge became more complicated, and terrain – more rugged. In fact, the ridge here looks like an intricate combination of roughly parallel, or en-echelon table-top uplifted fault blocks separated by narrow grabens, troughs and gorges. This segment has double-crested top with narrow sediment-filled trough at 2600 m b.s.l. The "bend" of the Lomononsov Ridge is reflected in configuration of magnetic anomalies (Brozena et al. 2003) and repeat the configuration of the active Gakkel Ridge (Cochran et al. 2003).

Siberian segment of the "bend" is 150 km long with two-tear system of terraces, merging on the North American side into single terrace 40–60 km wide at depth 1100 m. Southward, the Lomonosov Ridge gets shallower and its top rises to 950 b.s.l. at 87°N. On the Amerasina side the ridge slope abruptly drops down by 1000–1500 m to 40–50 km wide trough separating the ridge from the Marvin Spur of the same amplitude (1500 m). According to seismic data, this trough contains more than 2000 m of undisturbed sediments.

Further south, the Lomonosov Ridge gets wider and shallower (≥ 500 m) deep 50 km wide Klenova trough and wide Klenova Basin (or Lincoln Sea Basin) separate

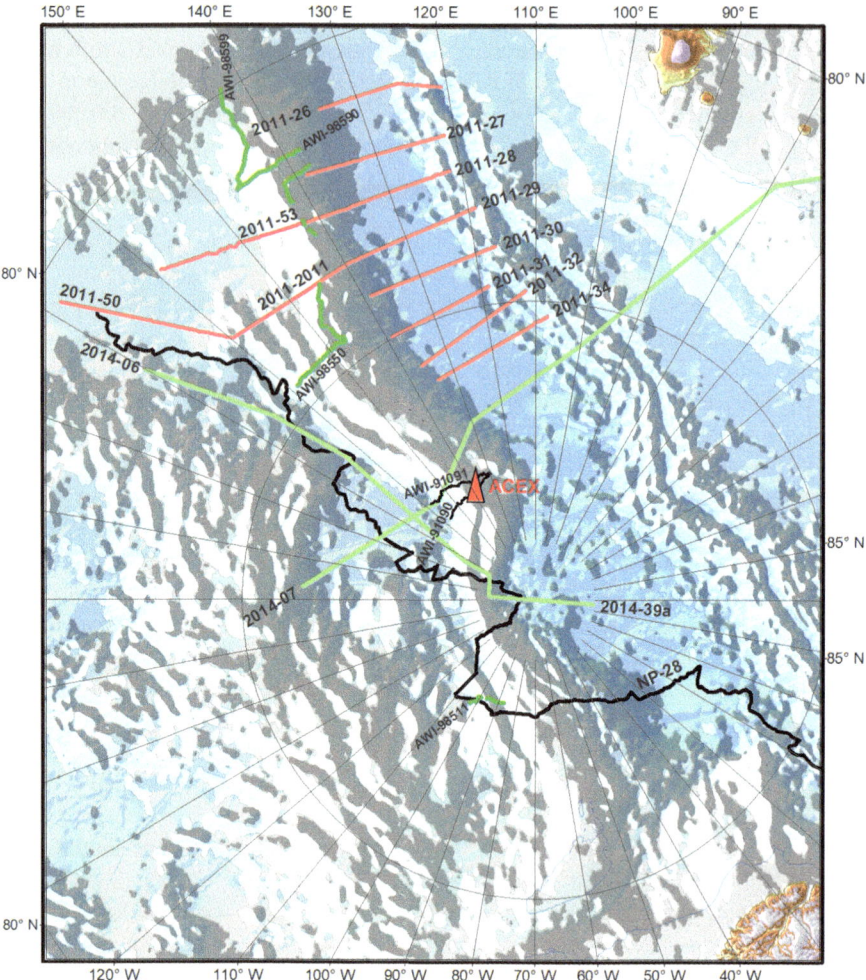

Fig. 4.6 Seismic surveys at the Lomonosov Ridge

the Lomonosov Ridge from the Ellesmere-Greenland and Linkoln Sea shelves (Grantz et al. 2009; Døssing et al. 2014). According to seismic and gravity data, the Klenova trough follows significant strike-slip fracture zone on drastically reduces continental crust (Døssing et al. 2014).

As a whole, the western, Amerasian, slope of the Lomonosov Ridge is classified as a complex slope (with multitude of terraces and intermediate slopes) and vertical drop of 2400–3000 m. Lower slope with its base at 2600–4200 m is practically undisturbed along the entire length of the Lomonosov Ridge.

4.2   Potential Fields Anomalies

Greenland-Canadian segment of the Lomonosov Ridge displays the pattern of strong potential anomalies oriented along its strike (Fig. 4.3). Magnetic anomalies of this sector are similar to those of Ellesmere Island and have positive correlation with terrain (Vogt et al. 1982, Piskarev 2004). Numerical interpretation and iterative modeling of gravity data, supported by bathymetry and seismic information expedition LOREX, (Weber and Sweeney 1990) revealed significant thickness of consolidated crust (around 28 km) with predominantly felsic-metamorphic composition of the crust in near-Pole region and mainly mafic – in vicinity of the Alpha Ridge (with possible HALIP trapp formations on surface).

In the near-Pole region, intensity of potential fields, trending here at oblique angle to the meridian trend of the ridge, increases significantly. Calculated magnetization of sources (2–5 A/m), is high enough to classify them as mafic, however no correlation between magnetic anomalies and terrain forms was reported. Upper crust density of 2.79 g/cm^3, estimated by gravity modeling, also indicates mostly basic composition of crystalline crust.

Gravity and magnetic anomalies of Siberian segment of the Lomonosov Ridge are weak, except the western part of the Lomonosov Ridge – Laptev shelf junction along the Eurasian Basin border with strong magnetic anomalies caused by basalt flows visible on seismic inside the northern Laptev Sea sedimentary section (Franke et al. 2001). Positive correlation between magnetic anomalies and terrain forms was also observed. Computed magnetization (1–2 A/m) suggests that the basement of this part of the Lomonosov Ridge composed mainly by felsic-methamorphic formations.

4.3   Sedimentary Cover

The Lomonosov Ridge is relatively well studied, having been, and still being, a subject of many expeditions and studies. Figure 4.6 illustrates seismic coverage and drilling locations in this region.

Stratigraphy of the Lomonosov Ridge sedimentary cover is based on information from wells drilled under the project ACEX at locations making correlation with seismic possible (Jokat 2005; Moran et al. 2006; Backman et al. 2008; Backman and Moran 2009; Langinen et al. 2009; Bruvoll et al. 2010; Rekant and Gusev 2012; Jokat et al. 2013; Poselov et al. 2014) (Fig. 4.7).

Two major lithostratigraphic complexes - upper, Miocene-Pleistocene (18,2 Ma–present) and lower, Eocene (56.2–44.4 Ma), are separated by erosional (stratigraphic) unconformity. Miocene-Pleistocene complex was formed in cool ("ice house") climatic conditions which replaced warm climatic maximum ("green house"), prevailing in Eocene Period, at about 44.8 Ma ago (Moran et al. 2006; Backman et al. 2008). A horizon with spores of freshwater fern Azolla was found in

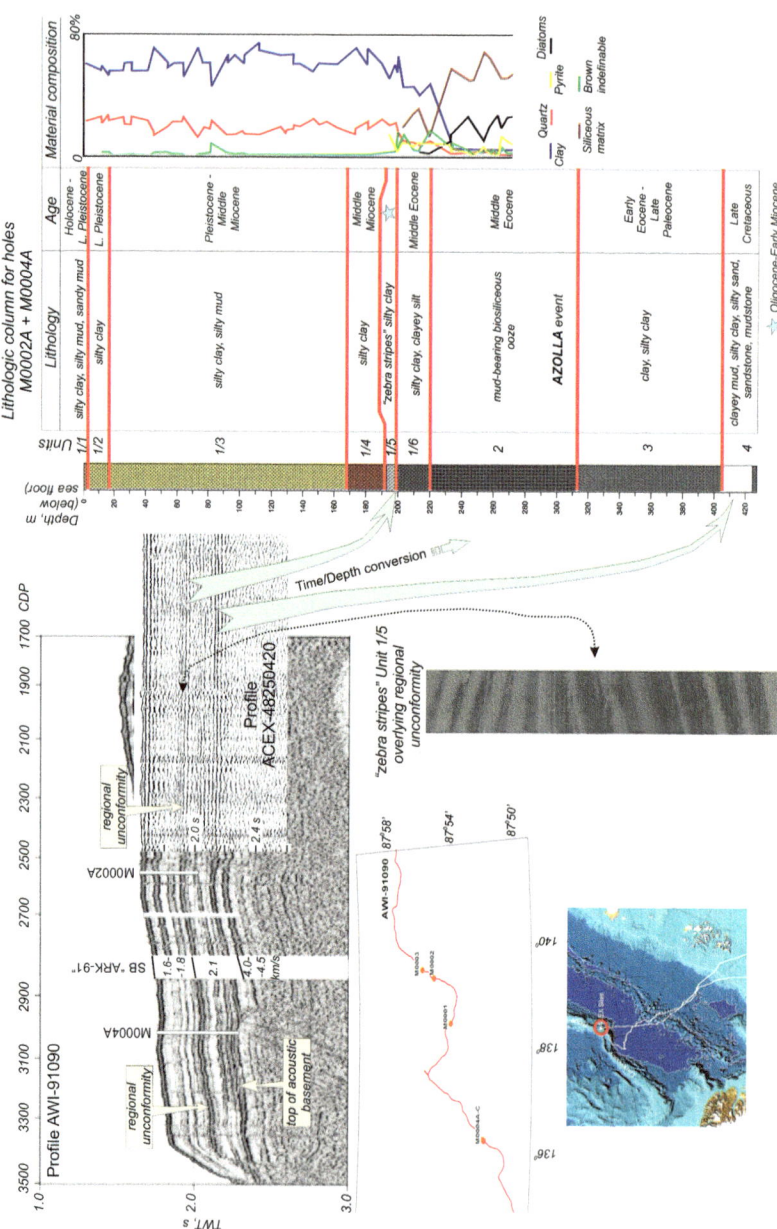

Fig. 4.7 The Lomonosov Ridge sedimentary cover - stratigraphy and correlation of ACEX data with seismic

the lower middle Miocene (~ 49 Ma). High concentration of fern Azolla spores (potential oil source material) can be attributed to massive accumulation of sediments during extreme climate warming at the end of Early Eocene climatic maximum.

Syn-rift deposits, presumably Cretaceous and/or Paleocene, were identified in tilted fault blocks. The interval from upper part of Mid-Eocene to lower pat of Lower Miocene initially was considered as non-depositional hiatus (Moran et al. 2006; Backman et al. 2008). But later ((Poirier and Hillaire-Marcel 2011), Fig. 4.8), 50 m-thick sequence of very fine-grained sediments was referred to this interval.

Incidentally, the newest chronology based on Os (osmium) isotops analysis, is close to one based on diatom algae (36.7 M), but disagrees with age determined by dinofalgellates cysts.

Pre-Eocene, or Post-Campanian, unconformity may mark the onset of spreading of the Eurasian Basin if we assume that before Eocene the region of the future Lomonosov Ridge went trough continental rifting. Erosional discontinuity present inside Lower Miocene sequence marks the erosional event still not accurately dated.

There are several geological models attempting to explain the specifics of geology of the Lomonosov Ridge and its junctions with Amundsen and Makarov Basins. Main difference between them lies in number of seismic-stratigraphic complexes and their correlation to the ACEX 302 well. Some researchers (Poselov et al. 2014) consider the reflector RU (for Regional Unconformity, related to Oligocene hiatus between neritic an hemipelagic deposits) as a principal regional marker horizon in the Lomonosov Ridge and Podvodnikov Basin. At the same these authors relate the cardinal change in the sedimentation and climate regime at the Paleogene-Neogene boundary with opening of Fram Gap (LMA 13) bringing massive inflow of Atlantic waters into the Arctic Basin. Another group (Grantz et al. 2001; Moran et al. 2006) think that the most important marker is strong seismic horizon LU – erosional AND angular unconformity present at the bottom of the well core and micropaleontologicaly dated as Late Paleocene. (Fig. 4.9).

At the same time Jokat (2003, 2005), Rekant and Gusev (2012) relate LU unconformity to the time of rigorous normal faulting of the Lomonosov Ridge creating numerous fault blocks subjected to active erosion of the crestal parts The reflector can be traced trough significant part of the Amerasian Basin and at the Lomonosov Ridge - Amundsen Basin boundary.

Five seismic lines AWI-98550-98,599 shot in 1998 by R/V (Jokat 2005) across the Lomonosov Ridge clearly demonstrate the results of Cenozoic rifting on the structure of the ridge. For more than 100 km, the Amerasian side of the Lomonosov Ridge is an agglomerate of continental blocks, broken off the ridge, and deep grabens filled by syn- and post-rift sediments. Rekant and Gusev (2012) interpreted Polarstern lines using the above mentioned seismic model. Their founding can be summarized as follows:

1. Upper seismic-stratigraphic complexes SSC-I and SSC-II (Eocene-Pleistocene) form well stratified, and almost completely undisturbed, sequence of hemipelagic sediments.

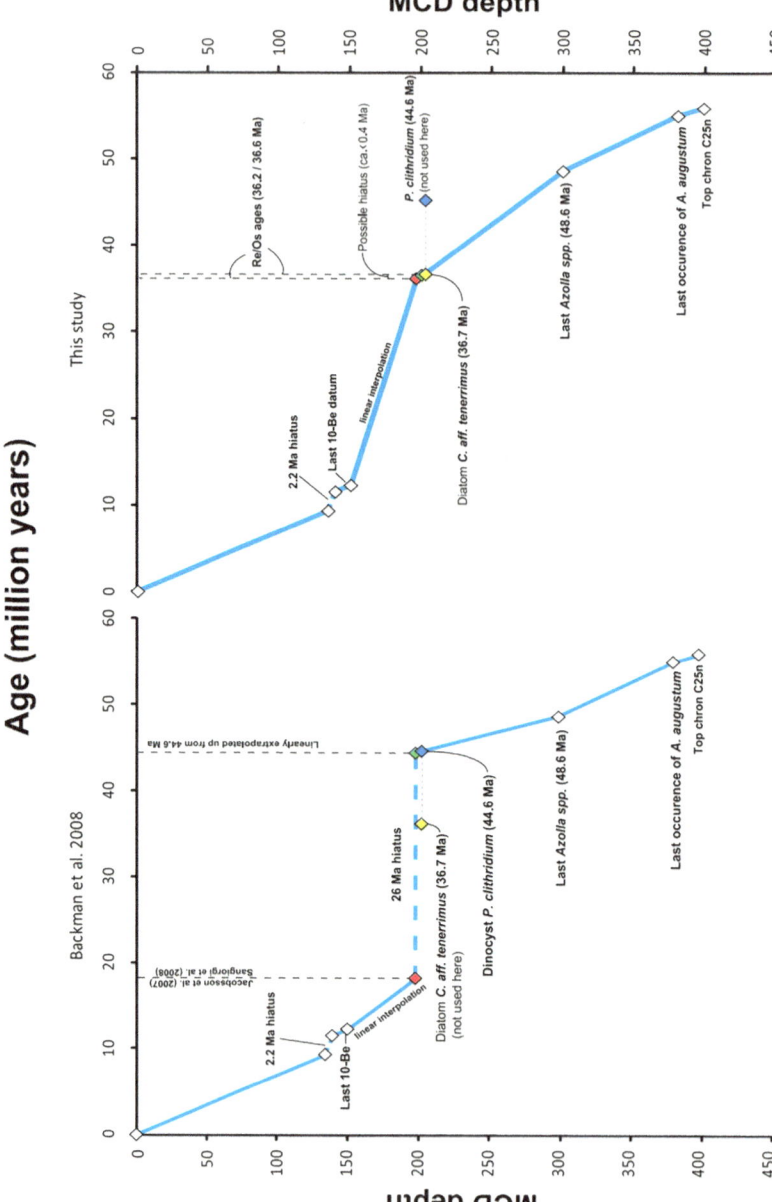

Fig. 4.8 Filling the Gap – comparison of age models for the ACEX core from Backman et al. (2008) and based on Re/Os isochrones (Poirier and Hillaire-Marcel 2011)

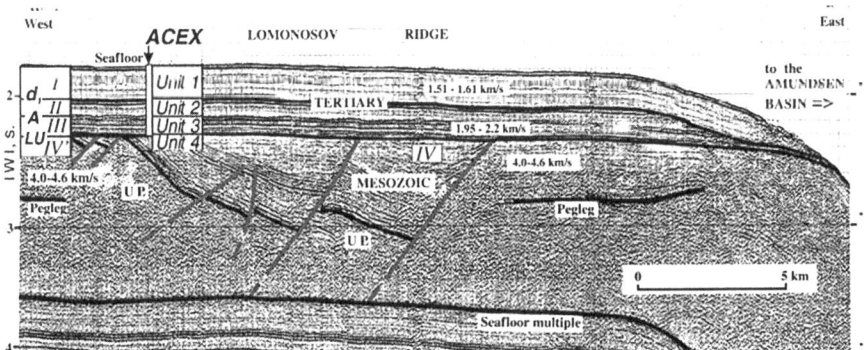

Fig. 4.9 Seismic line AWI-91090, interpreted section (Grantz et al. 2001) with well ACEX super-imposed. Seismic stratigraphy by (Moran et al. 2006)

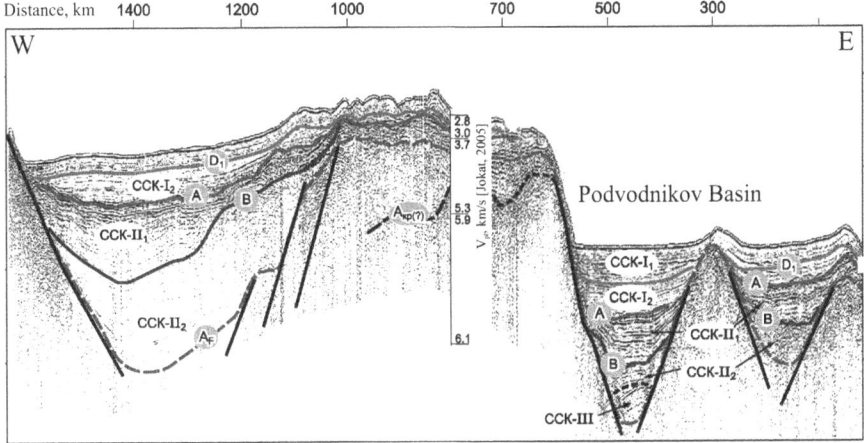

Fig. 4.10 Eeasten part of line AWI-98550 (Jokat 2005). Interpretation by (Rekant and Gusev 2012)

2. SSC-II$_1$ (Lower Paleocene), SSC-II$_2$ (Upper Cretaceous) and SSC-III (Lower Cretaceous, Aptian-Albian) – seismic-stratigraphic complexes fill deep graben-type depressions not only on the flanks, but also in the crestal parts and of the Lomonosov Ridge; they presumably consist of littoral and bathial sediments. (Figs. 4.10–4.12).

The complex SSC- II$_1$ almost completely levels the acoustic basement topography and has all the characteristics of an infill formation – sharp increase of thickness (up to 1000 m) inside the depressions, onlap, truncation at, and draping over, the uplifted parts. Apparently, the complex was formed during periods of intensive erosion of the crestal parts of the Lomonosov Ridge and filling tectonic depressions in the ridge and slope with eroded material.

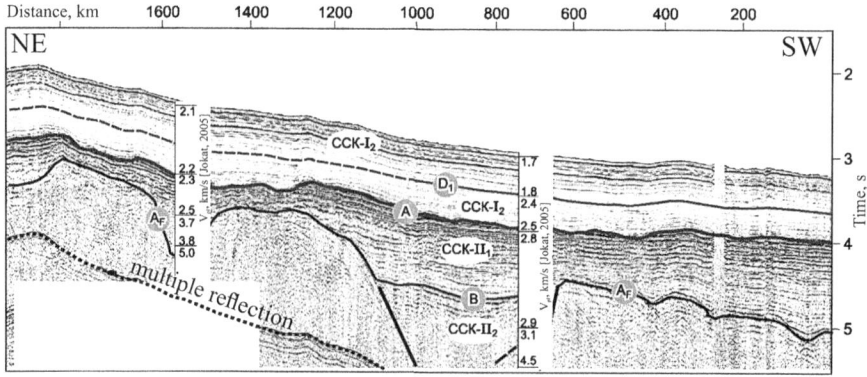

Fig. 4.11 Fragment of line AWI-98595 (Jokat 2005). Interpretation by (Rekant and Gusev 2012)

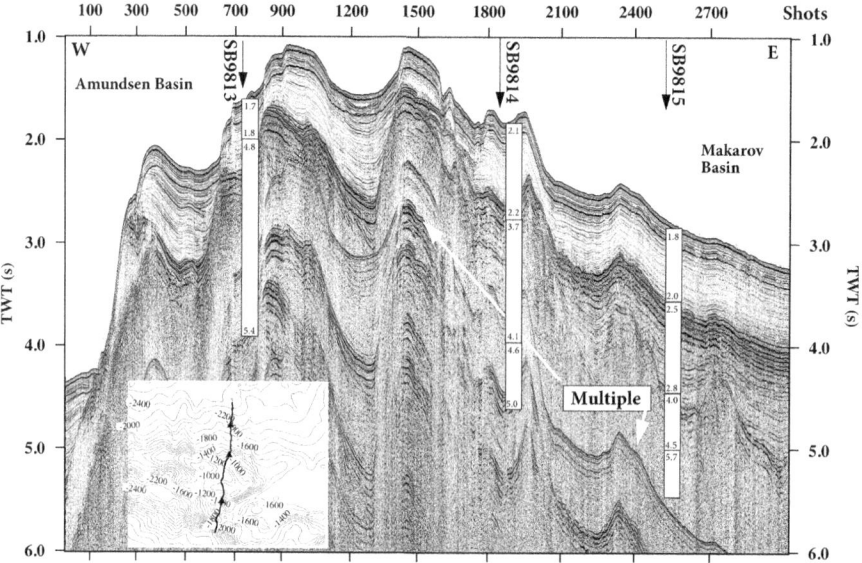

Fig. 4.12 Fragment of line AWI-98590 (Jokat 2005) across Lomonosov Ridge crest. Interpretation by (Rekant and Gusev 2012)

Seismic data from the Russian expedition ARCTICA -2011 (Fig. 4.13–4.14) – also confirm normal faulting of the ridge acoustic basement demonstrating numerous half-grabens at the base of the sedimentary cover not only in the ridge proper, but in the trough between the Lomonosov Ridge and the Geofizikov Spur (actually, the whole trough can be considered as complex system of fault blocks with different amplitudes formed on the continental crust). The half-grabens roughly follow the general trend of the Lomonosov Ridge and filled with Cretaceous, perhaps, even older sediments. Some of them might be reactivated in Paleocene.

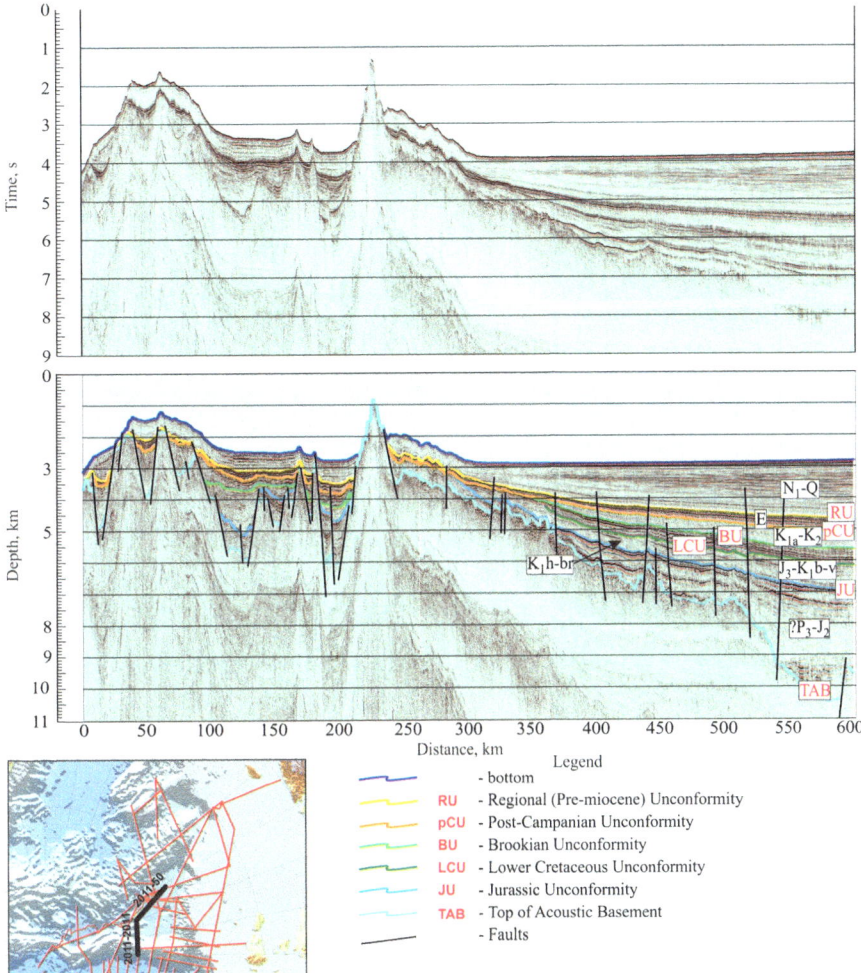

Fig. 4.13 Composite MCS line ARC-2011_2011–050

Further south, the Geofizikov Spur is buried under sedimentary sequence up to 3 km thick. Here hemipelagic undisturbed Pleistocene-mid Miocene sediments above pre-Miocene unconformity RU overlay Paleogene littoral deposits which, in turn, are separated from the underlying Lower-Upper Cretaceous sediments by post-Campanian unconformity pCU (Poselov et al. 2014). Lower parts of sedimentary sequence may contain Cretaceous, Jurassic and even Upper Paleozoic (below the JU unconformity) deposits, filling tectonic depressions at the base of the Lomonosov Ridge.

Line ARC-2011_053–065 demonstrates that two sub-horizontal stratigraphic units which authors identify with the units mentioned above (with erosional boundary in-between), are present at the Lomonosov Ridge. It also shows thin sedimentary wedge of Late Eocene — Early Miocene sediments (44,4–18,2 Ma) below

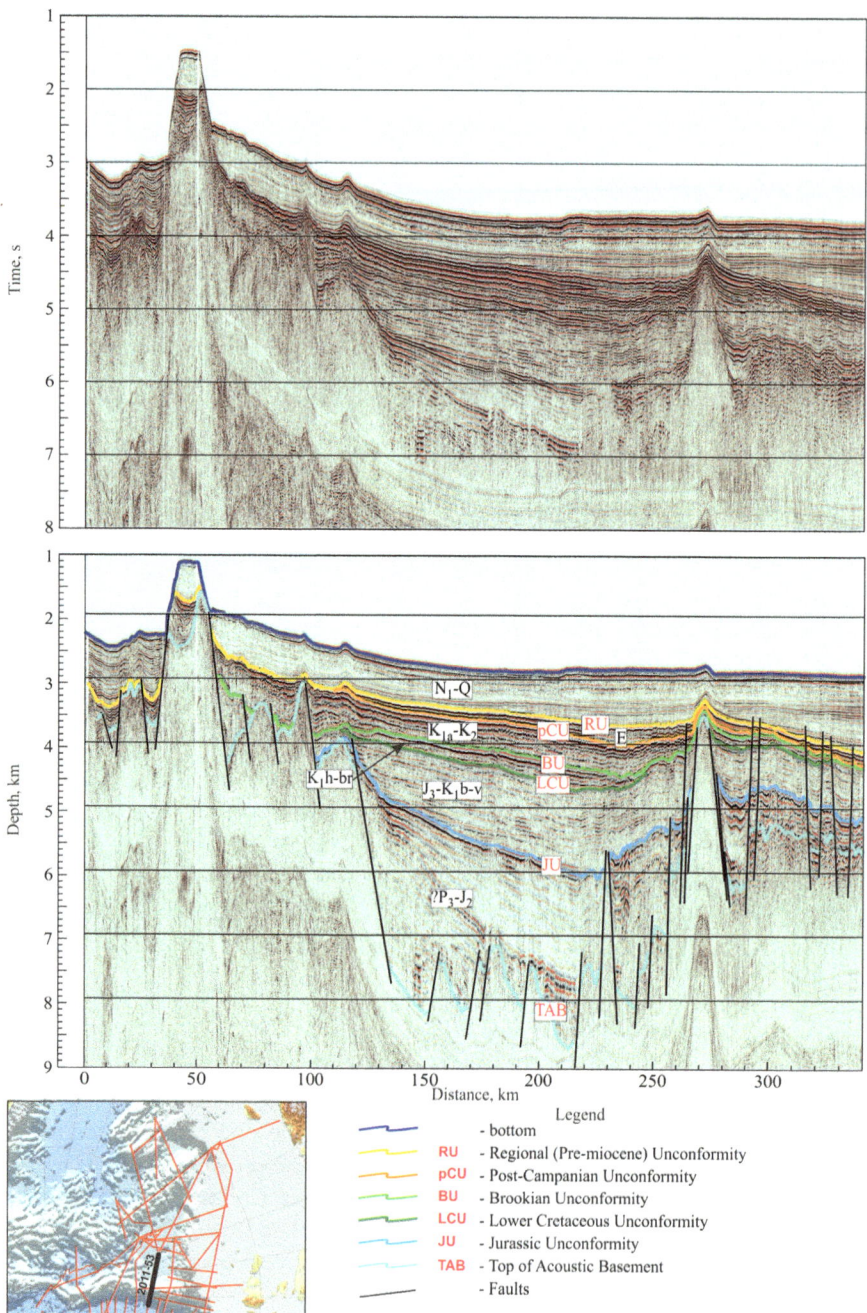

Fig. 4.14 MCS line ARC-2011_053–065

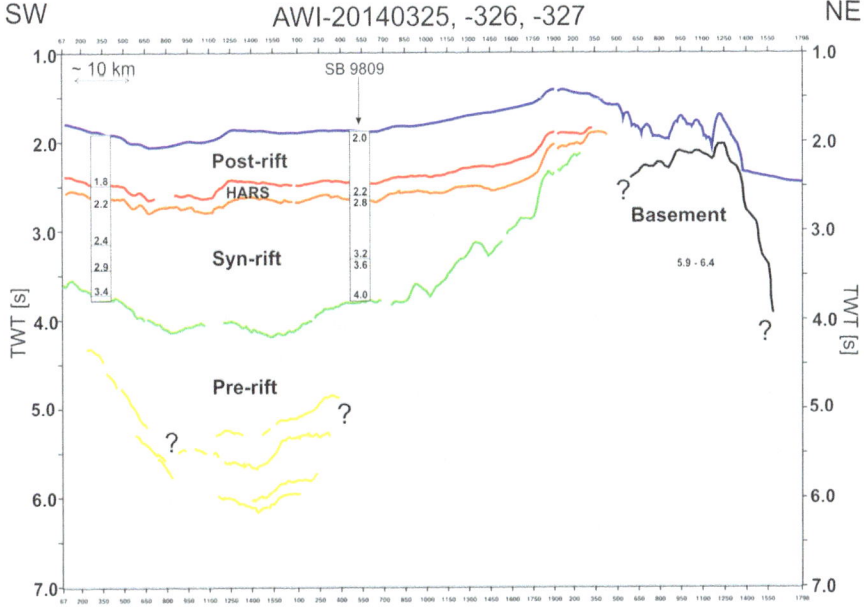

Fig. 4.15 Seismic units of along-ridge profiles (refer to Fig. 4.6 for locations)

Middle-Late Miocene layers. This Late Eocene — Early Miocene wedge is absent, or drastically thinner, in the wells drilled at the ridge crest but clearly visible on its slopes towards the Podvodnikov Basin.

Rather peculiar sedimentary pattern of the Lomonosov Ridge is illustrated by seismic lines AWI-20140325-327 orientd along the ridge's crest (Sauermilch and Jokat 2015) (Fig. 4.15).

Here the following sequences were identified (from top down):

– Post-rift homogenous hemipelagic sediments;
– High Amplitude Reflector Sequence (HARS- present along the entire ridge and marks regional drastic change in depositional environment (Jokat et al. 1992);
– Syn-rift sediments with rift-related deformations;
– Pre-rift depression.

Numerous post-Paleogene normal faults, most of which are re-activated faults of Cretaceous age, indicate that the Lomonosov Ridge was under dominant tension regime during the period of ultra-slow spreading of the Eurasian Basin.

The seismic acquired during the ice station NP-28 drift over the near-Greenland part of the Lomonosov Ridge is presented on Fig. 4.16 (Langinen et al. 2009). Seismostratigraphic complex of grabens infill correlates with SSC-II2 and SSC –III complexs of near Siberian section of the Lomonosov Ridge (Rekant and Gusev 2012).

Seismic from certain sections of IS 28 drift line closer to the North Pole clearly shows unconformity within Cenozoic sequence of well stratified sub-horizontal

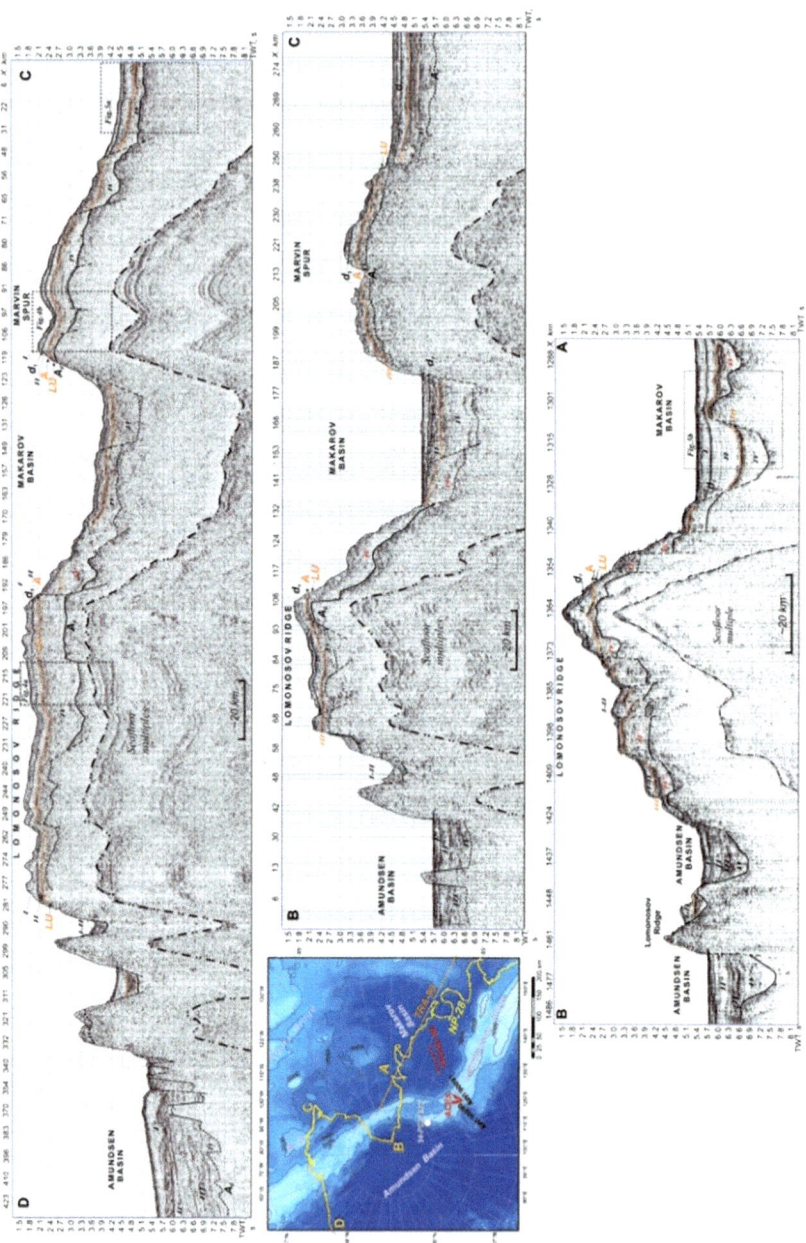

Fig. 4.16 Sesimic sections along drift line of IS NP-28 (Langinen et al. 2009)

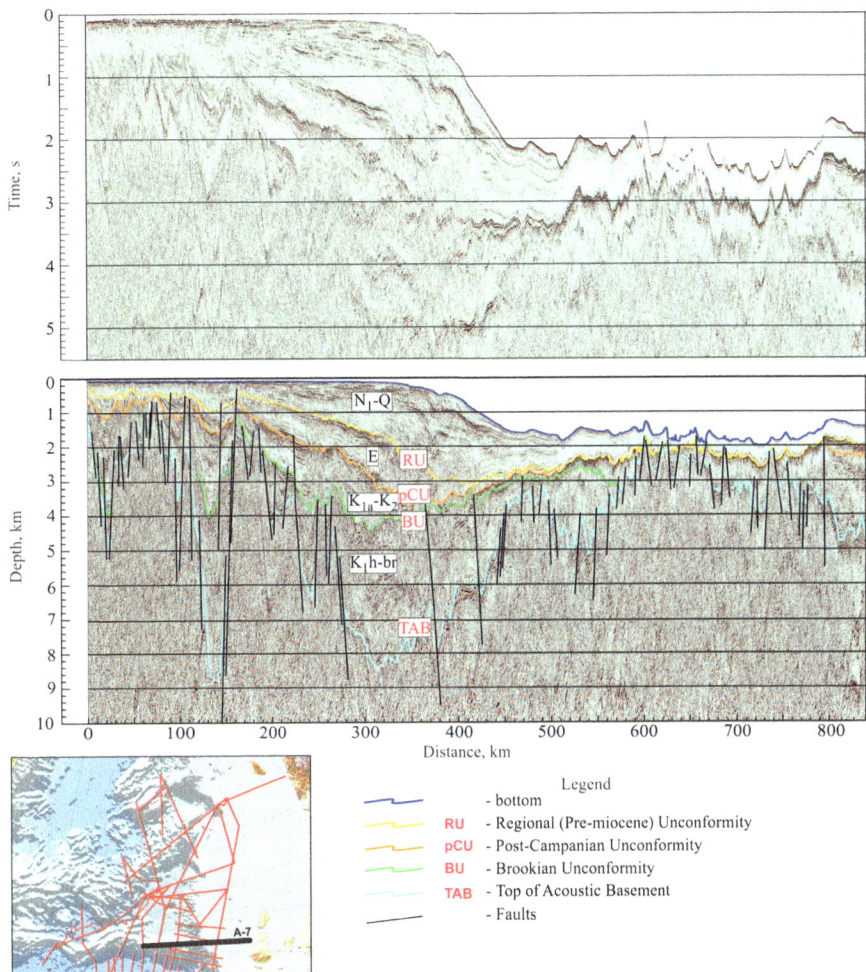

Fig. 4.17 MCS seismic line A7 along the near-Siberian segment of the Lomonosov Ridge and adjacent shelf

reflectors few hundred meters thick (LU unconformity), which correlates well with pCU unconformity in (Poselov et al. 2014)).

Near-Siberian segment of the Lomonosov Ridge abuts to the shelf sedimentary basin of the Laptev and East- Siberian Seas. The details of this junction are illustrated by MCS line A-7 (Fig. 4.17) showing that the base of shelf syn-rift Cretaceous formations become the base of Cenozoic sedimentary cover of the Lomonosov Ridge. The entire assemblage of seismostratigraphic complexes is present in the depocenter of the Vilkitsky Trough confidently outlined as a depression of the acoustic basement occupying the whole transition zone from the outer shelf (southern wing) to the Lomonosov Ridge (northern wing). At the ridge crest, hemipelagic

Miocene-Pleistocene deposits lie on top of Cretaceous separated by RU unconformity and Paleogene complex is either completely absent or only a few hundred meter thick (~200 m at the ACEX well location). Therefore, up to Early-Middle Miocene, the near-Siberian segment of the Lomonosov Ridge was above, or close to, sea level (Poselov et al. 2014).

It can be noted that the dating of the Laptev shelf sedimentary complexes based on presented seismic-stratigraphic interpretation agrees with earlier stratigraphic model in (Franke et al. 2001).

Therefore, it can be stated that the Cretaceous and Cenozoic sedimentary complexes continue uninterrupted from Eurasian Shelf via transitional zone to the Lomonosov Ridge without any significant changes of their seismostratigraphic and seismic-facial characteristics, which would be impossible if there was a significant strike-slip displacement of the ridge relative to the East-Siberian and/or Laptev continental margins. There also are no significant normal faulting within the transitional zone, as the configuration of the acoustic basement in the Vilkitsky Trough may attest (Poselov et al. 2014).

W. Jokat (Jokat et al. 2013) came to the similar conclusion postulating that the Lomonosov Ridge kept its position relative to North American, East Asian continental masses and other Amerasian Basin structures fixed since separation from Barents-Kara Continental Margin moving with the plate.

4.4   Acoustic Basement

The Lomonosov Ridge acoustic basement is thought to contain a sedimentary complex analogous to modestly dislocated Paleozoic-Early Mesozoic formations of the New Siberian Islands Archipelago. So far, only a single core was extracted from the basement at 89° N. It contains continental aleurolites, in composition and age similar to Upper Triassic-Lower Jurassic layers of Frantz-Joseph Land (Grantz et al. 2001).

Coarse material in sea floor samples collected in close proximity to the well and at the near-Siberian segments consists of argillites (70–80%), dolomites, quartz sandstones, aleurolites and limestones (20–25%) and occasional mafic lavas and crystalline rocks. Single sample from western slope of the Geofizikov Spur contains approximately equal amount of sedimentary and crystalline (metamorphic and volcanic) rocks.

Danish geological expedition dredged two locations on Amerasian slope of the Lomonosov Ridge in 2012. Collected material contained brekchiated clastic rocks (10%), meduim-grained pryable light-gray quartz sandstone (10%), inter-bedding dark grey slightly altered sandstones and aleurolites. Meta-sediments tentatively dated as Lower Paleozoic, quartz sandstone - Mesozoic. Other samples contain approximately equal amount of fine-middle grained cemented quartz sandstone (>20%) and fine inter-bedding slightly metamorphased gray sandstone and black argillite (>21%), preliminary dated by X. Markussen as Early Paleozoic. This infor-

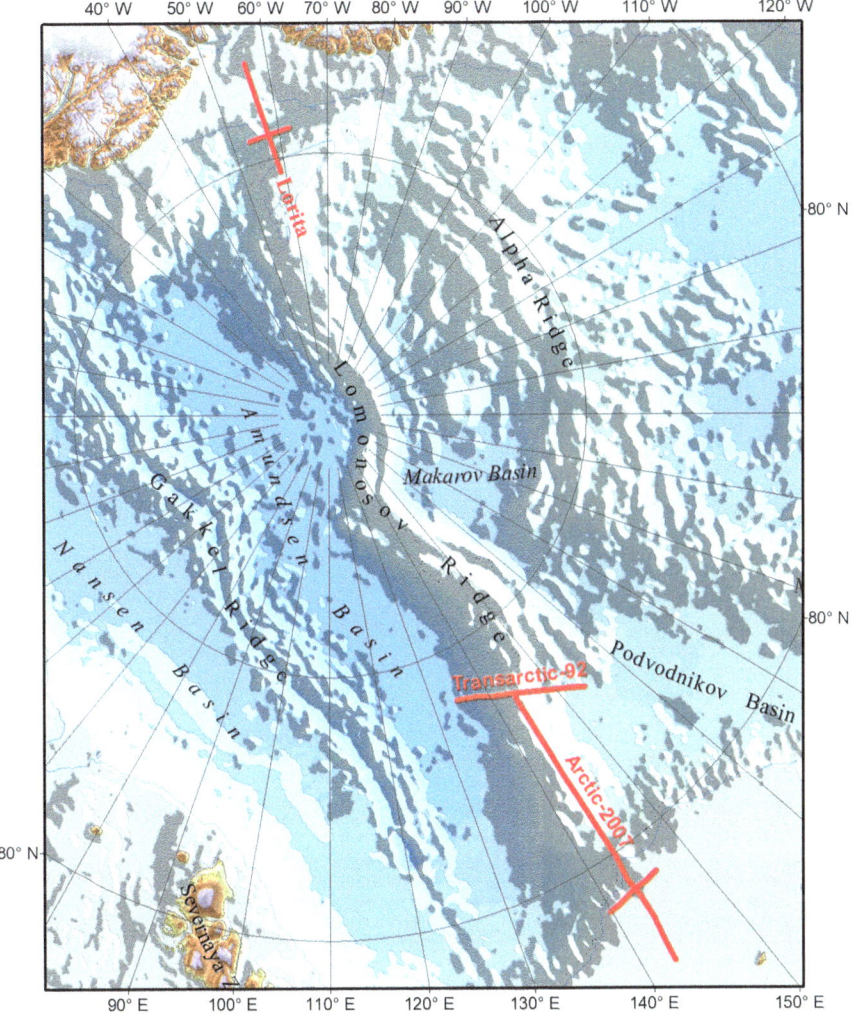

Fig. 4.18 Deep Seismic soundings at the Lonosov Ridge

mation support our notion of the Lomonosov Ridge basement as an agglomerate of Caledonian fold belts fragments.

4.5   Crustal Structure

Crustal structure of the Lomonosov Ridge was investigated by Deep Seismic Sounding (DSS) by Russian expeditions "TransARCTIC-1992" and "ARCTIC-2007" in near-Siberian sector and by Danish-Canadian - in the near-Greenland sector ("LORITA-2006") - Fig. 4.18. The resulting crustal velocity

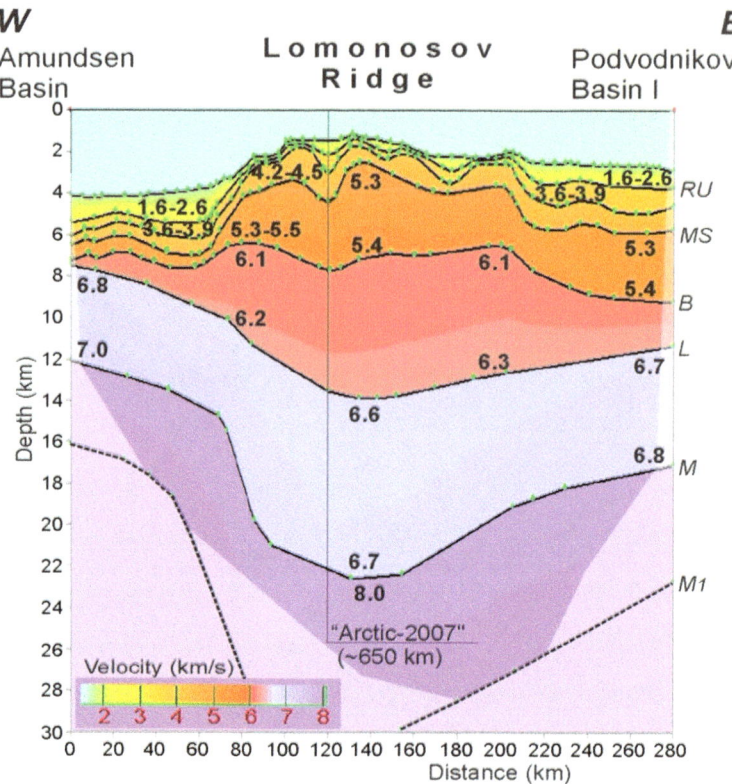

Fig. 4.19 "TransARCTIC-1992" DSS velocity model (P-wave, km/sec) (Poselov et al. 2014): RU — regional pre-Moicene unconformity; MS — top of meta-sedimentary complex; B, L - top of upper and lower crust, accordingly; M-Moho; M1- inner mantle interface

models for "TransARCTIC-1992" and "ARCTIC-2007" and "ARCTIC-2007" seismic programs are shown on Figs. 4.19, 4.20.

The major crustal units on this models are described as follows (Poselov et al. 2014):

– Three sedimentary complexes between RU (pre-Miocene) and pCU (post-campanian) unconformities with Vp 1.6–2.6 km/sec (upper), 3.6–3.9 km/sec (middle) and 4,2–4,5 km/sec (lower) with total thickness increasing from 1.5 km on the ridge to 2–2.5 km in adjacent basins;
– Meta-sedimentary complex (MS) below acoustic basemen with Vp 5.3–5.5 km/sec and maximum thickness ≈ 5 km in the axial part of the ridge (it drops to 4–4.5 km towards the Podvodnikov Basin and close to 1.5 km - on the Amundsen Basin slope).
– Upper crust with Vp 6.1–6.3 km/sec; its thickness reaches 6–7 km under the central Lomonosov Ridge. It thins to 2–2.5 km towards the Podvodnikov Basin and almost completely pinches out (or thins beyond the survey resolution) on the Amundsen Basin slope.

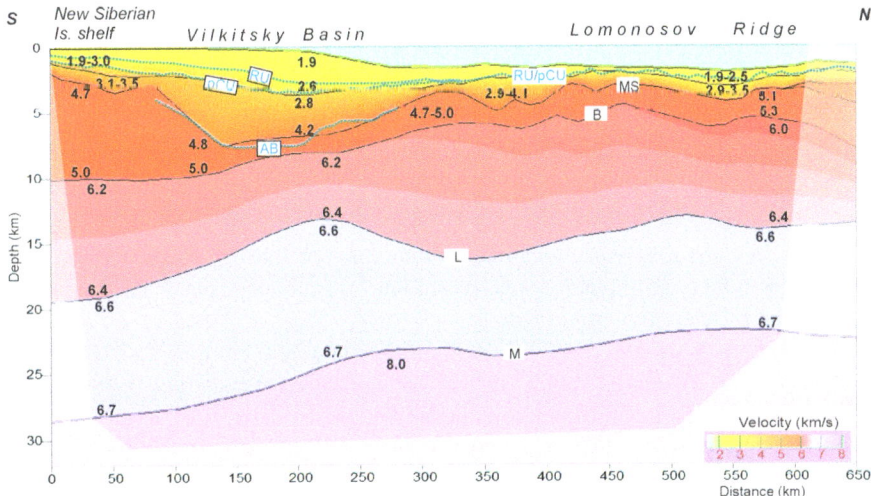

Fig. 4.20 "ARCTIC-2007" DSS velocity model (P-wave, km/sec) (Poselov et al. 2014): RU — regional pre-Moicene unconformity; pCU — post-Campanian unconformity; AB — acoustic basement; MS — top of meta-sedimentary complex; B, L — top of upper and lower crust, accordingly; M-Moho; blue dotted lines— major unconformities from depth converted MCS A-7 seismic line

– Lower crust with indirectly determined Vp (by PmP waves) around 6.7 km/sec (6.7–6.8 km sec on Podvodnikov slope, 6.8–7.0 km/sec - on Amundsen slope, all derived from refraction velocities); thickness varies from ≈ 8 km in the axial part to ≈ 6 km on Podvodnikov slope and ≈ 4 km – on Amundsen slope.

– Upper mantle with Vp 8 km/sec, reliably determined only on the slopes of the Lomonosov Ridge and only approximately estimated from PmP hodographs at the axial part; Moho plunges from ≈ 12–17 km under the slopes to ≈ 22 km under the axis of the Lomonosov Ridge. Pm1P synthetic wave pointed to existence of inner-mantle boundary with velocity 8.2 km/sec plunging steeply under the Lomonosov Ridge from ≈ 23 on the Podvodnikov Basin slope and from ≈ 15 km on Amundsen Basin slope.

In this model (Poselov et al. 2014) we see:

– Two sedimentary complexes divided by unconformity RU + pCU (within the Lomonosov Ridge proper, regional pre-Miocene and post-Campanian unconformities merge, at the scale of depth model, into one): upper with Vp 1.9–2.6 km/sec on shelf and 1.9–2.5 km/sec within the Lomonosov Ridge, lower - with 3.1–3.5 km/sec on shelf, 2.8–4.2 km/sec in the Vilkitsky Trough and 2.6–3.8 km/sec within the Lomonosov Ridge; total thickness drops from ≈ 7 km to ≈ 3 km, or less, in the Lomonosov Ridge;

– Meta-sedimentary complex (MS) with velocities changing laterally from 4.7–5.0 km/sec on shelf to 5.1–5.3 km/sec on the Lomonosov Riridge and thickness - from ≈ 7 km on shelf, to ≈ 1.5 km in the Vilkitsky Trough and 2–3 km in the Lomonosov Ridge;

– Upper crust, 6–9 km thick, with Vp 6.0–6.4 km/sec;
– Lower crust with velocities (mostly derived from fragments of PmP waves) of no
  more than 6.7 km/sec, 9–12 km thick; total thickness of crystalline crust lies
  between 16–18 km for the whole section length;
– Upper mantle velocity of 8.0 km/sec was extracted from rare examples of Pn
  waves records and fragments of PmP waves; Moho raises from 28 km on shelf to
  22–23 km under the Vilkitsky Trough and the Lomonosov Ridge.

These models clearly demonstrate continuation of the Cretaceous- Cenozoic
sedimentary, meta-sedimentary and crystalline crust complexes from the outer
Laptev Shelf to the Lomonosov Ridge.

Information from DSS ARCTIC-2007 profile about 600 km long and MCS and
DSS surveys by MAGE and other organizations was used to construct an integrated
3D seismic-density model of the Lomonosov Ridge.

Free Air gravity field input consisted of 10 × 10 km grid covering are a 360 ×
950 km and actual gravity modeling process was run along computational profiles
30 km apart using software Grav-3D (Piskarev and Tchernyshev 1997). Iterative
adjustment of density contrasts and geometry of causative boundaries continued
until calculated and observed gravity merged within 2–3 mGal (Piskarev and Savin
2010). Fig. 4.21 illustrates the resulting crustal density model along the central

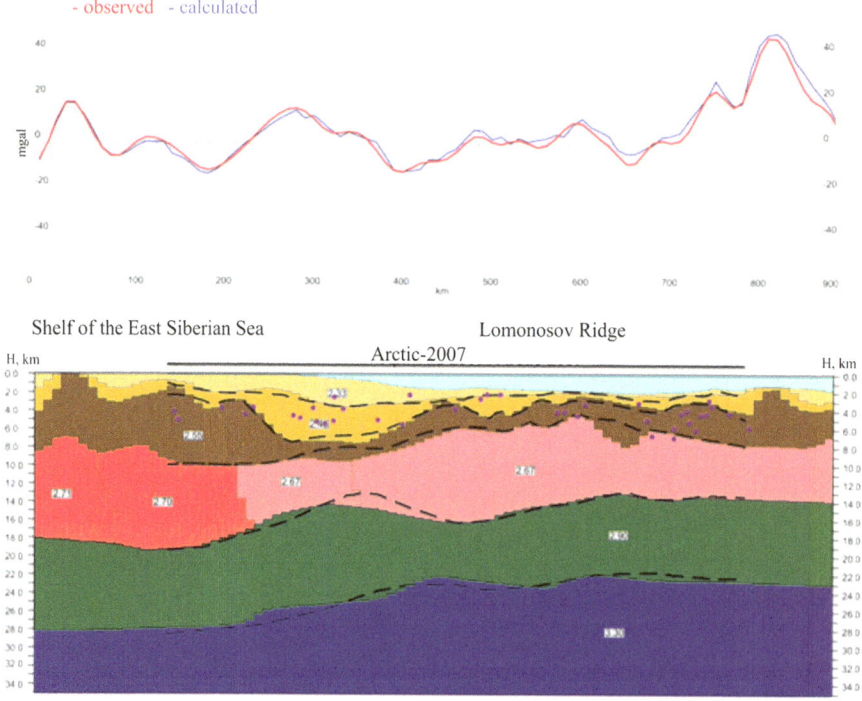

Fig. 4.21 Crustal model along DSS Arktika −2007. Red dots- calculated tops of magnetic sources;
dotted lines- interpreted seismic boundaries from seismic line ARCTIC -2007

computational profile. This profile was selected because it runs along the seismic line ARCTIC-2007 and interpreted seismic boundaries from this section were overlaid on gravity model (dotted lines).

This model clearly shows the continuation of geological boundaries and integrity of the crustal structure from the shelf to the Lomonosov Ridge with total crust thickness gradually changing from 28 km to 20–22 km in that direction.

Based on integrated seismic interpretation and core data from wells drilled at the Lomonosov Ridge (Backman et al. 2006; Kim and Glezer 2007), sedimentary section with maximum thickness of about 7 km at the shelf edge, was subdivided into upper (Neogene-Pleistocene, Vp 1.7–2.4 km/sec, density 2.33 g/cm^3) and lower (Late Cretaceous – Oligocene, Vp 3.2–4.6 km/sec, density 2.33 g/cm^3) parts, divided by regional unconformity.

The sedimentary section is underlain by acoustic basement with Vp 4.4–5.2 and computational density 2.55 g/cm^3, composed, most probably, by folded Mesozoic formations. Both sedimentary and acoustic basement formations can be traced through the entire length of the Lomonosov Ridge.

The crystalline crust, between acoustic basement and Moho, is divided into upper crust (Vp 6–6.4 km/sec, density 2.67–2.71 g/cm^3)· and lower (Vp 6.7–6.9 km/sec, density 2.9 g/cm^3).

Similar seismic-density crustal model also was created for MCS line 2014–07 (Fig. 3.1 b). At this location the axial part of the Lomonosov Ridge, 2–3 km higher than then adjacent Amundsen and Makarov Basins, is marked by 90 mGal-strong positive gravity anomaly. Sedimentary section here is 1–4 km thick, with density 2.0–2.3 g/cm^3. Consolidated crystalline crust is 22 km thick, with 9–15 km of upper crust (density 2.7 g/cm^3) and 5–6 km of lower (2.9 g/cm^3).

Seismic crustal velocity model of Greenland-Ellesmere Shelf (inner segment) and near-Greenland portion of the transitional zone and the Lomonosov Ridge (outer segment) along Danish-Canadian DSS line LORITA-2006 is presented in Fig. 4.22 (Jackson et al. 2010).

The sedimentary cover with maximal thickness up to 12 km was subdivided into three complexes (2.1–2.2 km/sec, 3.1–3.2 km/sec and 4.3–5.2 km sec) with the lowest one identified as Palesoic-Mesosoic (base of the second complex is marked by regional unconformity). Underlying complex with Pv 5.9–6.5 km/sec is equated with older meta-sediments. All four units continue uninterrupted from shelf to bathyal portion of the Lomonosov Ridge.

On the other section of the LORITA-2006 program, orthogonal to the illustrated above, meta-sedimentary complex is shallower and underlain by horizons with higher (6.2–6.7 km/sec) velocities. Radial magnetic anomalies within the Lomonosov Ridge proper are interpreted as having volcanic origin (Jackson et al. 2010); Moho depth varies from 17 to 27 km.

In general, the crustal velocity models of near Siberian and near-Greenland segments of the Lomonosov Ridge derived from different sources by the Russian, Canadaian and Danish teams are in good agreement by all key parameters and lead to conclusion that the Lomonosov Ridge is a block of typical continental crust in rifting environment.

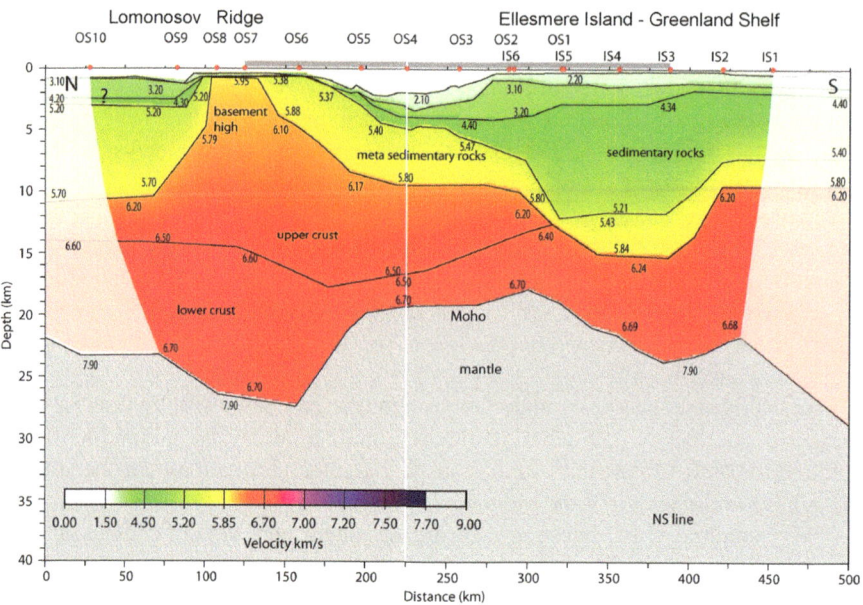

Fig. 4.22 Crustal velovity model along LORITA-2006. (Jackson et al. 2010)

It was also noted that regions of the Lomonosov Ridge-Eurasian Shelf junction are practically a-seismic – see Fig. 1.29. No contemporary seismic activity was recorded not only at the ridge-shelf junction, but also further east along the East-Siberian Sea continental slope. This unequivocally points out on absence of any movements of the Lomonosov Ridge in relation to the shelf.

But anothers question – were there any such movements in the past? – remains. If we answer this question affirmatively, we also have to admit that seismicaly active strike-slip, or transform, fault should exists at the ridge-shelf junction. Given the fact that the Eurasian Basin is still spreading, it is difficult to understand the cessation of tectonic movements and subsequent seismic activity along such fault's plane. The hypothesis of plate boundary jumping from abyssal areas to the Laptev shelf does not solve the problem of modern a-seismicity – earthquakes would continue, but as an intra-plate. If modern earthquakes are recorded in insignificant fault zones far removed from the spreading center, they should've been generated in large fracture zone of significant high-order fault. Another contradiction was noted in (Poselov et al. 2012): if the Lomonosov Ridge moves to the east relatively to other parts of the Eurasian Basin, the collision zones would form on its Eurasian slope, instead of abyssal basins. Therefore, the seismology confirms the unity of the Lomonosov Ridge and the adjacent shelf.

The general conclusions regarding structure of the Lomonosov Ridge are as follows:

1. The Lomonosov Ridge basement is thought to consist of Baikal - Caledonian fold belts with possible Cimmerians at the near-Siberian part. Both systems are outcropping at the adjacent continental and insular land masses in the periphery of the Arctic Basin.
2. The basement is overlain by meta-sedimentary acoustic basement, transitional by its geophysical parameters. It may include a variety of geological structures – slightly metamorphic Caledonides on North American side, or strongly consolidated and sometimes gently deformed Paleosoic (?) cratonic formations - on Siberian.
3. Sedimentary cover includes lower (syn-rift) Mesozoic and blanketing upper (Cenozoic) complexes. Apparent variations of their thickness notwithstanding, these formations can be confidently traced through the whole length of the Lomonosov Ridge and neighboring shelves of Greenland-Ellesmerian and Siberian margins. Towards the Podvodnikov Basin, a wedge of Middle - Late Eocene-Oligocene (?) deposits appears under the Miocene horizons; Cretaceous sediments may be possibly present in numerous grabens, visible here and at the Eurasian slope of the Ridge on seismic sections below Paleocene-Eocene deposits.
4. During Cenosoic, the Lomonosov Ridge, as a part of the Amerasian Plate and moving with it, maintained its fixed position among relative to the assemblage of the Amerasian Basin structures, including the East-Asian continental massif.
5. The crust within the Lomonosov Ridge proper is 20–22 km thick, with aproximately constant upper-lower crust thickness ratio.

According to seismic stratigraphy analysis and deep water drilling (ACEX-302), the sedimentary cover of the Arctic Ocean consists of unconsolidated terrigenous (mostly clays and aleuritic silts) sediments accumulated from Late Cretaceous (at places, from Apt-Alb) to Holocene. Drilling shows that no oceanic sediments were deposited on the Lomonosov Ridge until Miocene and that from Late Cretaceous to Early Eocene it was an area of shallow water pools and subtropical climate. Sizable areas of chemical weathering within Amerasian Basin shelf margins also points on prevailing lacustrine and/or littoral conditions during Paleogene. The non-depositional and erosional period lasted from Late Eocene to late Oligocene; Mid–Miocene marks the Lomonosov Ridge subsidence and appearance of the first ice sheets.

References

Backman J, Moran K (2009) Expanding the Cenozoic paleoceanographic record in the Central Arctic Ocean: IODP expedition 302 synthesis. Cent Eur J Geosci 1(2):157–175
Backman J, Moran K et al (2006) Sites M0001-M0004. Expedition 302 Scientists. Proceedings of the integrated ocean drilling program 302, p 169

Backman J, Jakobsson M, Frank M et al (2008) Age model and core-seismic integration for the Cenozoic Arctic coring expedition sediments from the Lomonosov ridge. Paleoceanography 23(1)

Brozena JM, Childers VA, Lawver LA et al (2003) New aerogeophysical study of the Eurasia Basin and Lomonosov ridge: implications for basin development. Geology 31(9):825–828

Bruvoll V, Kristoffersen Y, Coakley B et al (2010) Hemipelagic deposits on the Mendeleev and northwestern Alpha submarine ridges an the Arctic Ocean: acoustic stratigraphy, depositional environment and an inter-ridge correlation calibrated by ACEX results. Mar Geophys Res 31:149–171

Coakley BJ, Cochran JR (1998) Gravity evidence of very thin crust at the Gakkel ridge (Arctic Ocean). Earth Planet Sci Lett 162:81–95

Cochran JR, Kurras GJ, Edwards MH et al (2003) The Gakkel ridge: bathymetry, gravity anomalies, and crustal accretion at extremely slow spreading rates. J Geophys 108:2116–2137

Cochran JR, Edwards MH, Coakley BJ (2006) Morphology and structure of the Lomonosov ridge, Arctic Ocean. Geochem Geophys Geosyst 7(5):1525–2027

Døssing A, Hansen TM, Olesen AV et al (2014) Gravity inversion predicts the nature of the Amundsen basin and its continental borderlands near Greenland. Earth Planet Sci Lett 408:132–145

Franke D, Hinz K, Oncken O (2001) The Laptev Sea rift. Mar Pet Geol 18:1083–1127

Glebovsky VY, Verba VV, Kaminsky VD (2008) Potential fields of the Arctic Basin: history of investigations, older analog and modern digital compilations. In: 60 years in the Arctic, Antarctic and World Ocean. VNIIOkeangeologia, Saint Petersburg, pp 93–110

Grantz A, Pease VL, Willard DA et al (2001) Bedrock cores from 89° North: implications for the geologic framework and Neogene paleooceanology of the Lomonosov ridge and a tie to the Barents shelf. Geol Soc Am Bull 113(10):1272–1284

Grantz A, Scott RA, Drachev SS et al (2009) Map showing the sedimentary successions of the Arctic region (58°-64° to 90°N) that may be prospective for hydrocarbons. American Association of Petroleum Geologists GIS-UDRIL Open-File Spatial Library, Tulsa

Jackson HR, Dahl-Jensen T, the LORITA working group (2010) Sedimentary and crustal structure from the Ellesmere Island and Greenland continental shelves onto the Lomonosov ridge, Arctic Ocean. Geophys J Int 182(1):11–35

Jokat W (2003) Seismic investigations along the western sector of alpha ridge, Central Arctic Ocean. Geophys J Int 152:185–201

Jokat W (2005) The sedimentary structure of the Lomonosov ridge between 88°N and 80°N. Geophys J Int 163:698–726

Jokat W, Kristoffersen Y, Rasmussen TM et al (1992) ARCTIC 91:Lomonosov ridge – a double-sided continental margin. Geology 20:887–890

Jokat W, Ickrath M, O'Connor J (2013) Seismic transect across the Lomonosov and Mendeleev ridges: constraints on the geological evolution of the Amerasia Basin, Arctic Ocean. Geophys Res Lett 40(19):5047–5051

Kim BI, Glezer ZI (2007) Sedimentary cover of the Lomonosov ridge (stratigraphy, the history of the cover and structure formation, age dating of seismic units). Stratigr Geol Correl 15(4):63–83

Langinen A, Lebedeva-Ivanova N, Gee D et al (2009) Correlations between the Lomonosov ridge, Marvin spur and adjacent basins of the Arctic Ocean based on seismic data. Tectonophysics 472:309–322

Moran K, Backman J, Brinkhuis H et al (2006) The Cenozoic palaeoenvironment of the Arctic Ocean. Nature 441:601–606

Piskarev AL (2004) The basement structure of the Eurasia Basin and central ridges in the Arctic Ocean. Geotektonics 38(6):443–448

Piskarev AL, Savin VA (2010) Gravitational simulation of Earth's crust in Lomonosov range. Karotazhnik 9(198):41–54

Piskarev AL, Tchernyshev M (1997) Magnetic and gravity anomaly patterns related to hydrocarbon fields in northern West Siberia. Geophysics 62(3):831–841

Poirier A, Hillaire-Marcel C (2011) Improved Os-isotope stratigraphy of the Arctic Ocean. Geophys Res Lett 38(14):L14607

Poselov VA, Avetisov GP (ed) (2014) Russian Arctic Geotransects (results of geological and geophysical studies) FSUE "I.S.Gramberg VNIIOkeangeologia", St. Petersburg

Poselov VA, Avetisov GP, Butsenko VV et al (2012) The Lomonosov ridge as a natural extension of the Eurasian continental margin into the Arctic Basin. Geol Geophys 53(12):1662–1680

Poselov VA, Butsenko VV, Chernykh AA et al (2014) The structural integrity of the Lomonosov Ridge with the North American and Siberian continental margins. VI International conference on Arctic Margins (ICAM VI), Fairbanks, Alaska, 30 May–2 June 2011

Rekant PV, Gusev EA (2012) Seismic geological structure model for the sedimentary cover of the Laptev Sea part of the Lomonosov ridge and adjacent parts of the Amundsen plain and Podvodnikov Basin. Geol Geophys 53(11):1497–1512

Saltus RW, Miller EL, Gaina C (2011) Regional magnetic domains of the circum-Arctic: a framework for geodynamic interpretation. In: Spencer AM (ed) Arctic petroleum geology, vol 35. Geological Society of London, London, pp 49–60

Sauermilch I, Jokat W (2015) New multichannel seismic reflection data along the eastern part of Lomonosov ridge, Arctic Ocean European Geosciences Union General Assembly 2015, Vienna, 12–17 April 2015

Vogt PR, Taylor PT, Kovacs LC et al (1982) The Canada Basin: aeromagnetic constraints on structure and evolution. Tectonophysics 89:295–336

Weber JR, Sweeney JF (1990) Ridges and basins in the Central Arctic Ocean. In: The Arctic Ocean Region Vol. L. The Geological Society of America, Boulder, pp 305–336

Zinchenko AG (2004) Orographic zoning and general characterization of seafloor topography. In: Geology and mineral deposits of Russia. VSEGEI, Saint Petersburg 5(1)15–25

Chapter 5
Podvodnikov Basin

Oleg E. Smirnov, Victor V. Butsenko, Yury G. Firsov, Vladimir Yu. Glebovsky,
Evgeny A. Gusev, Valery D. Kaminsky, Gennady S. Kazanin,
Alexey L. Piskarev, and Victor A. Poselov

Abstract The vast bathymetric depression of the Podvodnikov Basin lies between the Lomonosov Ridge and Mendeleev - Alpha Ridges, bounded by the East-Siberian Sea shelf from the south and by the Makarov Basin - from the north. Sea depth in the Podvodnikov Basin varies from 800 m to 2700 m in the same direction. The Podvodnikov Basin has a benched structure and is considered to be a part of the terraced continental slope.

The seismic data divide the southern terrace of the Podvodnikov Basin into western and eastern parts, separated by the prominent Geofizikov Spur and its southern continuation as a deeper structure of the acoustic basement. Total thickness of the sedimentary cover here reaches 8 km.

The sedimentary cover of the Podvodnikov Basin is subdivided into six seismic-stratigraphic complexes, dated from late Permian to Holocene by correlation with ACEX well on the Lomonosov Ridge and exploration wells on the Alaska shelf. Integrated geophysical data interpretation and modeling classify the Basin's crust as stretched continental, with thickness of the crystalline crust from 10 to 23 km.

Keywords Podvodnikov Basin · Geofizikov Spur · Bathymetric depression · Extension structure · Sedimentary basin · Extended continental crust

O. E. Smirnov (✉) · V. V. Butsenko · Y. G. Firsov · V. Y. Glebovsky · E. A. Gusev
V. D. Kaminsky · V. A. Poselov
All-Russian Research Institute of Geology and Mineral Resources of the World Ocean
(VNIIOkeangeologia), Saint Petersburg, Russia
e-mail: vicb@vniio.nw.ru; gleb@vniio.nw.ru; okeangeo@vniio.ru

G. S. Kazanin
Marine Arctic Geological Expedition, Murmansk, Russia
e-mail: info@mage.ru

A. L. Piskarev
All-Russian Research Institute of Geology and Mineral Resources of the World Ocean
(VNIIOkeangeologia), Saint Petersburg, Russia

Saint Petersburg University, Saint Petersburg, Russia

© Springer International Publishing AG, part of Springer Nature 2019 187
A. Piskarev et al. (eds.), *Geologic Structures of the Arctic Basin*,
https://doi.org/10.1007/978-3-319-77742-9_5

5.1   Morphology

The "Podvodnikov Basin" toponym is applied to the large bathymetric depression south from 85°N between the Lomonosov Ridge and Mendeleev - Alpha Ridges (Fig. 5.1); the name "Makarov Basin" is reserved for the northern abyssal enclave between near-polar segment of the Lomonosov and Alpha Ridges. Other variants of toponymic used in (http://www.ngdc.noaa.gov/gazetteer/; Orographic Map of the Arctic Basin 1995; Central Arctic Basin, the Map 2002; Geomorphological Aspects of the Russian Continental Shelf Exterior Boundary in the Arctic 2005; Kaminsky et al. 2014), are disregarded hereby.

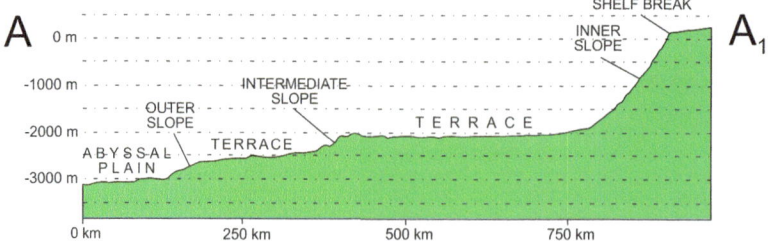

Fig. 5.1   The Podvodnikov Basin. Top – 3D rendition (grid IBCAO v. 3.0), looking south; bottom - bathymetry along track A$_1$-A

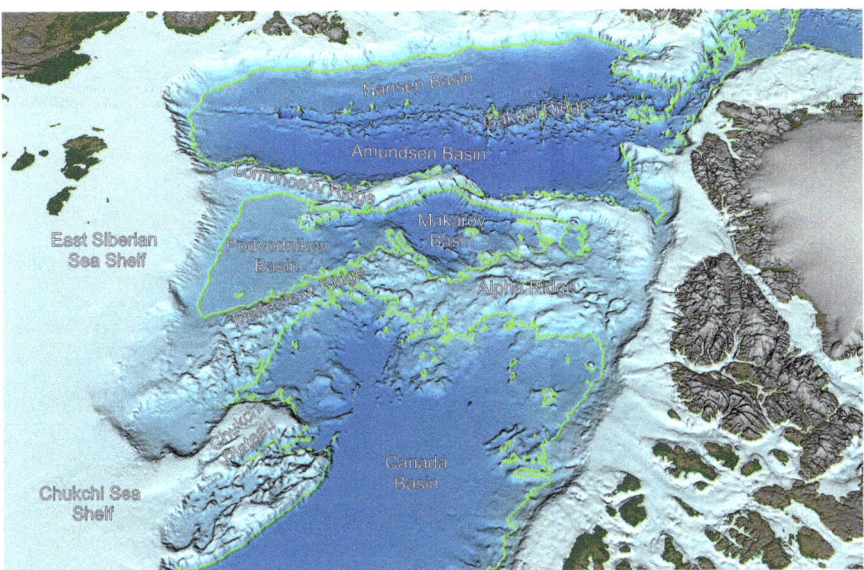

Fig. 5.2 The seafloor topography of the Arctic Basin (grid IBCAO v. 3.0, GeoCap processing); Green line marks 2500 m isobath

The international grid IBCAO with cell sizes 2 × 2 km or 0,5 × 0,5 km and multibeam data processed with GeoCap software gave adequate representation of the regional morphology (Fig. 5.2). In general, bathymetry and underwater topography of the Podvodnikov Basin differ from that of the Canada, Amundsen and Nansen Basins with their oceanic seafloor.

The Podvodnikov Basin has roughly triangular shape with 600 km-long base stretching parallel to the edge of the East Siberian Sea shelf and vertex – 650 km further north. Being a part of the submerged continental margin, the Podvodnikov Basin displays typical features: slight, hardly perceptible, tilt of its floor to the north and terraced structure. There are two terraces: larger southern, 2600–2800 m b.s.l., bounded from the south by the inner slope, and smaller northern, 3100–3200 m deep, separated by intermediate slope about 400 m high. The lower terrace, in turn, is terminated by the outer slope visible at the narrow neck where the Mendeleev and Alpha Ridges getting close to each other, marking the border between the Podvodnikov Basin and the abyssal (3800–4000 m) Makarov Basin. (Fig. 5.1, 5.3).

Pictures above clearly illustrate the difference between eastern an western sides of the basin. The eastern side of the southern terrace in contact with the Mendeleev Ridge can be classified as an ordinary slope with vertical drop decreasing from 600 m in the south to 200 m in the north. The eastern slope of the northern terrace is similar to its southern counterpart, except being taller – from 400–1000 m, with higher elevation in the center.

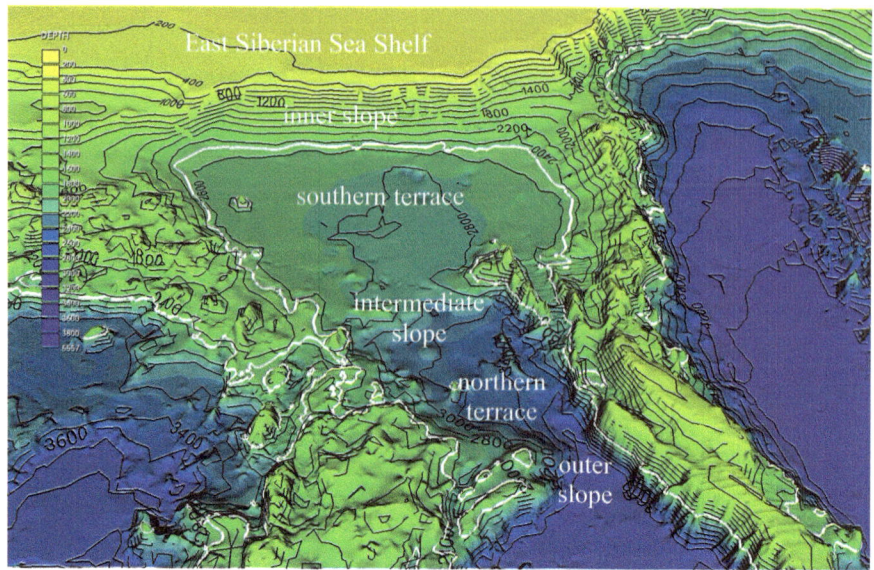

Fig. 5.3 Terraced morphology of the Podvoodnikov Basin (IBCAO grid processed by GeoCap)

The western side, at the basin junction with the Lomonosov Ridge, can be best described as a combination of the contrasting morphological forms – the Geofizikov and Senchura Spurs and similar unnamed feature along the edge of the northern terrace (Fig. 1.8). These ranges, divided by linear depressions, are considered to be a splinter-blocks of the Amerasian slope of the Lomonosov Ridge, tectonically separated from it. Geophysical data suggest that these blocks, buried under the sediments, continue into the Podvodnikov Basin. Therefore, being a part of the Lomonosov Ridge, they also shape the structure of the Podvodnikov Basin, thus confirming their genetic ties – they both suffered a neo-tectonic subsidence during formation of the Podvodnikov Basin.

Two GeoCap generated bathymetry profiles using IBCAO grid are shown on Fig. 5.4 and 5.5. The first one (Canada Basin – Mendeleev Ridge – Lomonosov Ridge – Amundsen Basin) demonstrates difference in depth of the Podvodnikov and Canada Basins (on the left) and Amundsen Basins (on the right).

The second profile (Fig. 5.5) runs north from the East-Siberian Sea shelf through Podvodnikov and Makarov Basins close to DSS traverse "Transarktika-1989–91" 1500 km long. It convincingly illustrates terraced morphology of the Podvodnikov Basin.

In light of the presented information we consider the Podvodnikov Basin assembly of the inner slope, southern terrace, intermediate slope, northern terrace and outer slope as a part of the continental slope of the corresponding continental margin.

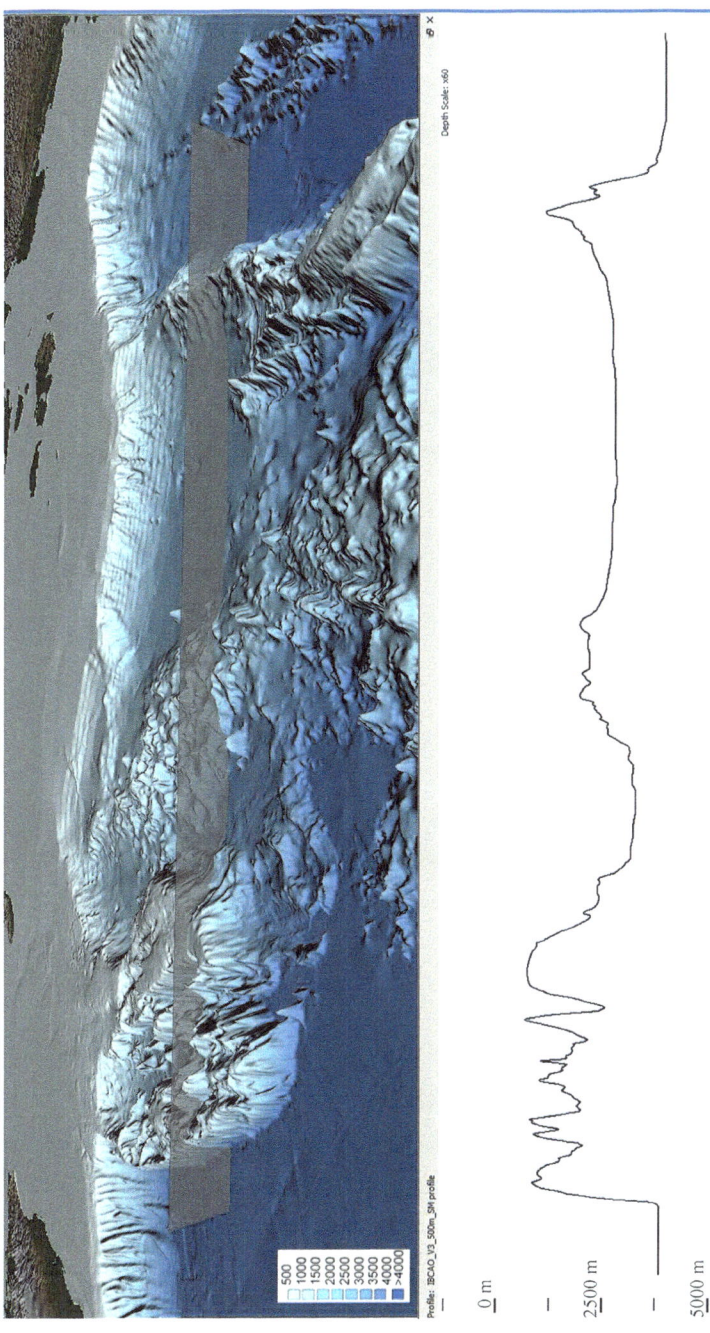

Fig. 5.4 Bathymetry profile Canada Basin – Mendeleev Ridge – Lomonosov Ridge – Amundsen Basin. From left to right: Canada Basin – Chukchi Plateau – Chukchi Basin –Mendeleev Ridge – Podvodnikov Basin – Lomonosov Ridge – Amundsen Basin

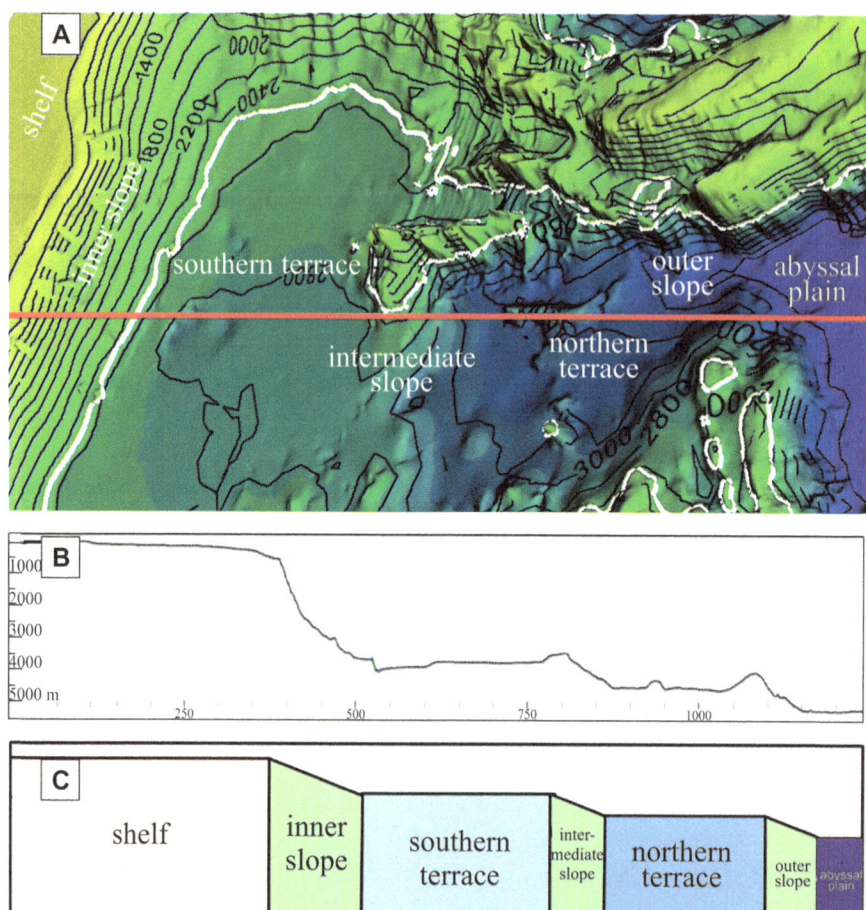

Fig. 5.5 Bathymetry of the Podvodnikov Basin
A — profile location; B-actual bathymetry; C — geomorphological interpretation

5.2 Potential Fields

As good regional indicators of composition and structure of geologically different regions, potential fields anomalies help identify the following crustal blocks inside the Podvodnikov Basin, surrounding uplifts and shelf structures in East-Siberian and Chukchi Seas (Fig. 5.6):

2d — The De Long Uplift - strong linear and circular (following the uplift shape) gravity/magnetic anomalies related to mafic Aptian-Albian (occasionally, Paleozoic and Cenozoic) magmatizm and relatively shallow, intensively faulted pre-Cambrian (most probably Baikalian) basement;

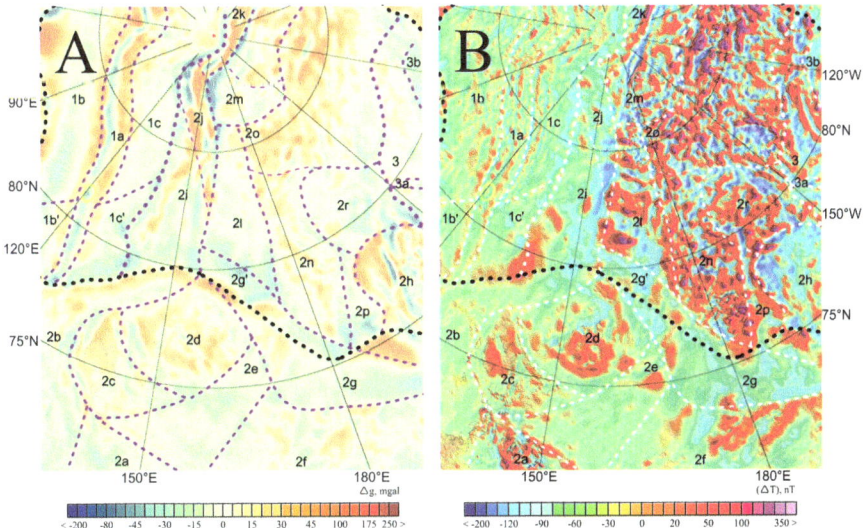

Fig. 5.6 Regionalization schematics of Free Air gravity (**A**) and magnetic (**B**) anomalies for the Central Arctic Uplifts and adjacent shelf. Black dots – deepwater basin boundary; dashed lines - crustal blocks borders (see text above)

2e —The Kolyma-Mendeleev Zone – wide north-east trending zone crossing East-Siberian Sea shelf form Kolyma Massif through the New Siberian depression to the Mendeleev Ridge, with weak gravity, and very weak magnetic, anomalies following trend.

2f — The East Siberian Uplift and the Wrangel Uplift – north-east and north-west trending anomalies of moderate strength, presumably related to folded and thrusted structures documented on the Wrangel Island;

2g — North-Chukchi Zone – clearly outlined by contrasting gravity and magnetic anomalies following southern (strong gravity and magnetic fields as well as seismic data) and northern (strong gravity, weak magnetic, anomalies) border of the North-Chukchi trough.

2l — Podvodnikov Basin – a complex assembly of Free Air gravity anomalies with gradually increasing amplitudes northward from 12 to 50 mGal and sub-linear magnetic anomalies trending north-west. These anomalies were interpreted as Cretaceous spreading-type linear anomalies (Gurevich and Mashenkova 2000), or as indicators of mafic magmatism related to continental rifting (Glebovsky et al. 1998) (numerical interpretation indicates mafic and ultra-mafic composition of the sources of the strong magnetic anomalies in the central parts of the region and seems to support the latter version); some positive magnetic anomalies correspond with basement uplifts expressed in seismic data, such as Geofizikov Spur.

Fig. 5.7 illustrates how superposition of all four major sources of information – bathymetry, gravity, magnetic and seismic – supplements and enhances interpretation of seismic data.

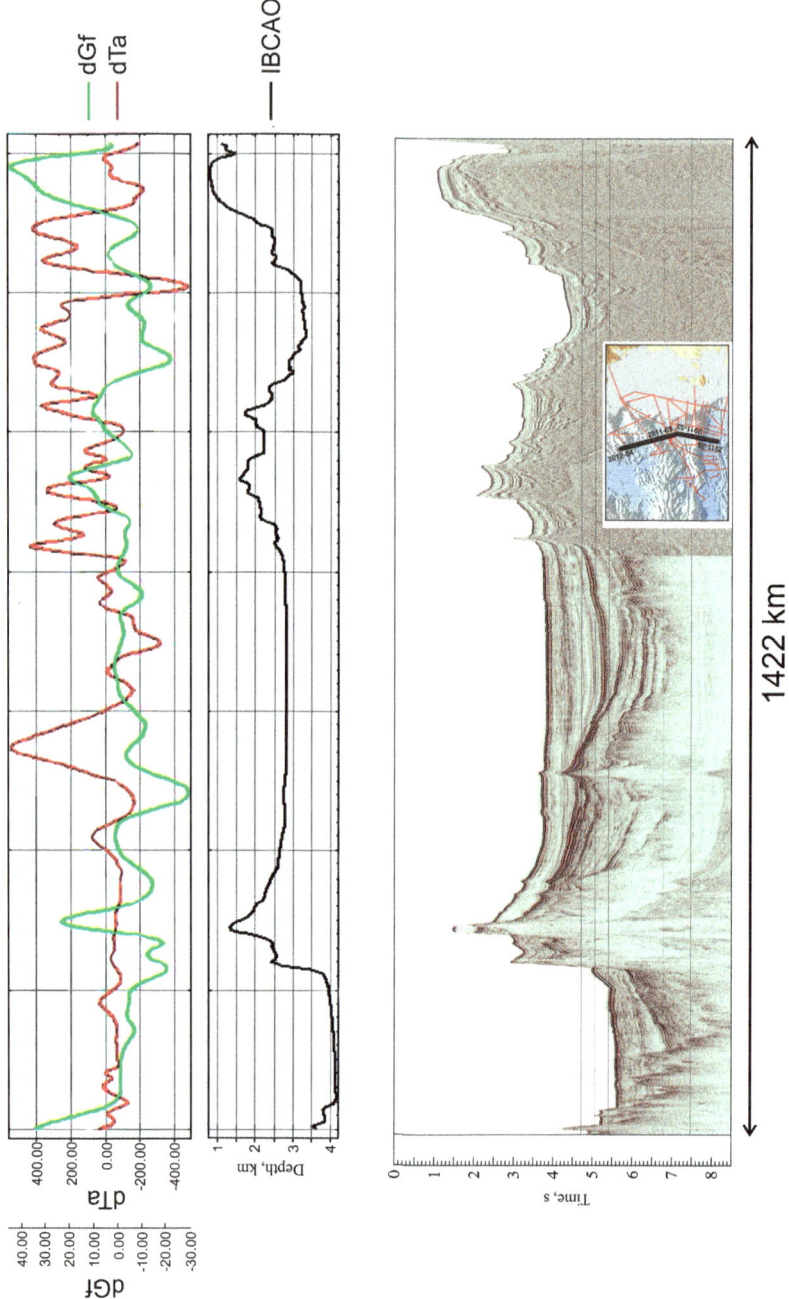

Fig. 5.7 Composite MCS line (2011–2012) from the Gakkel Ridge to the Chukchi Plateau; upper panel - Free Air gravity (dGf) and magnetics (dTa); middle panel – bathymetry (IBCAO grid)

5.3 Acoustic Basement

Several MCS lines illustrate differences in acoustic basement subsurface topography and structure within the Podvodnikov and Amundsen Basins (Fig. 5.8—5.11). The contrast between seismic signatures of the continental acoustic basement of the Podvodnikov Basin and oceanic crust of the Amundsen Basin, with its strongly differentiated surface and absence of coherent reflections, is striking.

Geological and geophysical data identify the De Long Uplift (north-west of the East-Siberian Sea, Fig. 5.6, 2d) as fragment of an ancient platform with basement of Baikal-Caledonian fold belts and Paleozoic sedimentary (or volcano-sedimentary in reactivated terranes) cover. One of these Caledonian belts with numerous diorite-porfiry sills and dykes, basalt and basalt andesite lava flows outcrops on the Henrietta Island (Vinogradov et al. 2004) with K-Ar age between 310 and 450 Ma. There also are some signs of even older consolidation: fragments of methamorphic rocks (gneisses, schists and qaurtzites) in gravellites and sandstones, or un-deformed Cambrian and Ordovician deposits of the Bennet Island. Lacking direct evidence, the geology of the De Long Islands may represent, to a certain degree, the composition and structure of the entire Podvodnikov Basin basement.

5.4 Sedimentary Cover

As can be seen from the Fig. 5.12, the Podvodnikov Basin is reasonably well covered by the Russian MCS programs completed between 2011 and 2014. We also include in our compilation the MCS traverse PS-2008 across the Podvodnikov Basin acquired by RV Polastern in 2008 (Alfred Wegener Institute expedition) and discussed in (Jokat et al. 2013).

All approaches to developing and correlation of the stratigraphic models of the Podvodnikov Basin sedimentary cover are based on information from ACEX wells drilled on the Lomonosov Ridge (Moran et al. 2006; Backman et al. 2008). Using not only the ACEX well data, but also information from wells drilled on the Chuckhi Sea shelf, and Alaska shelf as it was comprehensively described in Chap. 2, we subdivided the Podvodnikov Basin sedimentary cover into six seismic-stratigraphic complexes (Fig. 5.13).

SSC-1 (Lower Miocene-Pleistocene, 18.2–20 Ma) — undisturbed hemipelagic sediments between seafloor and regional unconformity RU;

SSC-2 (Upper Paleocene - Lower Oligocene) - marine shallow formation between unconformities RU and post-Campanian pCU;

SSC-3 (Lower Brooks, Lower Cretaceous and, partially, Upper Cretaceous, K_1a-K_2, 90–80 Ma) – aleurites and sandstones deposited during the latest stages of the HALIP magmatizm;

SSC- 4 (Lower Cretaceous, K_1h-br, 130—120 Ma) - associated with the initial stage of the HALIP magmatizm;

SSC-5 (Upper Jurassic – Lower Cretaceous, J_3-K_1b-v);

SSC-6 (Upper Ellesmere,? P_3-J_2).

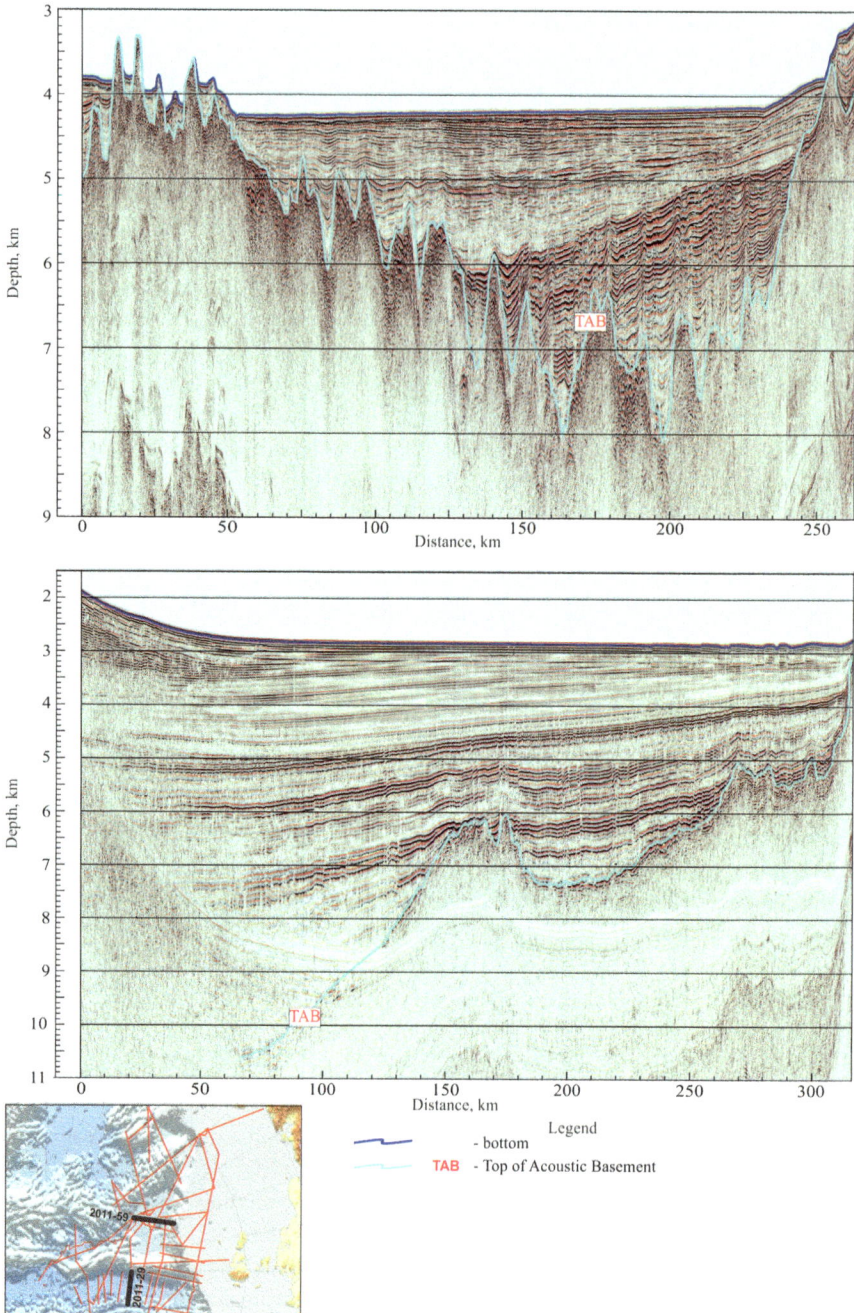

Fig. 5.8 Seismic signature of the Amundsen Basin oceanic acoustic basement (line 2011–29, upper panel) and continental acoustic basement of the Podvodnikov Basin (line 2011–59, lower panel)

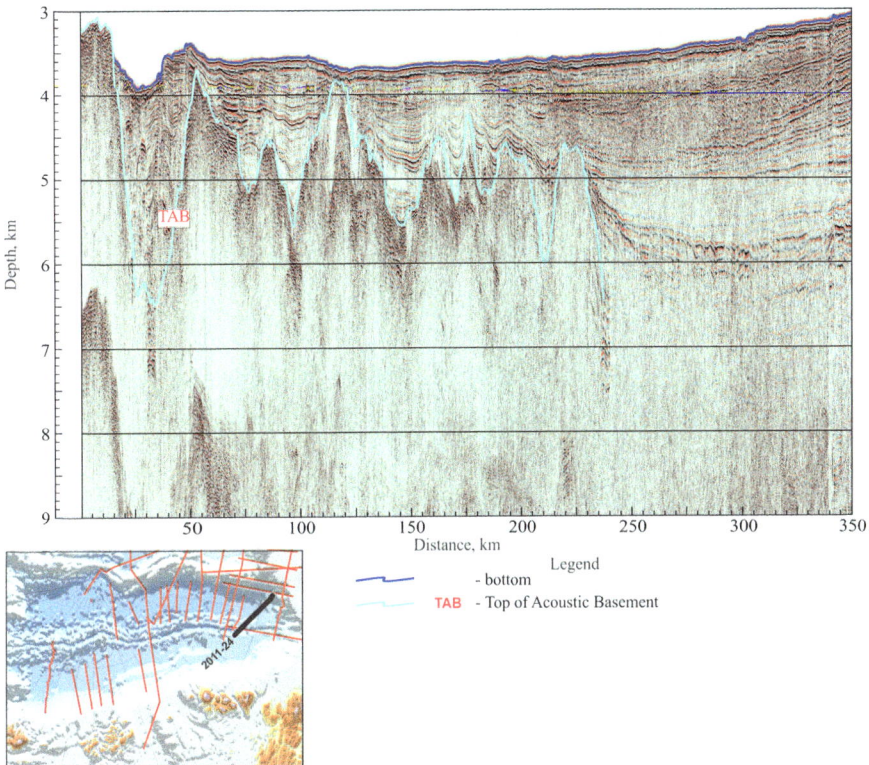

Fig. 5.9 MCS line 2011–24. Acoustic basement of the abyssal Amudsen Basin on the oceanic crust (note deep sediment-filled rift trough at the axial part of the Gakkel Ridge)

Several MCS lines crossing the Podvodnikov Basin – East-Siberian Sea shelf transition zone (Fig. 5.14, 5.15) reveal large basins with 7–8 km of sediments accumulated inside. According to (Nikishin et al. 2014) these structures are related to the East-Siberian Sea rifting in Early Cretaceous. The same age had been postulated for extension of the Podvodnikov Basin as a whole.

MCS lines crossing the Podvodnikov Basin – the Lomonosov Ridge boundary and the Geofizikov Spur, also show Cretaceous syn-rift sedimentary sequences at the base of the Basin. The same normal-fault block tectonics with down-throw towards the basin exists at the Podvodnikov Basin-Mendeleev Ridge junction.

Strong continuous reflections are typical for the seismic signature of the Podvodnikov Basin. They can be traced from the eastern flank of the Lomonosov Ridge (with minimal thickness of SSC-2 interval between merging RU and pCU unconformities) to the Mendeleev Ridge (Fig. 5.16).

Total thickness of the Podvodnikov Basin sedimentary section grows from 1000–1500 m on the Mendeleev and Alpha ridges edge to 6000 m in its central parts and drops back to 1000 m at the crest of the Lomonosov Ridge.

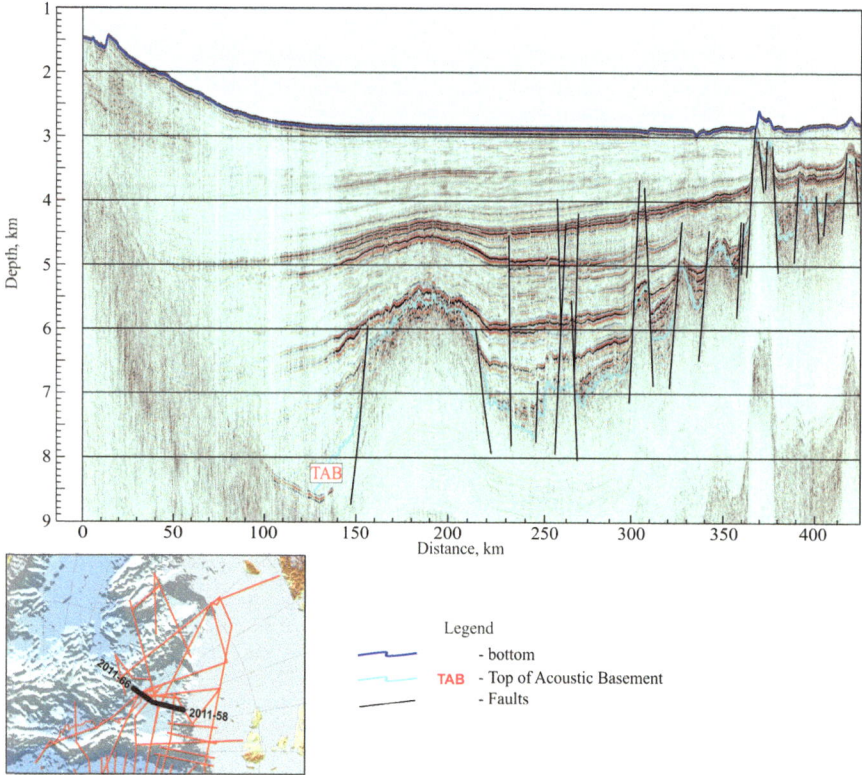

Fig. 5.10 MCS line 2011–58—66. Acoustic basement of the Podvodnikov Basin

The Geofizikov Spur and its southward continuation (as uplift of the acoustic basement buried under the sediments) divide the southern terrace of the Podvodnikov Basin into western and eastern parts. The acoustic basement of the western part, according to seismic, is moderately submerged continuation of the Lomonosov Ridge eastern slope with characteristic active normal faulting forming numerous rift half-grabens. In the eastern part the basement gradually plunges to 9–10 km b.s.l. and thickness of the sedimentary cover reaches 8 km.

Generally speaking, the structure of the Podvodnikov Basin southern terrace suggest two rifting episodes (Aptian –Albian and Late Cretaceous) leading to substantial stretching and thinning of the continental crust under the Basin. Several composite (seismic, gravity and magnetics) crustal models (Langinen et al. 2009; Lebedeva-Ivanova et al. 2011; Glebovsky et al. 2013; Jokat et al. 2013) support this hypothesis. They also demonstrate unity of crust under the East-Siberian Sea shelf and the Podvodnikov Basin, presenting the latter as a natural continuation of the former.

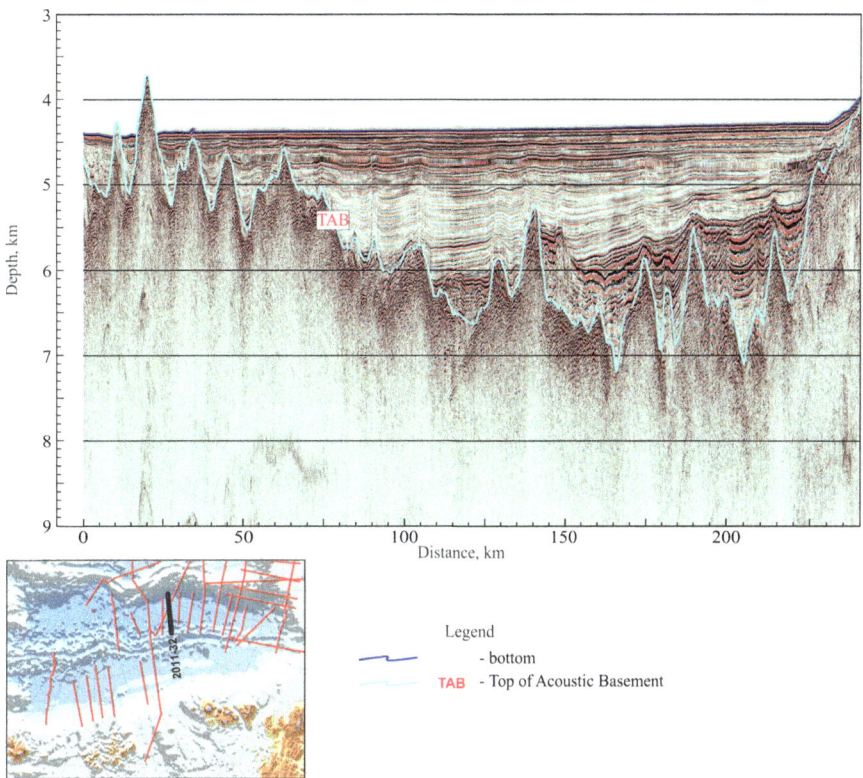

Fig. 5.11 MCS line 2011–32 Acoustic basement of the abyssal Amudsen Basin on the oceanic crust

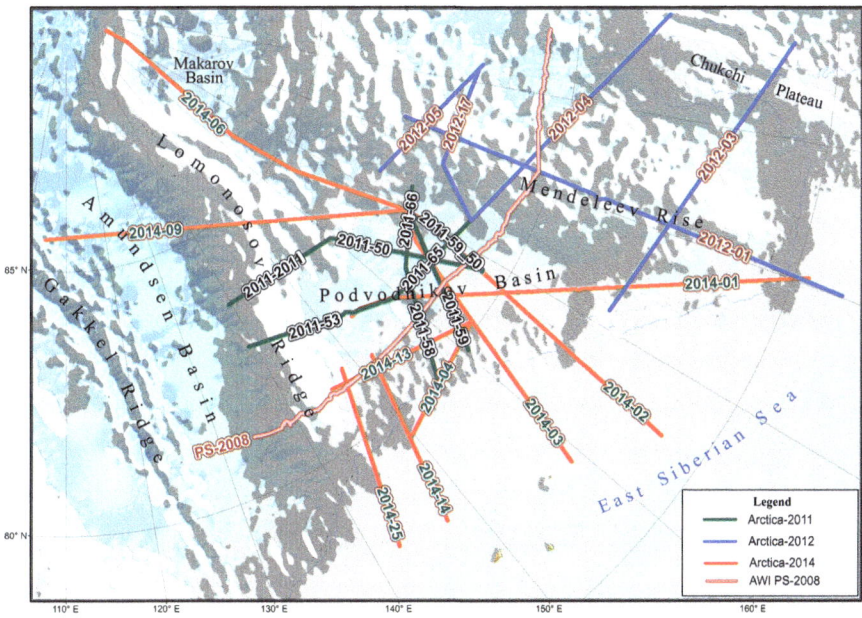

Fig. 5.12 Seismic programs in the Podvodnikov Basin, 2008–2014

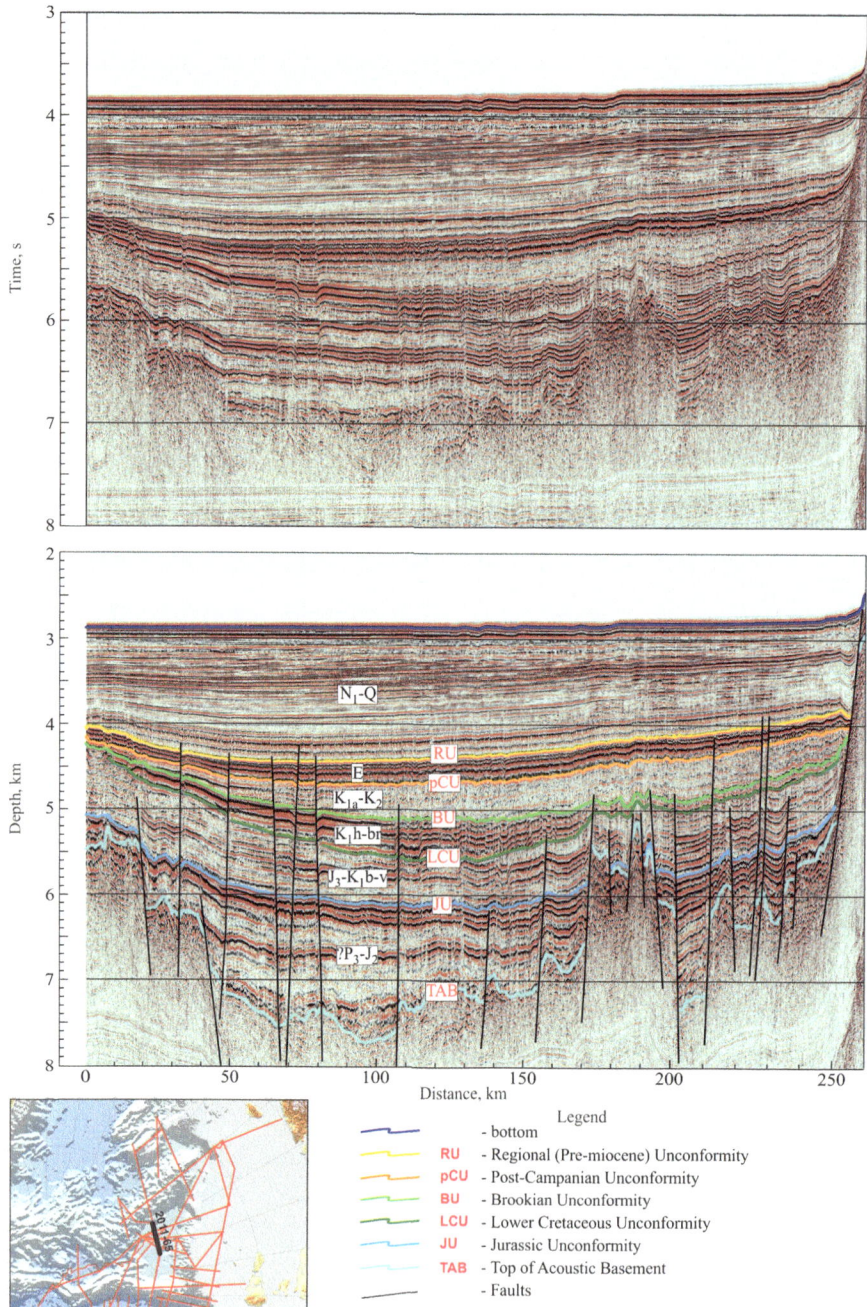

Fig. 5.13 Seismic-startigraphic interpretation of MCS line 2011–65

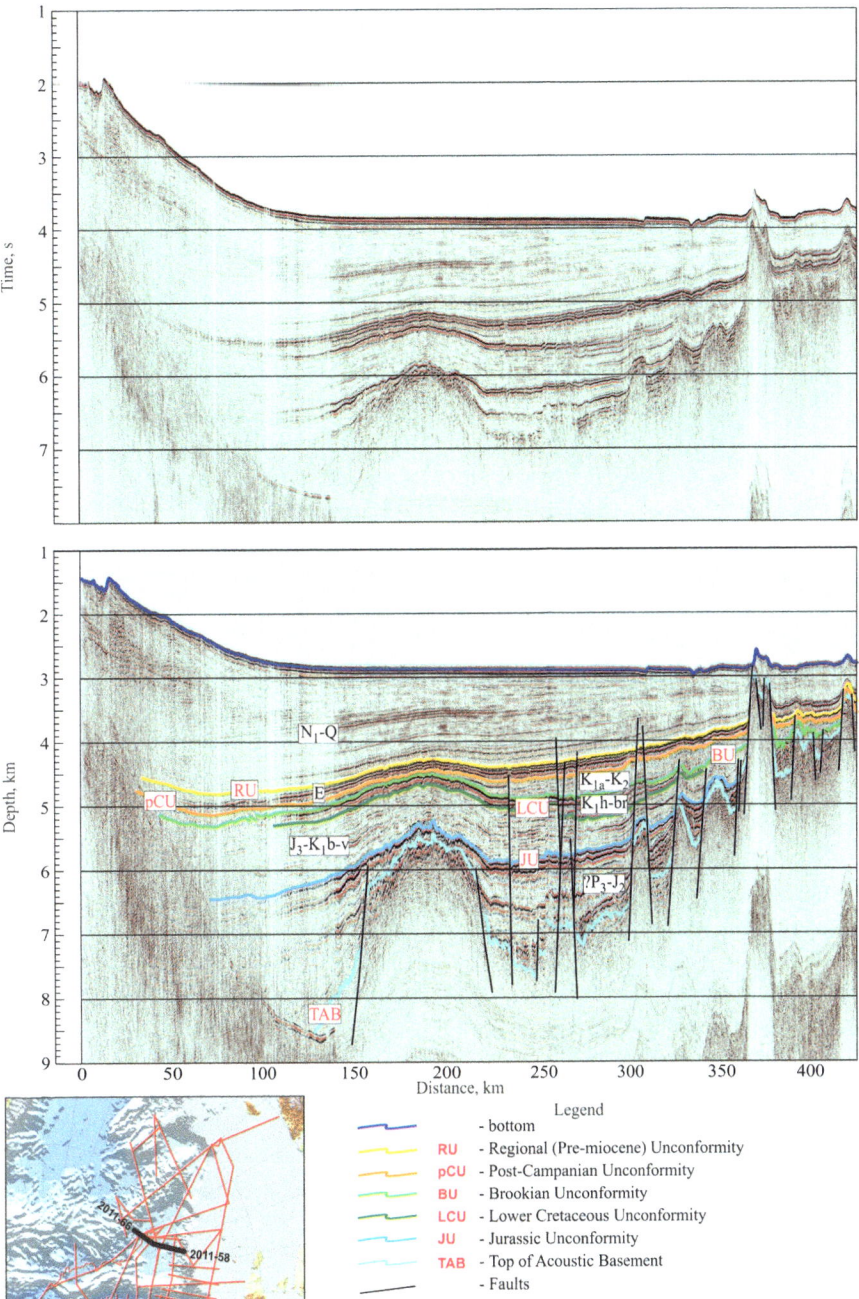

Fig. 5.14 Composite MCS line 2011–58-66

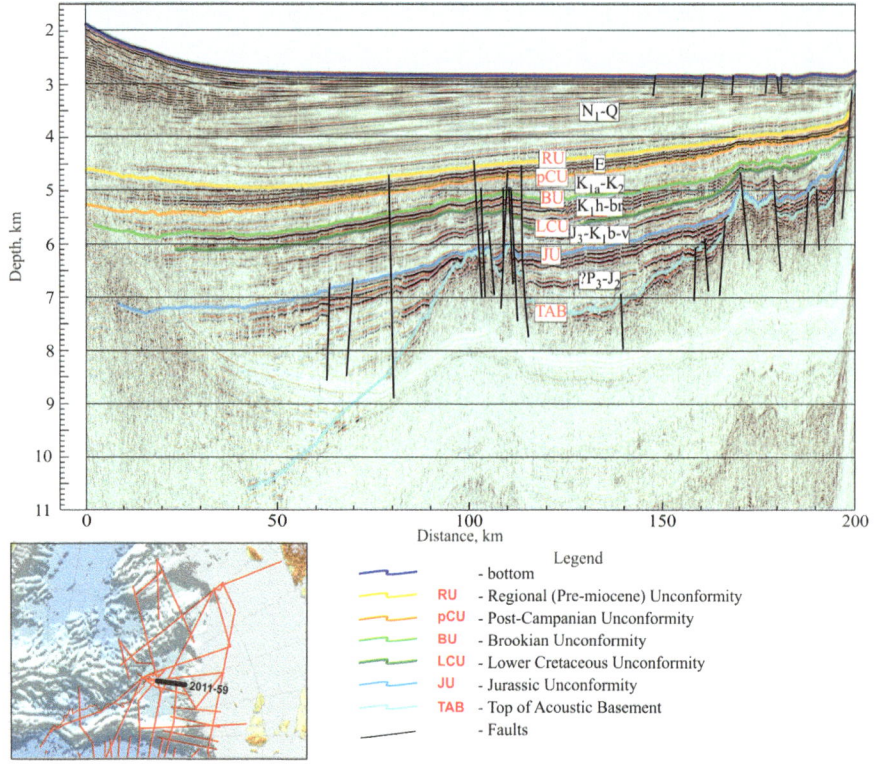

Fig. 5.15 MCS line 2011–59

The syn-rift sedimentary formations of the Podvodnikov Basin are overlain by post-rift Paleocene-Pleistocene sequence. MCS lines 2011–2011 and 2011–53 clearly show the sedimentary wedge between Mid Eocene and Miocene deposits ($\approx$ 44.4 Ma) thinning towards the Lomonosov Ridge eventually indicating non-depositional interval at the Ridge's crest.

Unconformities RU and pCU merge on the Lomonosov Ridge. East into the Podvodnikov Basin the RU and pCU unconformities separate again. Thickness of Paleogene complex ($\leq$ 300 m) is similar to that in the Lomonosov Ridge well (ACEX) where analysis of the core material indicates its neritic origin.

The most active accumulation of the Cretaceous bathial sediments (up to 2000 m) took place in the eastern and south-eastern parts of the Podvodnikov Basin and eastern parts of the Vilkitsky Trough, bordering the Basin from shelf side. Most of the Podvodnikov Basin domain can be considered as the Lomonosov Ridge sub-merged flank - the notion supported by (Jokat et al. 2013) analyzing the Polarstern MCS data.

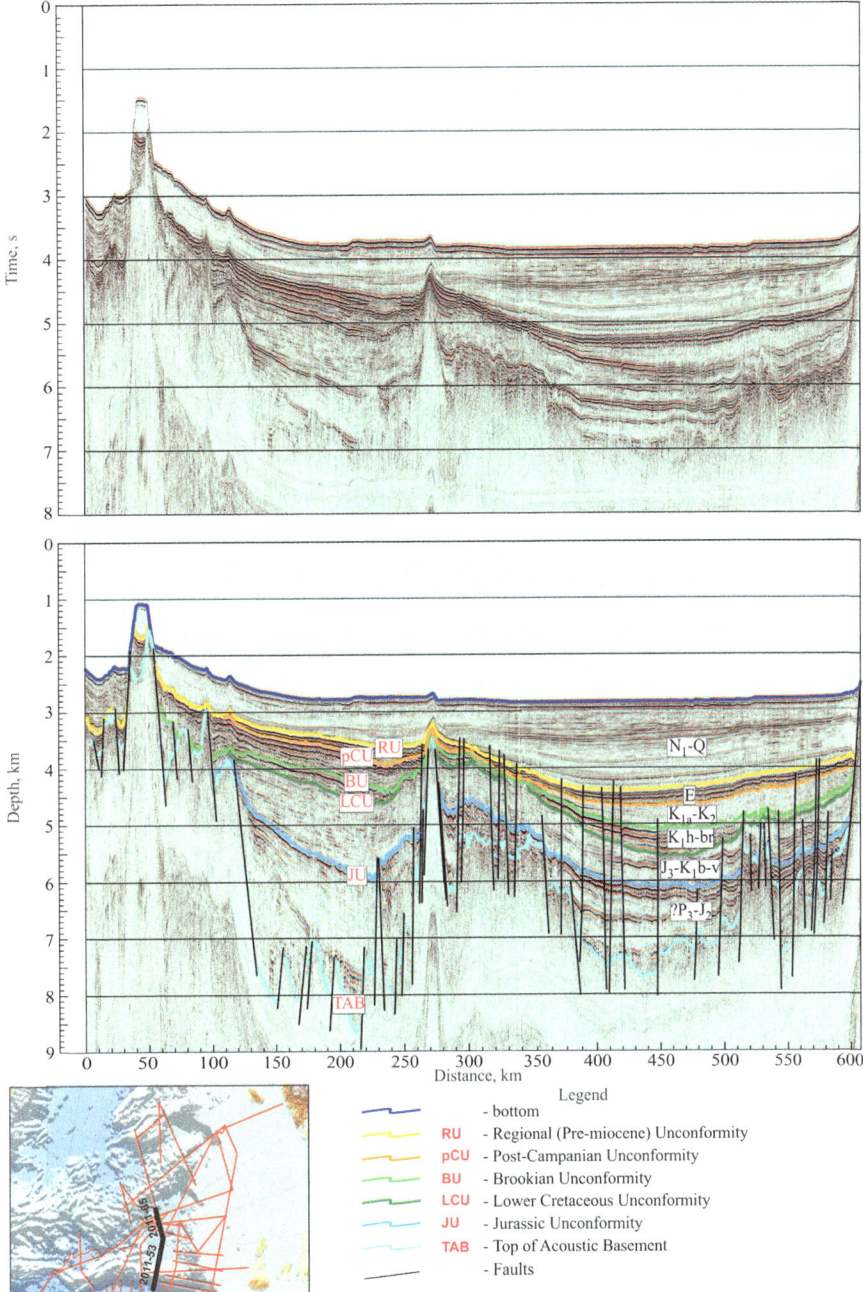

Fig. 5.16 Composite MCS line (2011–53 – 2011–65) Lomonosov Ridge - Podvodnikov Basin - Mendeleev Ridge

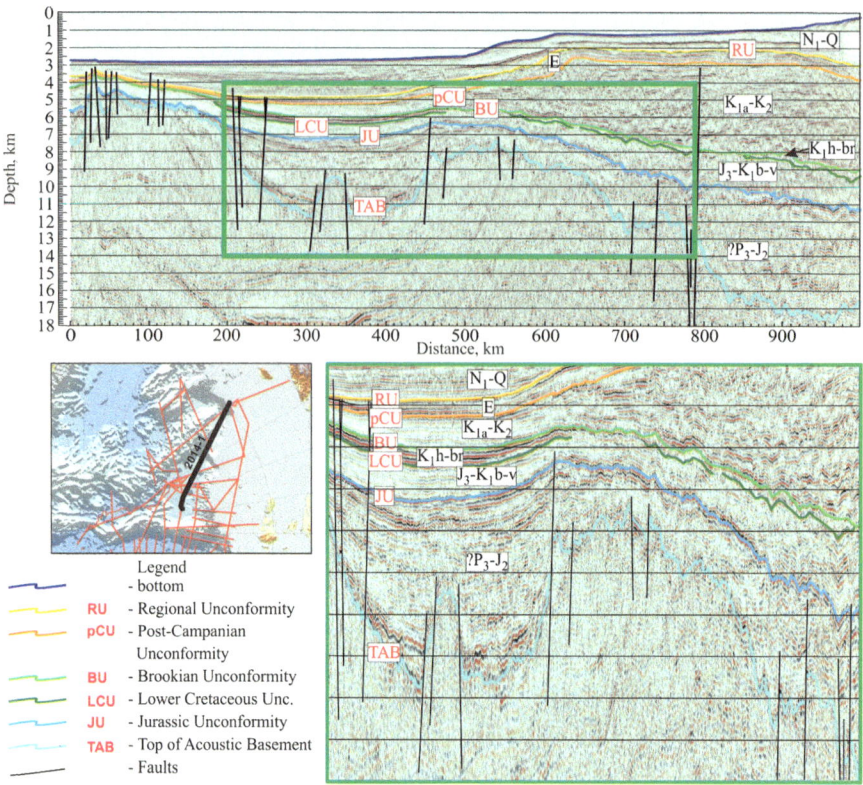

Fig. 5.17 MCS line 2014–01 (North-Chukchi Trough – Podvodnikov Basin)

Stratification of the Pre-Cenozoic sedimentary cover and correlation of the major seismic unconformities correspond with Popcorn well drilled on the Chukchi shelf. MCS sections show that three strong reflectors between the base of Cenozoic section (pCU) and acoustic basement (Fig. 5.14, 5.17) can be identified with Brooks (BU), Lower Cretaceous (LCU) and Jurassic (JU) unconformities, accordingly. The very basal sedimentary complex between JU and acoustic basement (TAB) is thought to contain Upper Paleozoic - Lower Mesozoic sediments.

The wide-angle sonobuoy refraction seismic in the Vilkitsky Trough - Podvodnikov Basin estimates the interval velocities of Upper Cretaceous (between pCU and BU) complex at 3.5–3.9 km/s, Pre-Cretaceous (between JU and TAB) – 4.1–4.4 km/s.

The latest seismic surveys made it possible to correlate the major seismic unconformities on the East-Siberian and Chukchi shelf with those established in the abyssal Arctic Basin. Fig. 5.18 presents the composite seismic profile connecting ACEX well on the Lomonosov Ridge with wells drilled on the Chukchi Sea shelf.

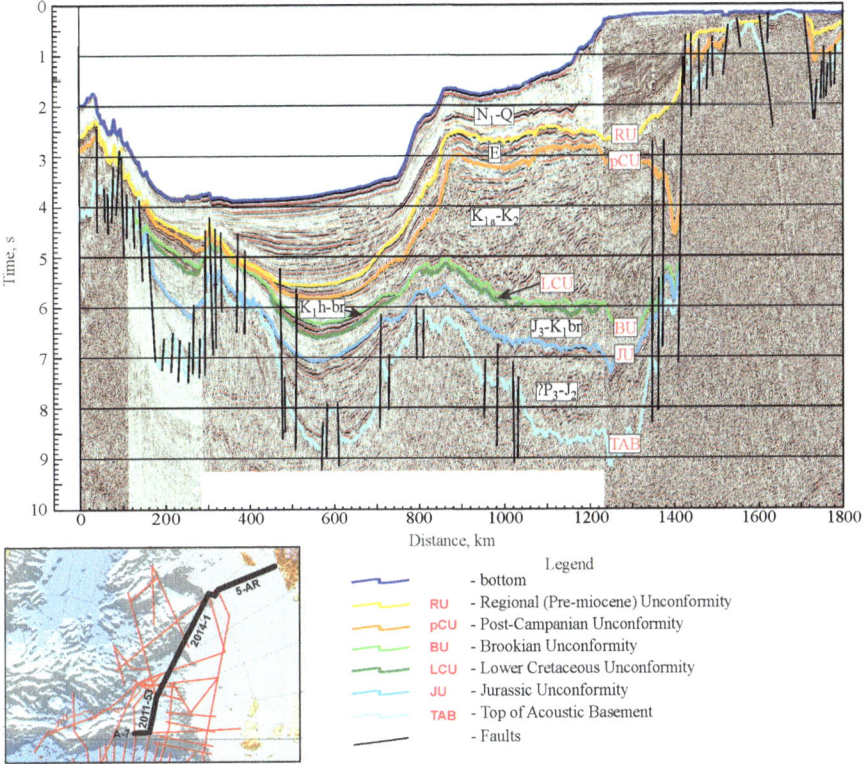

Fig. 5.18 Composite (A-7—2011-53—2014-1—5-AR) MCS traverse Lomonosov Ridge - Podvodnikov Basin - East-Siberian Sea shelf

This correlation convincingly associates the onset of sedimentation in the Podvodnikov Basin to Late Permian in the center or Early Cretaceous periods in the west.

The regional MCS sub-meridian transect 2014–02—06 connect the East-Siberian Sea shelf, the Podvodnikov Basin and the Lomonosov Ridge (at junction with the Makarov Basin). Interpreted seismic section clearly demonstrates not only the terraced character of the Podvodnikov Basin, but also un-interrupted continuation of the Paleogene sedimentary complexes from the East-Siberian shelf into the Podvodnikov Basin (Fig. 5.19). The configuration of the acoustic basement confirms that shelf was subjected to continental rifting, presumably in Cretaceous.

The large-scale acoustic basement uplift separates the southern and northern terraces of the Podvodnikov Basin (MCS lines 2014–02-06, Fig. 5.19) while actual contact between the two coincides with high-amplitude normal fault. Similarity between TAB seismic signature in the northern terrace and the Mendeleev Ridge suggests that the northern terrace might be a downthrown segment of the Mendeleev Ridge (gravity and magnetic data (Fig. 5.6) also point out in the this direction). Alternatively, it can be just a saddle between the Lomonosov and Mendeleev Ridges.

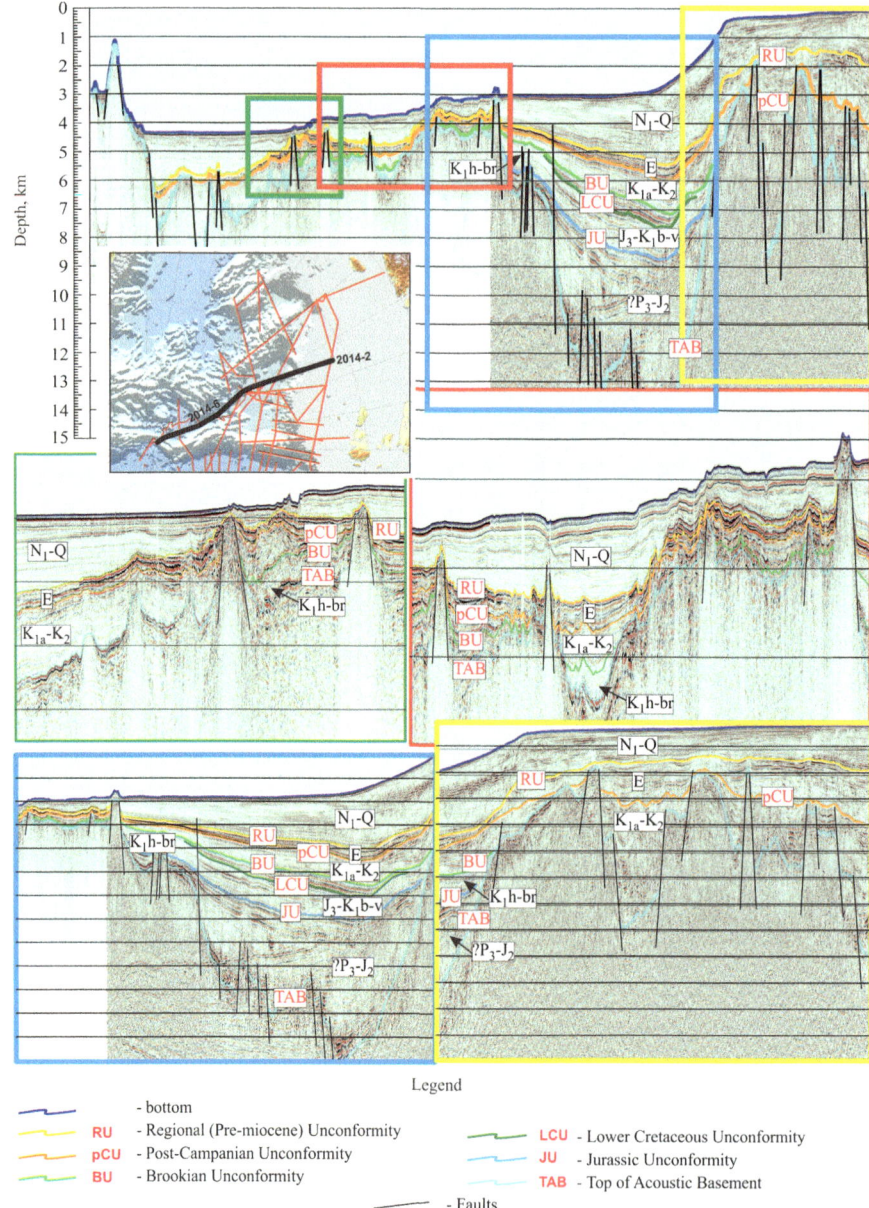

Fig. 5.19 Composite MCS transect 2014–02—06 (East-Siberian shelf - Podvodnikov Basin – Makarov Basin – Lomonosov Ridge)

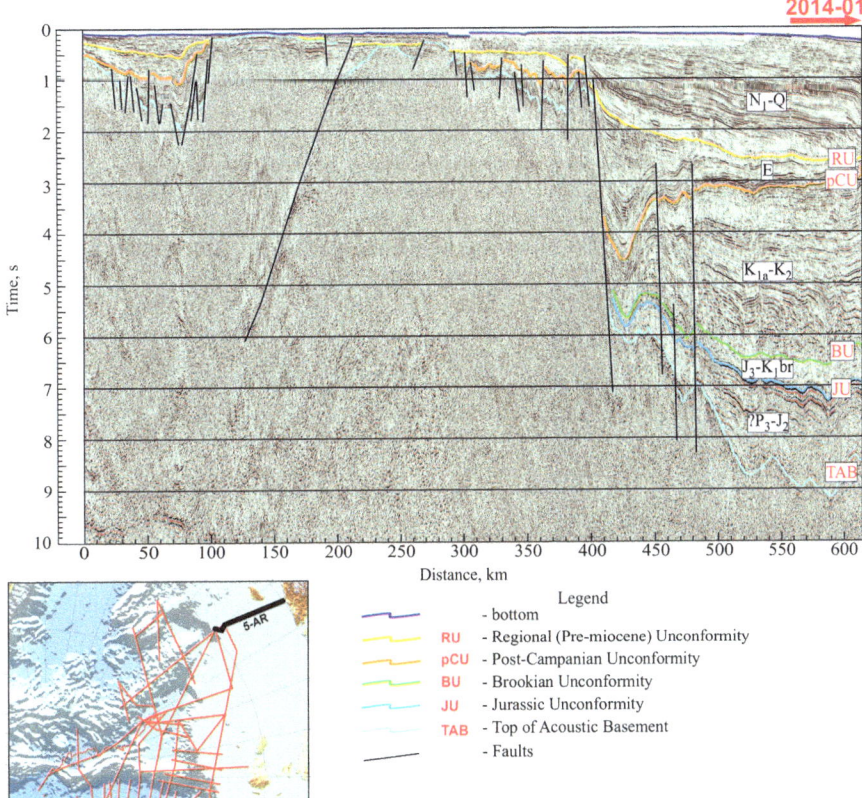

Fig. 5.20 MCS line 5-AR and unconformities identification corresponding with MCS line 2014–01

As we noted before, the MCS line 2014–01 (Fig. 5.17) most vividly illustrates the continuity of pre-Cenozoic seismic unconformities. The south-eastern end of this line joins the northern end of the MCS line 5-AR (Fig. 5.20) on East-Siberian Sea shelf, thus allowing tracing the JU (Jurassic) unconformity from shelf to the deepwater part of the Podvodnikov Basin.

North Chukchi Basin has up to 18 km of sediments and the lowest sedimentary complex is dated as Upper-Permian(?) - Mid Jurassic. Detailed description of correlation and dating of the sedimentary complexes in the North Chukchi Basin is given in Chap. 2 ("Seismic stratigraphy of the sedimentary cover").

Another important regional MCS traverse PS-2008 (Fig. 5.21) was acquired by RV "Polarstern" across the Podvodnikov Basin (for traverse location - see Fig. 5.12) in 2008 (Jokat et al. 2013; Weigelt et al. 2014). It was concluded (Jokat et al. 2013) that at least half of the Podvodnikov Basin resides on the stretched continental crust supporting considerably thick turbidite sequence no younger that Late Oligocene

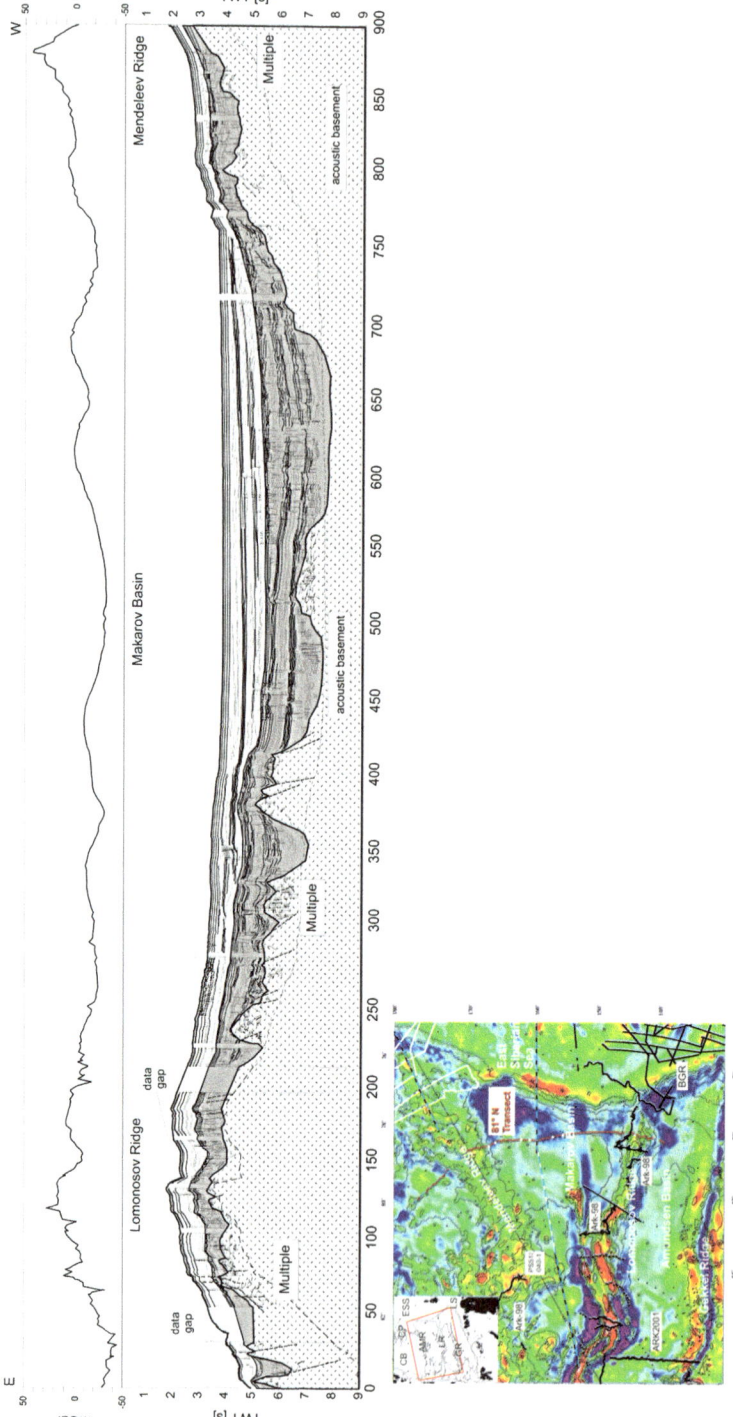

Fig. 5.21 Seismic traverse along 81° N Lomonosov Ridge – Podvodnikov Basin – Mendeleev Ridge (Jokat et al. 2013) Upper graph - Free Air Gravity; Light gray section – post- Fram Strait opening sediments; The "Multiple" line denotes the lower boundary of informative reflection data

(Hegewald and Jokat 2013a, b) (this dating agrees with pre-Miocene dating (Poselov et al. 2014) of the regional unconformity terminating this sequence). Overall, the sedimentary section thickness varies from 1000 m at the top of the Lomonosov Ridge to 6000 m in the central part of the Podvodnikov Basin (incidentally, in (Jokat et al. 2013), it was named "Makarov Basin").

Seismic program under the Russian expedition "Arktika-2014" delivered a large amount of new information which greatly improved our undrestanding of the Podvodnikov Basin geology in general, and its sedimentary cover, in particular. Some of the seismic cross-sections from this program are discussed below.

MCS line 2014–03 (Fig. 5.22), 650 km long, runs from the De Long Uplift through the southern Podvodnikov Basin to the western flank of the Mendeleev Ridge. Near the shelf edge, thickness of sedimentary cover inside the Vilkitsky Basin reaches 11 km. Apparent intrusion of igneous material (450–500 km) can be related to the Cretaceous magmatic activity, because the formations above the pCU unconformity are undisturbed. In contrast, the igneous intrusions visible at the foot-hills of the Mendeleev Ridge (630–650 km), penetrate the entire sedimentary section and expressed in the seafloor relief, placing this magmatic event in Pliocene – Pleistocene.

MCS line 2014–04 (Fig. 5.23) (250 km) obliquely crosses the shelf edge and terminates in the Podvodnikov Basin at depth $\approx$ 2500 m. The eye-catching trough, 10-km wide and 500 m deep, is interpreted as erosional channel carved into the sea bottom by high-energy turbidite flows, rather than tectonic structure. This conclusion is supported by continuous reflectors on both sides and under the bottom of the trough, as well as by presence of side mounds.

Main portion of the MCS line 2014–12 (Fig. 5.24) ($\approx$350 km) covers the De Long Uplifts demonstrating sharply reduced thickness of sediments. Identification of the acoustic basement with confidence on this section is difficult without correlation with well established markers in proximity, which may come in the future.

MCS line 2014–13 (Fig. 5.25), slightly more than 300 km long, starts at the eastern flank of the Lomonosov Ridge and crosses southern section of the Podvodnikov Basin. The seismic section clearly demonstrates draping of the ragged acoustic basement by overlaying formations and gradual increase of Cenozoic sediments toward the center of the Podvodnikov Basin. The uplifted block of acoustic basement in 210–230 km interval may represent submersed continuation of the Geofizikov Spur, also recognizable on seismic lines north from, and parallel to, the present one.

MCS line 2014–14 (Fig. 5.26) ($\approx$380 km), runs from the East-Siberian Sea shelf through the continental slope into the Podvodnikov Basin along the eastern flank of the southern Lomonosov Ridge. Here, at the shelf edge, another turbidite-related feature, similar to that of the line 2014–04, is present. The slope-basin transition is marked by dramatic increase of the sedimentary cover thickness.

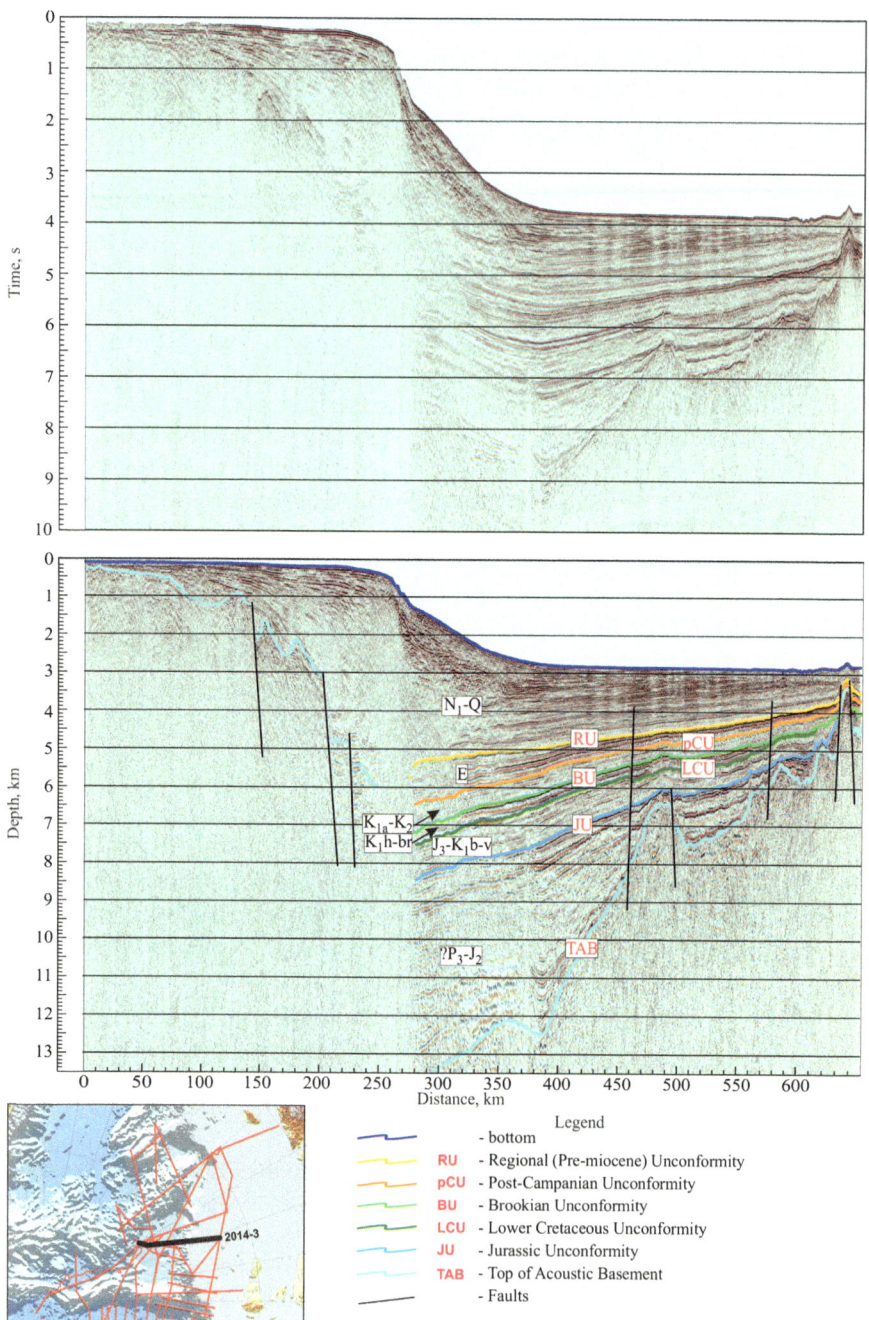

Fig. 5.22 MCS line 2014–03

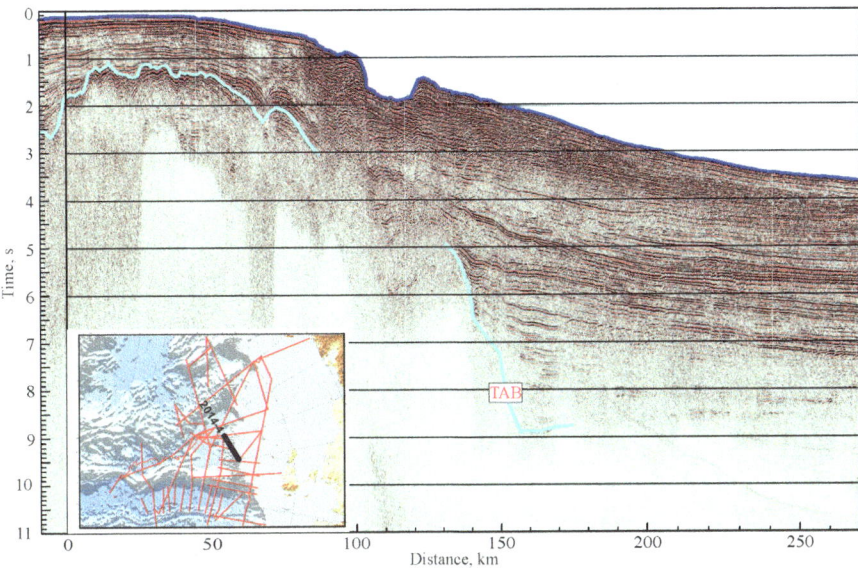

Fig. 5.23 MCS line 2014–04

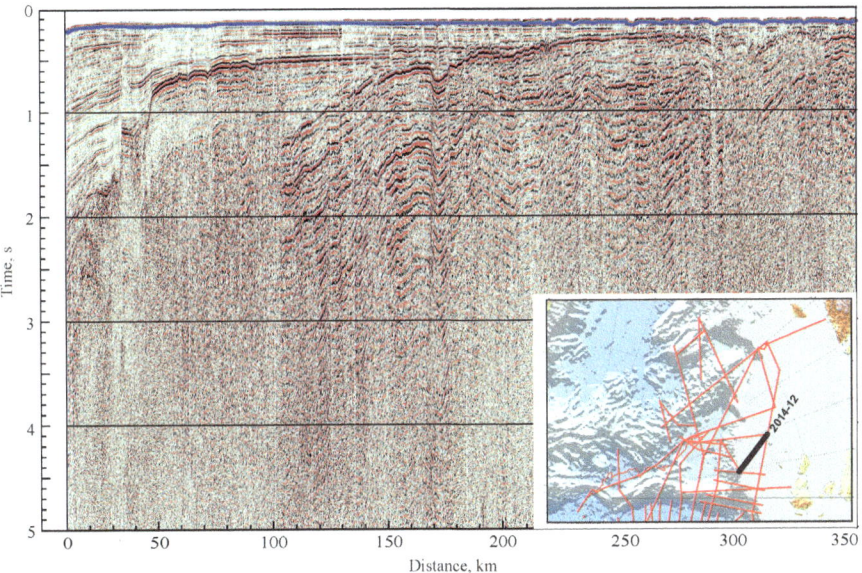

Fig. 5.24 MCS line 2014–12

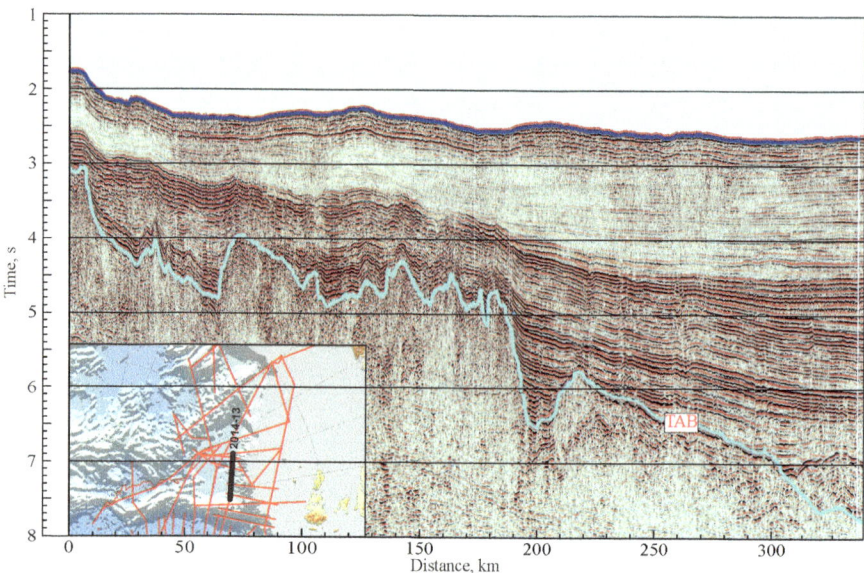

Fig. 5.25 MCS line 2014–13

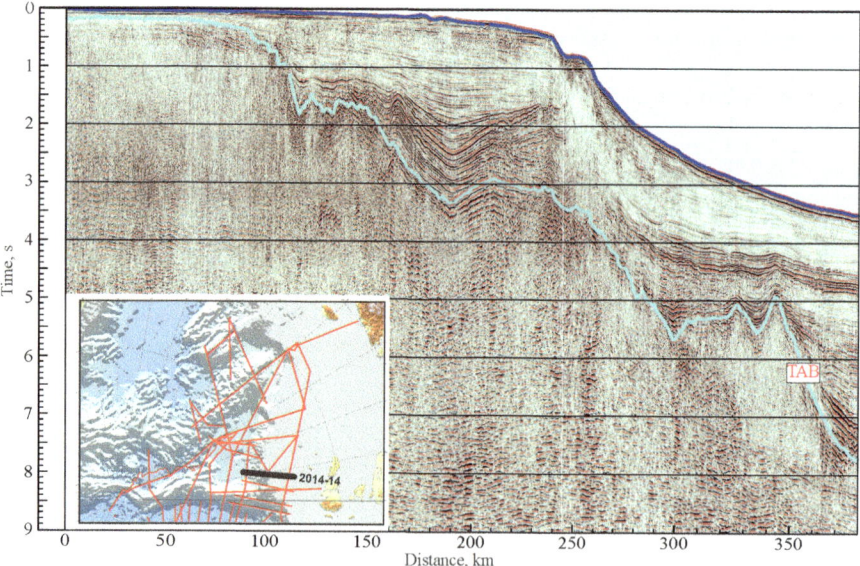

Fig. 5.26 MCS line 2014–14

5.5   Crustal Structure

The Podvodnikov Basin deep crustal investigations started after completion of the regional polar transect "Transarktika-89-91" (refraction deep seismic sounding or DSS, for short). It runs from the East-Siberian Sea shelf through the Podvodnikov and Makarov Basins to the near-polar segment of the Lomonosov Ridge foothills (Fig. 5.27). The obtained data were repeatedly interpreted and the results - published (Arctic Ridges: Results and Planning 1994; Verba 1996; Pavlenkin et al. 1996; Lebedeva-Ivanova et al. 2011).

In addition to the "Transarktika-89-91" DSS information, data from earlier geophysical programs (MCS line 90801, ice-borne seismic surveys "Transarktika-91" and SP-28-91) were also used. Integrated interpretation of all the above mentioned sources is presented on Fig. 5.27.

Total thickness of the sedimentary cover reaches its maximum of 7 km in the Vilkitsky Trough and then drops down in north-western direction to 4 km and even less, to few hundred meters.

Four-layered crust model of the Podvodnikov and Makarov Basins is postulated:

Layer I ($V_p = 1,7$—$3,8$ km/s) — Meso-Cenozoic sedimentary complexes;
Layer II ($V_p = 5,0$—$5,4$ km/s) — older sedimentary (and possibly volcano-sedimentary in basins) formations;
Layers III and IV ($V_p = 5,9$—$6,5$ km/s and $6,7$—$7,3$ km/s, accordingly) – crystalline crust.

Moho discontinuity is identified at depth of 20 km under the southern and northern parts of the Podvodnikov Basin plunging down to 30 in center, under the Arlis Gap. The authors associate the formation of the Podvodnikov basin to the Mesozoic stretching of the continental crust. The jointed Mendeleev—Alpha Ridges reside on the 25 km thick crystalline crust of layers III and IV. The abyssal Makarov Basin was suggested to have an oceanic crust of pre-spreading stage, 8–10 km thick, with possible inclusions of continental blocks split from the Lomonosov Ridge.

Computed seismic-gravity crustal model of the "Transarktika-89-91" transect (Piskarev 2004) is presented on Fig. 5.28.

The total thickness of the continental crust in the Podvordnikov basin (300–950 km marks) varies from 15 to 19 km. Upper consolidated crust (3–5 km) is classified as felsic intrusive-methamorphic complex with average density 2.70–2.72 gcm³. Under the Mendeleev Ridge (950–1080 km interval) the crust looks like standard continental one, only with noticeably reduced lower layer. Further north-west, in the Makarov Basin (1080–1260 km), the crust acquired all characteristics of the oceanic type, with total thickness close to 8 km. Transition back to continental crust of the Lomonosov Ridge is visible at the very end (1260–1350 km) of the transect.

Ray modeling of DSS data from MCS lines (Arctic-2014-1 and Arctic-2014-2) acquired by expedition "Arktika-2014", running from the De-Long Uplift across the entire Vilkitsky Trough into the Podvodnikov Basin, produced the detailed and reliable crustal velocity information (Fig. 5.29).

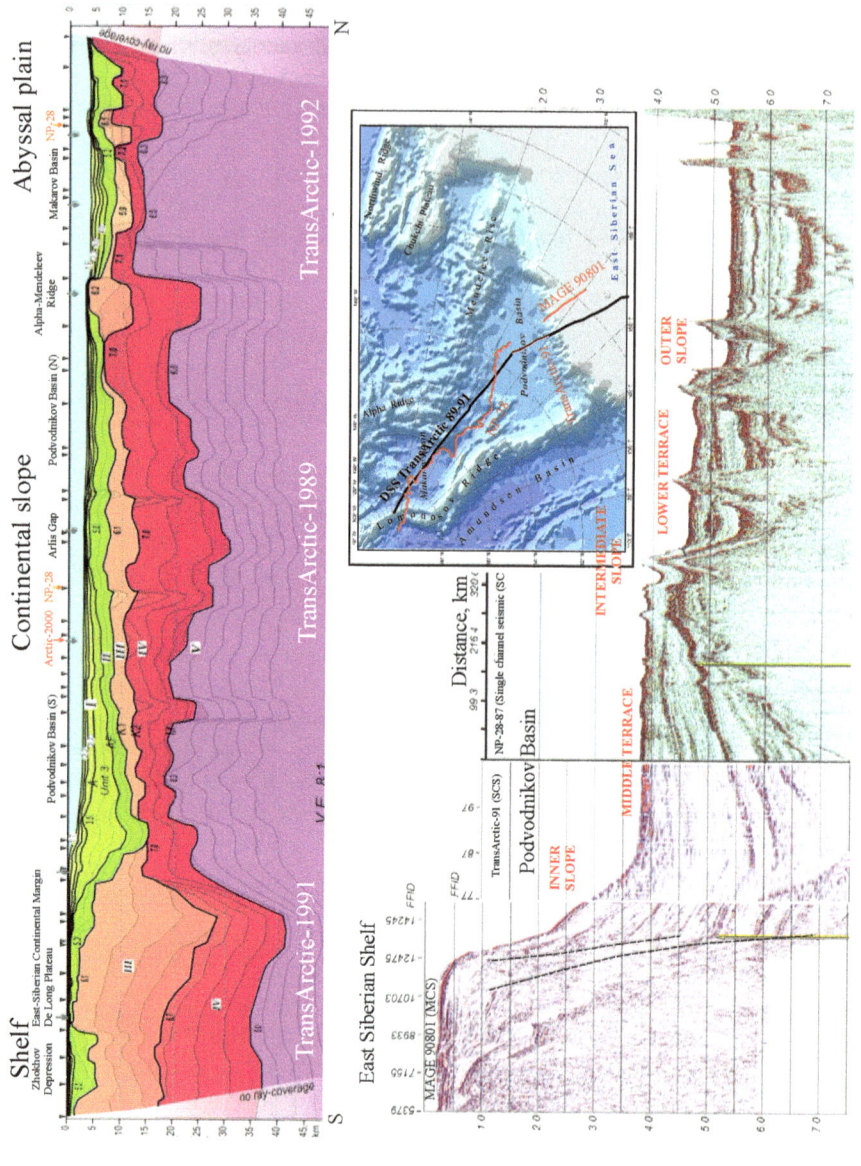

Fig. 5.27 Crustal velocity model of the Podvodnikov Basin, DSS "Transarktika-89-91" (Lebedeva-Ivanova et al. 2011)

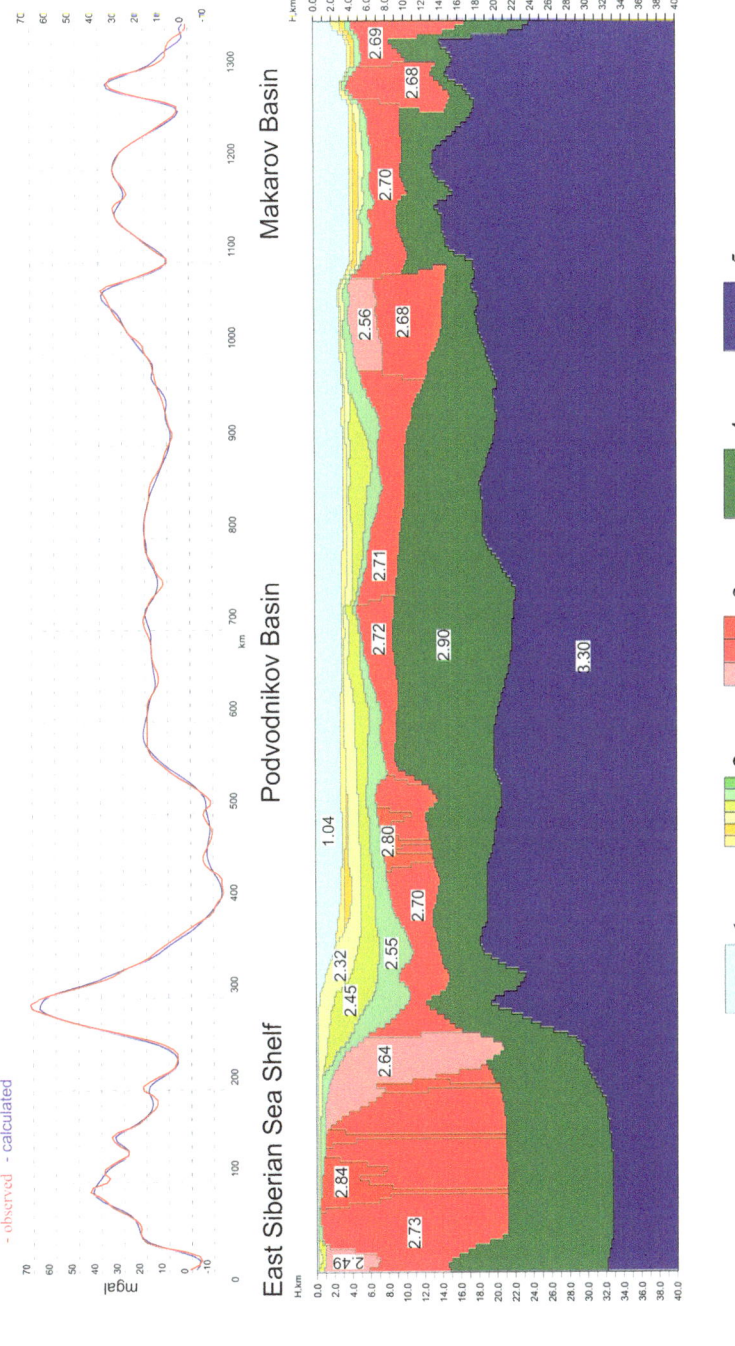

Fig. 5.28 The Podvodnikov Basin crustal model by seismic and gravity data ("Transarktika-89-91" transect) 1 — water; 2 — sedimentary comlexes; 3 — upper crust; 4 — lower crust; 5 — mantle

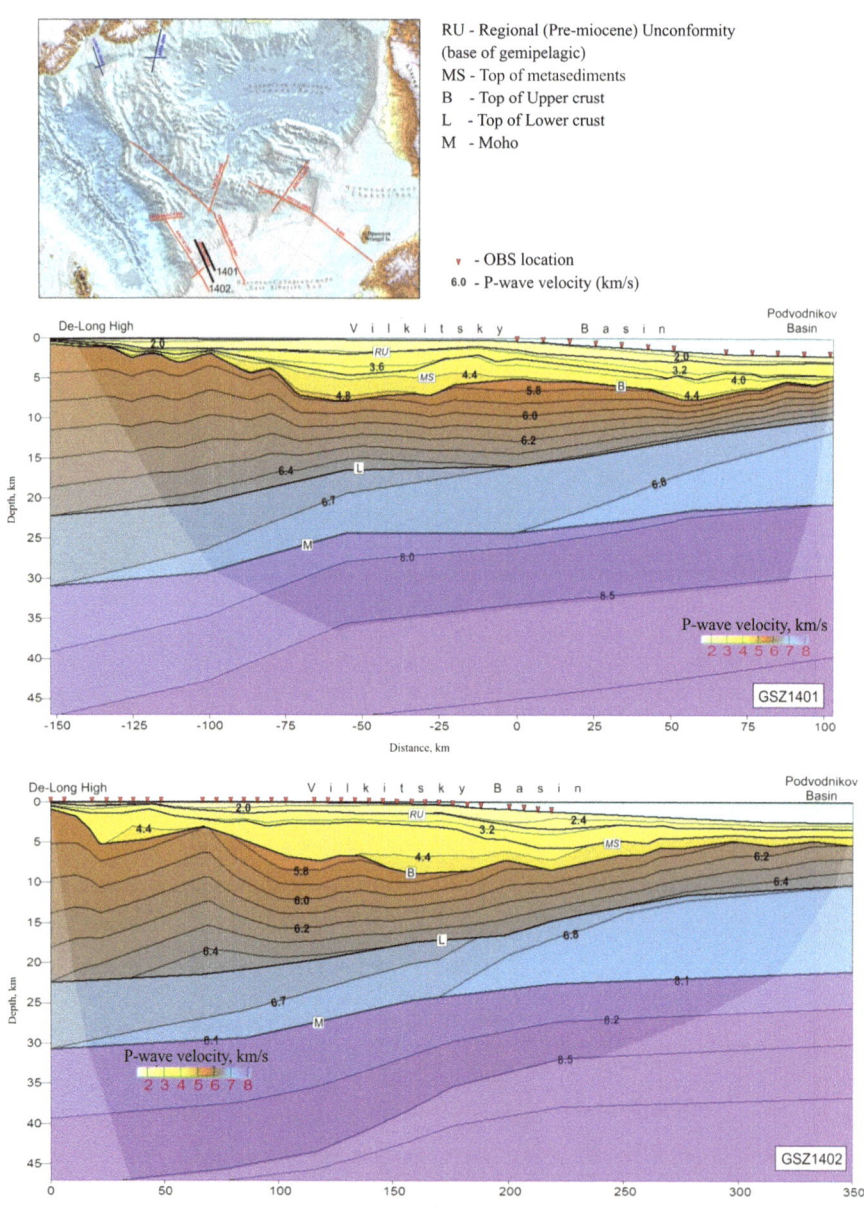

RU - Regional (Pre-miocene) Unconformity
(base of gemipelagic)
MS - Top of metasediments
B - Top of Upper crust
L - Top of Lower crust
M - Moho

▾ - OBS location
6.0 - P-wave velocity (km/s)

Fig. 5.29 Crustal velocities in the East-Siberian continental margin (DSS lines Arctic-2014-1 and Arctic-2014-2)

The principal crustal units (from top down) are identified as follows:

- unconsolidated hemi-pelagic sediments (Pv =1.9–2.4 km/s) between the sea floor and pre-Miocene unconformity RU, with thickness varying from 1.5–2.5 km (shelf and basins) to almost 0 (De-Long Uplift);
- shelf/neritic deposits below RU unconformity (Pv = 3.1–3.6 km/s) and thickness varying from 2–3 km (shelf and basins) to few hundred meters (Arctic-2014–02) or 0 (Arctic-2014-1) at the De-Long Uplift;
- more consolidated and, perhaps, metamorphic formations (Pv =4.0–4.8 km/s (4,0—4,4 km/s in the Vilkitsky Trough - Podvodnikov Basin transition slope); thickness, in average in the range of 1.5–4 km, increases to 6.5 km in the Vilkitsky Trough depocenter (Arctic-2014-2), or drops to few hundred meters at the De-Long Uplift (Arctic-2014-1);
- upper crust (P-wave velocity increasing from 5.8 to 6.4 km/s with depth); thickness drops from maximum around 22 km at the De Long Uplift to ≈5 km in the Podvodnikov Basin;
- lower crust (P-wave velocity increases in northerly direction from 6.6–6,7 km/s at the De Long Archipelgo shelf to 6.8–6.9 km/s in the Podvodnikov Basin) with thickness growing from 8 to 10–11 km in the same direction.
- upper mantle with Pv ≈ 8.0–8.1 km/s and Moho discontinuity fixed at ≈31 km depth under the De Long Uplift rising to ≈21 km in the Podvodnikov Basin, bringing the overall crustal thickness changing from 31 to 18–19 km, accordingly.

Total thickness of the sedimentary complexes, including possible metamorphic complexes, reaches ≈8 km in the shelf depocenter of the Vilkitsky Trough and ≈6.0 km – in its slope depocenter.

Presented information unambiguously shows the continuation of the stratified sedimentary complexes from the De Long shelf into the Vilkitsky-Podvodnikov depression, thus establishing structural and genetic commonality between them.

The crustal seismic-gravity model along the seismic MCS line 2014–01 (Fig. 5.30) was constructed using the subsurface geometry of crustal complexes from depth-converted interpretation of this line (Fig. 2.19) and their velocity-derived density.

The model stretches from the North-Chukchi Trough through the Podvodnikov Basin to the Geofizikov Spur.

In the North Chukchi Trough the total crustal thickness approaches 27 km, considerable part of which consists of sedimentary cover – 15-17 km in the first 170 km of the section; further northwest, towards the Podvodnikov Basin, the sedimentary cover thins to 8–12 km. The iterative modeling process produces the best fit between observed and computed gravity with densities of

2.0–2.55 g/cm3 for SSC1-SSC4 sedimentary complexes;
2,60—2,65 g/cm^3 – for SSC5-SSC6 sedimentary complexes;
2,72—2,86 g/cm^3 –for upper consolidated crust;
2,95—3,0 g/cm^3 – for lower consolidated crust.
3,3 g/cm^3 – for mantle.

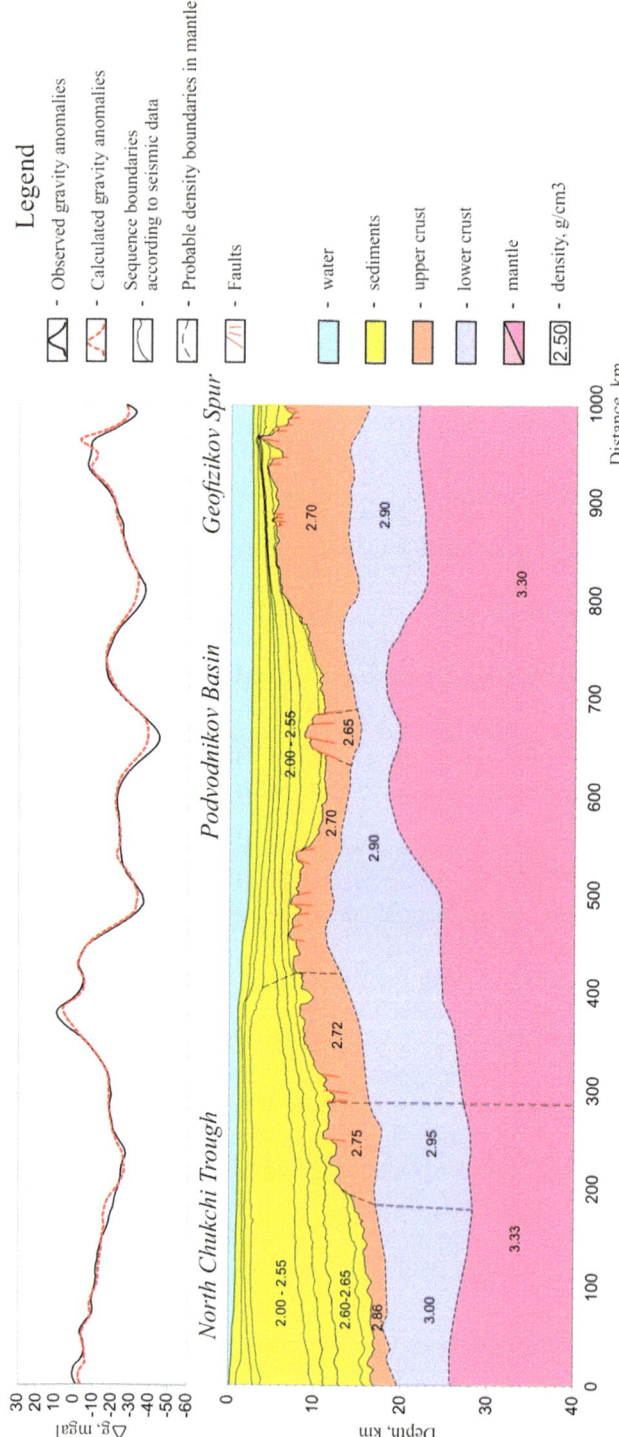

Fig. 5.30 Crustal seismic-gravity model along MCS line 2014–01

The consolidated crust tends to increase thickness toward the Podvodnikov Basin from 10 to 13 km (upper crust – from 2 to 7 km, lower - 7 to 12 km). The steady increase of gravity field intensity towards the deepest part of the North-Chukchi Trough (0–170 km) can only be explained by rising Moho and thinning of the consolidated crust in this direction.

In the Podvodnikov Basin crustal thickness is reduced to 16–17 km (6–8 km - sedimentary cover, 4–6 km – upper crust, 5–7 km – lower). Densities are in the same range as for the North Chukchi Trough, the only difference is absence of lower sedimentary complexes. The local gravity minimum at the 630–680 km interval coincides. This seemingly conflicting situation was resolved by reducing density of the basement block to 2.65 g/cm^3 (felsic intrusion?) and deepening Moho by 2 km.

At the Geofizikov Spur the total crustal thickness increases to 20 km along with sedimentary cover thinning to 3 km on average (around 1 km on uplifts, and up to 5 km in depressions). Consolidated crust is getting thicker mainly due to the upper crust almost doubling its thickness compare to adjacent parts of the Podvodnikov Basin - up to 9 km. Densities of the upper and lower crusts are laterally constant 2.70 and 2.90 g/cm^3, accordingly.

The composite MCS profile 2014–02—2014-06 (location and seismic section (Fig. 5.19) most clearly illustrates the deeply rooted geological connections between the East Siberian Sea shelf and abyssal depressions of the Arctic ocean. The crustal seismic-gravity model of this important profile is shown on Fig. 5.31.

In the East-Siberian Sea shelf Cenozoic sedimentary complexes SSC1 and SSC2 with density 2.0–2.3 g/cm^3 and 3 km thick on average, together with Mesozoic formation (2–6 km thick with density 2.50 g/cm^3), filling basement depressions, reside on the continental crust 28–33 km thick. The upper crust is 6–17 km thick, lower - 17–20 km.

In the Podvodnikov Basin the crustal thickness drops to 17 km with slight increase towards the Geofizikov Spur. The sedimentary cover (2.00–2.55 g/cm^3) shrinks in the same directions from 9 to 3 km. As in the previous model the crustal block with reduced density (2.65 g/cm^3) is introduced into the upper crust at the depocenter of the Podvodnikov Basin. At the crest of the Geofizikov Spur there is only 1–1.5 km of sediments; underneath it the crustal thickness reaches 20–22 km.

In Makarov Basin sedimentary cover (1–4 km) consists of first three complexes with densities from 2.00 to 2.40 g/cm^3, upper and lower crusts are 3–6 km and 3–4 km thick, accordingly. Total crustal thickness by modeling estimates (10–12 km) is in agreement with DSS data ("Ttransarktika 89–91", Fig. 5.27). Transition to the Lomonosov Ridge domain is manifested by substantial increase of crustal thickness up to 21 km and strong (100 mGal) gravity anomaly marking sharp, 2.5 km high, rise of the ridge above the seafloor. The sediments thickness here is negligent (first hundred meters).

The similar seismic-gravity crustal models were computed for DSS lines Arctic-2014-1 and Arctic-2014-2 (Fig. 5.29) using their density-converted velocities (Fig. 5.32, 5.33).

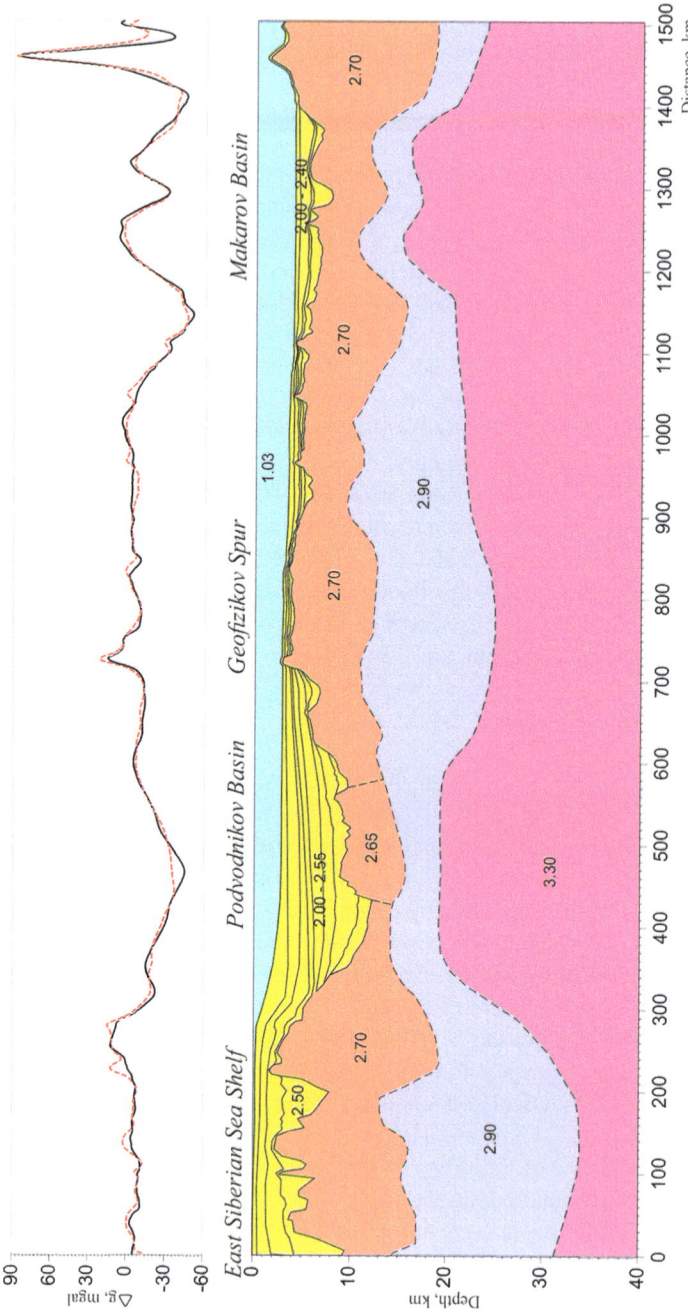

Fig. 5.31 Deep crustal model along composite MCS line 2014-02—2014-06

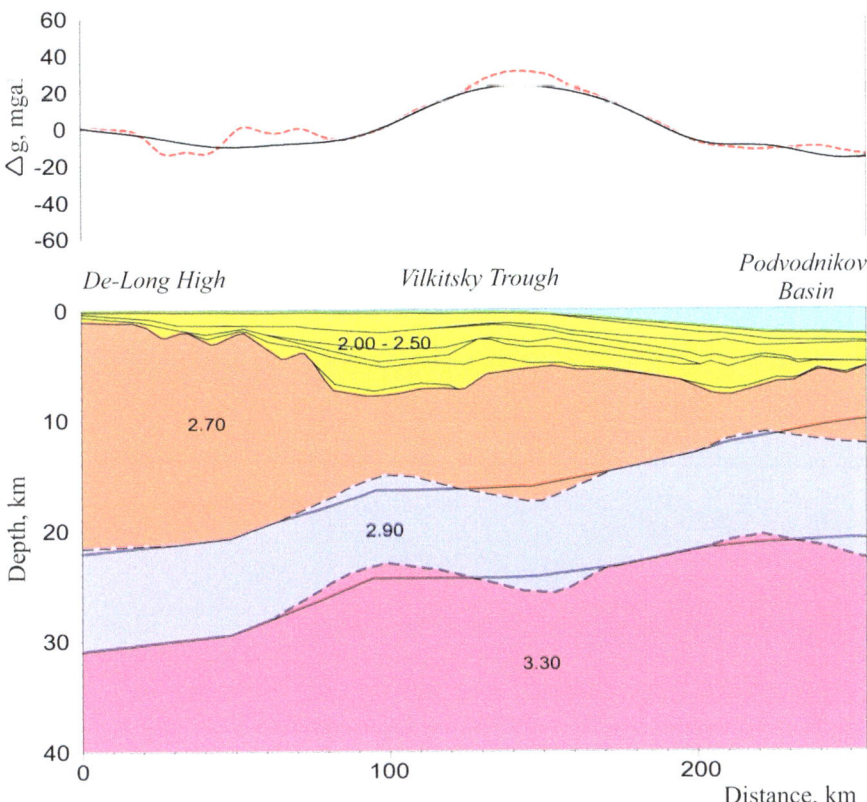

Fig. 5.32 Crustal model along DSS line Arctic-2014-1

The modeling demonstrates good agreement between gravity and seismic data. Slight discrepancy (1–2 km) in configuration of upper - lower crusts and Moho (see solid lines on Fig. 5.32, 5.33) are well within the accuracy of the method. The uncertainty of boundaries interpolation over 250–300 km interval, is not supported by seismic rays, and averaged nature of grid–to profile interpolated gravity values, may also be the cause.

Additional model (Fig. 5.34), was computed using seismic data from MCS line 2014–03 (Fig. 5.23) for location and seismic section.

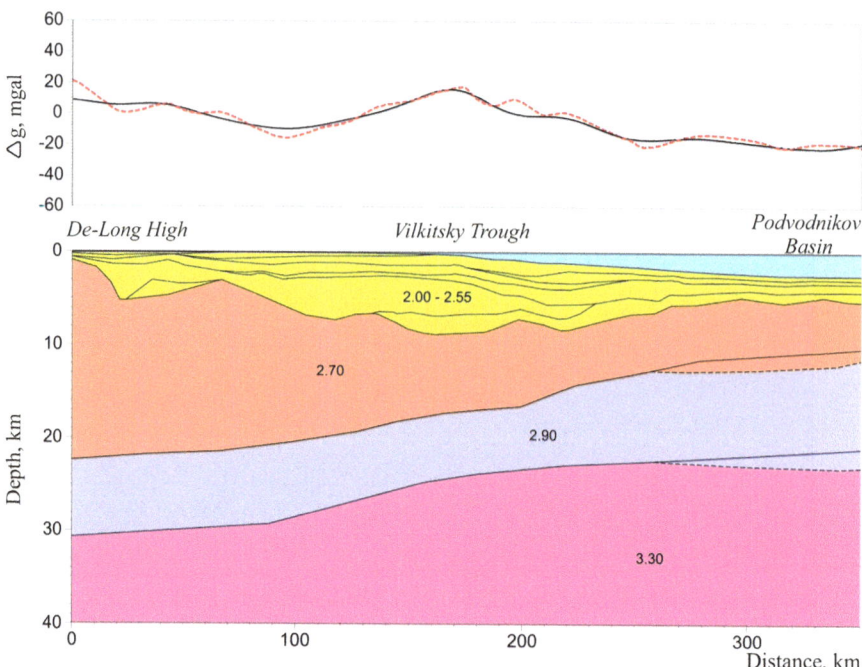

Fig. 5.33 Crustal model along DSS line Arctic-2014-2

5.6 Conclusions

1. The Podvodnikov Basin resides on the continental crust, stretched and thinned
   during the Creataceous rifting process. (Nikishin et al. 2014, 2015; Langinen
   et al. 2009; Lebedeva-Ivanova et al. 2011; Kashubin et al. 2011, 2013; Glebovsky
   et al. 2013; Laverov et al. 2013; Jokat et al. 2013). Crustal thickness is between
   19 and 25 km. The Basin subsidence and development supposedly took place
   between 118 and 56 Ma (Franke et al. 2004). The Geofizikov Spur separate the
   Podvodnikov Basin into western and eastern sub-basins distinguished mostly by
   stratigraphy of the basal sedimentary formations. The space between the
   Lomonosov Ridge and the Geofizikov Spur contains several half-grabens filled
   with syn-rift formations, presumably of Cretaceous age. In the eastern sub-basins
   seismic reflection data may suggest the presence of the older sedimentary complexes. The transitional slopes between the Lomonosov and Mendeleev Ridges
   are disturbed by multitude of north-south trending normal faults.
2. Total thickness of the Podvodnikov Basin sedimentary cover exceeds 6 km. The
   structure of the acoustic basement is radically different from the oceanic acoustic
   basement of the Amundsen basin.

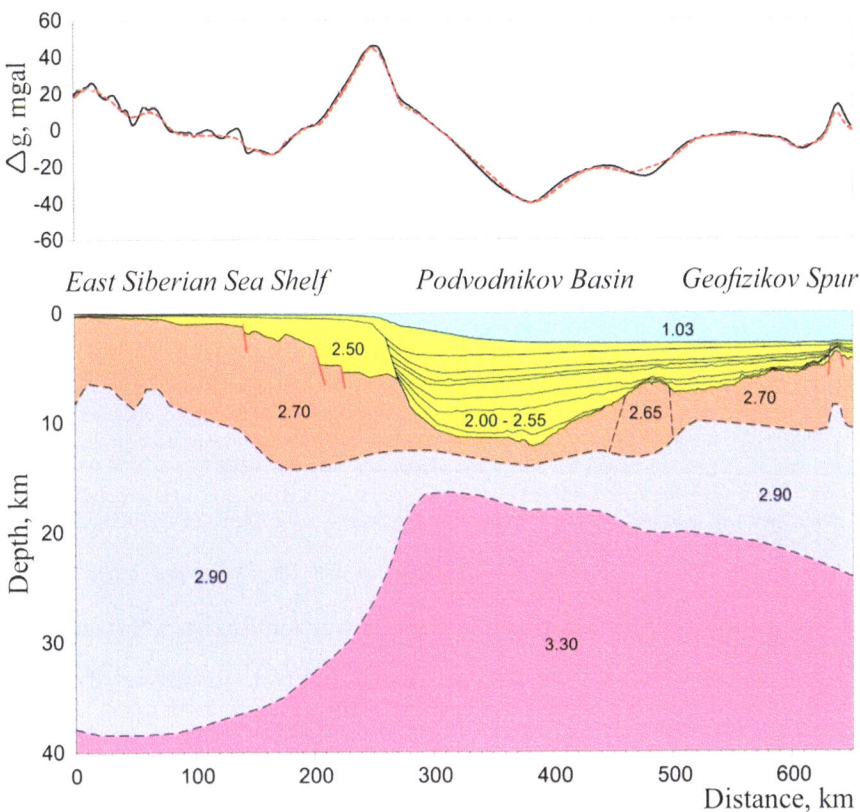

Fig. 5.34 Crustal model along MCS line 2014–03

3. Two strong seismic reflectors between the Post-Campanian unconformity pCU and the acoustic basement can be traced throughout the entire Podvodnikov Basin. These reflectors are correlated with Brooks and Jurassic unconformities identified at the base of Lower Brooks complex. Accordingly, the basal sedimentary complexes of the Podvodnikov Basin are dated as Pre- and Early Cretaceous.

4. Morphologically, the Podvodnikov Basin represents a complex terraced slope subjected to neo-tectonic normal faulting under extensional crustal regime causing stepped differential subsidence.

5. The Podvodnikov Basin can be classified as a part of the continental margin on the stretched continental crust. The stretching and thinning of the Podvodnikov Basin crust is presumed to be multiphase covering a certain period of time; possibility of the periodic changes of the tensional forces orientation also cannot be excluded.

References

Arctic Ridges: Results and Planning (1994) Workshop Report. AWI. Germany

Backman J, Jakobsson M, Frank M et al (2008) Age model and core-seismic integration for the Cenozoic Arctic coring expedition sediments from the Lomonosov ridge. Paleoceanography J 23(1)

Central Arctic Basin, the Map (2002) Scale 1:2500000 (at 75°N). GUNiO MO RF Saint Petersburg, №91115

Franke D, Hinz K, Reichert Ch (2004) Geology of the east Siberian Sea, Russian Arctic from seismic images: structures, evolution and implications for the evolution of the Arctic Ocean basin. J Geophys Res 109(7):1–19

Geomorphological Aspects of the Russian Continental Shelf Exterior Boundary in the Arctic (2005) Naryshkin GD (ed) GUNIO MO RF Saint Petersburg, p 58

Glebovsky VY, Kovacs LC, Maschenkov SP et al (1998) Joint compilation of Russian and US navy aeromagnetic data in the Central Arctic Seas. Polarforshung J 68:35–40. erschienen 2000

Glebovsky VY, Astafurova EG et al (2013) Thickness of the Earth's crust in the deep Arctic Ocean: results of a 3D gravity modeling. Russ Geol Geophys J 54(3):247–262

Gurevich NI, Maschenkov SP (2000) The crustal types of geological structures in the deep-sea Arctic basin in: collected articles. Geology and geophysics of the Arctic region lithosphere, vol 3. VNIIOkeangeologia, Saint Petersburg, pp 9–32

Hegewald A, Jokat W (2013a) Relative Sea level variations in the Chukchi region – Arctic Ocean – since the late Eocene. Geophys Res Lett 40:1–5

Hegewald A, Jokat W (2013b) Tectonic and sedimentary structures in the northern Chukchi region, Arctic Ocean. J Geophys Res 118(7):3285–3296

Jokat W, Ickrath M, O'Connor J (2013) Seismic transect across the Lomonosov and Mendeleev ridges: constraints on the geological evolution of the Amerasia Basin, Arctic Ocean. Geophys Res Lett 40(19):5047–5051

Kaminsky VD, Poselov VA, Avetisov GP et al (2014) Russian Arctic Geotransects (results of geological and geophysical studies). VNIIOkeangeologia, Saint Petersburg, p 164

Kashubin SN, Petrov OV et al (2011) Crust thickness in the circumpolar Arctic. Regionalnaya Geologiya i Metallogeniya J 46:5–13. (in Russion)

Kashubin SN, Pavlenkova NI, Petrov OV et al (2013) Types of the earth's crust the circumpolar Arctic. Regionalnaya Geologiya i Metallogeniya J 55:5–20. (in Russian)

Langinen A, Lebedeva-Ivanova N, Gee D et al (2009) Correlations between the Lomonosov ridge, Marvin spur and adjacent basins of the Arctic Ocean based on seismic data. Tectonophysics J 472:309–322

Laverov NP, Lobkovsky LI, Kononov MV et al (2013) A geodynamic model of the evolution of the Arctic basin and adjacent territories in the Mesozoic and Cenozoic and the outer limit of the Russian continental shelf. Geotectonics J 47(1):1–30

Lebedeva-Ivanova NN, Gee DG, Sergeyev MB (2011) Crustal structure of the east Siberian continental margin, Podvodnikov and Makarov basins, based on refraction seismic data (TransArctic 1989–1991). Geol Soc Lond 35(26):395–411

Moran K, Backman J, Brinkhuis H et al (2006) The Cenozoic palaeoenvironment of the Arctic Ocean. Nature J 441:601–606

National Centers for Environmental Information http://www.ngdc.noaa.gov

Nikishin AM, Malyshev KN et al (2014) Arctic region: new model of geodynamic history. Geophys Res Abstr 16:5575–5575

Nikishin AM, Petrov YB et al (eds) (2015) Highlights from the tectonic evolution of the Arctic Ocean revealed by new data from the Arktika-2011, Arktika-2012, and Arktika-2014 Russian expeditions and kinematic and numerical modeling. AAPG 3P Arctic polar petroleum potential Conference & Exhibition, Stavanger, Norway, 29 Sep – 2 Oct 2015

Orographic Map of the Arctic Basin (1995) [Maps] Naryshkin GD (ed) Scale 1 : 5 000 000. Karttakeskus, Helsinki

Pavlenkin AD, Poselov VA, Butsenko VV (1996) DSS surveys in the Arctic and the structure of the lithosphere. Geology and geophysics of the Arctic region lithosphere, vol 2, pp 145–155

Piskarev AL (2004) Petrophysical models of the Earth crust in the Arctic Ocean. VNIIOkeangeologia, Saint Petersburg

Poselov VA, Butsenko VV, Chernykh AA et al (2014) The structural integrity of the Lomonosov Ridge with the North American and Siberian continental margins. VI International Conference on Arctic Margins (ICAM VI), Fairbanks, Alaska, 30 May – 2 June 2011

Vinogradov VA, Gusev EA, Lopatin BG (2004) Age and structure of the sedimentary cover of the east Russian Arctic shelf. Geol-geophys Features Lithosphere Arctic Region J 5:202–212

Verba ML (1996) Gradual process of the oceanic-contninental crusts transformatrion in the Arctic Ocean. Geology and geophysics of the Arctic region Lithosphere, vol 1, pp 172–182. (in Russian)

Weigelt E, Jokat W, Franke D (2014) Seismostratigraphy of the Siberian sector of the Arctic Ocean and adjacent Laptev Sea shelf. J Geophys Res 119(7):5275–5289

Chapter 6
Makarov Basin

Alexey L. Piskarev, Yury G. Firsov, Victor A. Poselov, and Oleg E. Smirnov

Abstract Makarov Basin - abyssal plain, 3800 m bsl on average (occasionally–over 4000 m, very close to the depth of the Nansen and Canada Basins), is mostly flat, disturbed only by 800 m-high linear ridge – abyssal continuation of the Marvin Spur. The potential fields anomalies illustrate the variety of structural plans of the Makarov Basin and surrounding regions. Gravity anomalies inside the Makarov Basin display symmetrical pattern relative to the linear positive anomaly parallel to 120°E, flanked by gravity lows on both side.

The total thickness of the sedimentary cover varies from 2 to 4 km. The regional pre-Miocene unconformity RU at the base of the hemipelagic Miocene deposits is traceable through the whole Makarov Basin. The reduced Paleogene complex between the RU and post-Campanian unconformity pCU overlays the Upper Cretaceous and, possibly, older formations. Upper and lower crusts are estimated to be 3–6 km and 3–4 km, accordingly. Total crust thickness is 10–12 km.

Keywords Makarov basin · Gravity · Magnetic anomalies · Sedimentary cover · Earth's crust

6.1 Morphology

The Makarov Basin is a part of the Amerasian Basin of the Arctic Ocean, separated from the Eurasian Basin by the Lomonosov Ridge (Fig. 6.1).

The basin is enclosed from all sides by outer slopes which, in turns, are parts of complex tectonically affected continental slopes. The Makarov Basin is the deepest

A. L. Piskarev (✉)
All-Russian Research Institute of Geology and Mineral Resources of the World Ocean (VNIIOkeangeologia), Saint Petersburg, Russia

Saint Petersburg University, Saint Petersburg, Russia

Y. G. Firsov · V. A. Poselov · O. E. Smirnov
All-Russian Research Institute of Geology and Mineral Resources of the World Ocean (VNIIOkeangeologia), Saint Petersburg, Russia

© Springer International Publishing AG, part of Springer Nature 2019
A. Piskarev et al. (eds.), *Geologic Structures of the Arctic Basin*,
https://doi.org/10.1007/978-3-319-77742-9_6

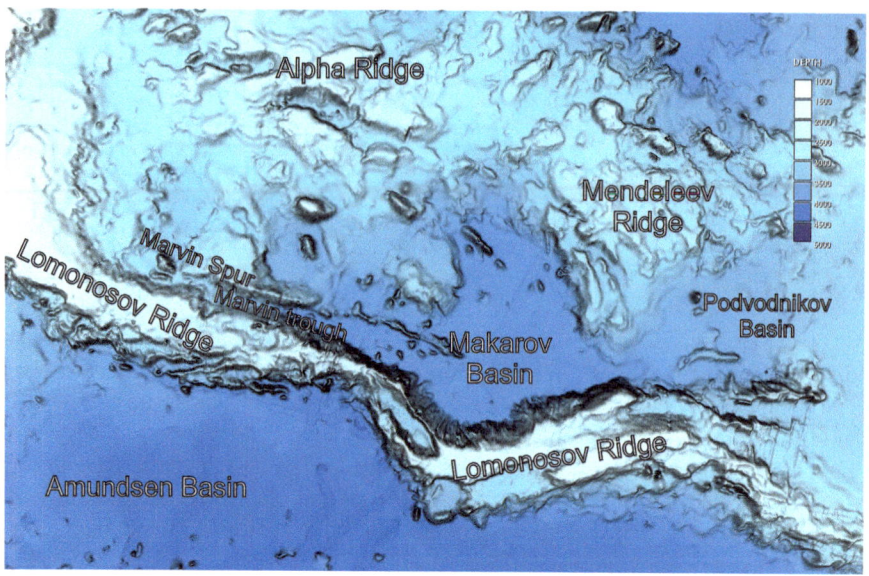

Fig. 6.1 The Makarov Basin (grid IBCAO 3.0, isobath interval-200 m)

of several depressions between the Lomonosov Ridge and Mendeleev-Alpha uplifts which includes the Podvodnikov Basin and unnamed depression outlined by −2600 m isobath between the Marvin Spur and the Alpha Ridge (Explanatory Notes to the Arctic Basin Maps 1999; Geomorphological Aspects of the Russian Continental Shelf Exterior Boundary in the Arctic 2005). Its abyssal plain, 3800 m b.s.l. on average (occasionally–over 4000 m) is mostly flat, disturbed only by 800 m-high linear range – abyssal continuation of the Marvin Spur. The common Lomonosov Ridge - Makarov Basin slope is known as Shmakov escarpment (http:// www.ngdc.noaa.gov/gazetteer/). It is much higher and steeper than the opposite slope of the Lomonosov Ridge facing the Amundsen Basin.

US project SCICEX data (submarine- based Side-Scan Sonar and gravimetry, (Cochran et al. 2006)) demonstrated that the Makarov Basin seafloor depth practically coincides with that of the Nansen and Canada Basins – the fact supported by the trans-oceanic bathymetry profile produced by GeoCap from the IBCAO grid (Fig. 6.2).

Compare to the Podvodnikov Basin, the Makarov Basin is not only deeper, but also has "rhomboidal" configuration similar to the generalized configuration of the "pull-apart" basins (Fig. 6.3).

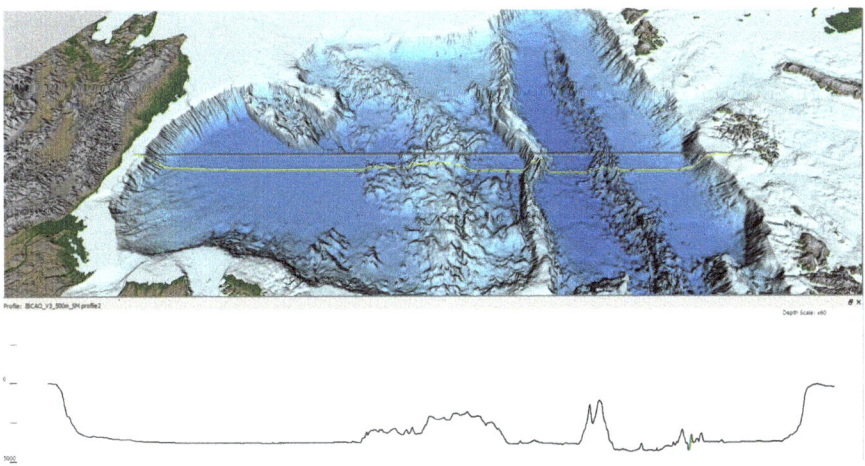

Fig. 6.2 Transoceanic bathymetry profile by GeoCap Project

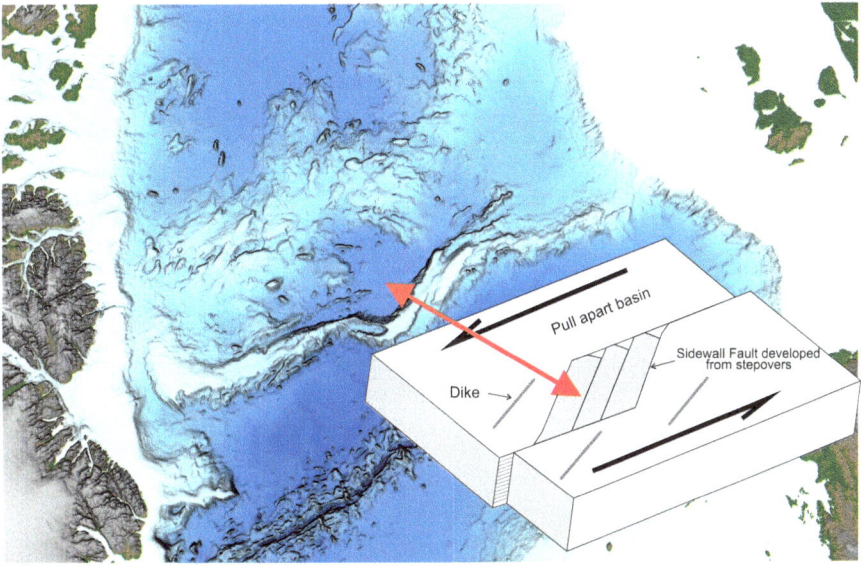

Fig. 6.3 The Makarov Basin as possible "pull-apart" basin - comparison with classical model

6.2   Potential Fields

As shown by the magnetic and Free Air gravity anomalies maps of the Central Arctic Uplifts inside the Amerasian Basin (Fig. 4.3), the potential fields anomalies graphically illustrate the variety of structural plans of the Makarov Basin and surrounding regions. Moreover, physical properties of rocks composing the structural elements can be estimated using special numerical methods.

Gravity anomalies inside the Makarov Basin display symmetrical pattern relative to the linear positive anomaly parallel to 120°E, flanked by gravity lows on both side. Sub-latitudinal magnetic anomalies (Fig. 4.3) originally were classified as related to spreading (Taylor et al. 1981), but later were related to faulting. Only small part of the Makarov Basin adjacent to the Lomonosov Ridge contains magnetic anomalies resembling the spreading type. There is no visible correlation between magnetic anomalies and underwater terrain. Judging by the high amplitude of the magnetic anomalies, the Makarov Basin basement has mainly mafic (basalt) composition, with some ultra-mafic material present.

6.3 Sedimentary Cover

The Podvodnikov and Makarov Basin were investigated by the Soviet ice station "North Pole - 28" (NP-28) (Fig. 6.4). Seismic data defined the Marvin Spur as a block separated from the Lomonosov Ridge and established that blocks of the continental crust are present in the Makarov Basin basement (Langinen et al. 2009).

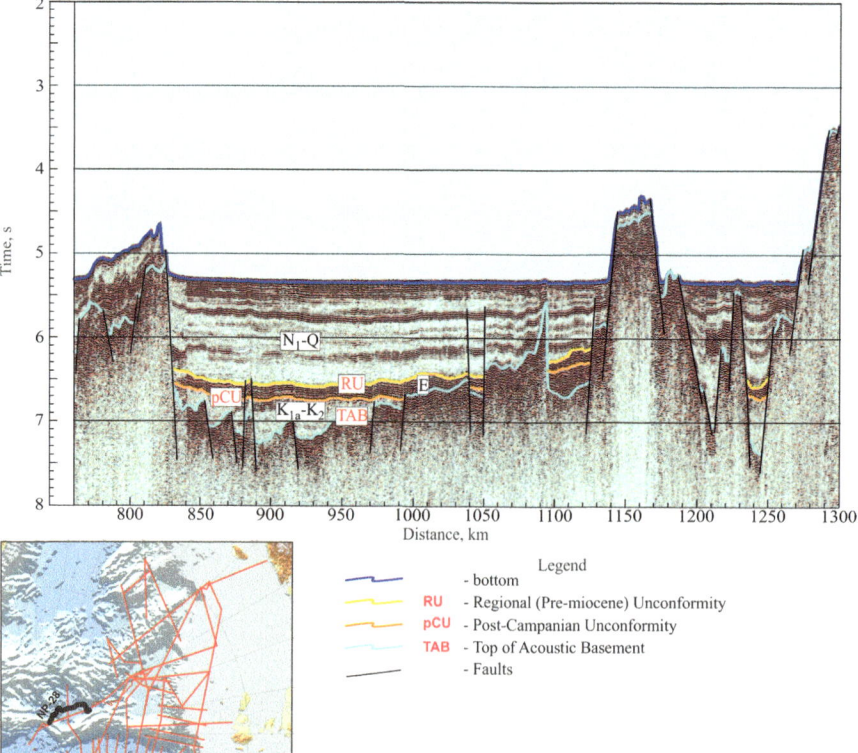

Fig. 6.4 MCS line along NP-28 drift in the Makarov Basin; RU and pCU (Poselov et al. 2014) correlate with A and LU (Langinen et al. 2009)

The total thickness of the sedimentary cover varies from 2 to 4 km (Jokat 2005; Langinen et al. 2009; Lebedeva-Ivanova et al. 2011). Both western (the Lomonosov Ridge) and eastern (Alpha Ridge) flanks of the Makarov Basin are formed by system of down-stepping normal faults. Top of the basement bears the signs of active, presumably rift-induced, faulting creating sub-latitudinal graben-horst system (Langinen et al. 2009).

Important trans-regional seismic program in the Central Arctic Ocean was completed in 2014. MCS lines 2014–06 and 2014–07 crossed the Makarov Basin from west to east and from north to south (Fig. 1.26).

The regional pre-Miocene unconformity RU at the base of the hemipelagic Miocene deposits, is traceable through the whole Makarov Basin. The Paleogene complex between the RU and post-Campanian unconformity pCU is also present but with reduced thickness. It overlays the Upper Cretaceous and, possibly, older formations and, lacking continuous reflectivity, may have syn-rift origin. Total thickness of the sedimentary cover in the Makarov Basin reaches 2 km. Buried under the Miocene-Pleistocene sediments, the rift valley in the central part of the Makarov Basin is filled by undisturbed Paleogene sediments. (Fig. 6.5).

Fig. 6.6 shows the crustal velocity model of the MCS line 2014–06 produced by seismic ray modeling.

In this model the crust consists of (from top down):

– Miocene–Pleistocene hemipelagic complex above the RU unconformity with $V_p = 1.8$–2.0 km/s in the Podvodnikov Basin and 1.9—2.3 km/s in the Makarov Basin; thickness increases laterally from 0.5 km to 1.5 in the same direction;
– Cretaceous deposits below regional pCU unconformity (here practically merged with RU unconformity); velocity changes from 2.2–2.3 km/s to 2.5 at the Podvodnikov Basin upper terrace (Arlis Uplift) to 2.5–2.7 km/s at its lower terrace and in the Makarov Basin depocenter - to 2.8–3.3 km/s; thickness - ~0.7 km, ~1 km and ~2,1 km accordingly (except the local uplifted blocks, where thickness can drop to few hundred, or even tens, of meters);.
– Acoustic basement, including metasediments, with $V_p = 4.$-4.6 km/s (zone of anomalously high velocity – 4.8-5.2 km/s - marks the transition between the Podvodnikov and Makarov Basins); thickness varies from 2 to 3 km, dropping to several hundred meters in the Makarov Basin depocenter;
– Crystalline basement confidently defined by refracted waves with typical for upper continental crystalline crust interval velocity of 5.8–6.2 km/s.

The velocity information from the above model was used for depth conversion of this line.

In Chap. 3 (Fig. 3.5) we presented the crustal velocity model for whole MCS line 2014–07 which spans both Eurasian and Amerasian Basins. In the Amerasian (1100–1500 km) part the velocity model identifies:

– Miocene–Pleistocene hemipelagic complex above the RU unconformity with Vp = 2.1–2.3 km/s in the Makarov Basin and 1.8–1.9 km/s on the Lomonosov Ridge; thicknesss - ~1.0 km in the Makarov Basin, decreasing to ~ 0.6 km in th Lomonosov Ridge;

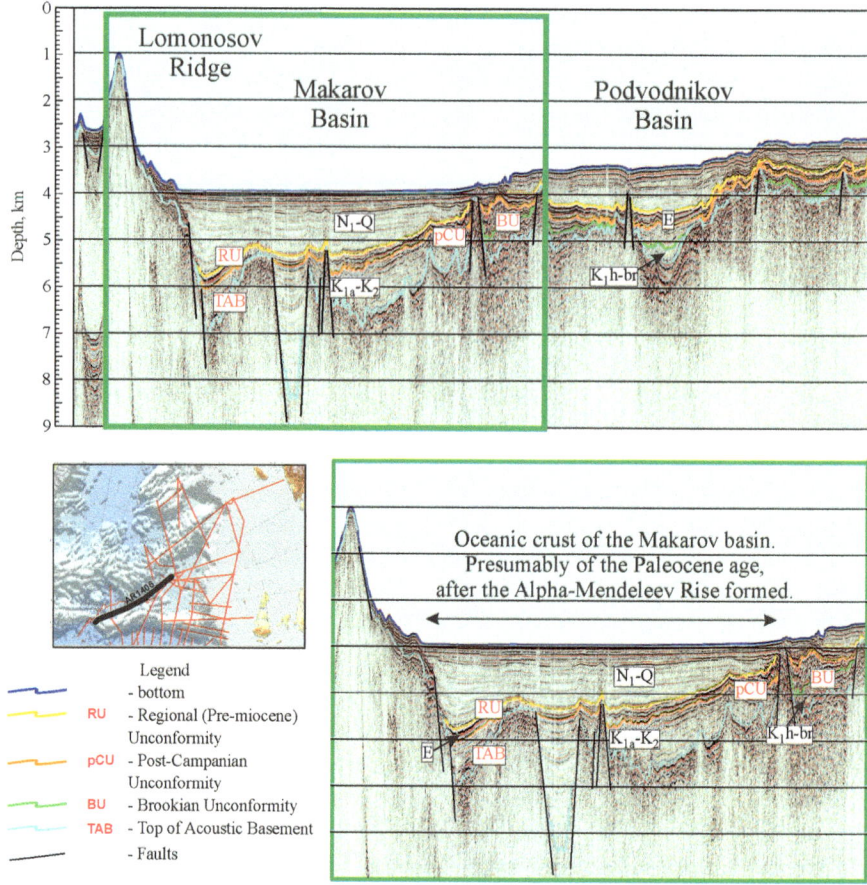

Fig. 6.5 MCS line 2014–06

– Cretaceous deposits below regional RU + pCU unconformities with Pv = 3.1–
 3.4 km/s and thickness, about 3 km, present only in the graben at the foot of the
 Lomonosov Ridge;
– Acoustic basement, including metasediments, with Vp = 4.-4.6 km/s in the
 Makarov Basin and 4.1—4.3 km/s in the Lomonosov Ridge; ~ 2.5 km thick in
 the eastern part of the Makarov Basin, thinning to ~ 1.3 km under the rift graben
 and reaching 1.5–2.1 km in the Lomonosov Ridge;
– Crystalline basement (by refracted waves) with typical for upper continental
 crystalline crust interval velocity of 5.8–6.2 km/s.

The TWT section of the eastern (Amerasian) part of that line (1100—1500 km)
is shown on Fig. 6.7).

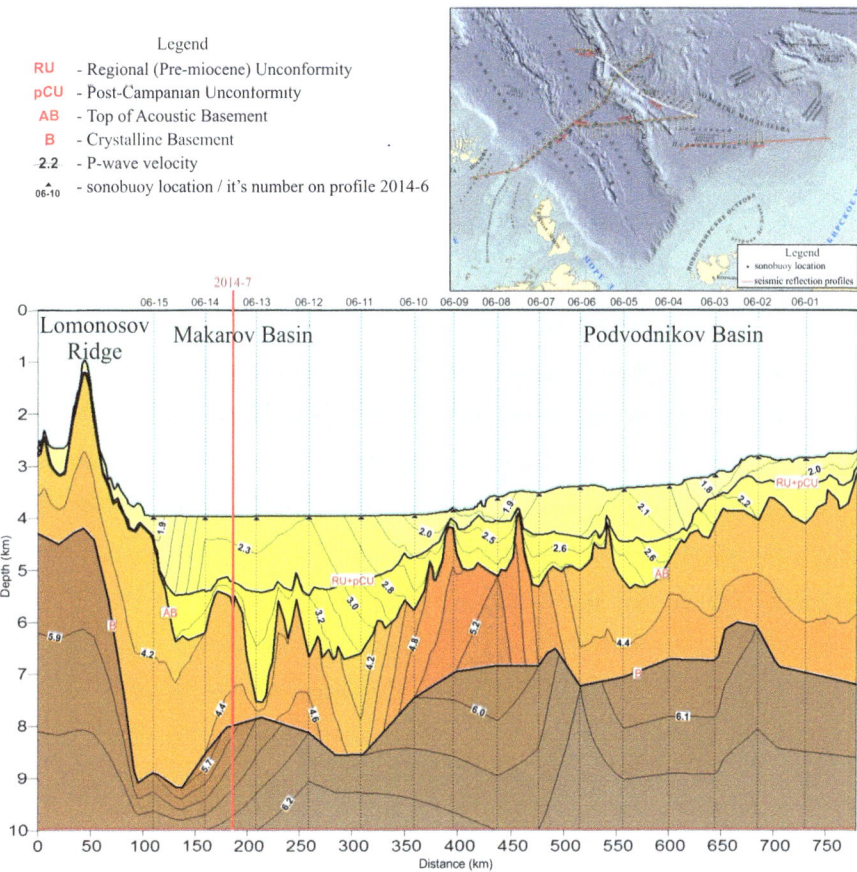

Fig. 6.6 Crustal velocity model (MCS line 2014–06)

(note the deep graben at the foot of to Lomonosov Ridge (1230–1250 km) filled with deformed, assumingly Cretaceous, sediments. The main part of the Makarov Basin (Fig. 6.7) is filled by Paleogene formations discordantly overlain by the Miocene-Pleistocene deposits.

MCS and refraction seismic data were acquired in 2011 along the 650 km long line from the Alpha Ridge across the Makarov Basin to the Lomonosov Ridge (Evangelatos and Mosher 2016). The thicknes of the sedimentary cover increase from 1.9 km in the Makarov Basin to almost 5 km at Lomonosov Ridge slope. Interval velocities inside the sedimentary cover vary from 1.6 to 4.3 km/s (Fig. 6.8).

The authors attribute relatively high velocity of the deeper sedimentary horizons to the presence of volcanic formations.

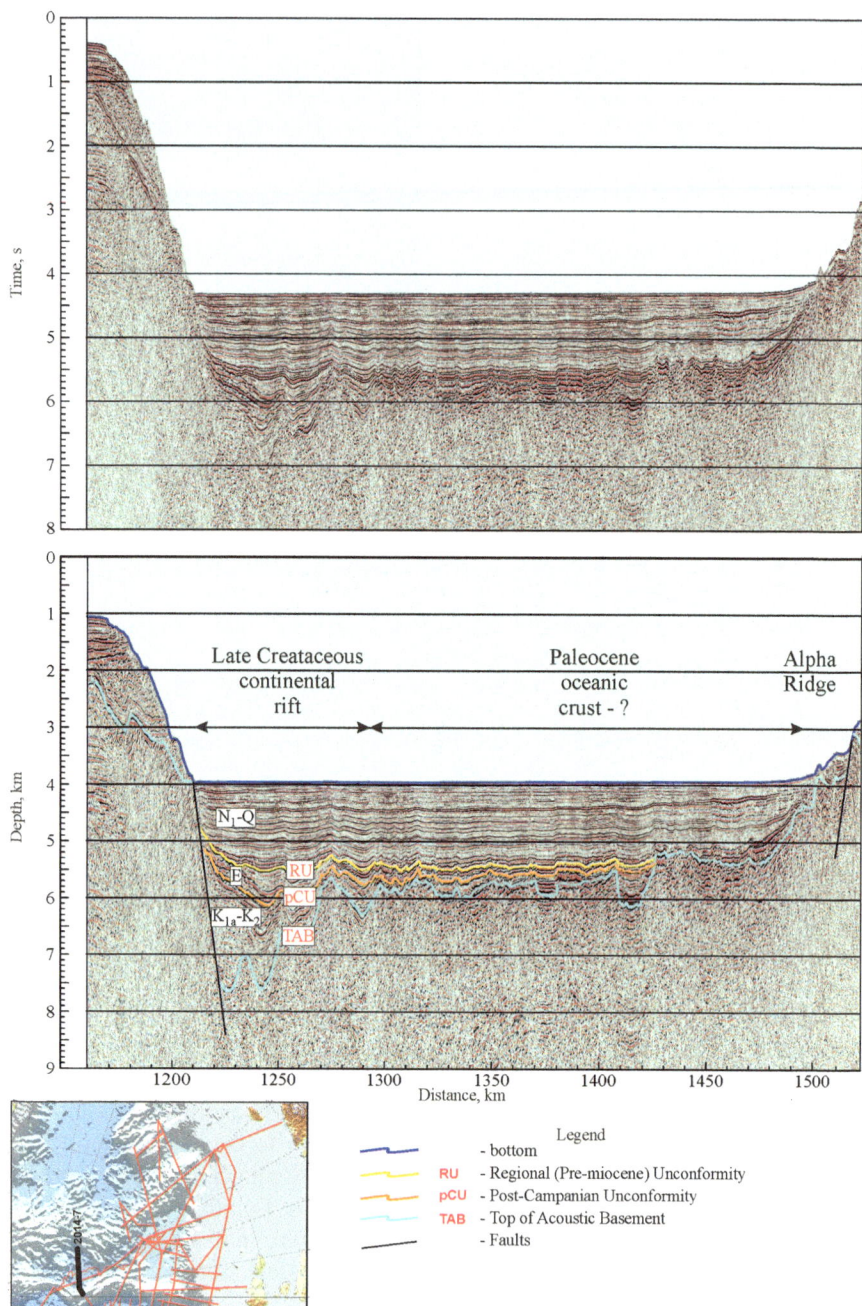

Fig. 6.7 MCS line 2014–07 (Makarov Basin)

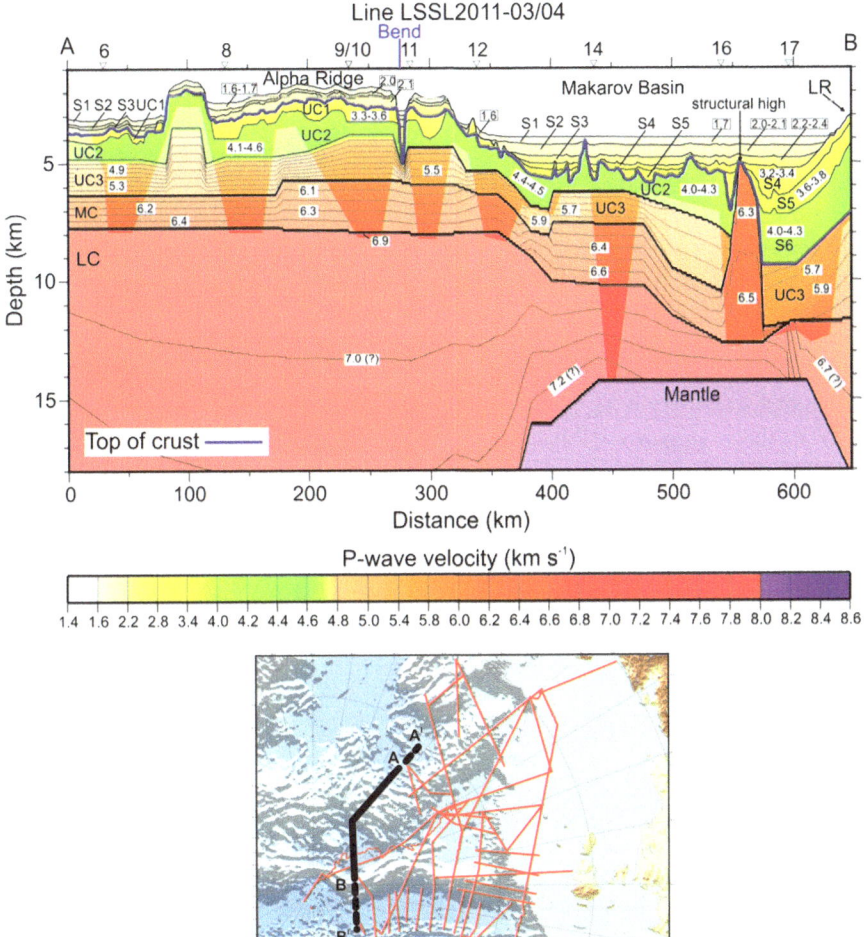

Fig. 6.8 P-wave velocity model for line LSSL2011–03/04 (A'-B'). Pale colours indicate sections unconstrained by CDP or refraction data. "Bend" - location of the line orientation change (Evangelatos and Mosher 2016)

6.4   Deep Crustal Structure

Based on DSS survey (1989–1991) along the regional profile from the De Long uplift into the Podvodnikov and the Makarov Basin, four-layered model of the Podvodnikov and Makarov Basins crust was postulated:

Layer I (V_p = 1,7—3,8 km/s) — Meso-Cenozoic sedimentary complexes;
Layer II (V_p = 5,0—5,4 km/s) — older sedimentary and, possibly, volcano-sedimentary formations (so called meta-sedimentary complex);

Layers III and IV (V_p = 5,9—6,5 km/s and 6,7—7,3 km/s, accordingly) – crystalline crust.

Moho discontinuity is identified at depth of 20 km under the southern and northern parts of the Podvodnikov Basin plunging down to 30 in center, and rising to 12–16 km under Makarov Basin (Fig. 5.27, 5.28). The spur of Mendeleev Ridge along the southern boundary of Makarov Basin is characterized by 25 km thick crystalline crust made of layers III and IY. It was suggested (Langinen et al. 2009; Lebedeva-Ivanova et al. 2011) that the abyssal part of the Makarov Basin is supported 8–12 km oceanic crust with possible inclusions of continental blocks split from the Lomonosov Ridge.

Two more gravity models highlighting the deep crustal structure of the Makarov Basin and their descriptions can be found in Chap. 3 (Fig. 3.6, 900–1200 km interval) and Chap. 5 (Fig. 5.31, 1200–1500 km interval). Transition from the Makarov Basin to the Lomonosov Ridge domain is manifested by substantial increase of crustal thickness up to 21 km (mostly at the expense of the upper crust thickening to 14–17 km) and strong (100 mGal) gravity anomaly marking sharp, 2.5 km above the seafloor, rise of the ridge. The sediments thickness here is negligent (first hundred meters).

Gravity model of profile LSSL2011–03/04 (Evangelatos and Mosher 2016) is presented on Fig. 6.9:

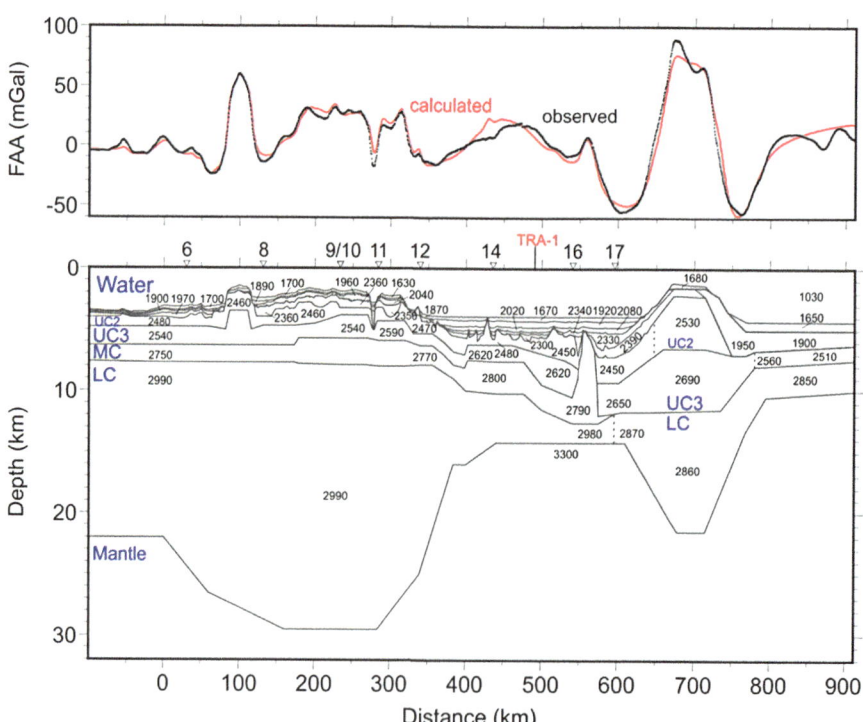

Fig. 6.9 Gravity model for line LSSL2011–03/04 (location -Fig. 6.8; densities - in kg/m³)

It was noted that:

(a) sharp increase of crustal thickness at the Makarov Basin - Lomonosov Ridge interface may indicate its transform nature;
(b) prominent basement uplift at the Makarov Basin depocenter with maximaum thickness of the sedimentary cover may signify the spreading activity in Late Cretaceous and/or Paleogene (Evangelatos and Mosher 2016).

6.5 Conclusions

The Makarov Basin can be described as micro-oceanic basin with oceanic crust, formed within the Central Arctic Uplifts as a "pull-apart" basin possibly synchronous to opening of the Canada Basin. According to N. Bogdanov (2004), the initial pull-apart depression was wider than the modern Makarov Basin and its floor consequently was segmented by active normal and strike-slip faulting.

The Makarov Basin contains (from top down):

- Hemipelagic Miocene Pleistocene complex above the pre-Miocene unconformity with P-wave velocity between 1.9–2.3 km/s and thickness 0.5–1.5 km;
- Cretaceous sedimentary complex under the RU + pCU regional unconformity, Pv from 2.2–2.3 to 2.8–3.3 km/s, thickness 0.7–2.1 km, (a few hundred or tens of meters at the local uplifted blocks);
- Acoustic basement with P-waves velocity 4.2–4.6 km/s (locally reaching 4.8–5.2 km/s) and thickness 2–3 km.

Upper and lower crusts are estimated to be 3–6 km and 3–4 km, accordingly. Total crust thickness is 10–12 km.

References

Bogdanov NA (2004) Tectonics of the Arctic Ocean. Geotectonics 3:13–30

Cochran JR, Edwards MH, Coakley BJ (2006) Morphology and structure of the Lomonosov ridge, Arctic Ocean. Geochemistry Geophysics Geosystems 7(5):1525–2027

Evangelatos JE, Mosher DC (2016) Seismic stratigraphy, structure and morphology of Makarov Basin and surrounding regions: tectonic implications. Marine Geology 374:1–13

Explanatory notes to the Arctic Basin maps: Orographic map of the Arctic Basin (1999) The Arctic ocean seafloor topography (Text) Saint Petersburg, p 39

Geomorphological aspects of the Russian continental shelf exterior boundary in the Arctic. (2005) Naryshkin GD (ed) GUNIO MO RF Saint Petersburg, p 58

Jokat W (2005) The sedimentary structure of the Lomonosov ridge between 88°N and 80°N. Geophys J Int 163:698–726

Langinen A, Lebedeva-Ivanova N, Gee D et al (2009) Correlations between the Lomonosov ridge, Marvin spur and adjacent basins of the Arctic Ocean based on seismic data. Tectonophysics 472:309–322

Lebedeva-Ivanova NN, Gee DG, Sergeyev MB (2011) Crustal structure of the east Siberian conti-
    nental margin, Podvodnikov and Makarov basins, based on refraction seismic data (TransArctic
    1989-1991). Geological Society of London 35(26):395–411
National Centers for Environmental Information http://www.ngdc.noaa.gov
Poselov VA, Butsenko VV, Chernykh AA et al (2014) The structural integrity of the Lomonosov
    Ridge with the North American and Siberian continental margins. VI International conference
    on Arctic Margins (ICAM VI), Fairbanks, Alaska, 30 May–2 June 2011
Taylor PT, Kovacs LC, Vogt PR et al (1981) Detailed aeromagnetic investigation of the Arctic
    Basin. J Geophys Res 86:6323–6333

Chapter 7
Mendeleev and Alpha Ridges

Victor V. Butsenko, Yury G. Firsov, Evgeny A. Gusev, Valery D. Kaminsky, Sergey P. Kashubin, Alexey L. Piskarev, and Victor A. Poselov

Abstract The Mendeleev Ridge was formed on the continental crust with total thickness 27–32 km of which the upper crust takes 4–7 km. Continuous chain of progressively seaward deepening bathymetric terraces demonstrate the morphological connection of the Mendeleev Ridge with shallow water regions of the Siberian-Chukchi continental margin. Seismic data reveal that the structure and stratigraphy of the Mendeleev Ridge are affected by intensive normal faulting of acoustic basement creating complex system of grabens and half-grabens. It depicts the Mendeleev Ridge as an extensional structure of Cretaceous age illustrating tectonic evolution of the Central Arctic region. The Miocene-Pleistocene complex, with continuous undisturbed layers of hemipelagic sediments and regional erosional unconformity at its base, drapes the entire Mendeleev Ridge and marks the completion of contemporary morphological modern system of the Mendeleev-Alpha Ridge System.

Keywords Mendeleev Ridge · Alpha Ridge · Sedimentary cover · Extensional structure · Continental crust · Normal faults

V. V. Butsenko (✉) · Y. G. Firsov · E. A. Gusev · V. D. Kaminsky · V. A. Poselov
All-Russian Research Institute of Geology and Mineral Resources of the World Ocean
(VNIIOkeangeologia), Saint Petersburg, Russia
e-mail: vicb@vniio.nw.ru; okeangeo@vniio.ru

S. P. Kashubin
A.P. Karpinsky Russian Geological Research Institute (VSEGEI), Saint Petersburg, Russia
e-mail: sergey_kashubin@vsegei.ru

A. L. Piskarev
All-Russian Research Institute of Geology and Mineral Resources of the World Ocean
(VNIIOkeangeologia), Saint Petersburg, Russia

Saint Petersburg University, Saint Petersburg, Russia

© Springer International Publishing AG, part of Springer Nature 2019   239
A. Piskarev et al. (eds.), *Geologic Structures of the Arctic Basin*,
https://doi.org/10.1007/978-3-319-77742-9_7

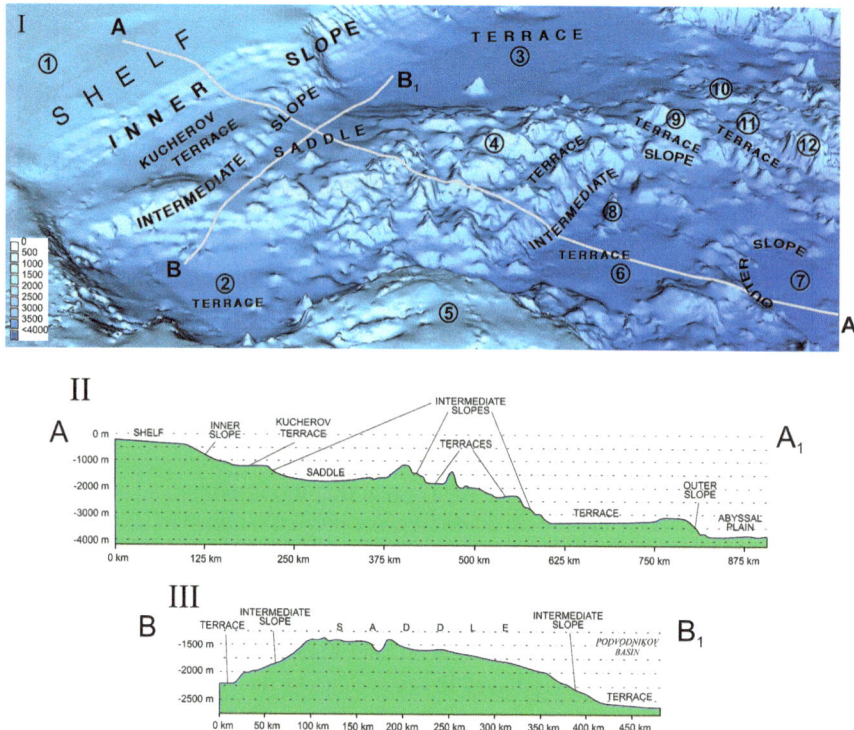

Fig. 7.1 Geomorphology of the Mendeleev Ridge region: (grid IBCAO v.3): I – general view, looking from the Chukchi Plateau; II and III – bathymetry along lines A-A$_1$ and B-B1, accordingly; Circled numbers: 1 – East-Siberian shelf; 2 – Chukchi Basin terrace; 3 – Podvodnikov Basin; 4 – Mendeleev Ridge; 5 – Chukchi Plateau; 6 – Mendeleev Basin terrace; 7 – Nautilus Basin; Seamounts: 8 – Karusev; 9 – Shamshura; 10 – Trukshin; 11 – Kovrigin; 12 – Rogotsky

7.1 Morphology

Mendeleev Ridge is the largest uplift of Central Arctic seafloor topography. (Fig. 7.1).

At the East-Siberian shelf the Mendeleev Ridge is 450 km wide. Stretching north for 700 km, it narrows down to 150–200 km at the Alpha Ridge junction. For its entire length, the Mendeleev Ridge is sculptured by the intermediate slopes; western, straight and sloping, with vertical drop decreasing from 800 m in the south to 200 m northward, and eastern - steeper, taller (600–800 m) and more complex. In the southern termination the Mendeleev Ridge intermediate slopes become inner slopes on the fringes of the Podvodnikov and Chukchi basins. Intersecting linear topographical features with sub-meridional (prevailing) and north-eastern (secondary) orientation define complex morphology of the Mendeleev Ridge (Fig. 7.2).

The large linear table-top plateaus, so typical for the Lomonosov Ridge at 1000 m b.s.l, are absent in the Mendeleev Ridge. Instead, only isolated remnants of these plateaus, 400–800 tall are present here. They are, on average, 400–800 m tall,

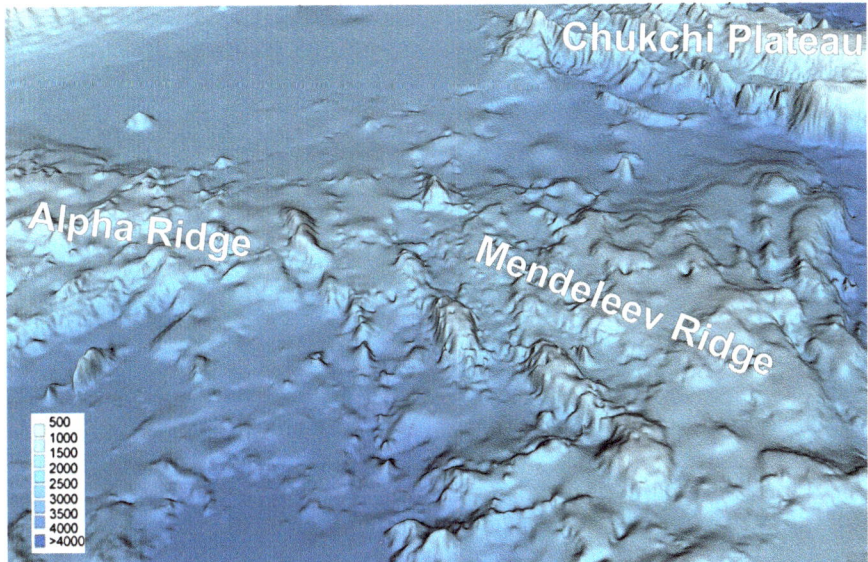

Fig. 7.2 Horst-like terrain of the central Mendeleev-Aplha Ridge System (grid IBCAO v.3)

but some of them reach 1000–1400 m (Shamshura, Rogotsky and Trukshina seamounts). Few linear troughs are visible on the surface, but geophysical data indicate that more of these extensional tectonic features are buried under the sediments. In contrast, terraces of different shape and size are widely spread. The ridge northern boundary is marked by large linear 10–20 km wide valley trending north-west. With depth of 2600–2800 m, this valley marks the deepest part of the Mendeleev Ridge at the junction with the Alpha Ridge.

One of the apparent features of the Mendeleev Ridge morphology is its stepped character defined by several terraces and saddles at distinct bathymetric intervals, deepening northward. The first such interval (1000–1200 m) is represented by the Kucherov Terrace between shelf inner slope and the intermediate slope (Fig. 7.1). The subsequent intervals are fixed at 1400–1800 (saddle, terrace), 2200–2400 m and 2600–2800 m (terraces). The A-A₁ profile shows the lowest interval is represented by Mendeleev Basin terrace, bounded seawards by outer slope – the last bathymetric gradient of the Mendeleev Ridge separating it from the abyssal plain.

If we concentrate our attention to the shelf-Mendeleev Ridge transition zone, we will find first signs of stepped block-wise terrain subsidence at the outer shelf. Here several seaward tilted and 200–250 km wide terraces are present at 300–350 m depth, heavily eroded at the shelf edge. Below the shelf edge, the inner slope (up to 1000 m deep) joins the Kucherov Terrace. The Kucherov Terrace is, in turn, terminated by intermediate slope with amplitude 400–500 m. This intermediate slope brings us to the wide saddle, opening and tilting to the west.

This continuing sequence of regularly deepening bathymetric intervals convincingly demonstrates the morphological ties between the Siberia-Chukchi continental margin and the Mendeleev Ridge. Morphology of the Mendeleev Ridge suggests that it was subjected to intensive tensional stresses leading to tectonic fragmentation of initially flat-topped plateau into present system of differently displaced fault blocks, which distinguishes the Mendeleev Ridge among other major morphologic structures of the Central Arctic region.

7.2 Potential Field Anomalies

Both the Mendeleev Ridge and Alpha Ridge are characterized by very strong an highly irregular magnetic fields with no easily recognizable patterns of anomalies (Fig. 4.3). Initialy, Karasik (1980) and later Gurevich N.I. and Maschenkov S.P. (2000) considered as prevailing anomalies following the axial line, but later north-eastern trending and latitudinal anomalies were also identified (Fig. 1.16, 4.3). Lack of linearity handicaps identification and tracing of faults, however, some signs of it can be discerned in the deep depressions north from Chukchi Plateau and Makarov Basin. In any case, it must taken into account that aeromagnetic maps of 60th–70th vintage are unreliable due to low positioning accuracy (Fig. 1.13).

Despite the visual similarity of anomalous magnetic field in the Mendeleev Ridge and Alpha Ridge, the satellite magnetic data, recorded at much higher altitude, show significant difference: The Mendeleev Ridge has practically no magnetic signature, while the Alpha Ridge is accompanied by strong magnetic anomaly possibly indicating the shallower sources in the former.

The magnetic fields similar with similar characteristics are observed in large magmatic provinces. The comparative spectral analysis by V.V. Verba (2006) showed similarities between magnetic fields of the Mendeleev-Alpha and Siberian trap province. Recent analysis of gravity (Glebovsky et al. 1998; Kenyon and Forsberg 2001) and magnetic (Verhoef et al. 1996) anomalous fields identified Mendeleev Ridge and Alpha Ridge as a part High Arctic Large Igneous Province (HALIP) (Grantz et al. 2009, 2011; Saltus et al. 2011) and combined them in to one geological province.

Anomalous Free Air gravity field (Fig. 4.4) of the Mendeleev-Alpha Ridge System consists of alternating highs and lows, generally reflecting its the bathymetry and gradually decreasing in intensity towards adjacent plateaus and depression, especially Canada Basin.

7.3 Sedimentary Cover

The grid of MCS lines over the Mendeleev Ridge was acquired during the Russian expedition "Arktika-2012" (Fig. 1.26). The interpretation of seismic sections was based on the seismic-stratigraphic benchmark model developed for the Central Arctic uplifts (see Chap. 2) which, in turn, relied on correlation with the results of the "Arktika-2011" program in the Podvodnikov Basin and, consequently, with

ACEX well (Backman et al. 2006). All that made possible direct uninterrupted tracing of the major regional unconformities - Pre-Miocene (RU), Post-Campanian (pCU) and Brookian (BU) - from the North-Chukchi Trough to the Mendeleev Ridge (Poselov et al. 2017). Bottom sampling information (Morozov et al. 2013) also was utilized.

Three seismic-stratigraphic complexes are identified in the in the Mendeleev Ridge sedimentary cover:

N_1-Q – between seafloor and RU, Lower/Mid-Moicene – Plestocene (18.2–0 Ma); correlates with Unit 1 in ACEX well (Moran et al. 2006; Backman et al. 2008); hemipelagic "ice house" glacial-marine clays unconformably overlaying Lower-Mid Eocene/Oligocene clays;

E - (between RU and pCU); correlated with ACEX Unit 2 ("green house" biosilicate clays and silts, Lower/Mid-Eocene) and Unit 3 (compact silty clays-argillites, Mid Paleocene - Eocene on top of shallow-water Cretaceous sands) (Moran et al. 2006). Lower-Mid Eocene sediments containing fresh-water Azolla pollen were identified in the ACEX well. It was assumed that reduced Paleogene section in the Lomonosov Ridge points to either erosion or depositional hiatus. Admittedly, this interval may be more complete in the Podvodnikov Basin and at the Mendeleev Ridge, therefore this complex is interpreted as un-subdivided Paleogene.

K – syn-rift complex between pCU and acoustic basement (AB); present in grabens, half-grabens and sub-basins, often with strong reflectors close to AB, possible indicators of syn-rift basalt flows. The deepest tectonic depressions may contain Upper Jurassic deposits.

Several examples of "Arktika-2012" MCS lines covering the Mendeleev Ridge arc presented below.

7.3.1 MCS Line 2012–01

MCS line 2012–01 (1030 km) runs from East-Siberian - Chukchi continental margin along the Mendeleev Ridge axis to the junction with the Alpha Ridge (Fig. 7.3). The North Chukchi Basin (0–270 km), with AB 16–17 km deep, has maximum volume of sediments and the most complete stratigraphic section. Despite weak reflectivity below BU unconformity (the line was shot with streamer only 600 m long), it is obvious that Pre-Cretaceous horizons are truncated by rising northwestern flank of the North-Chukchi Trough and do not spread onto the Mendeleev Ridge. The Brookian unconformity BU (base of prograding Aptian-Albian – Upper Cretaceous sequence of the North-Chukchi Trough – see Chap. 2) is visible only in the North-Chukchi Trough and under the Kucherov Terrace. Cenozoic clinoforms (especially Paleogene clinoforms between RU and pCU) have stronger seismic signature under the Kucherov Terrace; Neogene clinoforms become more pronounced further south, closer to edge of the Chukchi Shelf.

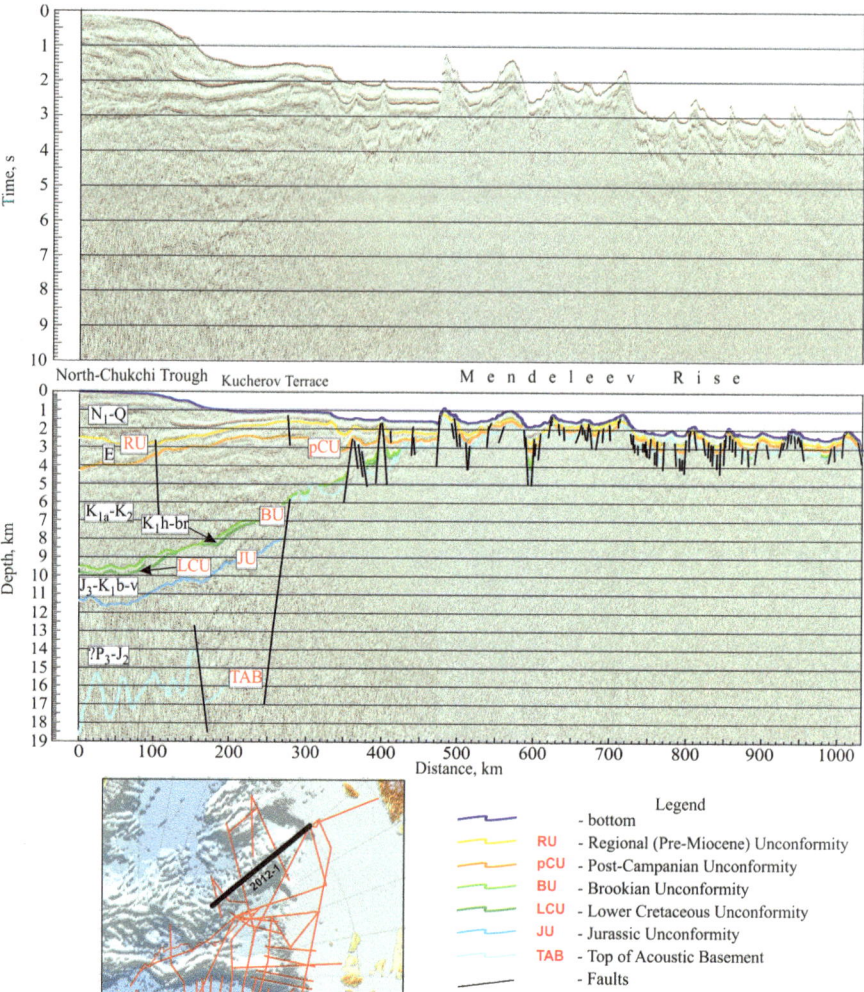

Fig. 7.3 MCS line 012–01

The central and northern parts (470–1030 km) highlights the structure of the Mendeleev Ridge as horst-graben system of differentiated fault blocks reflected in sea bottom topography, gradually merging into similar structure of the Alpha Ridge (at 920–945 km the line crosses eastern slopes of the Shamshura Seamount). Several local grabens were formed in the Cretacous and buried under Paleogene sediments.

Figure 7.4 illustrates the process of tracing of major regional unconformities RU and pCU from the North-Chukchi Trough into the Mendeleev Ridge and further into Podvodnikov Basin and the Lomonosov Ridge using the existing system of seismic lines. Incidentely, the seismic events representing these unconformities arrive at intersections not only at the similar TWT, but also in the same phase, thus confirming the veracity of presented "long distance" correlation and interpolation.

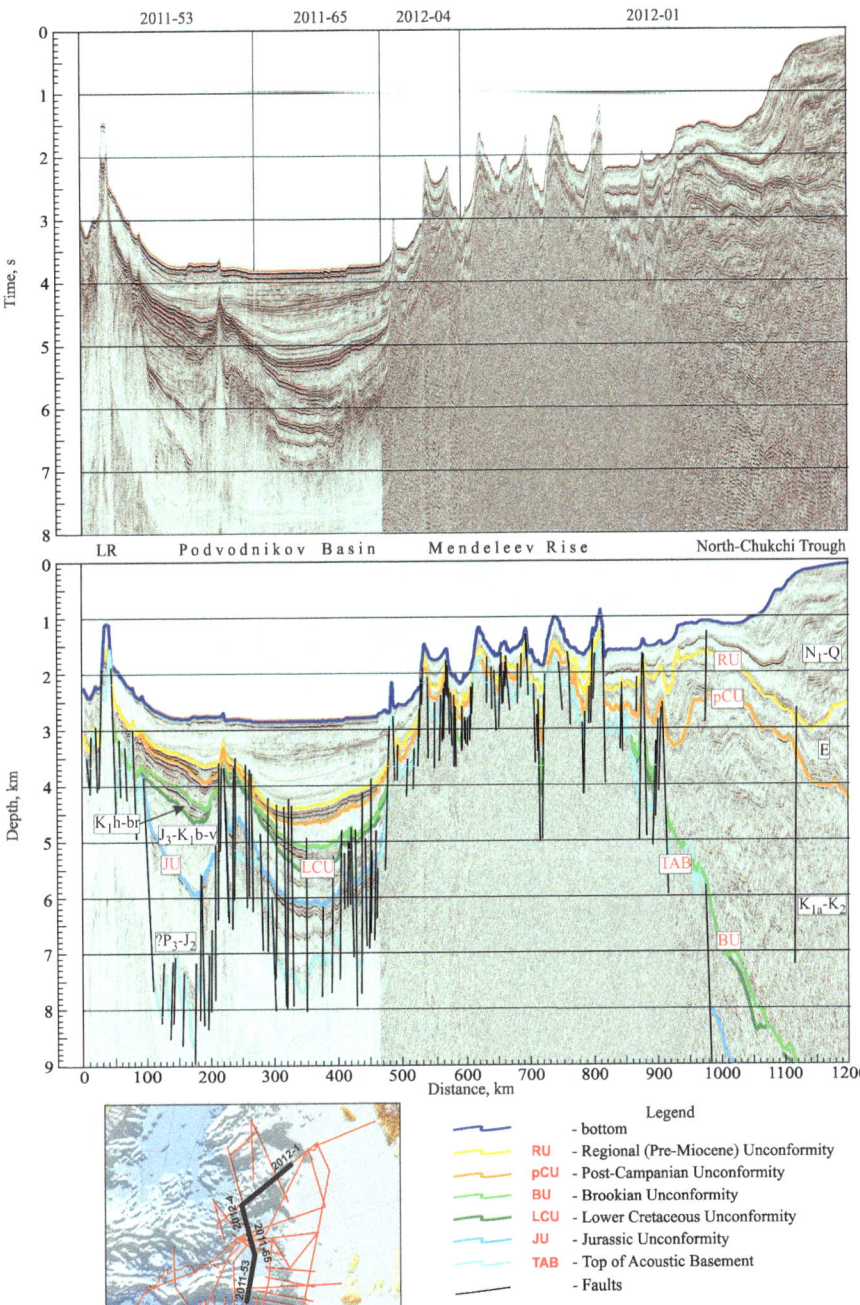

Fig. 7.4 Correlation of seismic horizons on the composite MCS profile (lines 2011–53, 2011–55, 2012–4 and 2012–1) from the Lomonosov Ridge to the North-Chukchi Trough

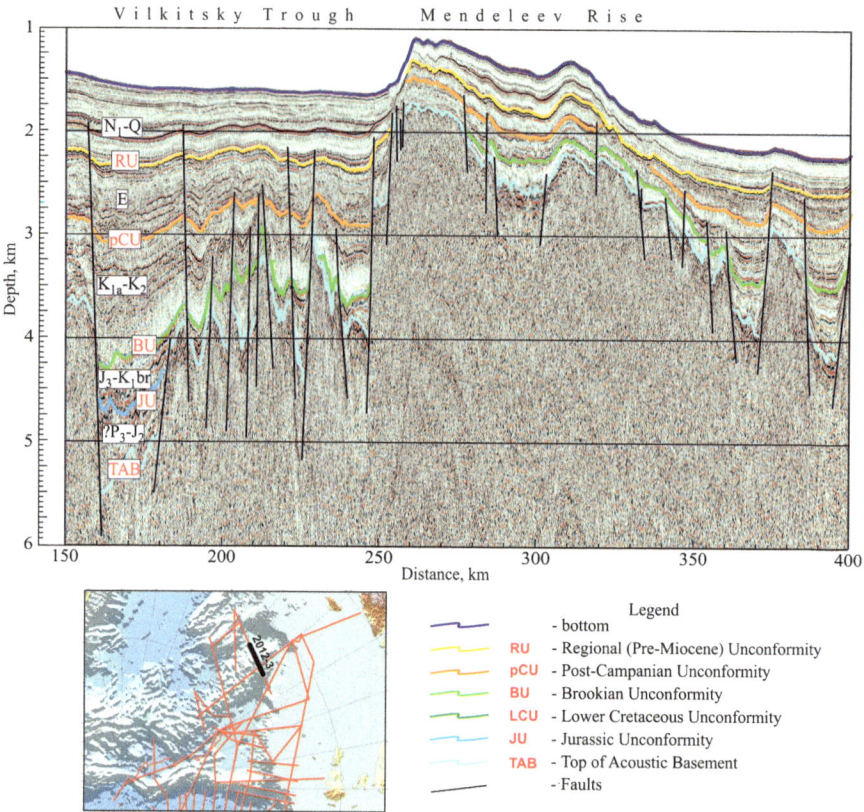

Fig. 7.5 Fragment of MCS line 2012–03

7.3.2 MCS Line 2012–03

MCS line 2012–03, shot with 4500 m streamer crosses southern part of the Mendeleev Ridge (Fig. 7.5). The western part (150–250 km) shows the south-western closure of the Vilkitsky Trough (or south-eastern closure of the North-Chukchi Trough) with AB at 4–5 km and 3–4 km of sediments. Numerous grabens and half-grabens, striking north-south or north-east, of Jurassic?- Early Cretaceous activation disturb the AB but are compensated by Paleogene sediments.

The line crosses the top of the Mendeleev Ridge at interval 250–320 km. Steep western slope formed by system of sub-meridional high-amplitude normal faults dispalcing AB by more than 2 km. More gradual eastern slope is complicated by another system of the normal faults active from Upper Jurassic?- Createcous to Paleogene.

7.3.3   MCS Line 2012–04

This 610 km long line crosses the northern part of the Mendeleev Ridge and shows it relations to the adjacent Podvodnikov and Mendeleev basins (Fig. 7.6).

The central part of the Mendeleev Ridge is complicated by distinct uplifted fault block (90–143 km) with top at 1500–1700 m b.s.l. Another similar, but smaller, features exist west (30–35 km, 2200 m b.s.l.) and east (240–270 km, 2100–2500 m b.s.l.) from it. Thickness of the Cretaceous-Cenozoic sedimentary section varies from 300-500 m at the local uplifts to 1500–2000 m in the local depressions (150–200 km). Noticeable increase in thickness westwards to the Podvodnikov Basin is due to presence of the Jurassic? formations.

Important information regarding the structure of the upper parts of the acoustic basement can be deduced from the enlarged part of 100–140 km interval of the discussed line (Fig. 7.7). Here strong and continuous (up to 20 km) dipping to north-west reflectors are visible below TAB defining a specific seismic-stratigraphic sequence up to 1500 m thick and with interval velocity 2.9–3.0 km/s. This information let us to assume that at least upper part of the acoustic basement consist of deformed sedimentary platform formations.

7.3.4   MCS Line 2012–05

316 km long line 2012–05 highlights the details of the Mendeleev Ridge-Alpha Ridge and Mendeleev Ridge-Mendeleev Basin transition zones. It also crosses some locations where bottom samples were collected Fig. 7.8–7.9).

Trukshin Seamount (31–55 km) stands out as uplifted block formed by normal faults with amplitudes close to 2000 m. Those faults strike north-east, the orientation more typical for the Alpha Ridge, rather than the Mendeelev Ridge (which, perhaps, is what can be expected in the transition zone between the two ridges). Those faults, synchronous with formation of the Mendeleev–Alpha in-block structure, were initiated in the Cretaceous and periodically re-activated during Paleogene till modern times. Several weaker faults are present on the Mendeleev Ridge eastern slope, reflected by the drape structures inside the N_1-Q complex. Thickness of the Cretaceous-Cenozoic formations changes from 200–300 m at the summit of the Trukshin Seamount to 2000 m in local depressions east and west from it.

The acoustic basements outcrops on the flanks of the Trukshin Seamount. Strong reflections observed in the upper parts of the acoustic basement can be interpreted as seismic signature of the interbedding basalt flows and tuffs.

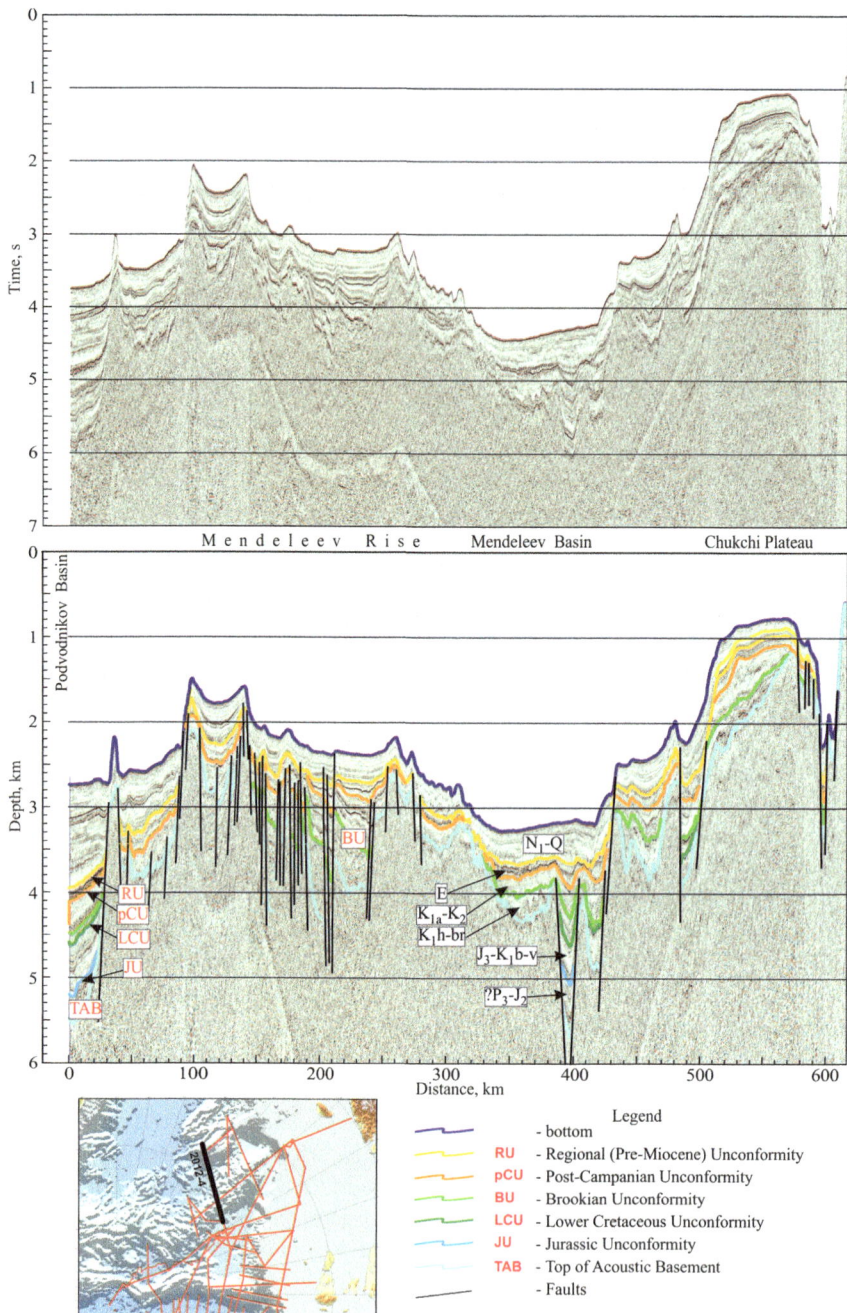

Fig. 7.6 MCS line 2012–04

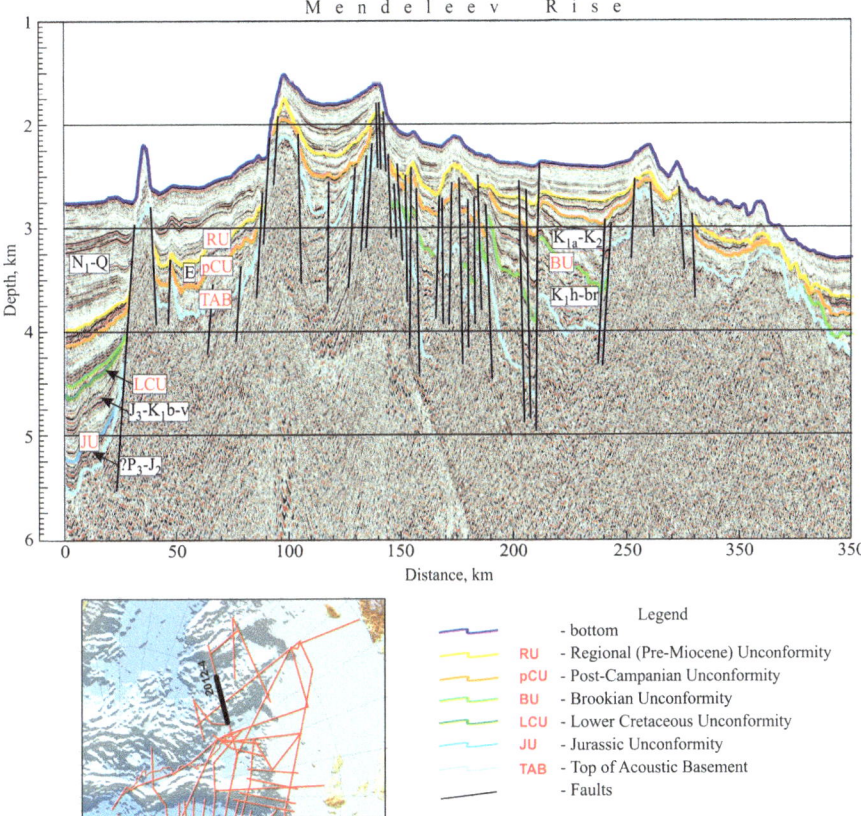

Fig. 7.7   MCS line 2012–04 - fragment

7.3.5   *MCS Line 2012–17*

MCS line 2012–17 illustrates the structural relations of the northern Mendeleev Ridge and its neighbors from the east (Mendeleev Basin) and west (Podvodnikov Basin (Fig. 7.10–7.11).

The Shamshura Seamount is located on the axial part (155–200 km) of the Mendeleev Ridge. It is formed by faults with vertical displacement close to 1000–1200 m, originated in Cretaceous and re-activated in Paleogene-Miocene. The acoustic basement deepens step-wise by series of low-amplitude normal faults towards the Podvodnikov and Mendeleev basins. The thickness of Cretaceous-Cenozoic complex increases from 200–300 m on the Shamshura Seamount to 3000 m in the Podvodnikov Basin owing to Jurassic? sequence.

Drill core extracted from the outcrop during expedition 'Arktika-2012" sampling program (Fig. 7.9, 32 km mark) contains brecciated trachybasalt with clayish matrix, typical for bathyal explosive volcanism.

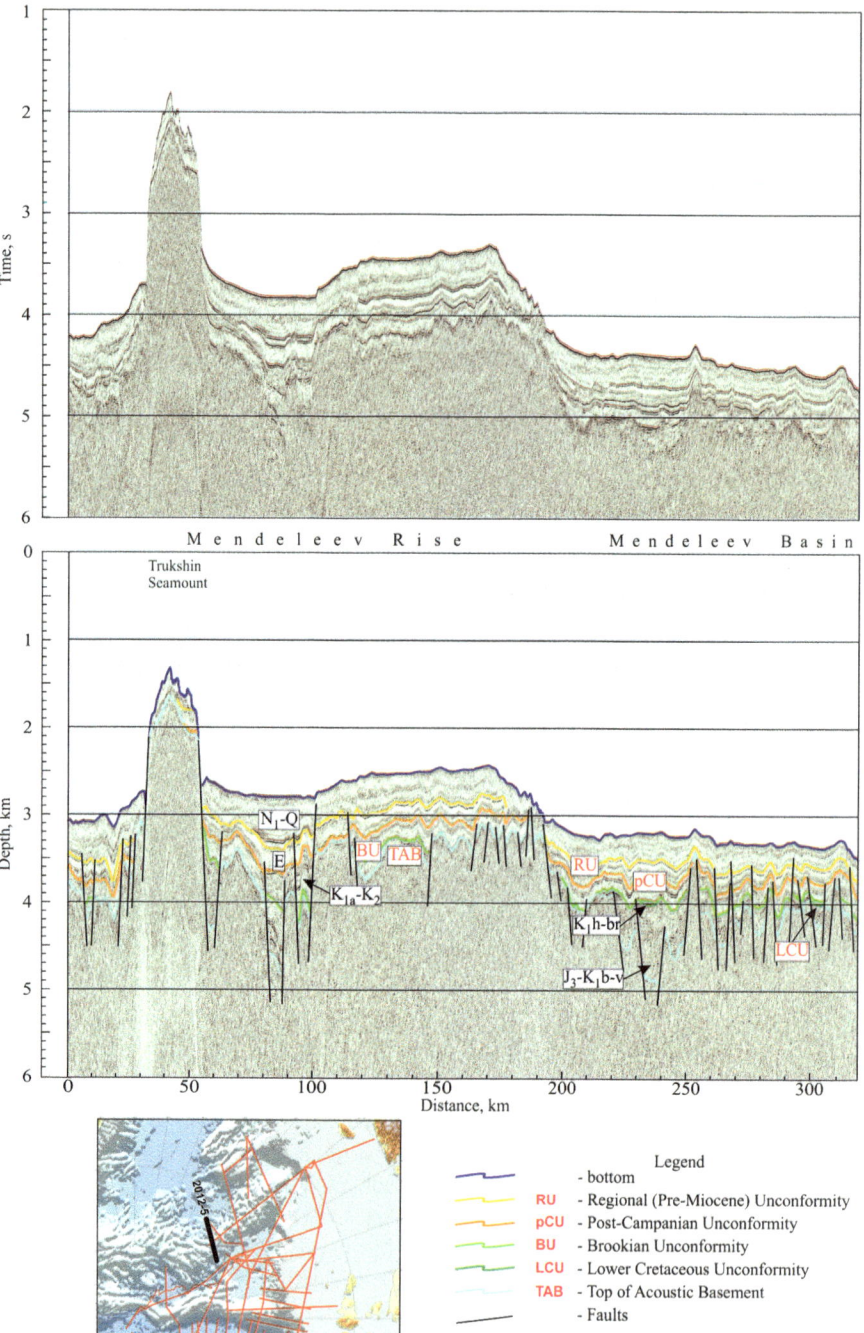

Fig. 7.8 MCS line 2012–05

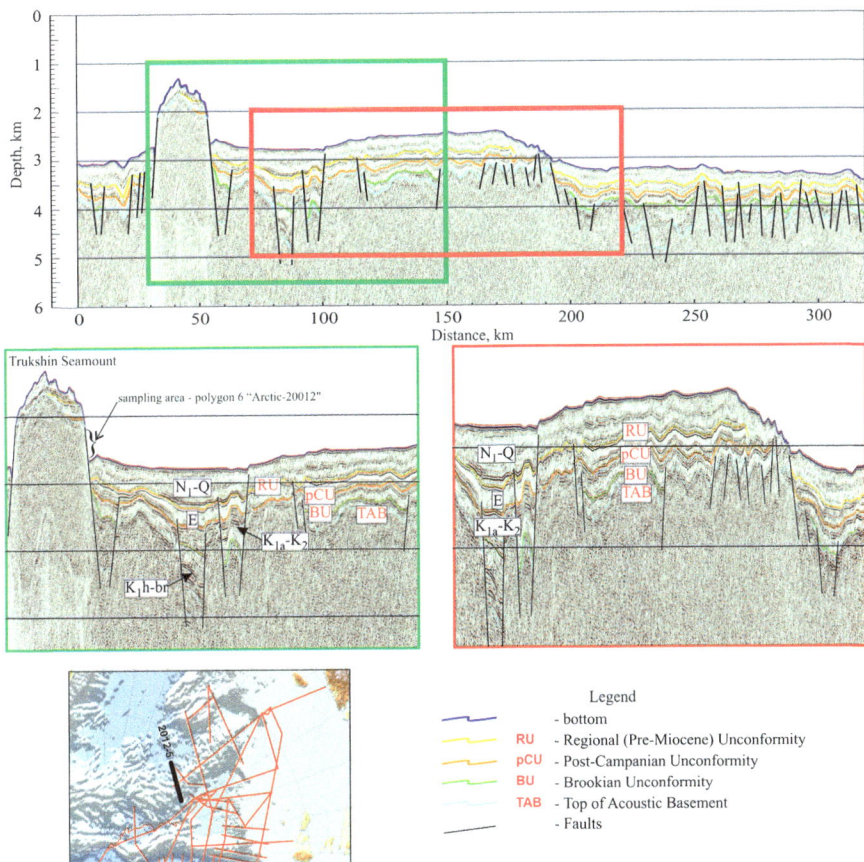

Fig. 7.9 Fragments of MCS line 2012–05

The acoustic basent formations outcrops only on the Shamshura Seamount eastern and western slopes (Fig. 7.11). For the most of the line only irregular chaotic reflections are visible below the TAB, except 130–155 km (eastern slope), 155–185 km (Shamshura Seamount) and 220–235 km (wetern slope) interval. Short but traceable dipping reflections can be seen there.

Ignoring some minor differences in seismic signature of the Meso-Cenozoic formations, all discussed seismic profiles have several important things in common:

– The continuous sedimentary complex N_1-Q drapes the whole modern system of the Central Arctic Uplifts, connecting it to the Siberian-Chukchi continental margin;
– N_1-Q complex has thickness varying from few hundred meters to 700–900 m, stable interval velocity of 1.6–1.8 km/s and regional erosional unconformity at its base. By these and other parameters the complex can be confidently correlated

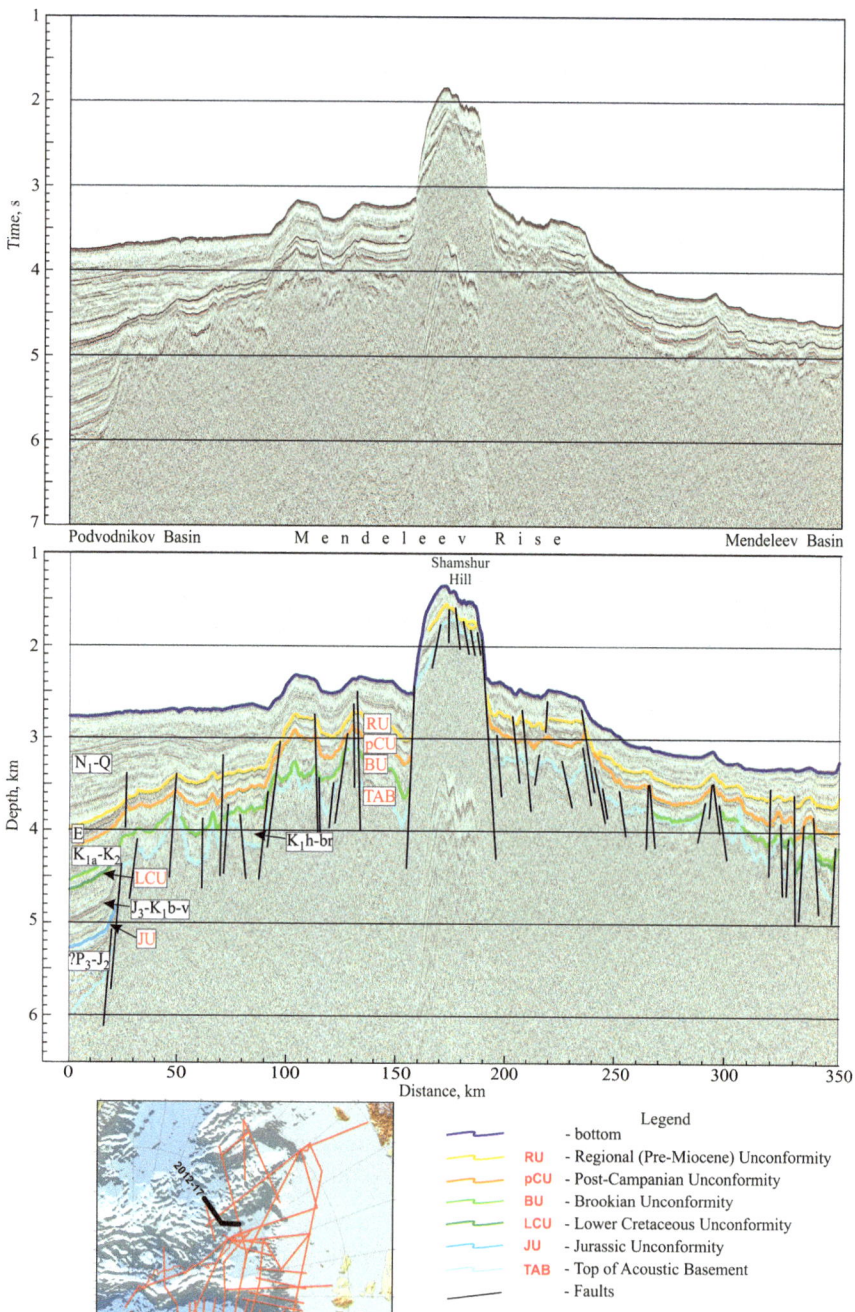

Fig. 7.10 MCS line 2012–17

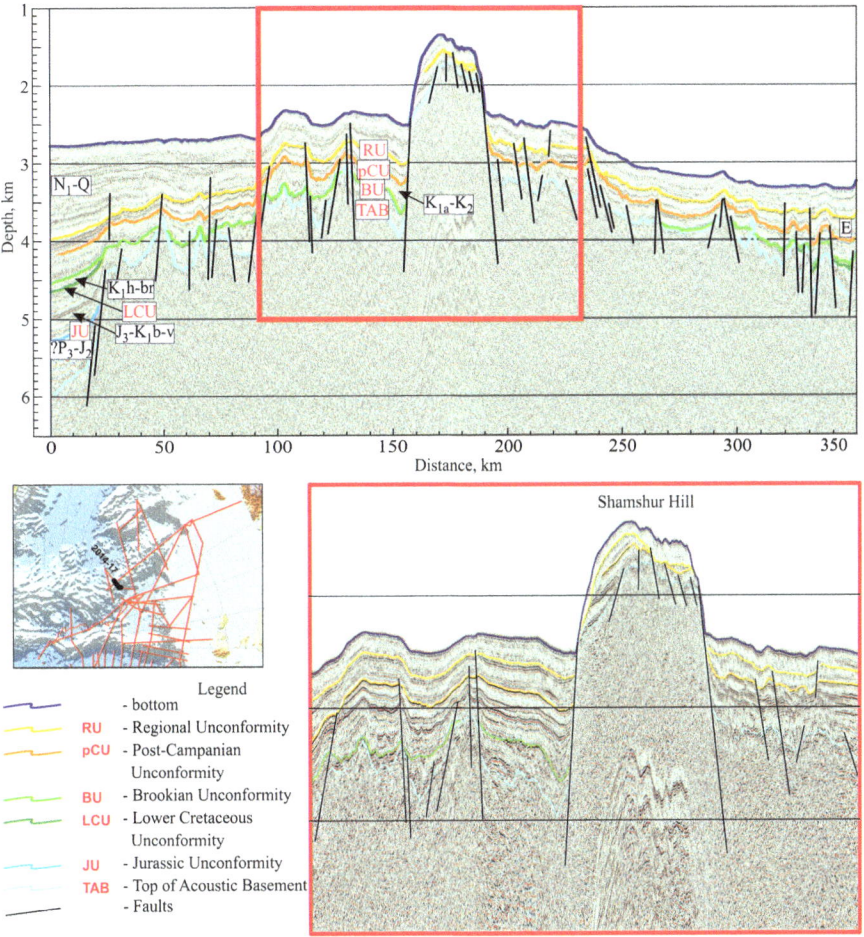

Fig. 7.11 MCS line 2012–17 – fragment including Shamshura Seamount

with Miocene-Pleistocene hemipelagic horizons in ACEX well at the Lomonosov Ridge;

– Onset of N_1-Q accumulation marks the completion of contemporary morphological structure of the Mendeleev-Alpfa Ridge System;

– The Paleocene and Cretaceous sedimentary complexes, despite their varying thickness, belong to the same sedimentation system of depositional basins in the Podvodnikov Basin, Mendeleev-Alpa Ridge and Chukchi Sea shelf; local grabens with sedimentary infill were originated in Early Cretaceous and were active till Paleogene.

– Sporadic strong seismic events in the upper parts of acoustic basement are thought to represent HALIP volcanic rocks, the presence of which is confirmed by seafloor drilling and sampling. There are geochemical parallels between the exposed here bedrocks and Upper Cretacous formation of the Siberian platform. According to paleontological and petrographic analysis, the material collected from the deeper parts of the acoustic basement ((Fig. 7.9, 51 km mark) correlate with terrigenous formations of the Upper Ellesmere (Triassic) and carbonate or carbonate-terrigenous Lower Ellesmere (Carboniferous-Permian) complexes in Popcorn well drilled on Alaska shelf (Morozov et al. 2013).

Seismic signature of the sedimentary formations in the Medeleev Ridge:

Neogene-Pleistocene (N_1-Q) This complex contains long continuous parallel low-amplitude reflections typical for hemipelagic facies forming uninterrupted blanket over the entire Mendeleev Ridge.

Unconformity RU This unconformity is represented by coherent reflection marking the erosional surface at the base of transgressive sequence, almost parallel to the overlain beds (disconfomity). Widely distributed over large areas, it is interpreted as regional Pre-Miocene unconformity.

Paleogene (E) Reflectivity patterns of this complex are affected by basement tectonics: over flat or slightly deformed basemen the reflections are parallel and continuous, but in the areas of strongly dislocated and faulted basements reflectors acquire more complex geometry, either onlapping on the steep slopes, or being truncated by faults.

Unconformity pCU This unconformity is interpreted as also regional angular unconformity between Cretaceous and transgressive Paleogene sequences.

Cretaceous – Well stratified complex with strongly variable thickness – from many hundred meters in grabens to complete pinch-out on uplifted blocks – and reflectivity also affected by basement structure, as was described above in ***Paleogene (E).***

7.3.5.1 Seismic Velocities

Refraction soundings at sono-buoys locations along the MCS lines provided the interpreters with additional valuable information. The resulting velocity models, in combination with depth-converted interpreted MCS line 2012–04 seismic section, are presented on Fig. 7.12.

7.4 Acoustic Basement

Basalts from outcropping bedrocks of the Mendeleev–Alpha Ridge were described in (Andronikov et al. 2008; Brumley et al. 2008; Bruvoll et al. 2012; Morozov et al. 2013; Jokat et al. 2013). The basalts, geochemically varying from tholeite to alkali

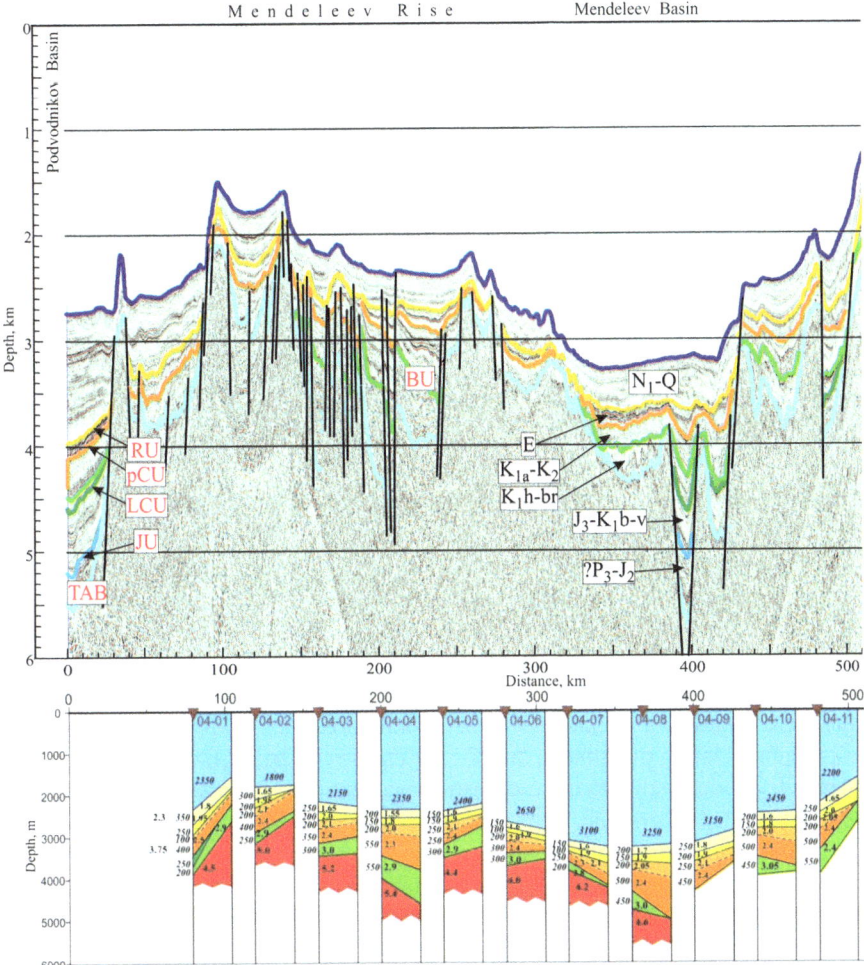

Fig. 7.12 Velocity models at sono-buoys locations along MCS line 2012–04

and from oceanic to cratonic, were dated as 89 Ma (Polarstern-1998), 76–100 and 112–115 Ma, 82 Ma and 127–260 Ma ("Arktika-2012"); basalts as old as Paleozoic and even Pre-Cambrian (U-Pb, zircons) were dredged from the Mendeleev Ridge escarpments.

There still is no universally accepted explanation of such variance of age and geochemistry. Accepting dating of basalts from the Frantz-Joseph Land, Svalbard and De-Long Archipelago as the most reliable and consisting, we place the onset of HALIP magmatism at around 123 Ma (Aptian). If the youngest known basalts were erupted around 80 Ma ago, then the HALIP magmatism within the Mendeleev–Alpha Ridge System lasted 43 Ma.

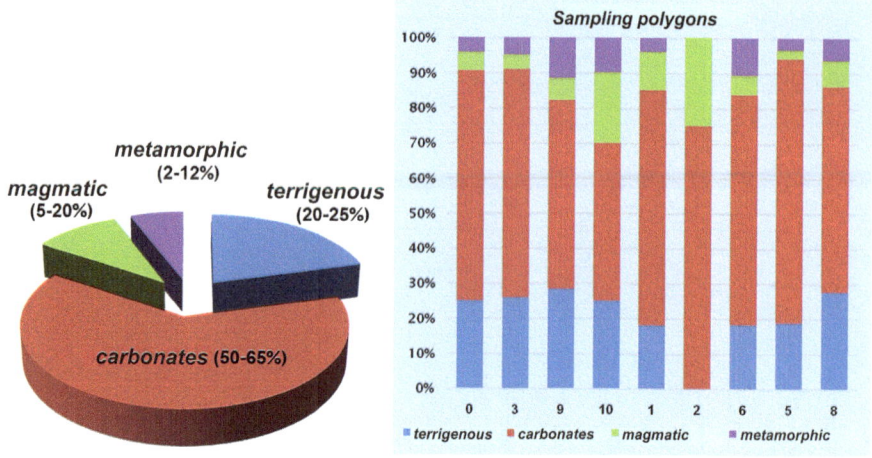

Fig. 7.13 Composition of the bottom-sampled material from the Mendeleev Ridge

The Russian Arctic expeditions collected large volume of sedimentary, igneous and metamorphic material dredged from the Mendeleev Ridge escarpments (Morozov et al. 2013) (Fig. 7.13). Detrital zircons from sandstones have ages from Pre-Cambrian to Mesozoic with the youngest population from 450–350 and 250–200 Ma, suggesting Paleozoic and Triassic-Early Cretaceous ages of the source rocks. Carbonate rocks (limestones and dolomites) contain Mid- and Late Paleozoic microfauna. Detail quantitative analysis of clastic material from the slopes of the Mendellev Ridge shows that the acoustic basement may consist of Early Paleozoic (perhaps, Caledonian) fold belts with magmatic inclusions and Late Paleozoic – Early Mesozoic carbonate-terrigenous cratonic complexes.

Basing on joint analysis of seismic and gravity data collected by research ice-breaker "Healy" in 2005, the specific seismic facies in the upper parts of the acoustic basement were interpreted as volcanic-sedimentary complex (at least 1000 m thick) of intercalating basalts, sills, volcaniclastics and sediments (Bruvoll et al. 2010, 2012). Relatively low interval velocity of this complex (Vp = 2.3–4.0 km/s) may indicate on considerable volume of tuffs which, in turn, require their sources (underwater volcanoes) to be present at the neritic depth. The cessation of HALIP magmatism and creation of volcanic-sedimentary formations in the Mendeleev-Alpha Ridge System is dated by 80 Ma (Campanian).

7.5 Earth's Crust

Deep Seismic Sounding (DSS) of the Mendeleev Ridge (see Chap. 1 for technical details) were acquired by expeditions "Arktika-2000", "Arktika-2005" and "Arktika-2012" along the namesake profiles (Fig. 1.27). The velocity models were

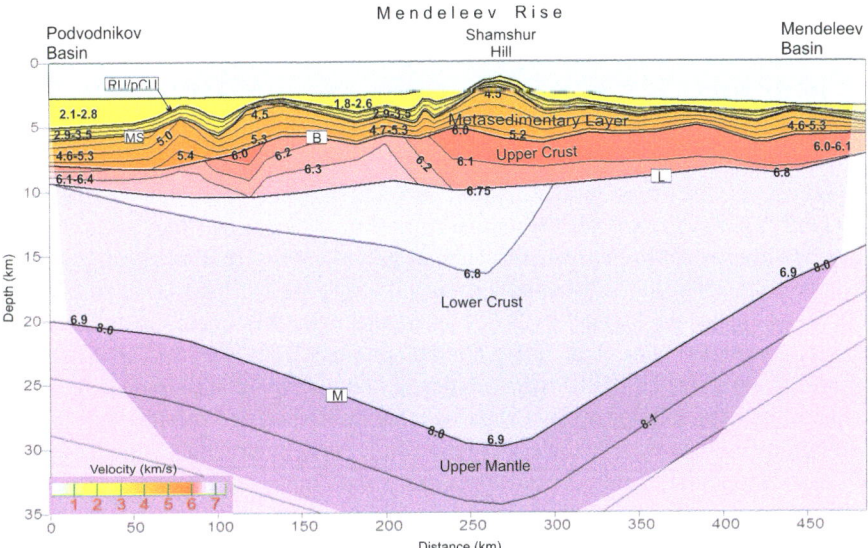

Fig. 7.14 Mendeleev Ridge crustal velocity model ("Arktika-2000") (Poselov et al. 2012)
RU – Pre-Miocene regional unconformity; pCU – Post-Campanian unconformity; MS
– MetaSedimentary complex; B – upper crust; L – lower crust; M – Moho; 6.0 - P-wave velocities,
km/s

constructed by ray-tracing and synthetic modeling based on iterative alterations of
the geometry and velocity of seismic interfaces until the desired fit between observed
and calculated time-distance curves of all refraction/reflection P-waves is
achieved (Poselov et al. 2012).

7.5.1   "Arktika-2000" (Fig. 7.14)

The model shows:

– Two sedimentary complexes, upper, (Vp = 1.8–2.8 km/s) and lower (2.9–
  3.5 km/s) separated by merged RU and pCU unconformities; total thickness –
  from 3.5 km in the Podvodnikov Basin to ≈ 0.5 km on the Shamshura Seamount;
– Metasedimentary complex (acoustic basement, Vp = 4.5–5.3 km/s) with thickness varying from 1 km to 4 km under the Shamshura Seamount;
– Upper crust with interval velocities from 6.0 to 6.4 km/s; thickness gradually
  drops from 4–5 km under the Mendeleerv Ridge to 1,5–2.0 km in the adjacent
  basins;
– Lower crust with Vp from 6.7 (confidently derived from the first breaks of the P_L
  waves) to 6.9 (only approximated by fragments of the PmP waves); lower crust
  thickness reaches its maximum of ≈ 20 km under the Mendeleev Ridge and

declines to ≈ 10 km under the Podvodnikov Basin and ≈ 7 km under the Mendeleev Basin;

- Upper mantle (Vp = 8.0 km/s); Moho depth, derived from Pn and PmP waves, subsides to ≈ 30 km under the Mendeleev Ridge; it is at depth of ≈20 km and ≈ 15 km under the Podvodnikov and Mendeleev basins, accordingly.

With total thickness growing to 24 km and interval velocities from 1.8 to 6.9 km/s, we may classify the crystalline crust of the Mendeleev Ridge as a standard continental crust. Proven presence of the basalt volcanism (LIP-thickening which may be responsible for substantial increase of the lower crust thickness relative to the upper) makes the Mendeleev Ridge analogous to the Ellesmere-Greenland continental margins (Funck et al. 2004). On the other hand, uninterrupted continuation of crustal units (including the metasedimentary complex with its carbonate/terrigenous formations and upper crust with $V_P = 5.9$–6.5 km/s) and morphological features from shelf to the Mendeleev–Alpha Ridge System structurally ties it to the Eurasia continental margin (Lebedeva-Ivanova and Zamansky 2006).

7.5.2 "5-AR - Arktika-2005"

The composite profile "5-AR - Arktika-2005" runs from Chukchi coast through the East-Siberian – Chukchi seas shelf onto the Mendeleev Ridge for 1400 km (Fig. 1.27).

The deep water part ("Arktika-2005") was covered by landing air-borne operations using wide-angle DSS system on drift ice recording refracted and reflected P-waves at offsets 200–250 km. On shelf (line 5-AR) it was ship-borne operations with OBS and air-guns. The final velocity model (Poselov et al. 2012) based on DSS and MCS data is shown on Fig. 7.15. The model shows:

- Two sedimentary complexes, upper, (Vp = 1.8–2.5 km/s – North-Chukchi Trough, 1.6–1.9 km/s – Mendeleev Ridge) and lower (Vp = 3.9–4.4 km/s – North-Chukchi Trough, 3.1–3.3 km/s – Mendeleev Rise) separated by merged RU and pCU unconformities; (the third sedimentary complex with $V_P = 4.7$– 5.9 km/s and ≈4 km thick is defined only inside North-Chukchi Trough); Total thickness of sedimentary complexes is close to 16 km in the North-Chukchi Trough depocenter and is reduced to 2. 5 km at the Mendeleev Ridge.
- Metasedimentary complex (Vp = 4.8–5.1 km/s. thickness 2–3 km) tracing from the Mendeleev Ridge to the northern flank of the North-Chukchi Trough;
- Upper crust (North-Chukchi Trough - Vp = 6.1–6.3 km/s, thickness 2–3 km, Mendeleev Ridge - Vp = 6.2–6.3 km/s, thickness 4–7 km);
- Lower crust (North-Chukchi Trough - Vp = 6.6–6.8 km/s, thickness 9–10 km, Mendeleev Rise - Vp = 6.7–6.9 km/s, thickness 20–22 km); total thickness of the continental crust almost doubles from 13 km under the shelf edge to 26 km under the Mendeleev Ridge.

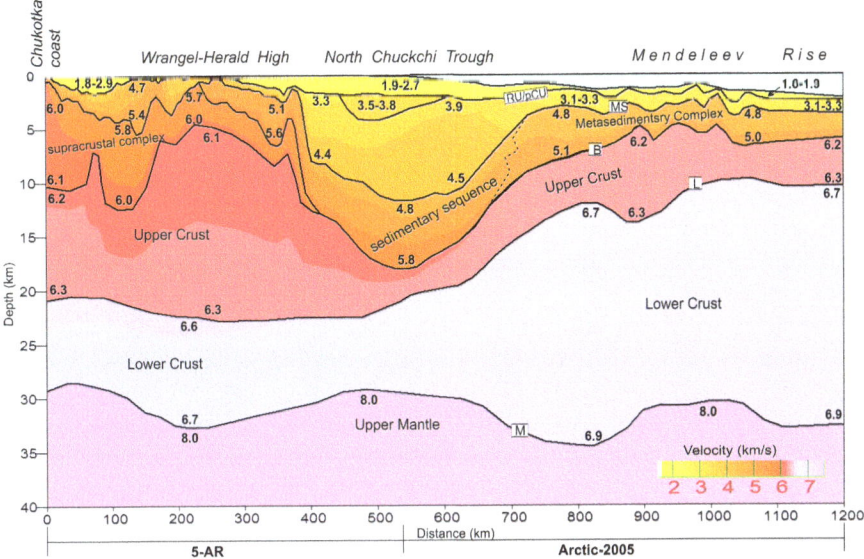

Fig. 7.15 Crustal velocity model for the composite profile "5-AR - Arktika-2005" (Poselov et al. 2012)
RU – Pre-Miocene regional unconformity; pCU – Post-Campanian unconformity; MS – MetaSedimentary complex; B – upper crust; L – lower crust; M – Moho; 6.1 - P-wave velocities, km/s

– Upper mantle with Vp ≈ 8.0 (from Pn and PmP waves). The P-wave velocity in the upper mantle is ~8.0 km/s. Moho depth changes fom 28–29 km under the North-Chukchi Trough to 31–34 km under the Mendeleev Ridge.

Again, the consolidated crust and sedimentary cover complexes continue from the East-Siberian-Chukchi continental margin to the Mendeleev Ridge with its attenuated upper crust. More pronounced attenuation is visible under the deep depositional center of North-Chukchi Trough indicating it formation under prevailing extensional forces.

In the North-Chukchi Trough depocenter (with ≈17 km thickness of sediments) the 5–6 km thick lowest sedimentary sequence with Vp = 4.8–5.8 km/s lies directly on top of the upper crust. At the Mendeleev Ridge the thickness and velocities of this sequence drops to 3–4 km and 4.8–5.1 km/s, accordingly, and it is interpreted there as a metasedimentary complex.

Thickness of the upper crust increases dramatically from ≈3 km under the North-Chukchi Trough depocenter to ≈ 20 km under the Wrangel-Herald uplift and very gradually to 5–9 km – towards the Mendeleev Ridge. Upper crust velocity is rather stable (6.1–6.3 km/s) but slows down to 5.7–6.0 km/s under the Wrangel-Herald High and Chukchi coast, hinting on possible presence of supra-crustal folded formations.

Striking similarity in velocity spectrum, deep structure and thickness of crustal units exists between the above described model and that of the Greenland continental margin (Funck et al. 2004). Both models demonstrate doubling of lower crust (V_P = 6.7–6.9 km/s) thickness relative to the upper one (V_P = 6.15–6.3 km/s) (Fig. 7.16).

The phenomena of the lower crust thickening in the Mendeleev Ridge, Ellesmere-Greenland margin (or, for that matter, any other structures inside HALIP) can be explained by constant addition of igneous material from the mantle upward into the crust during periods of active magmatism - so-called LIP-thickening, or magmatic underplating.

The crustal density model for composite profile "5-AR - Arktika-2005 was created using its interpreted MCS sections and above described velocity model (Fig. 7.17).

According to this model, the sedimentry infill of the 17 km deep North-Chukchi Trough starts with presumably Jurassic (or older?) formation with density 2.62 g/cm^3 and ends with Miocene – Pleistocene deposits with density 2.34 g/cm^3. The computed densities for the upper crust, lower crust and mantle were accepted as 2.80 g/cm^3, 2.96 g/cm^3 and 3.34 g/cm^3, accordingly. Lower-upper crust interface under the Mendeleev Ridge is fixed at ≈16 km. Moho is consistently 28–30 km deep under the Mendeleev Ridge rising to 25–26 km under the North-Chukchi Trough depocenter.

Lacking the surface gravity observations, the satellite data, "naturally" deprived of short-wave anomalies, were used as observed values. This makes attempts to achieve very close fit between "observed" and "simulated"gravity anomalies and explains why the former looks like smoothed version of the latter.

7.5.3 "Arktika-2012"

Sub–latitude DSS profile "Arktika-2012" crosses the Central Arctic Uplifts from south-eastern closure of the Vilitsky Trough thought Mendeleev Ridge and Chukchi Basin into the Chukchi Plateau (Fig. 1.27).

Figure 7.18 presents the P-waves crustal model, Fig. 7.19 – S-waves (after seafloor conversion). Both models demonstrate uninterrupted continuation of structural complexes of the consolidated crust between the Mendeleev Rise and Chukchi Plateau.

The Mendeleev Ridge is distinguished by maximal crustal thickness and the deepest Moho (up to 34 km). The P-wave velocity of the lowest parts of the consolidated crust may reach anomalously high values (7.2–7.3 km/s) in the deepest.

Attenuation of the upper crust to 7–9 km and enlargement of the lower to18–19 km are most likely related to HALIP magmatism (Kashubin et al. 2016).

The crustal density model for DSS profile "Arktika-2012" (Fig. 7.20) was constructed using its interpreted MCS sections and above described velocity models. For instance, position and geometry of Moho and upper-lower crust interface were

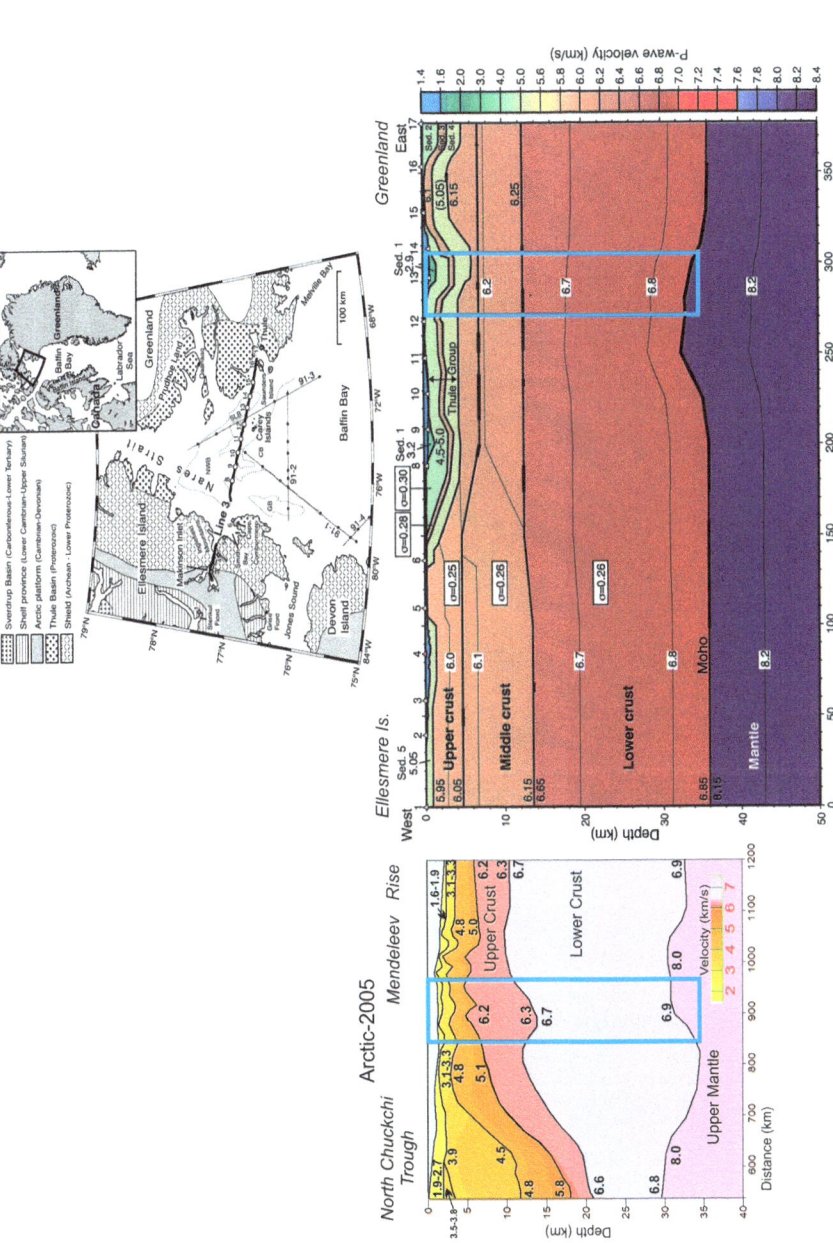

Fig. 7.16 Comparison of crustal velocity models for the Mendeleev Ridge (Poselov et al. 2012) and the Greenland Continental Margin (Funck et al. 2004)

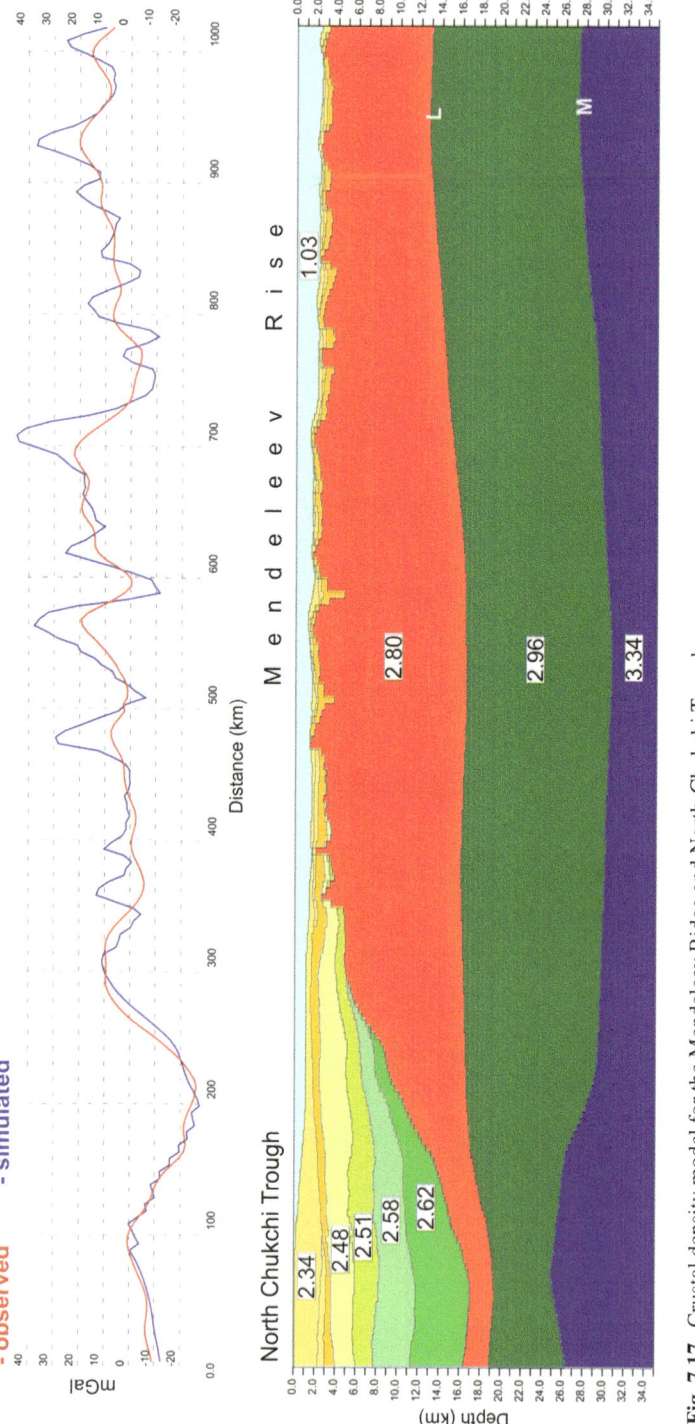

Fig. 7.17 Crustal density model for the Mendeleev Ridge and North-Chukchi Trough

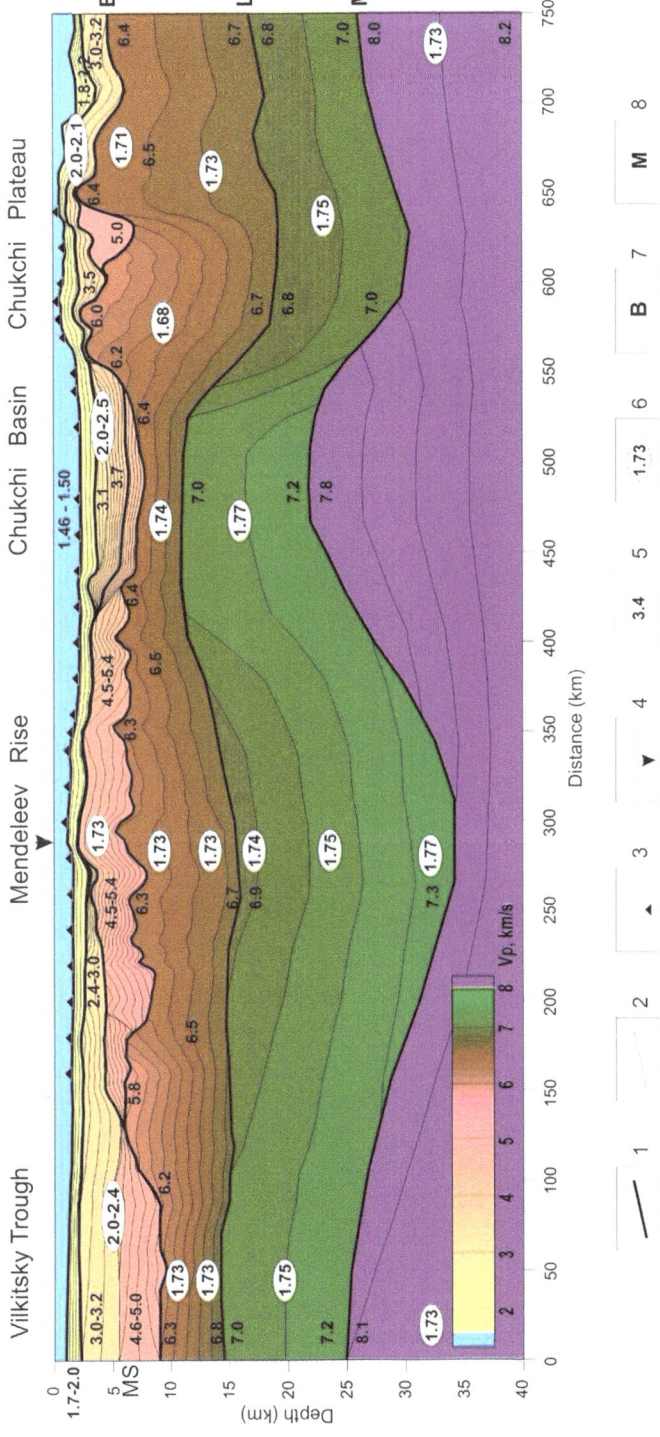

Fig. 7.18 Crustal Vp model for DSS profile "Arktika-2012" (Kashubin et al. 2016): 1 – seismic interfaces; 2 – Vp contours; 3 – ocean bottom stations (OBS); 4 – DSS "Arktika-2005" intersect; 5 – Vp; 6 –Vp/Vs; 7 – sedimentary cover base; 8 – Moho

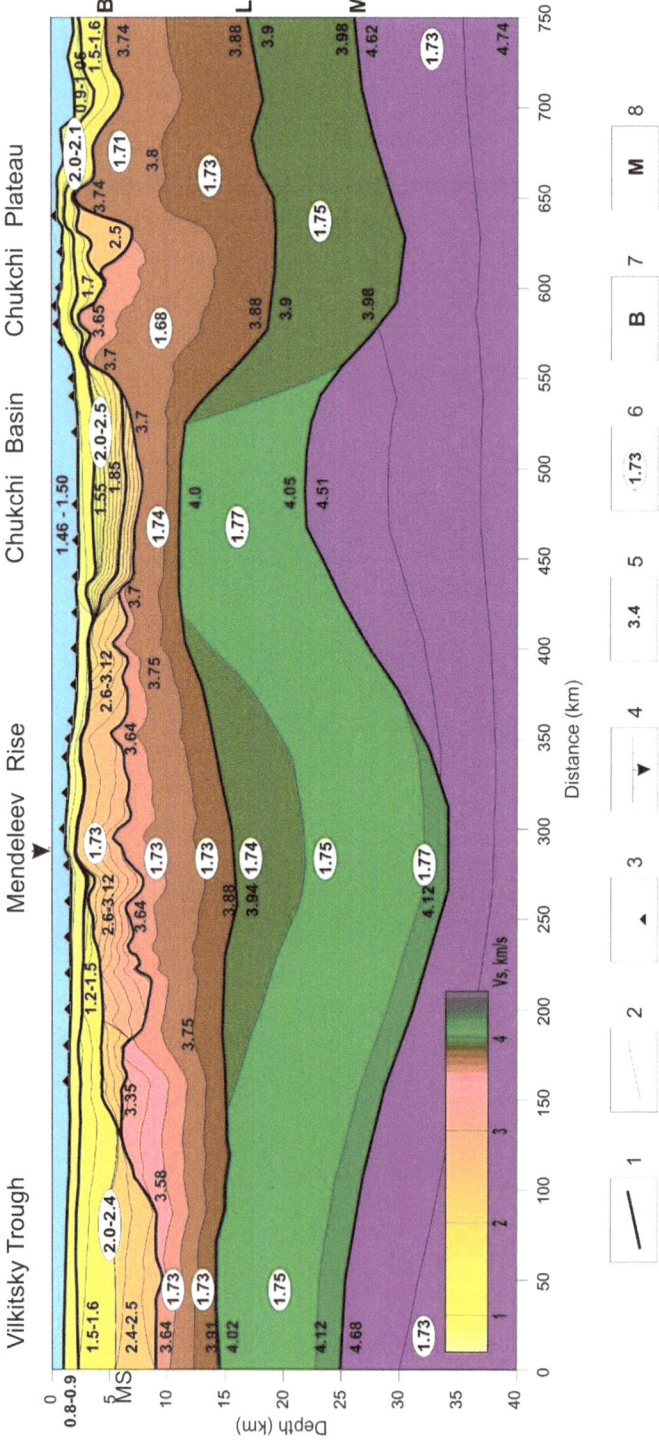

Fig. 7.19 Crustal Vs model for DSS profile "Arktika-2012" (Kashubin et al. 2016): 1 – seismic interfaces; 2 – Vs contours; 3 – ocean bottom stations (OBS); 4 – DSS "Arktika-2005" intersect; 5 – Vs; 6 – Vp/Vs; 7 – sedimentary cover base; 8 – Moho

copied from velocity models and remained unaltered by the density modeling process (compare Fig. 7.20 and Fig. 7.17), Therefore the depth to Moho and lower crust thickness in the density model are the same as in velocity models in Fig. 7.17 and Fig. 7.18. Mantle density is estimated at 3.34 g/cm^3.

Following lateral P-wave velocity change in the lower crust from 7.1 km/s to 6.8 km/s from west to east, its density were estimated as 2.96 g/cm^3 under the Vilkitsky Trough and the Mendeleev Ridge, 2.90 g/cm^3 under the Chukchi Basin and 2.84 g/cm^3 - under the Chukchi Plateau.

Density 2.80 g/cm^3 was assigned to the upper crust and 2.73 g/cm^3 - to the metasedimentary complex.

Several seismic anomalies were identified on the MCS section (120, 200, 380 km) in the metasedimentary complex. Accompanied by strong magnetic anomalies with no visible gravity effect, these structures were tentatively interpreted as volcanic centers with its highly magnetic mafic lavas, unconsolidated lower density volcaniclastics and hydrothermal activity.

Introduction of high density (2.95 g/cm^3) block at 240 km was necessary to compensate for the negative gravity effect of increasingly thick sedimentary section on the Mendeleev Ridge flank.

Stratification of the upper sedimentary cover was taken from the interpreted MCS sections. Velocity-density conversion was based on empirical functions established for the region (Franke et al. 2001; Piskarev 2004; Kazanin et al. 2006; Malyshev and Obmetko 2009).

7.6   Conclusions

1. The Mendeleev Ridge was formed on the continental crust with total thickness 27–32 km of which the upper crust takes 4–7 km. Terriggenous Meso-Cenozoic sedimentary cover is underlain by metasedimentary complex of Paleozoic cratonic, mostly littoral carbonate formations, as shown by bottom sampling ("Arktika-2000" and "Arktika-2012").

2. Continuous chain of progressively seaward deepening bathymetric terraces defines the most specific feature of the Mendeleev Ridge - its cross-wise or diagonal hypsometric zoning – and demonstrate the morphological connection of the Mendeleev Ridge with shallow water regions of the Siberian-Chukchi continental margin.

3. Seismic data reveal that the structure and stratigraphy of the Mendeleev Ridge are affected by intensive normal faulting of acoustic basement creating complex system of grabens and half-grabens. It depicts the Mendeleev Ridge as an extensional structure of Cretaceous age and as an example of tectonic evolution of the Central Arctic region.

4. Cretaceous-Cenozoic sedimentary complexes continue uninterrupted from the continental shelf (North-Chukchi Trough) to the Mendeleev Ridge. According to

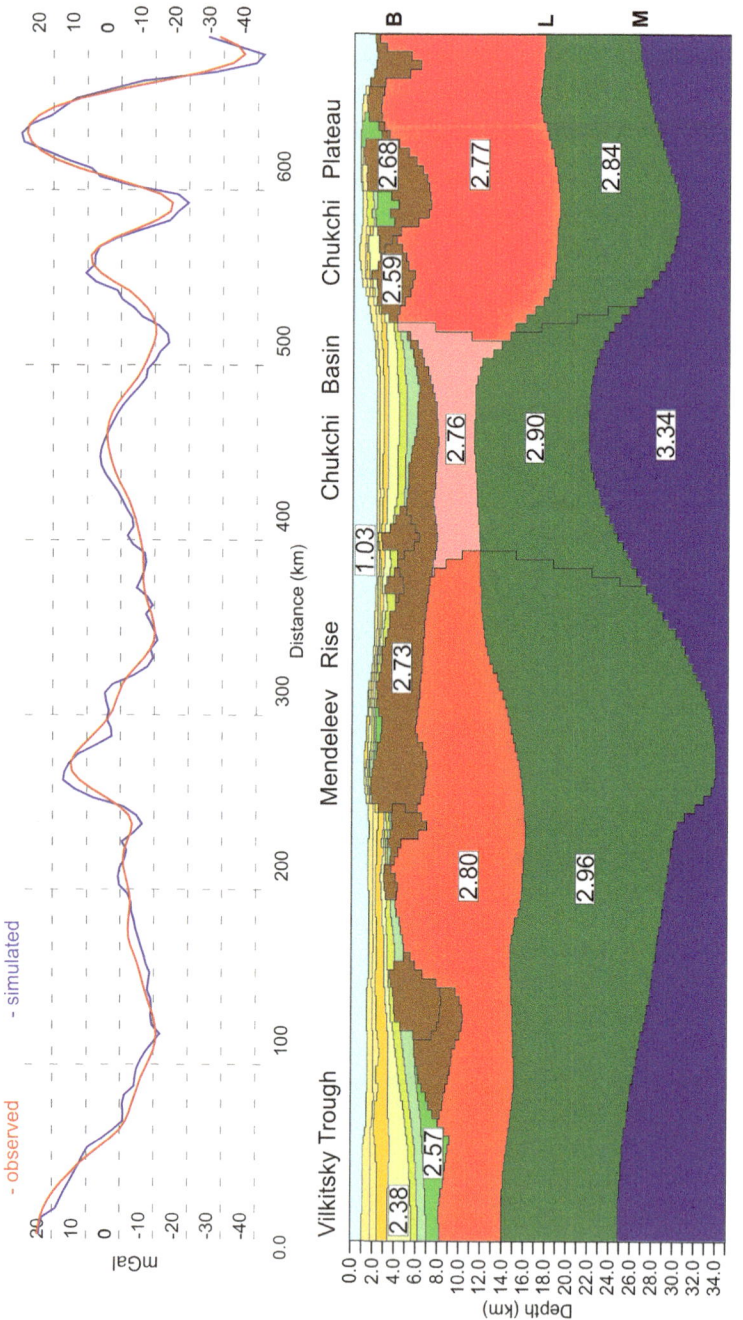

Fig. 7.20 Crustal density model for DSS profile "Arktika-2012" (Kashubin et al. 2016)

seismic data, no major normal or strike-slip faults exist between the shelf and the Mendeleev Ridge.

5. The Miocene-Pleistocene complex, with continuous undisturbed layers of hemipelagic sediments and regional erosional unconformity at its base, drapes the entire Mendeleev Ridge and marks the completion of contemporary morphological modern system of the Mendeleev-Alphfa Ridge System.

References

Andronikov A, Mukasa S, Mayer LA et al (2008) First recovery of Submarine Basalts from the Chukchi Borderland and Alpha. Eos. Trans.AGU 89(53). Abstract V41D-2124

Backman J, Moran K et al (2006) Sites M0001-M0004. Expedition 302 Scientists. Proceedings of the integrated ocean drilling program 302, p 169

Backman J, Jakobsson M, Frank M et al (2008) Age model and core-seismic integration for the Cenozoic Arctic coring expedition sediments from the Lomonosov ridge. Paleoceanography J 23(1)

Brumley K, Mayer LA, Miller EL et al (2008) Dredged rock samples from the alpha ridge, Arctic Ocean: implications for the tectonic history and origin of the Amerasian Basin. Eos Transactions American Geophysical Union, 15–19 December 2008

Bruvoll V, Kristoffersen Y, Coakley B et al (2010) Hemipelagic deposits on the Mendeleev and northwestern alpha submarine ridges an the Arctic Ocean: acoustic stratigraphy, depositional environment and an inter-ridge correlation calibrated by ACEX results. Mar Geophys Res J 31:149–171

Bruvoll V, Kristoffersen Y, Coakley B et al (2012) The nature of the acoustic basement on Mendeleev and northwestern alpha ridges, Arctic Ocean. Tectonophysics J 514:123–145

Franke D, Hinz K, Oncken O (2001) The Laptev Sea rift. Mar Pet Geol J 18:1083–1127

Funck T, Jackson HR, Dehler SA et al (2004) A refraction seismic transect from Greenland to Ellesmere Island, Canada: the crustal structure in Southern Nares Strait. Polarforschung J 74(1–3):97–112

Glebovsky VY, Kovacs LC, Maschenkov SP et al (1998) Joint compilation of Russian and US navy aeromagnetic data in the Central Arctic Seas. Polarforshung J 68:35–40. (erschienen 2000)

Grantz A, Scott RA, Drachev SS et al (2009) Map showing the sedimentary successions of the Arctic region (58°-64° to 90°N) that may be prospective for hydrocarbons. American Association of Petroleum Geologists GIS-UDRIL Open-File Spatial Library, Tulsa

Grantz A, Hart PE, Childers VA (2011) Geology and tectonic development of the Amerasia and Canada basins, Arctic Ocean. Geol Soc Lond Mem 35:771–800

Gurevich NI, Maschenkov SP (2000) The crustal types of geological structures in the deep-sea Arctic basin in: collected articles. Geology and geophysics of the Arctic region lithosphere, vol 3. VNIIOkeangeologia, Saint Petersburg, pp 9–32

Jokat W, Ickrath M, O'Connor J (2013) Seismic transect across the Lomonosov and Mendeleev ridges: constraints on the geological evolution of the Amerasia Basin, Arctic Ocean. Geophys Res Lett 40(19):5047–5051

Karasik AM (1980) Strtucture and evolution of the of the Arctic Ocean floor in light of aeromagnetic data. Marine geology, sedimentology, petrography and ocean geology, pp 178–193. (in Russian)

Kashubin SN, Petrov OV et al (2016) Deep structure of crust and the upper mantle of the Mendeleev rise on the Arktic-2012 DSS profile. Regionalnaya geologiya i metallogeniya J 65:16–36. (in Russian)

Kazanin GS et al (2006) Some of the results of seismic exploration of the MCS survey in the Laptev sea. Abstracts of the international conference "Oil and gas of the Arctic shelf". Murmansk. CD-R

Kenyon S, Forsberg R (2001) Arctic gravity Projecta status. International Association of Geodesy Symposia 123:391–395

Lebedeva-Ivanova NN, Zamansky YY (2006) Seismic profiling across the Mendeleev ridge at 82°N: evidence of continental crust. Geophys J Int 165(2):527–544

Malyshev NA, Obmetko VV, Borodulin AA at al (2009) The Laptev Sea shelf sedimentary cover according to the newest information Geology of the Earth Polar Reigons: Proceedings of XLII Tectonic conference II:32–37

Moran K, Backman J, Brinkhuis H et al (2006) The Cenozoic palaeoenvironment of the Arctic Ocean. Nature J 441:601–606

Morozov AF, Petrov OV, Shokalsky SP et al (2013) New geological data are confirming continental origin of the Central Arctic rises. Reg Geol Metallog J 52:2–25

Piskarev AL (2004) The basement structure of the Eurasia Basin and central ridges in the Arctic Ocean. Geotektonics J 38(6):443–448

Poselov VA, Avetisov GP, Butsenko VV et al (2012) The Lomonosov ridge as a natural extension of the Eurasian continental margin into the Arctic Basin. Geol Geophys J 53(12):1662–1680

Poselov VA, Butsenko VV, Zholondz SM et al (2017) Seismic stratigraphy of sedimentary cover in the Podvodnikov Basin and North Chukchi trough. Dokl Earth Sci J 474(2):688–691

Saltus RW, Miller EL, Gaina C (2011) Regional magnetic domains of the circum-Arctic: a framework for geodynamic interpretation. In: Spencer AM (ed) Arctic petroleum geology, vol 35. Geological Society of London, London, pp 49–60

Verba VV, Fedorov VI (2006) Anomalous magnetic field over the Central Arctic uplifts the Amerasian Basin of the Arctic Ocean. Geophys J 28(5):95–103. (in Russian)

Verhoef J, Roest WR, Macnab R et al (1996) Magnetic anomalies of the Arctic and North Atlantic oceans and adjacent land areas. Geol Surv Can., Open File 3125a:577

Chapter 8
Chukchi Plateau and Chukchi Basin

Victor V. Butsenko, Yury G. Firsov, Sergey P. Kashubin, Alexey L. Piskarev, and Sergey M. Zholondz

Abstract The Chukchi Plateau can be described as a fragment of the continental margin jutted into the abyssal part of the Arctic Ocean. The Chukchi Plateau has the most direct connection with the adjacent shelf of the East Siberian - Chukchi continental margin. The principal structural complexes of the consolidated continental crust continue uninterrupted between the Mendeleev Ridge and Chukchi Plateau with total thickness of 26–29 km and approximately equal thickness of the upper and lower crusts under the Chukchi Plateau.

In the Chukchi Basin the Earth's crust is 20 km thick with upper crust – 3 km. The Chukchi Basin, with its undisturbed sedimentary complexes, also includes 1200 m of pre- Upper Jurassic (Upper Ellesmerian) formations. Almost complete absence of folding and faults in the Chukchi Basin may point to active sedimentation, not the attenuation of the crust, as a principle cause of its subsidence.

Miocene-Pleistocene complex above the pre-Miocene unconformity continues from the Chukchi Basin to the Chukchi Plateau and marks the end of formation of these present-day morphological structures.

Keywords Chukchi Plateau · Chukchi Basin · East Siberian-Chukchi continental margin · Morphological structures · Sedimentary seaquences · Continental crust

V. V. Butsenko (✉) · Y. G. Firsov · S. M. Zholondz
All-Russian Research Institute of Geology and Mineral Resources of the World Ocean (VNIIOkeangeologia), Saint Petersburg, Russia
e-mail: vicb@vniio.nw.ru

S. P. Kashubin
A.P. Karpinsky Russian Geological Research Institute (VSEGEI), Saint Petersburg, Russia

A. L. Piskarev
All-Russian Research Institute of Geology and Mineral Resources of the World Ocean (VNIIOkeangeologia), Saint Petersburg, Russia

Saint Petersburg University, Saint Petersburg, Russia

© Springer International Publishing AG, part of Springer Nature 2019   269
A. Piskarev et al. (eds.), *Geologic Structures of the Arctic Basin*,
https://doi.org/10.1007/978-3-319-77742-9_8

8.1 Morphology

The Chukchi Plateau can be best described as a fragment of the continental margin jutted into the abyssal part of the Arctic Ocean. Two flat sub-horizontal abyssal terraces connected by the Charley Gap - the Chukchi and Mendeleev basins – separate the Chukchi Plateau from the Mendeleev Ridge (Figs. 8.1, 8.2).

With its flat almost undisturbed top, the Chukchi Plateau has the simplest morphology among other Central Arctic Uplifts and the most direct connection with the adjacent shelf.

At depth 600–800 m, very gentle inner slope imperceptibly transforms into 100–200 m-deep saddle marking the shelf-plateau transition zone, less pronounced than that of the Lomonosov Ridge or the Mendeleev Ridge (Fig. 8.3). The surface of the Chukchi Plateau (400–800 m b.s.l) is disturbed by north-south trending flat-bottom depression 200–400 m deep and almost 100 km long; farther north from its termination, there are two more similar "en-echelon" structures. The Egiazarov Trough, 200 km long, cuts deeply (> 2000 m) into the eastern Chukchi Plateau.

As profile BB1 across the Chukchi Plateau shows, there are deep troughs cutting into the Plateau flat top. These morphological features are usually oriented north-south and interpreted as tectonic depression (grabens) formed by normal faults under the extensional regime. The slopes of the Chukchi Plateau are steep (up to 45°), tall (1000 m in the south, 1600 in the west and 2500 m in the north), devoid of any substantial omplicating forms and with angular base contours outline. Long narrow spur reaches into the abyssal plain from base of the northern slope (Geomorphological Aspects of the Russian Continental Shelf Exterior Boundary in the Arctic 2005).

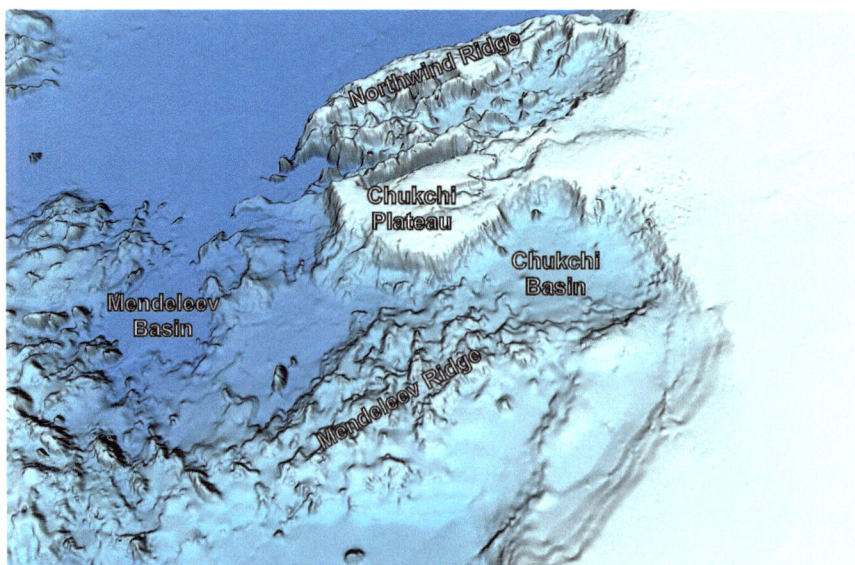

Fig. 8.1 Chukchi Plateau and Chukchi Basin

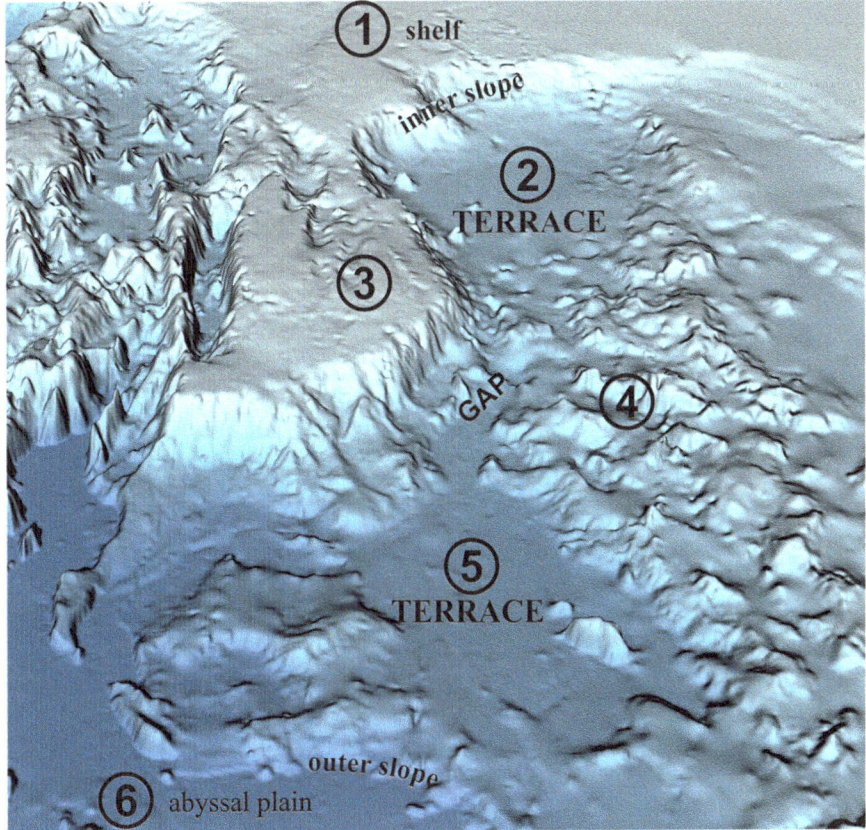

Fig. 8.2 Chukchi and Mendeleev abyssal plains (gird IBCAO v.3.0)
1 - Chukchi shelf; 2 – Chukchi terrace; 3 - Chukchi Plateau; 4 - Mendeleev Ridge; 5 - Mendeleev terrace; 6- Canada Basin

In the broad picture, they are parts of the complex continental slope and can be classified as intermediate slopes. The eastern slope of the Chukchi Plateau is more complex and rugged in contact with the Northwind Trough, but further south it looks like simple slope 1000 m high merging with the Northwind Plain (2000 m b.s.l.) The western slope merges into the terraces of the Chukchi and Mendeleev basins, while the northern – into outer slope bordering with abyssal plain of the Canada Basin. Some similarity exists between the continental slopes in the regions around the Chukchi and the Ermak plateaus.

8.2   Potential Field Anomalies

The Chukchi Plateau - Norhwind Ridge and the Mendeleev Ridge – Chukchi Basin regions have different patterns of the magnetic anomalies (Fig. 4.3). Interpretation of magnetic anomalies (Dove et al. 2010) excludes the Chukchi Plateau from

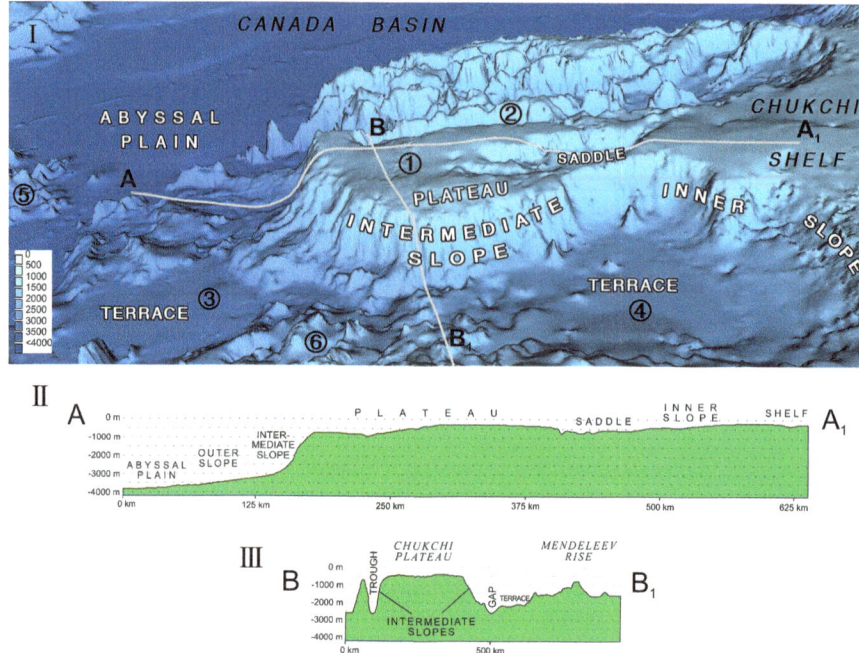

Fig. 8.3 Chukchi Plateau bathymetry
I – 3D view (gird IBCAO v.3.0), looking east from the Mendeleev Ridge; II and III – bathymetry profiles along AA₁ and BB₁ tracks, accordingly; 1 – Chukchi Plateau; 2 – Egiuazarov Trough; 3 – Mendeleev Basin; 4 – Chukchi Basin; 5 – Seamounts region; 6- Mendeleev Ridge

HALIP, contrary to the Mendeleev-Alpha Ridge System, the Chukchi Basin and part of the Canada Basin. Strong inverse correlation between surface topography and magnetic signature in the Chukchi Plateau combined with strong magnetizations of the morphlogical forms (5–10 A/m) signify the presence of Mesozoic plateau-basalts (Piskarev 2004).

Gravity anomalies of the Chukchi Borderland (Fig. 4.3) display direct correlation with surface topography and inverse – with magnetic anomalies.

Strong magnetic anomalies highlight not only uneven spacial distribution of magnetized formations in the basement of Chukchi Basin and the northern part of the Mendeleev Ridge, but also their structural and compositional similarity (Fig. 4.3). While the local gravity magnetic, anomalies directly correlate positively with each other the long-wavelength gravity seems to correspond mainly to the seafloor topography.

8.3 Sedimentary Cover

The Chukchi Plateau, Mendeleev and Chukchi basins were studied be several expeditions summarized in (Arrigoni 2008; Brumley 2009, 2014; Hegewald and Jokat 2013a, b; Dove et al. 2010; Bruvoll et al. 2010, 2012).

R/V Polarstern 2008 expedition collected valuable information regarding the Chukchi Plateau sedimentary cover synthesized by (Hegewald and Jokat 2013a, b). Their seismic stratitgraphy model includes Cretaceous sediments inside grabens, Upper Oligocene (RU by (Poselov et al. 2014)) and Upper Miocene unconformities. Post-Miocene normal faulting episodes were also postulated. Acoustic basement interval velocitie were estimated at 5.4 km/s, intermediate sedimentary complex – 3.6–4.1 km/ s and upper sedimentary complex – 1.6–2.3 km/s. Dating of sedimentary complexex was based on correlation with well drilled on Chukchi Sea Alaska shelf (Sherwood 2011).

Project HOTRAX-2005 (RV "Healy") was analyzed in (Dove et al. 2010; Bruvoll et al. 2010, 2012) describing four seismic-stratigraphic complexes above the acoustic basement (of which two upper ones were correlated to the Miocene-Pleistocene layers.

Icebreaker "Healy" 2010–2011 expedition covered the Chukchi Plateau and the Mendeleev Basin establishing presence of Miocene-Pleistocene, Paleogene and Upper Cretaceous sequences (Brumley 2014).

MCS seismic data acquired by "Arktika 2012" expedition (Fig. 1.17) made possible direct and uninterrupted tracing of all major unconformities – Pre-Miocene (RU), Post-Campanian (pCU), Brookian (BU) and Upper Jurassic (JU) (Poselov et al. 2017) - from the Vilkitsky Trough to the Mendeleev Ridge, through the Chukchi Basin into the Chukchi Plateau.

The interpretation of the seismic data is based on the seismic-stratigraphic model which was developed for the Central Arctic uplifts region tying the seismic data from "Arktika 2012 expedition with ACEX well drilled on the Lomonosov Ridge. The bottom sampling provided an additional information regarding ages of bedrock (Grantz and Hart 2012; Morozov et al. 2013).

Below we demonstrate several seismic lines from the "Arktika'2012" program in the Chukchi Plateau and the Chukchi Basin.

8.3.1   MCS Line 2012–03

The Chukchi Basin with full compliment ($\approx$ 4000 m) of undisturbed sedimentary complexes including 1200 m thick pre- Upper Jurassic (Upper Ellesmerian) formations, converges on the western slope of the Chuckhi Plateau through the sequence of high-amplitude normal faults (Fig. 8.4).

The Chukchi Plateau (530–660 km) stands out as uplifted block with several grabens and horsts inside. The most prominent graben in the center was formed in Jurassic?, most active in Cretaceous and fully compensated in Paleogene. The Miocene-Pleistocene hemipelagic complex drapes all structural elements present on this line.

The total thickness of the sedimentary cover changes from 200–300 at the highest point of the Chukchi Plateau (610–635 km) to 3000 m in the central graben (580–605 km.) Signs of erosion and angular discontinuity at the pCU

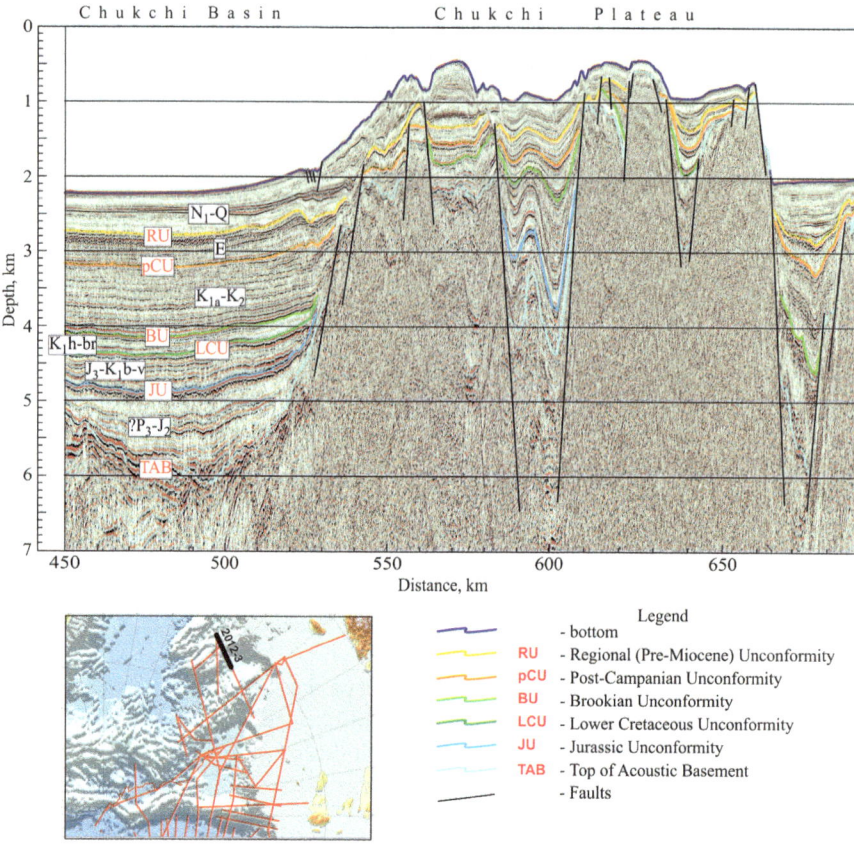

Fig. 8.4 MCS line 2012–03, fragment (Chukchi Basin – southern Chukchi Plateau)

(post-Campanian) unconformity on uplifted blocks (605–635 km) signify high stand and active denudation of these blocks in Paleocene.

Acoustic basement (AB) outcrops only at the steep, 1200 m high, eastern escarpment. Strong dipping reflectors close to the top of acoustic basement (TAB) (540–600 km) may represent the volcanic formations.

8.3.2 MCS Line 2012–04 (Northern Chukchi Plateau – Southern Mendeleev Basin)

The west slope of the Chukchi Plateau, sedimentary complexes and TAB deepen step-wise to north-west through the graben-horst system into the Mendeleev Basin and Charlie Trough (Fig. 8.5).

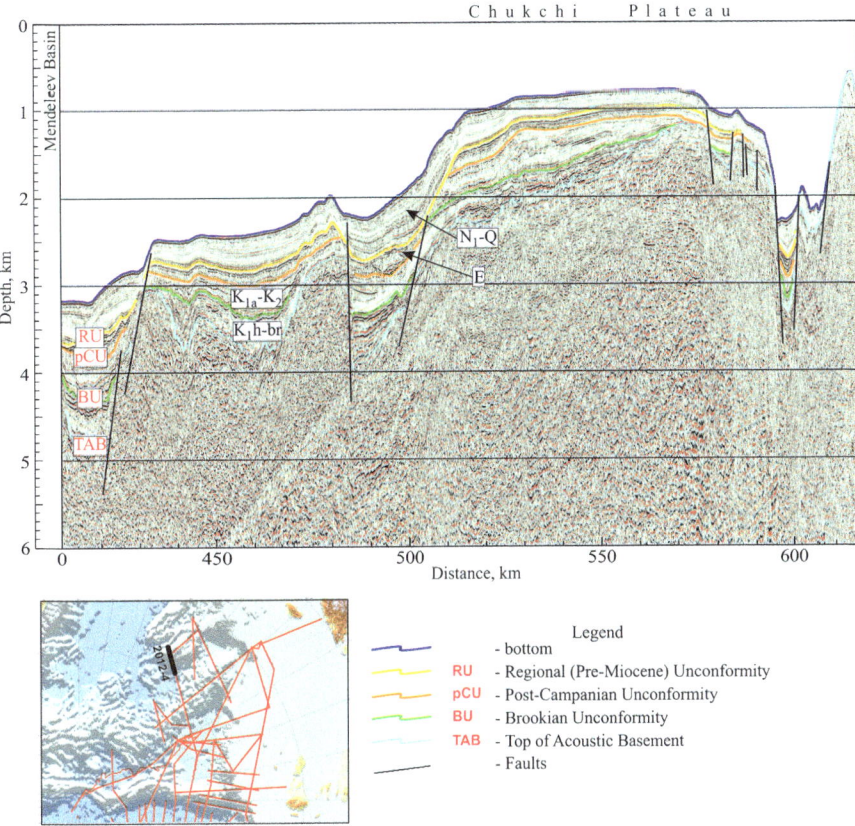

Fig. 8.5 MCS line 2012–04 (fragment, Chukchi Plateau)

The total thickness of the Meso-Cenozoic sedimentary cover varies from 300–500 m at the highest point of the Chukchi Plateau to 1600–1800 m in the local grabens with overall tendency to grow in the north-west direction.

Occurring everywhere in this section, the 200–600 m thick Miocene-Pleistocene complex above the RU unconformity marks the completion of the Central Arctic Uplifts forming process. The erosional unconformity at the base of this sequence and its interval velocities (1.6–1.8 km/s) convincingly identify it with the Miocene-Pleistocene hemipelagic sediments in the ACEX well on the Lomonosov Ridge. Initiated in the Early Cretaceous, the local grabens were completely filled (compensated) by Paleogene.

The acoustic basement outcrops only sporadically at the steep escarpments in the eastern parts of the Plateau.

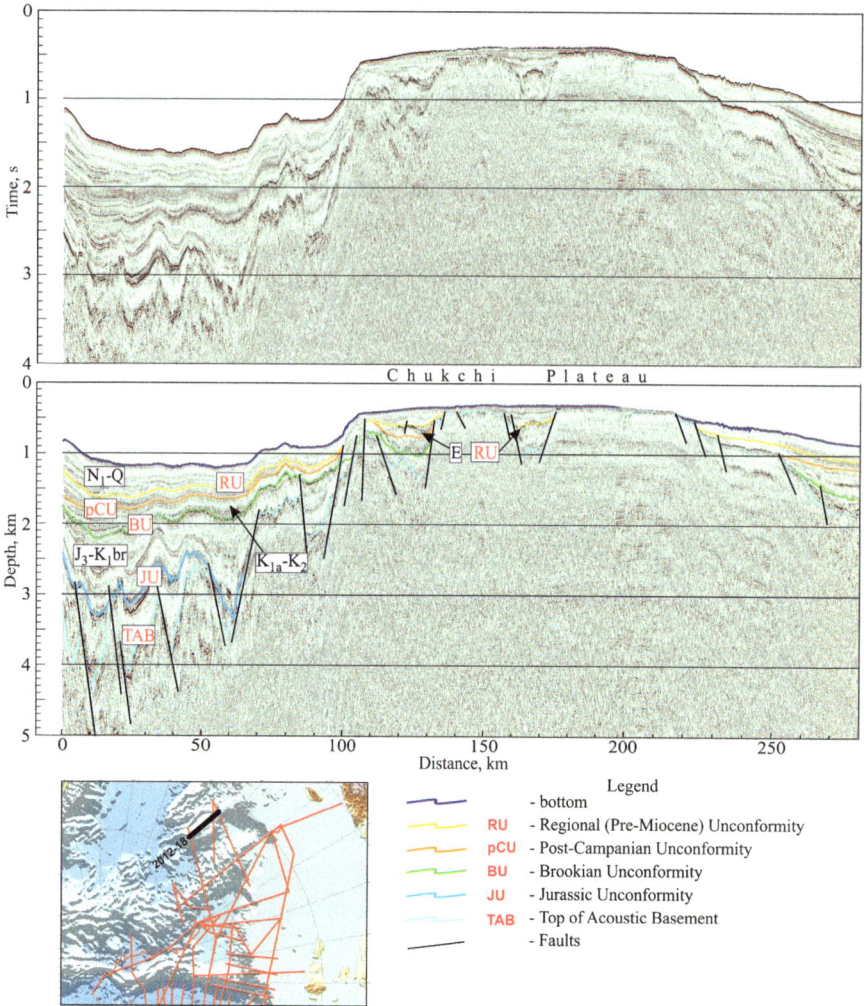

Fig. 8.6 MCS line 2012–18

8.3.3 MCS Line 2012–18 (Chukchi Plateau)

This sub-meridional line (281 km long) was shot along the axial part of the Chukchi Plateau (Fig. 8.6).

Southern part of the line (0–180 km runs along the central graben of the Chukchi Plateau, formed in Jurassic? and active till Late Cretaceous–Paleogene. Accordingly, the sedimentary sequences inside the graben, with total thickness approaching 3000 m, can be subdivided on syn-rift Mesozoic and post-rift Neogene–Pleistocene complexes. Judging by strong fragmentary reflectors, volcanic can be present in the upper parts of the acoustic basement.

The rest of the line (180–280 km) covers the highest section of the Chukchi Plateau formed by uplifted block of the acoustic basement, with two shallow grabens at 110–122 and 150–170 km. At the most of this block the sedimentary cover is drastically reduced and at 190–220 km completely absent exposing the acoustic basement at the sea bottom bearing signs of active erosion, perhaps, even in the sub-aerial conditions. North from these outcrops the thickness of the sedimentary cover increases again to 1000 m.

8.3.4   MCS Line 2012–19

The line (460 km long) stretches from the Eastern slope of the Mendeleev Plateau, over the northern Chukchi Basin, across the central Chukchi Plateau and into the Northwind Trough (Fig. 8.7).

Structurally, the line has three distinct segments. The north-western segment (0–60 km) represents eastern slope of the Mendeleev Ridge with topography closely following basement subsidence to south-east through step-wise system of sub-meridional normal faults.

The central segment (0–180 km) crosses the northern closure of the Chukchi Basin with complex system of the north-south trending Cretaceous normal faults.

The south-eastern segment illustrates the contrasting block structure of the Chukchi Plateau (200–420 km) with its distinctly different western and eastern parts. The western (200–310 km) part is affected by multiple normal faults culminating by the central graben (310–350 km) formed by Late Jurassic - Paleogene faults. In this part the Miocene-Pleistocene sediments are not displaced by normal faults.

The eastern part (350–420 km) can be described as an area of active high-amplitude normal faulting of Late Cretaceous activation (and still neotectonically active) creating a system of differentially displaced basement blocks. This area is bounded from south-east by steep, 1000 m high, fault escarpment of the Northwind Trough and it is only here where the acoustic basement is exposed.

Upper Cretaceous unconformity pCU with sings of active erosion places this part of the Chukchi Plateau above the sea level at some time during the Paleocene.

Despite the active normal faulting the Miocene-Pleistocene hemipelagic unit (180–300 m thick and Pv = 1.6–1.8 km/s) drapes over all above described structures and marks the completion of the Central Arctic uplifts forming process.

8.4   Acoustic Basement

Dredged material from steep escarpments with exposed acoustic basement consists of Lower-Upper Paleozoic shallow-water cratonic, mainly carbonate formations representing what was identified on crustal models described in previous Chapters

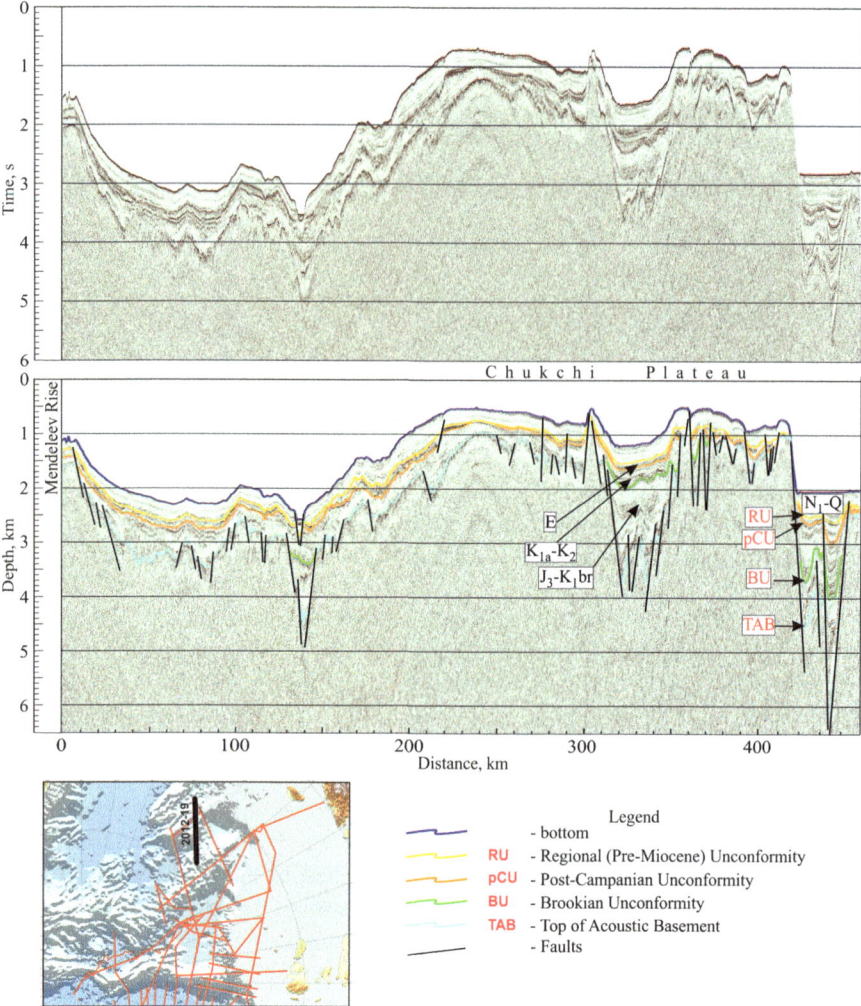

Fig. 8.7 MCS line 2012–19

as a metasedimentary complex, overlain by Mesozoic-Cenozoic sedimentary cover. Triassic aleurolites and sandstones, Late Jurassic argillites and Cretaceous argillites, sandstones and Cenomanian-Turonian felsic volcaniclastics (90.6 ± 2.1 Ma, 40Ar-39AR) – were also found.

Core sampling from flanks of the Northwind Ridge established presence of Paleozoic and Upper Jurassic sediments. Permian red beds in cores correlate with similar synchronous formations of the Sverdrup Basin in the Canadian Archipelago, suggesting that the Chukchi Plateau was attached to the Arctic Canada and Alaska and separated by rifting/spreading leading to formation of the Canada Basin (Grantz et al. 1998). The youngest syn-rift deposits on the Northwind Ridge are dated as Early Jurassic (Arrigoni 2008).

Large volume of material (? 500 kg) was dredged from the steep northern slope of the Chukchi Plateau ("Healy" expedition, 2009). It contains fragments of orto-gneisses, crystalline and green schists, granites and aplites. Zircons from orto-gneisses were dated at 428 ± 3.4 Ma and 500 Ma, with inner core of the single crystal giving 850–1000 Ma suggesting Caledonian re-activation of the Grenville crust.

8.5 Earth's Crust

The crustal models of the Chukchi Plateau presenting P-and S-waves velocities of the major crustal interfaces, their geometry and density (Figs. 7.18, 7.19 and 7.20) were based on DSS data acquired by the "'Arktika −2012" expedition (Fig. 1.27).

The models demonstrate that the principal structural complexes continue unin-terrupted between the Mendeleev Ridge and Chukchi Plateau, with total thickness of 26–29 km and approximately equal thickness of the upper and lower crusts under the Chukchi Plateau. The lower crust P-wave velocity does not exceed 7 km/s, char-acteristic for the continental type. Moho subsides from 22 km under the Mendeleev Ridge to 30.5 km under the Chukchi Plateau and rises again to 26.5 km under the Northwind Trough. Densities of the major structural complexes were computed as follows: mantle - 3.34 g/cm^3; lower crust – from 2.90 g/cm^3 under the Chukchi Basin to 2.84 g/cm^3 under the Chukchi Plateau; upper crust - 2.77 g/cm^3; metasedimentary complex – from 2.40 to 2.68 g/cm^3 (reflecting its complexity). The sedimentary cover of the Chukchi Plateau has significantly higher density then that of the Mendeleev Ridge suggesting older age of the former.

8.6 Conclusions

1. The Chukchi Plateau can be best described as a fragment of the continental mar-gin jutted into the abyssal part of the Arctic Ocean. With its flat almost undis-turbed top, the Chukchi Plateau has the simplest morphology among other Central Arctic Uplifts Complex and the most direct connection with the adjacent shelf of the East Siberian-Chukchi continental margin.
2. The principal structural complexes of the consolidated continental crust continue uninterrupted between the Mendeleev Ridge and Chukchi Plateau with total thickness of 26–29 km and approximately equal thickness of the upper and lower crusts under the Chukchi Plateau. In the Chukchi Basin the crust is 20 km thick with upper crust – 3 km.

3. Lower part of sedimentary cover consists of shallow-water cratonic, mainly carbonate, formations overlain by Cretaceous-Cenozoic (possible pre-Cretaceous in deep grabens) terrigenous sequences disturbed by normal faulting at the base.
4. The Chukchi Basin, with its undisturbed sedimentary complexes, also includes 1200 m of pre- Upper Jurassic (Upper Ellesmerian) formations. Almost complete absence of folding and faults in the Chukchi Basin may point to active sedimentation, not the attenuation of the crust, as a principle cause of its subsidence.
5. Miocene-Pleistocene complex above the RU unconformity continues from the Chukchi Basin to the Chukchi Plateau and marks the end of formation of these present-day morphological structures.

References

Arrigoni V (2008) Origin and evolution of the Chukchi borderland. Dissertation, Texas A&M University
Brumley K (2009) Tectonic geomorphology of the Chukchi borderland: constraint for tectonic reconstruction models. Dissertation, University of Alaska Fairbanks
Brumley K (2014) Geologic history of the Chukchi borderland, Arctic Ocean. Dissertation, Stanford University
Bruvoll V, Kristoffersen Y, Coakley B et al (2010) Hemipelagic deposits on the Mendeleev and northwestern alpha submarine ridges an the Arctic Ocean: acoustic stratigraphy, depositional environment and an inter-ridge correlation calibrated by ACEX results. Mar Geophys Res J 31:149–171
Bruvoll V, Kristoffersen Y, Coakley B et al (2012) The nature of the acoustic basement on Mendeleev and northwestern alpha ridges, Arctic Ocean. Tectonophysics J 514:123–145
Dove D, Coakley B, Hopper J et al (2010) HLY0503 geophysics team. Bathymetry, controlled source seismic and gravity observations of the Mendeleev ridge; implications for ridge structure, origin, and regional tectonics. Geophys J Int 183(2):481–502
Geomorphological Aspects of the Russian Continental Shelf Exterior Boundary in the Arctic (2005) Naryshkin GD (ed) GUNIO MO RF Saint Petersburg, p 58
Grantz A, Hart P (2012) Petroleum prospectivity of the Canada Basin, Arctic Ocean. Mar Pet Geol J 30:126–143
Grantz A, Clark DL, Phillips RL et al (1998) Phanerozoic stratigraphy of Northwind ridge, magnetic anomalies in the Canada Basin, and the geometry and timing of rifting in the Amerasia basin, Arctic Ocean. GSA Bull 110(6):801–820
Hegewald A, Jokat W (2013a) Relative Sea level variations in the Chukchi region – Arctic Ocean – since the late Eocene. Geophys Res Lett 40:1–5
Hegewald A, Jokat W (2013b) Tectonic and sedimentary structures in the northern Chukchi region, Arctic Ocean. J Geophys Res 118(7):3285–3296
Morozov AF, Petrov OV, Shokalsky SP et al (2013) New geological data are confirming continental origin of the Central Arctic rises. Reg Geol Metallogeny J 52:2–25
Piskarev AL (2004) The basement structure of the Eurasia Basin and central ridges in the Arctic Ocean. Geotektonics J 38(6):443–448
Poselov VA, Butsenko VV, Chernykh AA et al (2014) The structural integrity of the Lomonosov Ridge with the North American and Siberian continental margins. VI International Conference on Arctic Margins (ICAM VI), Fairbanks, Alaska, 30 May – 2 June 2011
Poselov VA, Butsenko VV, Zholondz SM et al (2017) Seismic stratigraphy of sedimentary cover in the Podvodnikov Basin and North Chukchi trough. Dokl Earth Sci J 474(2):688–691
Sherwood KW (2011) Geologic cross-section showing regional correlations among wells from the Chukchi shelf and western Arctic Alaska. Bureau of Ocean Energy Management

Chapter 9
Extensional Structures of the Central Arctic Uplifts Complex

Victor A. Poselov and Victor V. Butsenko

Abstract The whole volume of the contemporary information describes the Central Arctic Uplifts Complex as a composite block of continental crust. Rift-related stretching and attenuation of the continental crust is the principal factor dictating the tectonic evolution of this block and its two-phased HALIP magmatism. The most evident signs of the rift-induced strain, – systems of grabens and half-grabens, high-altitude and gently dipping normal faults – are present in the Lomonosov Ridge, Mendeleev Ridge, Chukchi Plateau and on the slopes of the uplifts into the western parts of the Podvodnikov and Chukchi basins. Depocenters of the Vilkitsky Trough (deep-water prolongation of the offshore North Chukchi Trough) and Chukchi Basin are filled with substantially thick Jurassic (or pre- Upper Jurassic) sequence, traceable from the North Chukchi Trough. Jurassic (or pre- Upper Jurassic) deposits are interpreted as relicts of the pre-oceanic Ellesmerian structural stage preserved in near-shelf tectonic depressions. They are strongly affected by rifting only at the elevated parts of the Central Arctic Uplifts Complex, and much less – in the depocenters of the sedimentary depressions.

Keywords Central Arctic Uplifts Complex · Lomonosov Ridge · Podvodnikov Basin · Vilkitsky Trough · Mendeleev Ridge · Chukchi Plateau · Chukchi Basin · East-Siberian shelf · Grabens and half-grabens · Normal faults · Continental crust

The Central Arctic Uplifts Complex (CAUC) occupies the largest portion of the Amerasian Basin. It includes not only the major underwater uplifts (Lomonosov Mendeleev, Alpha and Northwind ridges, Chukchi Plateau) but also large abyssal depressions - Podvodnikov, Makarov, Mendeleev, and Chukchi basins (Fig. 9.1).

For the last 50 years, origin of the Amerasian Basin (including the Canada Basin), in general, and the CAUC, in particular, was subject of numerous investigations and

V. A. Poselov (✉) · V. V. Butsenko
All-Russian Research Institute of Geology and Mineral Resources of the World Ocean (VNIIOkeangeologia), Saint Petersburg, Russia
e-mail: vicb@vniio.nw.ru

© Springer International Publishing AG, part of Springer Nature 2019
A. Piskarev et al. (eds.), *Geologic Structures of the Arctic Basin*,
https://doi.org/10.1007/978-3-319-77742-9_9

Fig. 9.1 Central Arctic Uplifts Complex

discussions (Carey 1958; Grantz et al. 1979, 2011; Scotese 2011; Miller and Verzhbitsky 2009; Brumley et al. 2008, 2011; Lobkovsky et al. 2013; Vernikovsky et al. 2013; ets).

From the period from 2005 to 2014 the Russian expeditions in this part of the Amerasian Basin acquired 35,000 km of bathymetry profiles, over 23,000 km of Multi-Channel Seismic (MCS), 4000 km of Deep Seismic Sounding (DSS) and 150 Refraction-Reflection Soundings (RRS).

Up-to-date processing, integration and interpretation of this dataset presents the CAUC as a large unified mega-block with contemporary structure created by rift-related extension of its continental crust. The regional composite seismic profile from the Amundsen Basin to the Chukchi Plateau (Fig. 9.2) clearly displays the signs of this process.

Most of normal faults oriented north-south, indicating sub-latitude orientation of extension vector, similar to orientation of the extensional structures on the adjacent shelf (see Chap. 3).

Below we include detail description of the extensional features in major CAUC structures.

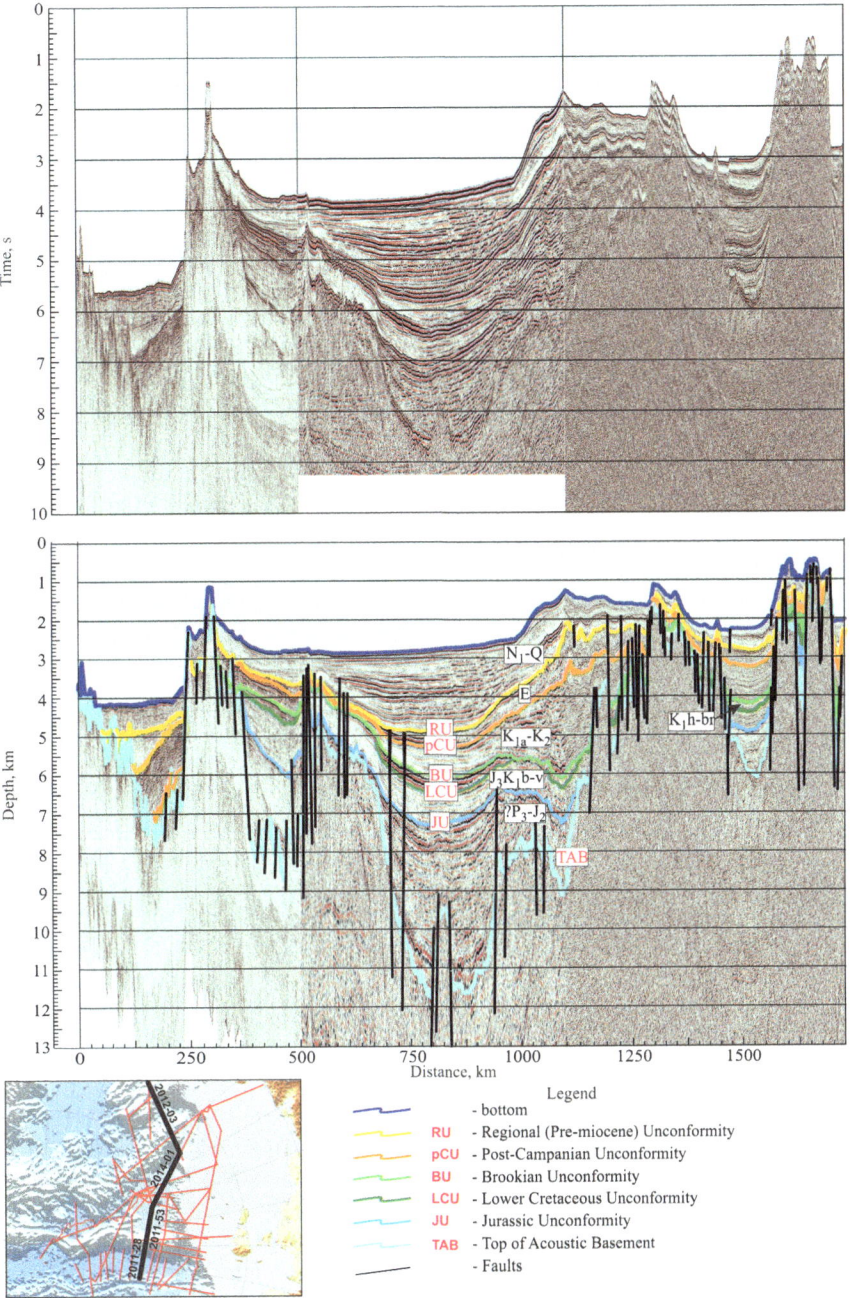

Fig. 9.2 Composite MCS profile from the Amundsen Basin to the Chukchi Plateau
Major unconformities: RU – regional pre-Miocene; pCU – post-Campanian; BU – Brookian; LCU – Lower Creatceous; JU –Upper Jurasic; TAB – top of acoustic basement

9.1 Lomonosov Ridge

According the modern reconstructions, the Lomonosov Ridge does not constitute a separate terrane. Instead, it was just a part of the CAUC and moved east together with it after CAUC separation from the Barents-Kara continental margin 55–53 Ma ago (Poselov et al. 2014; Jokat et al. 2013).

The Lomonosov Ridge crystalline crust, 18–21 km thick with approximately equal upper and lower parts, is undoubtedly continental. Transitional by its physical properties, the so called "metasedimentary" complex lies on top of the crystalline crust and represents the acoustic basement for the overlaying stratified sequences.

Above the acoustic basement, the lower (Cretaceous-Paleogene) syn-rift complex with neritic sediments and the upper (Miocene-Pleistocene) draping complex of hemipelagic deposits, are omnipresent throughout the Central Arctic Uplifts Complex.

The cardinal change of the sedimentation regime between Late Paleogene and Early Neogene might be triggered by opening of the Fram Strait - the important tectonic event (LMA 13, ≈33 Ma) allowing massive influx of the North Atlantic waters into the Arctic Basin. There are only few hundred meters (200 m at ACEX well) of Paleogene sequence at the Lomonosov Ridge. Therefore it was either above, or just below, the sea level. We may conclude that the Lomonosov Ridge existed as a mountain range in Cretaceous and became underwater uplift in Miocene.

Seismic section contains numerous signs of rift-related extensional regime within the Lomonosov Ridge proper. The most clearly they are visible on the Lomonosov Ridge slope towards the Podvodnikov Basin where array of normal faults with 100–2300 m displacement of Cretaceous syn-rift formations and fault planes dipping at ≈16° create system of grabens and half-grabens (Fig. 9.3). Faults displacing the Paleogene beds on the opposite slope of the Lomonosov Ridge into the Amundsen Basin are interpreted as Cretaceous reactivated during ultra-slow Cenozoic spreading of the Eurasian Basin.

9.2 Podvodnikov Basin

The eastern slope of the Lomonosov Ridge and the western slope of the Mendeleev Ridge forms western and eastern edges of the Podvodnikov Basin as a sedimentary depression and as a morphological entity. Both flanks are affected by sub-meridian normal faulting. To the north, the floor of the Podvodnikov Basin deepens stepwise toward the Makarov Basin. It classifies the floor as a part of complex continental slope subjected to the most recent extension of the continental crust and differential subsidence of separate blocks.

DSS data describe the Earth's crust of the Podvodnikov Basin as rift-strained continental margin type (total thickness 19–22 km, with 2–3 km for upper crust). The acoustic basement from the Lomonosov Ridge continues under the whole

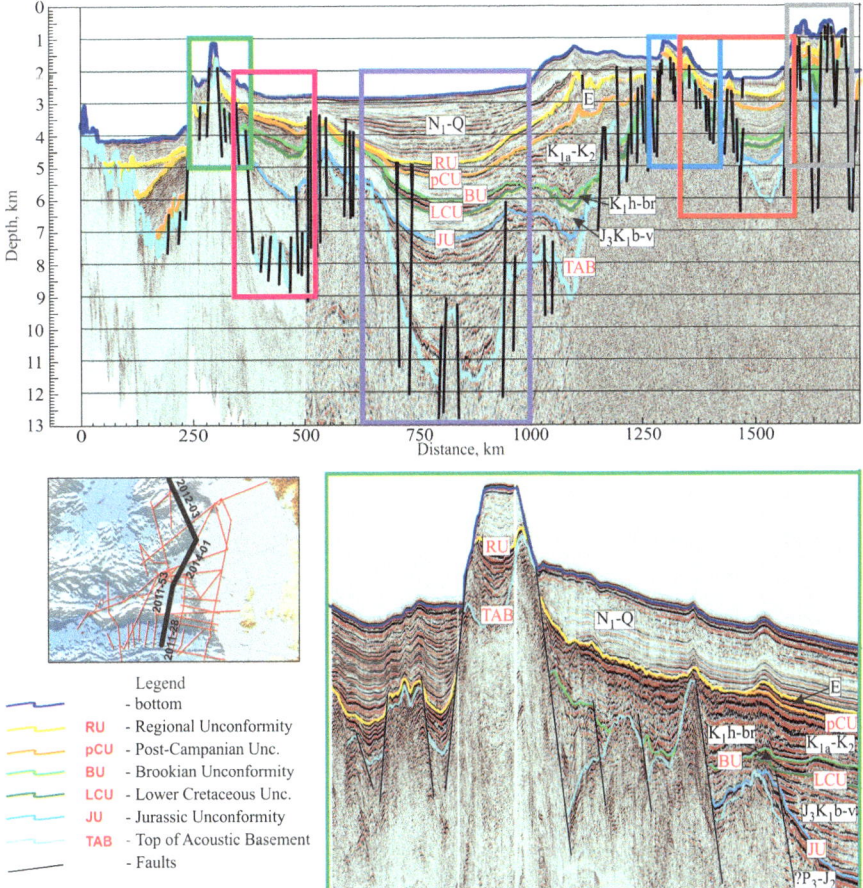

Fig. 9.3 The Lomonosov Ridge - fragment of the composite MCS profile (Fig. 9.2)

western part of the Podvodnikov Basin up to the border with the Vilkitsky Trough (also noted by W. Jokat in (Jokat et al. 2013)). The Geofizikov Spur, the western edge of the Vilkitsky Trough, is separated from the Lomonosov Ridge by deep graben with multitude of secondary half-grabens at its base, formed by normal faults with amplitude 800–1000 m and fault planes dipping at ≈30°. The half–grabens filled Jurassic beds which, in turn, are overlain by the thick Cretaceous sequences (Fig. 9.4). Paleogene neritic complex is considerably reduced here, similar to situation at the ACEX well at the near-pole section of the Lomonosov Ridge. The Miocene-Pleistocene hemipelagic complex forms the top of the sedimentary section.

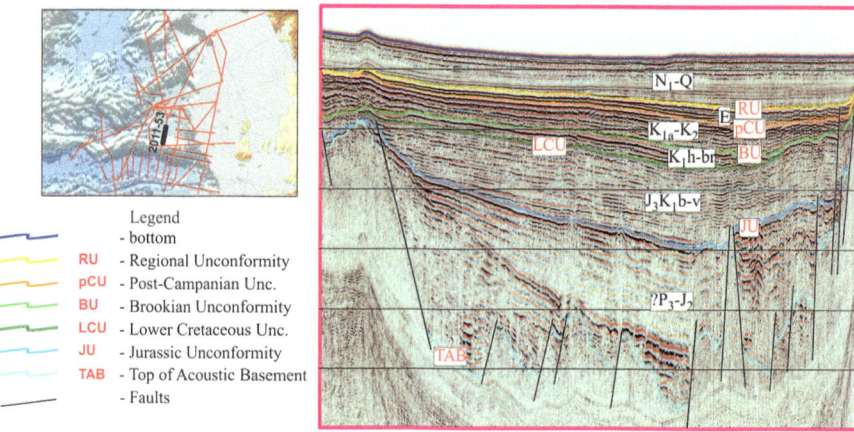

Fig. 9.4 Deep graben between the Lomonosov Ridge and the Geofizikov Spur – fragment of the composite MCS profile (Fig. 9.2)

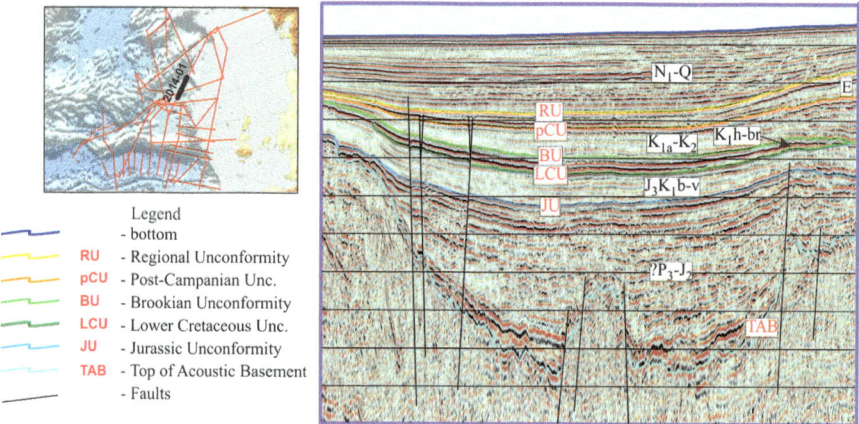

Fig. 9.5 The Vilkitsky Trough (deep-water prolongation of the offshore North Chukchi Trough) - fragment of the composite MCS profile (Fig. 9.2)

9.3 Vilkitsky Trough

Seismic depicts the Vilkitsky Trough (deep-water prolongation of the offshore North Chukchi Trough) as a rift depression with traced from the shelf thick Jurassic (or pre- Upper Jurassic) complex at its depocenter (Fig. 9.5).

The marine Upper Cretaceous complex is also characteristically thicker here while the thickness of the neritic Paleogene complex is similar to that in the graben between the Lomonosov Ridge and the Geofizikov Spur. The Miocene-Pleistocene hemipelagic complex forms the top of the sedimentary section.

There few visible extensional structures in the central part of the Vilkitsky Trough and all sedimentary complexes above the Upper Jurassic unconformity are practically undisturbed. Some normal faults displace Jurassic and Paleogene layers by only few hundred meters and have steep (up to 50°) planes. The depocenter of the Vilkitsky Trough is filled with substantially thick Jurassic (or pre- Upper Jurassic) sequence, traceable from the offshore North Chukchi Trough.

The subsidence of the Vilkitsky Trough can be explained not only by extension but also compensating regime of sedimentation, as it was discussed in (Artyushkov and Poselov 2010).

9.4 Mendeleev Ridge

In the west, the Mendeleev Ridge borders with the Vilkitsky Trough through the stepped fault zone, in the south – with the North Chukchi Trough, in the east - with the Chukchi Basin. Several stepping down northward bathymetric terraces tie the Mendeleev Ridge with Eurasian shelf.

In general, the Mendeleev Ridge 30–32 km thick Earth's crust can be classified as stretched continental margin crust. The major distinction of the Mendeleev Ridge crust is dramatic thickening of the lower crust compare to the upper (20–22 km against 4–8 km, accordingly, Fig. 7.15). This can be attributed to relatively stronger magmatizm in this particular region leading to phenomena of so called "LIP-thickening" or "magmatic underplating" (Poselov et al. 2012).

Seismic sections display variety of extensional grabens and half-grabens at the base of the Mendeleev Ridge sedimentary cover (Fig. 9.6), created by normal faults with amplitudes 160–220 m and moderately steep fault planes (15°–30°). Stronger amplitudes (to 500 m) and steeper fault planes (up to 40°) concentrate at the

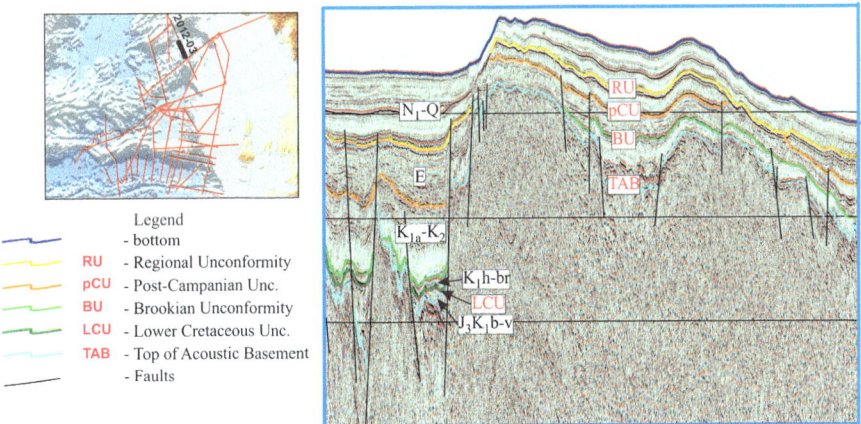

Legend
- bottom
RU - Regional Unconformity
pCU - Post-Campanian Unc.
BU - Brookian Unconformity
LCU - Lower Cretaceous Unc.
TAB - Top of Acoustic Basement
- Faults

Fig. 9.6 The Mendeleev Ridge - fragment of the composite MCS profile (Fig. 9.2)

Mendeleev Ridge - Vilkitsky Trough junction. Syn-rift microbasins with increased thickness of the Lower Cretaceous deposits often associate with half-grabens. Strong reflectors at the very top of the acoustic basement are interpreted as volcanics–related. The volcanic rock samples in the region associated with the acoustic basement were dated as Cretaceous and indicated on near surface environment at the extrusion location (Morozov et al. 2013). Similar indicators of the extension tectonics were documented on the Alpha Ridge (Bruvoll et al. 2012). The whole Mendeleev Ridge is draped by the Miocene-Pleistocene hemipelagic complex.

Existing opinion of purely volcanic origin of the Mendeleev Ridge (Bruvoll et al. 2012) was disproved by sample analysis of outcropping acoustic basement visually documented by the "Arktika-2012"expedition (Gusev et al. 2017).

9.5   Chukchi Basin

The Chukchi Basin flanks, both from the Mendeleev Ridge and the Chukchi Plateau sides, are formed by series of submeridional steep fault escarpments. Similar to the Podvodnikov Basin, the Chukchi Basin Earth's crust (total thickness 19–20 km, upper crust 2–3 km) also can be described as rift-strained continental margin type (Kashubin et al. 2016). The reduction of the upper crust is thought to be rift-induced.

From the structural point of view, the Chukchi and the Podvodnikov basins are also close to each other. Similar to the western part of the Podvodnikov Basin, underlain by the sloping acoustic basement of the Lomonosov Ridge, the western part of the Chukchi Basin is underlain by the sloping acoustic basement of the Mendeleev Ridge (Fig. 9.7). Therefore, the Chukchi Basin, like the Podvodnikov Basin, ought to be subdivided into western and eastern sub-basins (the western sub-basin can be

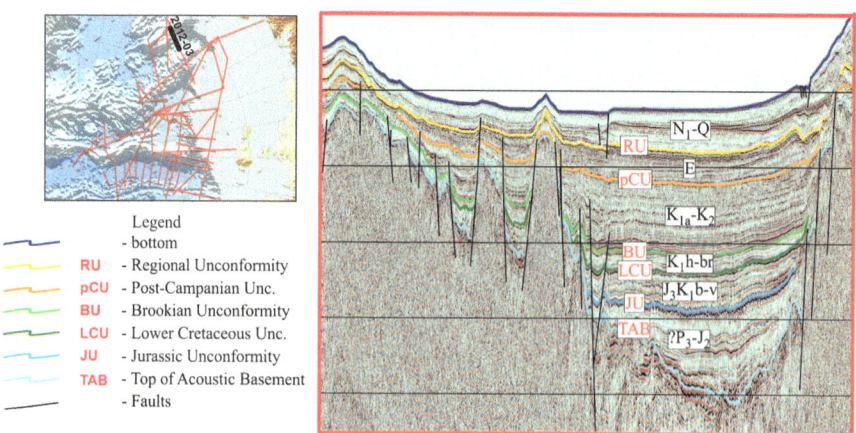

Fig. 9.7 The Chukchi Basin - fragment of the composite MCS profile (Fig. 9.2)

interpreted as a slope of the Mendeleev Ridge). The principal difference between western and eastern sub-basins lies in the different style of extension tectonics.

Moderate extensional activity is observed at the western sub-basin where a few normal faults with 300–400 m amplitude and plains dip ≈30° displacing the acoustic basement. The resulting grabens and half–grabens are filled with Lover Cretaceous sediments and compensated by thick Upper Cretaceous sequence. The Paleogene complex here is noticeably thicker than in the Podvodnikov Basin and also affected by normal faulting.

In the eastern sub-basin the sedimentary cover is almost completely undisturbed – the extensional normal faults with amplitude up 1100 m are only present at the Chukchi Basin - Chukchi Plateau junction forming the escarpment dipping at ≈34° (Fig. 9.7). Substantially thick Jurassic (or pre- Upper Jurassic) sedimentary complex fills the sub-basin depocenter; hemipelagic sediments of the Miocene-Pleistocene complex comprise the very top of sedimentary cover.

It is possible that the tectonic evolution and subsidence process of both the Vilkitsky Trough and the eastern Chukchi Basin was determined not only by stretching of the continental crust, but also compensating regime of sedimentation.

9.6 Chukchi Plateau

The Chukchi Plateau represents the slightly depressed block of the continental crust with total thickness of 28 km and approximately equal ratio of upper and lower crusts (Kashubin et al. 2016).

Active rift strain tectonics are clearly visible on the seismic section (Fig. 9.8), where the 28–43° steep normal faults with amplitudes up to 600–900 m displace not only acoustic basement, but also complete sedimentary cover and seafloor.

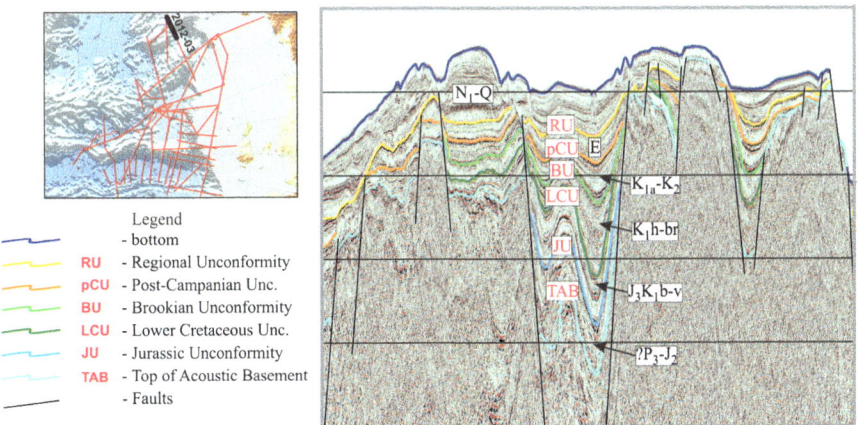

Fig. 9.8 The Chukchi Plateau - fragment of the composite MCS profile (Fig. 9.2)

9.7 East-Siberian Shelf

According to some estimates (Miller and Verzhbitsky 2009), lateral, east-west extension of the East Siberian Sea shelf reached 100%, leaving systems of typical extensional grabens and half-grabens as an unequivocal expressions of this process. As we mentioned it in Chap. 3, within the East Siberian and Laptev Seas shelves these structures are accompanied by strong local gravity and magnetic anomalies. Seismic data (Fig. 9.9) clearly demonstrate that the extensional structures are ubiquitous not only in the De Long Uplift but also in the adjacent (to west) Anisin Trough and (to east) - North Chukchi Trough, with high-amplitude (up to 300 m), steep (36° - 57°) faults, syn-rift sedimentary complex inside them and volcanics at the tops of acoustic basement. Age of the syn-rift complexes varies from Lower Cretaceous at the De-Long Uplift–Anisin Trough junction to the Upper Cretaceous on the De-Long Uplift proper and its slope into the North Chukchi Trough.

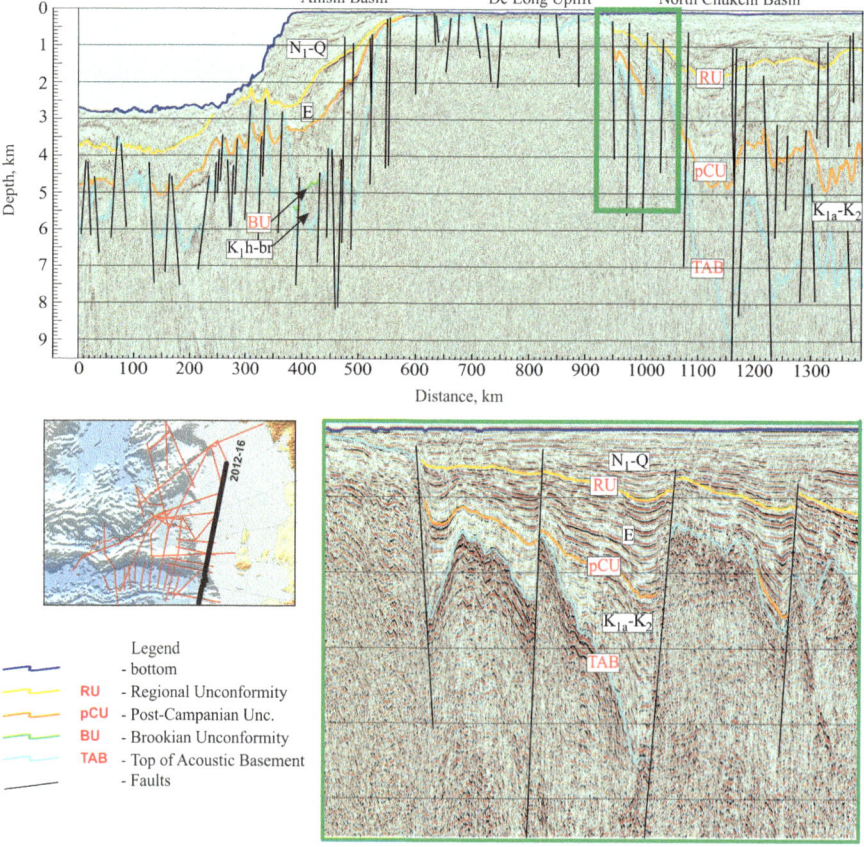

Fig. 9.9 De-Long Uplift and a fragment of the MCS profile 2012–16

9.8   Volcanism of the Central Arctic Uplifts Complex

Recent studies outlined the scope, composition and age of the entire High Arctic Large Igneous Province (HALIP) and its part within the CAUC. Many MCS sections demonstrate bright, short, but correlatable, reflectors at the top of the acoustic basement. These reflections are interpreted as seismic expression of volcano-sedimentary formations comprised of interbedding mafic sills, basalt lava flows, pyroclastics and sediments. Obvious affinity of these formations to grabens and half-grabens connects their origin to the rift-induced stretching of the continental crust (Figs. 9.6, 9.8, 9.9).

There are indications that the CAUC went through two stages of intensive long-lasting volcanic activity, typical for the HALIP. The first stage took place during Jurassic-Lower Cretaceous and was connected to the continental rifting of the Central Arctic Uplifts Complex and opening of the Canada Basin, 120–130 Ma ago (U-Pb 128 Ma, dating of volcanic seafloor samples, Mendeleev Ridge, "Arktika-2012", (Morozov et al. 2013)). The second, younger stage (80–90 Ma) was established by samples from the Mendeleev - Alpha Ridge System and Chukchi Plateau (Morozov et al. 2013) (see Fig. 1.38). The collected rocks were identified as:

- Subalkalic basalts (Mendeleev Basin);
- Alkali basalts (Northwind Ridge);
- Both alkali and subalkalic (axial Mendeleev Ridge);
- Transitional alkali-subalkalic - from the Chukchi Plateau (Mukasa et al. 2009), all with chemistry different from chemistry of basalts from typical mid-oceanic ridges.

In many aspects, basalt volcanism of the Central Arctic Uplifts Complex is close to that of the Spitsbergen, Frantz–Joseph Land, De-Long Islands, Ellesmere and other islands of the Canadian Arctic Archipelago (Morozov et al. 2013).

It becomes clear that there is substantial difference in morphology and deep crustal structure between the Amerasian and Eurasian basins of the Arctic Ocean. The Amerasian Basin can be described as a continental margin subjected to rifting and subsequent disintegrtaion into several blocks sunk to different depth (Gramberg 2001). L.I. Lobkovsky (Lobkovsky et al. 2013) and V.A. Vernikovsky (Vernikovsky et al. 2010, 2013) quite adequately explained causes and mechanism of rift-related stretching of the Central Arctic Uplifts Complex by upper mantle convection. According to this model, subduction of the South-Anuyi Ocean in Eraly Jurassic was augmented by permanently active subduction of the Pacific lithosphere. The resulting steady and intensive subduction of mantle into subduction zone of the north-west Pacific and the South-Anuyi oceans subjected the crust to tensile stress, causing opening of the Canada Basin and Late Jurassic-Early Cretaceous rifting in the CAUC.

Being in good agreement with existing multi-disciplinary data, this geodynamic model of upper mantle convection can be considered as a general mechanism defining Meso - Cenozoic tectonic evolution of the Arctic.

9.9 Conclusions

The whole volume of the contemporary information describes the Central Arctic Uplifts Complex as a composite block of continental crust.

Rift-related stretching and attenuation of the continental crust is the principal factor dictating the tectonic evolution of this block and its two-phased HALIP magmatizm. The most evident signs of the rift-induced strain, – systems of grabens and half-grabens, high-altitude and gently dipping normal faults – are present in the Lomonosov Ridge, Mendeleev Ridge, Chukchi Plateau and on the slopes of the uplifts into the western parts of the Podvodnikov and Chukchi basins. At the same time, rifting activity in the eastern parts is subdued leaving all sedimentary complexes, practically speaking, undisturbed.

Depocenters of the Vilkitsky Trough and Chukchi Basin are filled with substantially thick Jurassic (or pre- Upper Jurassic) complex, traceable from the offshore North Chukchi Trough. Jurassic (or pre- Upper Jurassic) deposits are interpreted as relics of the pre-oceanic Ellesmerian structural stage preserved in near-shelf tectonic depressions. They are strongly affected by rifting only at the elevated parts of the Central Arctic Uplifts Complex, and much less – in the depocenters of the sedimentary depressions.

References

Artuyshkov EV, Poselov VA (2010) Abyssal depressions in the Russain sector of the Amerasian Basin and eclogitisation of the lower continental crust. Dokl RAN J 431(5):680–684. (in Russian)

Brumley K, Mayer LA, Miller EL et al (2008) Dredged rock samples from the Alpha ridge, Arctic Ocean: implications for the tectonic history and origin of the Amerasian Basin. Eos Transactions American Geophysical Union, 15–19 December 2008

Brumley K, Miller EL, Mayer LA et al (2011) Petrography and U-Pb geochronology of Caledonian age orthogneisses dredged from the Chukchi Borderland, Arctic Ocean. Fall Meeting American Geophysical Union, 5–9 December 2011

Bruvoll V, Kristoffersen Y, Coakley B et al (2012) The nature of the acoustic basement on Mendeleev and northwestern Alpha ridges, Arctic Ocean. Tectonophysics J 514:123–145

Carey SW (1958) A tectonic approach to continental drift. In: Carey SW (ed) Continental drift: a symposium. University of Tasmania, Hobart, pp 177–355

Gramberg IS (2001) Comarative geology and mineralogy of the oceans and their stadial development. Geotektonika J 6:3–19

Grantz A, Eittreim S, Dinter DA (1979) Geology and tectonic development of the continental margin north of Alaska. Tectonophysics J 59:263–291

Grantz A, Hart PE, Childers VA (2011) Geology and tectonic development of the Amerasia and Canada Basins, Arctic Ocean. Geol Soc Lond Mem 35:771–800

Gusev E, Rekant P, Kaminsky V et al (2017) Morphology of seamounts at the Mendeleev rise, Arctic Ocean. Polar Res J 36:2–10

Jokat W, Ickrath M, O'Connor J (2013) Seismic transect across the Lomonosov and Mendeleev ridges: constraints on the geological evolution of the Amerasia Basin, Arctic Ocean. Geophys Res Lett 40(19):5047–5051

Kashubin SN, Petrov OV et al (2016) Deep structure of crust and the upper mantle of the Mendeleev rise on the Arktic-2012 DSS profile. Regionalnaya Geologiya i Metallogeniya J 65:16–36. (in Russian)

Lobkovsky LI, Shipilov EV, Kononov MV (2013) Geodynamic model of Meso-Cenozoic upper mantle convection and lithosphere transformation. Fizika Zemli 6:20–38

Miller EL, Verzhbitsky VE (2009) Structural studies near Pevek, Russia: implications for formation of the east Siberian shelf and Makarov Basin of the Arctic Ocean. EGU Stephan Mueller Publication Series 4:223–241

Morozov AF, Petrov OV, Shokalsky SP et al (2013) New geological evidence justifying the nature of the continental area of the Central Arctic elevations. Reg Geol Metallog J 53:34–55

Mukasa S, Andronikov A, Mayer L et al (2009) Geochemistry and geochronology of the first intra-plate lavas recovered from the Arctic Ocean. Portland GSA annual meeting 138: p 11

Poselov VA, Avetisov GP, Butsenko VV et al (2012) The Lomonosov ridge as a natural extension of the Eurasian continental margin into the Arctic Basin. Geol Geophys J 53(12):1662–1680

Poselov VA, Butsenko VV, Chernykh AA et al (2014) The structural integrity of the Lomonosov Ridge with the North American and Siberian continental margins. VI International conference on Arctic margins (ICAM VI), Fairbanks, Alaska, 30 May–2 June 2011

Scotese CR (2011) Paleogeographic reconstructions of the Circum-Arctic region since the Late Jurassic. Paleogeographic and Paleoclimatic Atlas. AAPG. Online Journal E&P Geoscientists. 2011 (www.searchanddiscovery.com/pdfz/documents/2011/30192scotese)

Vernikovsky VA, Dobretsov NL, Maminsky VD (2010) Geodynamics of the central and eastern Arctic. In: Proceedeing of the meeteing on RAS coordination on the study of the Arctic and Antarctica. UrO RAN, Yekaerinburg, pp 41–58. (in Russian)

Vernikovsky VA, Dobretsov NL, Metelkin DV et al (2013) Concerning tectonics and the tectonic evolution of the Arctic. Russ Geol Geophys 54(8):838–858

Chapter 10
Canada Basin

David C. Mosher and Deborah R. Hutchinson

Abstract Perennial sea-ice cover over much of Canada Basin of the Arctic Ocean has hampered geoscientific studies, but concerted efforts over the past decade– particularly with the use of two ice-breakers working collaboratively – has led to new seismic and sample acquisitions. These studies have revealed extensive non-oceanic basement beneath Canada Basin that coincides with proof of a central spreading axis and limited oceanic crust. Additionally, seismic reflection studies have shown its sedimentologic history and stratigraphic development. High resolution subbottom and multibeam detail have revealed its more recent geologic past, including the extent of ice margins during the Pleistocene and the role of submarine landslides and ocean currents within the basin. Despite this new information, there are still significant challenges in understanding the basin. These challenges result from the fact that the basin did not form by a simple rift/extension scenario, but rather more likely through a complexity of events that included variably oriented extension, trans-tension and transform tectonics. Additionally, emplacement of the high arctic magnetic domain (Alpha Ridge and Mendeleev Rise) masks underlying tectonic structures, and lack of age control inhibits correlation with global events.

Keywords Arctic ocean · Canada basin · Seismic-stratigraphy · Tectonics · High Arctic magnetic domain · Transitional crust · Ocean crust · Canada basin gravity low · Glaci-genic debris flows · Turbidites

D. C. Mosher (✉)
Geological Survey of Canada, Natural Resources Canada, Bedford Institute of Oceanography, Dartmouth, NS, Canada
e-mail: david.mosher@canada.ca

D. R. Hutchinson
United States Geological Survey, Woods Hole Coastal and Marine Science Center, Woods Hole, MA, USA
e-mail: dhutchinson@usgs.gov

© Springer International Publishing AG, part of Springer Nature 2019   295
A. Piskarev et al. (eds.), *Geologic Structures of the Arctic Basin*,
https://doi.org/10.1007/978-3-319-77742-9_10

10.1   Introduction

Canada Basin forms the largest, deepest, and oldest portion of the subregion of the Arctic Ocean basin known as Amerasia Basin (Fig. 10.1). It largely lies between about 110°W and 180° longitude and 70° to 80°N latitude. The geology of Canada Basin is among the least studied of any ocean basin in the World. Prior to the year 2006, less than 3000 line-km of seismic reflection data existed within the basin, and the vast majority of those data were close to its southern margins. Only a few seismic profiles acquired from ice camps sampled the central and northern portions of the basin. Additionally, there were few bathymetric data and few geologic samples except from the periphery or acquired from drifting ice camps. Between 2006 and 2016, however, more than a dozen ice-breaker expeditions were undertaken over abyssal plain and continental margin regions of Canada Basin. These expeditions were predominantly part of Arctic coastal States' efforts to map their margins for

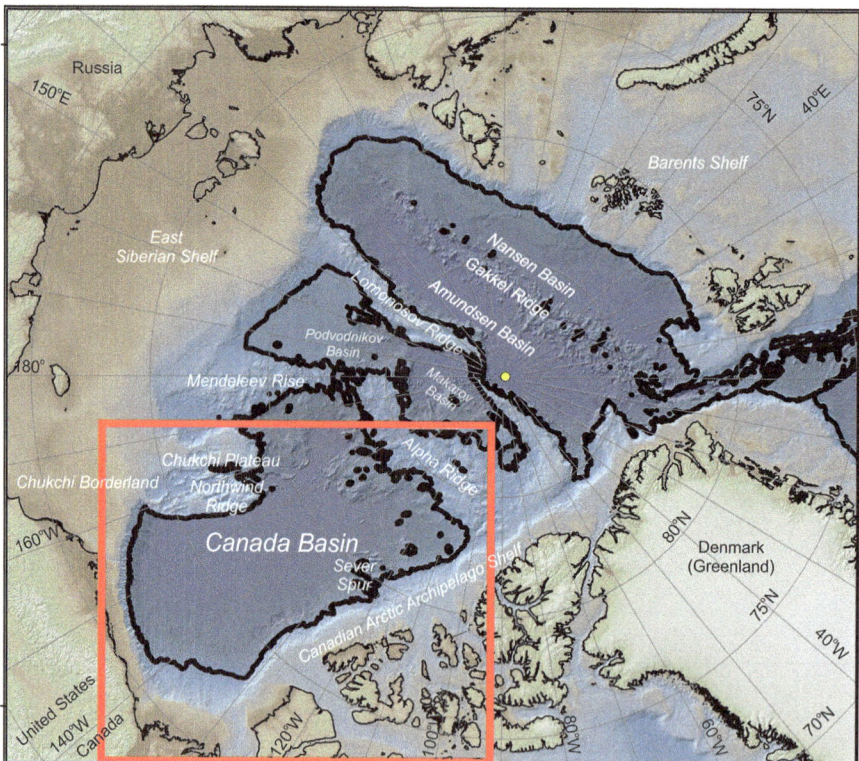

Fig. 10.1 Morphologic features of the Arctic Ocean with the major subbasins and surrounding land areas. The study area (Canada Basin) is outlined with the red box. The 2500 m isobath is shown with a thick black line. Amerasia Basin is the portion of the Arctic Basin between Lomonosov Ridge and Alaska and Canada to Russia. Eurasia Basin lies from Lomonosov Ridge to (and including) the Barents Shelf and from Greenland to Russia

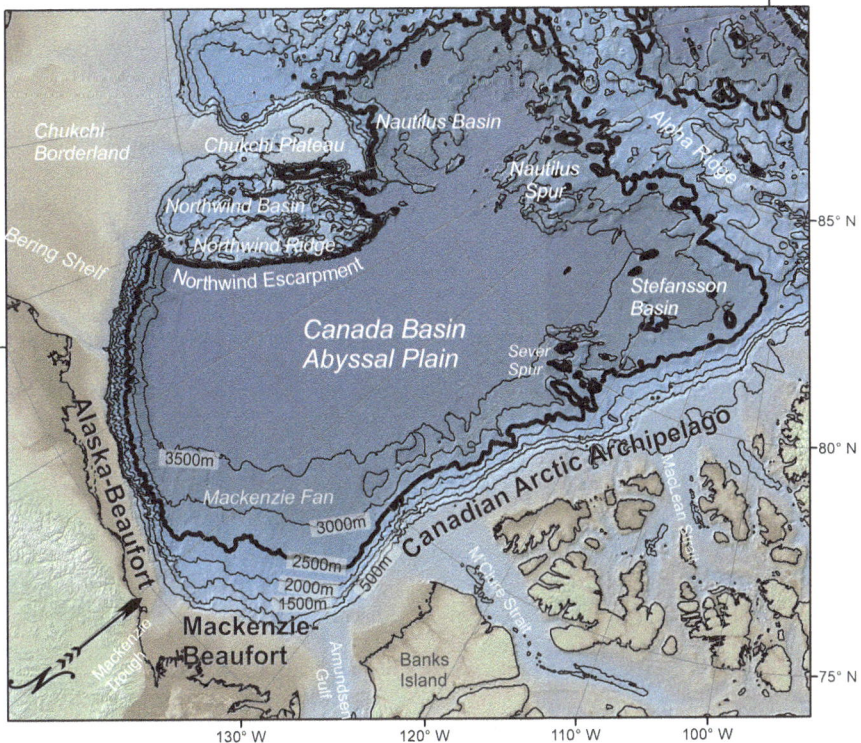

Fig. 10.2 Bathymetry and basemap of Canada Basin. Isobaths are in 500 m intervals; the thick black line is the 2500 m isobath, which broadly outlines the basin. The base map is the IBCAO version 3 (Jakobsson et al. 2012) with a − 90° central meridian. Each subsequent map is of this same extent and includes the 2500 m isobath for geographic reference

extended continental shelf purposes. These expeditions acquired marine multi-channel seismic reflection and refraction, subbottom profiler, marine gravity, and multibeam echosounder bathymetric data. This paper reports on some of the results of these expeditions, integrated with new compilations of potential field data, that provide a significant leap forward in our understanding of the tectonic and geological factors involved in the formation of Canada Basin.

Canada Basin is bound to the south by the Canadian and U.S. Alaska Beaufort margins, to the east by the Canadian Arctic Archipelago, to the west by Northwind Escarpment of the Chukchi Borderland and to the north by the Alpha Ridge (Figs. 10.1 and 10.2). It includes the Canada Basin Abyssal Plain, and Nautilus and Stefansson basins in the north as well as the continental margins of the adjacent Canadian and U.S. land masses (Fig. 10.2). It measures 1500 km north to south and 650 km west to east, totaling about 1,000,000 km² in area. This chapter describes Canada Basin first with regional potential field data (gravity and magnetics), showing the large-scale tectonic and geologic aspects of the basin, then reviews the status of knowledge on the crustal elements of the basin, and describes acoustic basement that was mapped during recent expeditions. The framework established from these overviews sets the stage to characterize the stratigraphy of the sedimentary section within the basin as well as the geomorphology and surficial geology, integrating multibeam data and the International Bathymetric Chart of the Arctic Ocean (IBCAO) bathymetric data sets as well as subbottom profiler data from international sources. The chapter finishes with a tectonic model for the evolution of the basin developed from many of the recent datasets.

10.2   Gravity and Magnetics

Recent compilations of gravity and magnetic anomaly data for the circum-Arctic have been completed for the International Polar Year (e.g., Anderson et al. 2010; Gaina et al. 2011). These compilations have been augmented with more recent gravity measurements taken during shipborne surveys and aeromagnetic and aerogravity surveys (e.g., Døssing et al. 2013, 2014; Oakey and Saltus 2016). These data provide important regionally extensive information concerning the deep structure of Canada Basin and the continuity of structures from onshore to offshore. They also provide a valuable tool for interpreting the regional seismic profiles and interpolating between individual seismic lines.

10.2.1   Gravity data

A significant component of free-air gravity anomalies in the Arctic Ocean is caused by bathymetric relief (i.e. in general, positive gravity anomalies correlate with bathymetric highs and gravity lows correlate with bathymetric lows). This correlation fails at continental margins, where the combined effects of changing water depths, changing sediment thicknesses and changing Moho depths combine to create a shelf-edge paired positive-negative gravity anomaly that is difficult to interpret. In order to compensate for bathymetric variations, a marine Bouguer gravity anomaly was calculated using bathymetry data from the IBCAO Version 3.0 gridded data set (Jakobsson et al. 2012). This Bouguer correction also compensates for the component of the shelf-edge paired anomaly caused by changing water depths. The correction was calculated by replacing the water column (density of 1030 kg/m^3) with material with a crustal density of 2670 kg/m^3, using a polynomial expansion algorithm (Nady 1966) in which the water layer was simulated by 3-D cubes, each with a size of 2 × 2 km and 100 m thick. Since the marine Bouguer anomaly field is dominated by high-amplitude long-wavelength components caused by deep water regions, a high pass filter was applied to remove all wavelengths greater than 400 km. This filtering focuses on Bouguer anomalies related to variations in sediment thickness, crustal thickness, and density variations within the crust. Fig. 10.3 shows Bouguer anomalies in the Canada Basin region.

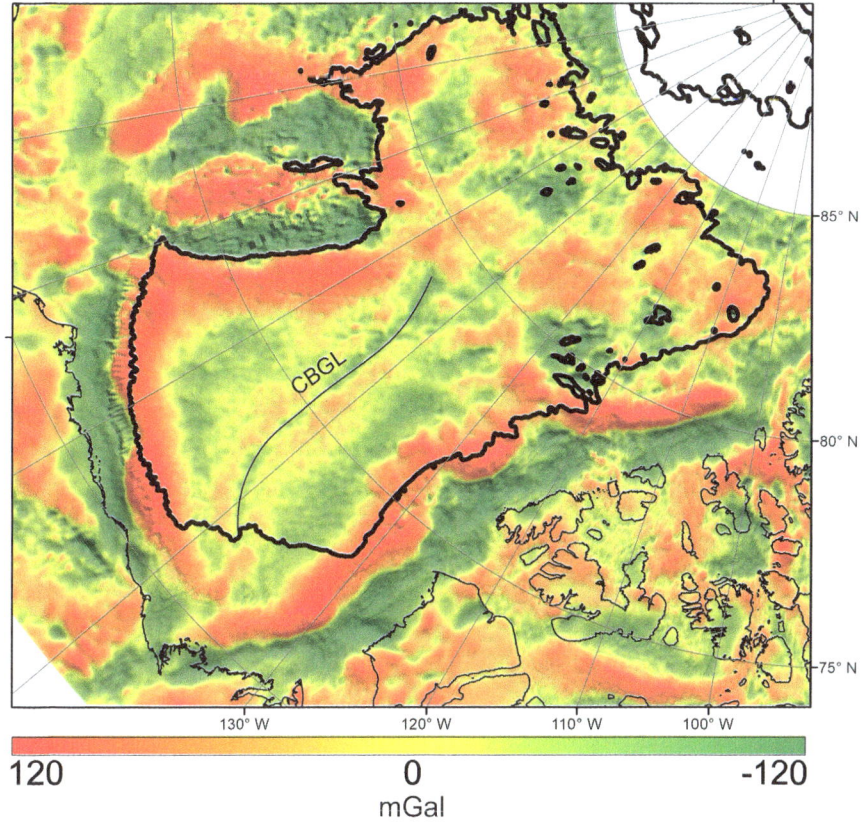

Fig. 10.3 Marine Bouguer gravity anomaly map of Canada Basin. Note the near linear Canada Basin Gravity Low (CBGL) marked with a thin curved line to indicate the axis of the low. The thicker black line is the 2500 m isobath

In general, gravity highs of Bouguer anomalies represent a localized mass surplus and gravity lows represent a mass deficit. Since the density effect of water has been removed, most residual gravity lows correlate with sedimentary basins, where a thick layer of low density sediments replace higher density crustal rocks. In contrast, gravity highs can be caused by either basement highs (buried beneath sedimentary cover), high density crustal blocks (e.g. dense volcanic intrusive in continental crust) or thick sedimentary sequences that are not in isostatic equilibrium (i.e., sediments are replacing water). Most prominent on the gravity grid shown in Fig. 10.3 are anomalies associated with margins of the basin, in part still an effect of the topographic change, but also due to deep sedimentary basins and changing Moho depths at the flanks of the margins. Within Canada Basin, the gravity anomalies are of small amplitude. A subtle narrow gravity low is apparent from south to north in the central basin (Fig. 10.3). This low is referred to as the Canada Basin Gravity Low (CBGL) and is flanked to either side by higher values in a symmetrical fashion. The low is about 40 km wide and extends from the Mackenzie-Beaufort margin toward the north for about 870 km. The CBGL has been hypothesized to represent a signature of an extinct spreading ridge (e.g., Grantz et al. 1979; Laxon and McAdoo 1994). It dissipates amongst more disparate gravity anomalies that form the northern portion of Canada Basin where anomalies associated with Alpha Ridge dominate. Anomalies show only slight variation throughout the bathymetric trend of Alpha Ridge and Mendeleev Rise, which implies minimal lateral variability in its density structure (Oakey and Saltus 2016).

10.2.2 Magnetic Data

Over the years, several regional compilations of the magnetic field in the Arctic were produced (e.g. Kovacs et al. 1985; Verhoef et al. 1996). Gaina et al. (2011) provided a regional description of the magnetic anomalies in the Arctic region, using the Circum-Arctic Mapping Program – Magnetics data (CAMP-M) compilation. The CAMP-M map was assembled from seven regional compilation grids: the 1 km gridded datasets for the regions of Canada (based on the Canadian Aeromagnetic Data Base), Alaska (based on Alaska USGS aeromagnetic database) and northwest Europe (Fennoscandiab compilation). These data sets were augmented with 5 km gridded data for the oceanic regions (Verhoef et al. 1996) and Russia (VSEGEI compilation). The final CAMP-M compilation has a 2 km grid resolution and the gridded data were upward continued to 1 km above ground or sea-level. A consistent regional long-wavelength component was introduced by a satellite-based lithospheric magnetic model MF6 (Maus et al. 2008).

Oakey and Saltus (2016) provide the most recent assessment of magnetic anomaly data in Canada Basin, although their focus is the region of Alpha Ridge and Mendeleev Rise. The region is characterized by a distinctive, irregular, high-amplitude magnetic anomaly pattern, with short wavelength highs and lows in no consistent orientation (Fig. 10.4). Oakey and Saltus (2016) refer to this area as the High Arctic Magnetic High domain (HAMH, Fig. 10.4). These anomalies are amongst the highest amplitude global long wavelength magnetic anomalies on earth (e.g., Coles and Taylor 1990) and Oakey and Saltus (2016) suggest they result either from deep volcanic crustal roots beneath the ridges or from volcanic intrusions into the crust. The distinctive magnetic anomalies have long been interpreted to be part of a High-Arctic Large Igneous Province (HALIP, e.g., Døssing et al. 2013).

These high amplitude magnetic anomalies extend further into Canada Basin than the bathymetric high associated with Alpha Ridge and Mendeleev Rise. This discrepancy suggests that the sediments of Canada Basin onlapped and buried components of this magnetic domain. South of this magnetic domain, central Canada Basin is characterized by four sub-parallel magnetic anomalies that are broadly symmetrical with the CBGL, i.e., two anomalies on either side of the CBGL (Figs. 10.3 and 10.4). The two anomalies furthest from the location of the CBGL are

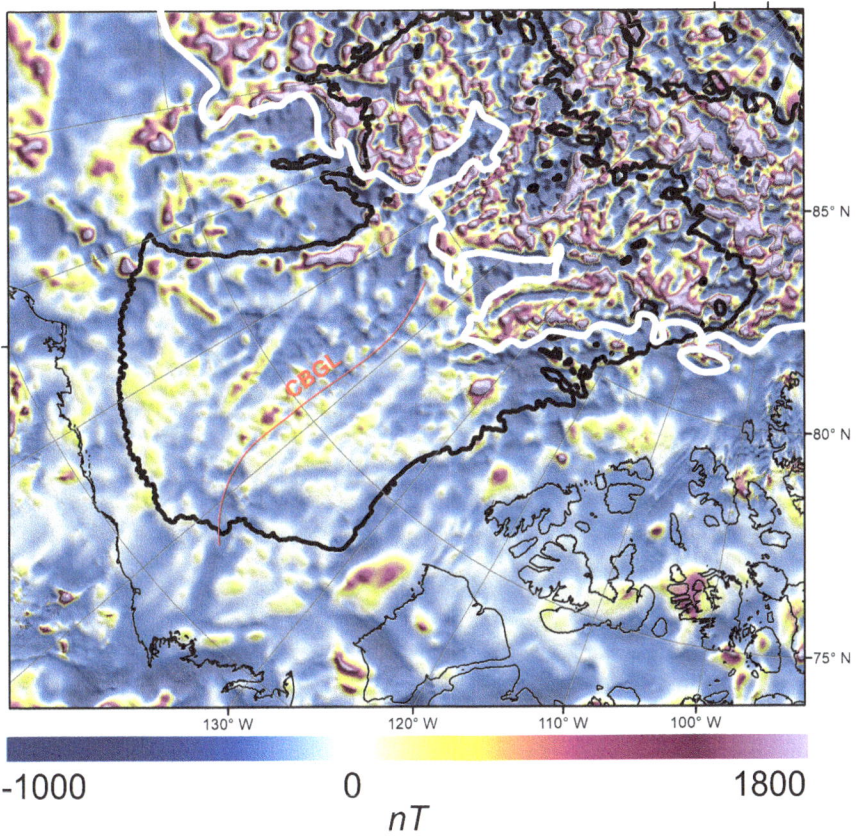

Fig. 10.4 Magnetic anomaly data of Canada Basin (see Oakey and Saltus 2016). The thick black line is the 2500 m isobath. The central axis of the CBGL is shown with a thin red curve. The heavy white line outlines the High Arctic Magnetic High domain (Oakey and Saltus 2016). Note the four near-linear magnetic anomalies in the central Canada Basin on either side of the CBGL

slightly wider apart to the north than the south (Fig. 10.4). Many studies interpreted these subtle linear trends within Canada Basin as oceanic magnetic lineations (e.g., Grantz et al. 1998).

Chian et al. (2016) proposed that these anomalies are located within a region mapped with velocities typical of oceanic crust (see Sect. 10.3 below). One unre-solved problem, however, is determination of the ages of these anomalies. In the analysis of Chian et al. (2016), there are two notable constraints in order for the positive-negative pairs of these magnetic anomalies to be created by seafloor spread-ing processes. First, they have to be created before or after the Cretaceous quiet period, i.e., M0 to C34 / 124 to 84 Ma. Second, they are probably created after M5 (~130 Ma) because the many short reversals prior to M5 are not likely to be preserved in a slow-spreading regime. Slow spreading is inferred in Canada Basin from the rough basement morphology, wide central spreading center valley, and anomalous seismic velocities within the central valley region. They propose the most likely age of spreading was 130–124 Ma although the resulting spreading rate is not consistent with slow spreading.

10.3   Crustal Structure

The crustal structure of Canada Basin has recently been interpreted by Chian et al. (2016) using velocity analyses of an extensive collection of sonobuoy wide-angle reflection and refraction data (Chian and Lebedeva-Ivanova 2015) in combination with seismic reflection (Mosher et al. 2015) and gravity and magnetic data (both shipborne and the grids presented above). Mantle refractions (Pn) and less commonly high-amplitude Moho reflections (PmP) with velocities in the range of 7.7–8.2 km/s were observed, providing constraints on crustal thickness.

The most significant result of the Chian et al. (2016) study is identification of oceanic crust limited to a 340-km-wide zone in the central Canada Basin (Fig. 10.5). Crustal velocities in this area were mapped based on velocities of 6.7–7.1 km/s, typical of oceanic layer 3. Mantle arrivals constrained the thickness of oceanic crust to be 5–6 km, consistent with slow spreading mid-ocean ridge settings (Bown and White 1994).

Chian et al. (2016), using refraction analysis, were also able to identify regions of transitional crust, with diagnostic velocities of 7.2–7.6 km/s. These transitional zones geographically surround the oceanic crust (Fig. 10.5). They interpret that these zones can be caused by either serpentinized mantle (e.g. Chian et al. 1995; Pickup et al. 1996) or with mafic intrusions associated with underplating of the Alpha Ridge and Mendeleev Rise (e.g., Funck et al. 2011). Using this geological framework, Chian et al. (2016) interpreted transitional crust in northern Canada Basin to most likely be underplated or magma-rich in composition (i.e., associated with the high Arctic large igneous province). In southern Canada Basin, these transitional velocities were interpreted to most likely be serpentinized mantle (i.e., associated with amagmatic rifting).

Continental crust was noted by Chian et al. (2016) in a number of regions on the periphery of Canada Basin. In the south, continental crust is interpreted by Chian et al. (2016) to extend seaward more than 300 km from Alaska, into deep-water, with basement velocities of 4.9 km/s underlain by crustal velocities in the range of 5.1–5.4 km/s above a clear PmP arrival. Further east, beneath the Mackenzie Delta, continental velocities were recorded on just two sonobuoys that lie close the coast, before reflecting more transitional velocities offshore. Continental crustal velocities are also observed on two sonobuoys along the northern tip of Northwind Ridge, with velocities ranging from 4.4 to 6.6 km/s (Chian and Lebedeva-Ivanova 2015; Chian et al. 2016). This result is consistent with geological samples of metamorphic rocks taken in this area (Brumley et al. 2015; O'Brien et al. 2016). Thinned continental crust is interpreted in one location offshore of the Canadian Arctic

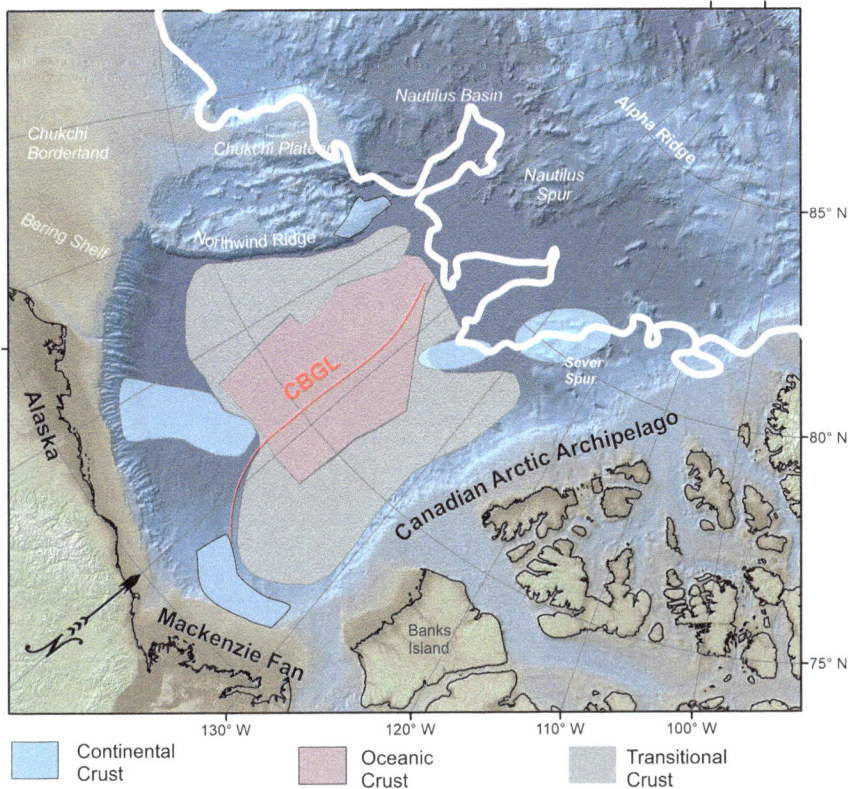

Fig. 10.5 Crustal types (from Chian et al. 2016) overlain on a morphology map of Canada Basin. The polygons represent areas interpreted based on refraction velocities. Areas of no polygon are because there were no measurements or results were ambiguous. The white line outlines the HAMH. The thin red curve through central Canada Basin marks the position of the CBGL

Archipelago, where a large rift graben was observed on seismic profile. Crustal thickness in this area is about 7 km, determined on strong PmP reflections.

10.4   Acoustic Basement

In excess of 16,000 line km of multi-channel seismic reflection data were acquired within Canada Basin between 2007 and 2014 by Canadian and American icebreaker expeditions. Acquisition and processing parameters are found in relevant expedition reports and publications (see for example, Mosher et al. 2012a; Mosher et al. 2015). These data were used to map acoustic basement and sediment thickness, as well as to assess structures along the margins. Acoustic basement is considered to be the low frequency, often high amplitude reflection or reflections below which little

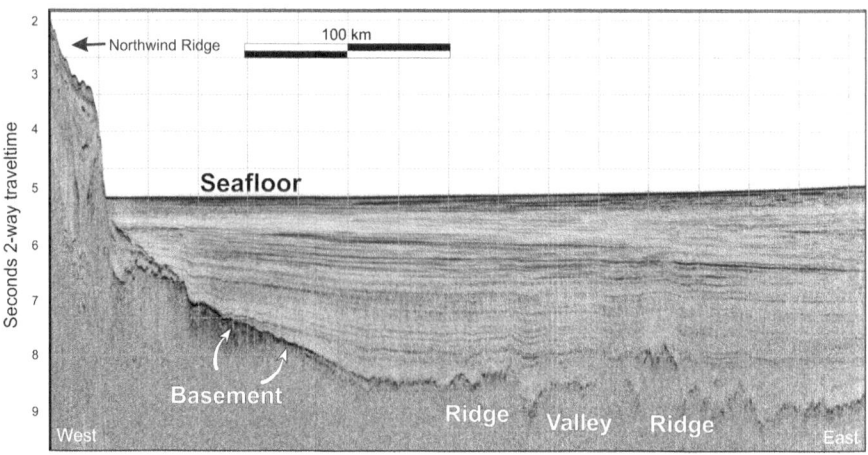

Fig. 10.6 Seismic reflection profile across Canada Basin from west to east showing acoustic basement and the sediment column. Note the structures and morphology of basement. The feature
labelled valley is where the seismic profile crosses the CBGL. Also note how the sediment section
thins to the west and abuts abruptly against Northwind Ridge. See Fig. 10.7 for the profile
location

coherent seismic reflection acoustic energy returns (Fig. 10.6). It is generally con-
trasted with well-stratified reflections that lie above. In large part, this contrast is
interpreted as the interface between unlithified or semi-lithified sediment above and
bedrock beneath.

Depth to acoustic basement from present sea level is shown in Fig. 10.7.
Published industry data help constrain basement depths in the near-shore regions
(e.g., Helwig et al. 2011). Basement rises to 5 km below sea-level close to Alpha
Ridge; resulting in a regional dip of 0.5° toward the south (Fig. 10.8). In the central
to south-central Canada Basin, basement is calculated at about 14 km deep in the
east and 5 km in the west, giving a dip of about 0.9° toward the east (Fig. 10.9a).
The principal cause of this dip is likely sediment loading in the south and east, with
sources of sediment shedding first from Alaska and later the Canadian Arctic
Archipelago and finally the Mackenzie (or proto-Mackenzie) River (Mosher et al.
2012a; Coakley et al. 2016).

Subsidence due to sediment loading does not account for all of the basement dip;
differences in crustal densities likely contribute. If Alpha Ridge and Mendeleev
Rise are underpinned by continental rocks as suggested by some researchers (e.g.,
Funck et al. 2011), then they will have lower densities and therefore higher elevation. The HALIP/HAMH extends well into northern Canada Basin, as evidenced by
the magnetic data (Oakey and Saltus 2016). Northwind Ridge and escarpment is
continental in origin as well (Brumley et al. 2015; Chian et al. 2016) and it too
would have a higher isostatic elevation, influencing the dip of basement.

A ridge-valley-ridge structure is suggested by depth-to-basement contours
(Fig. 10.7) and is illustrated in seismic data (Fig. 10.6). The depth of the axial valley

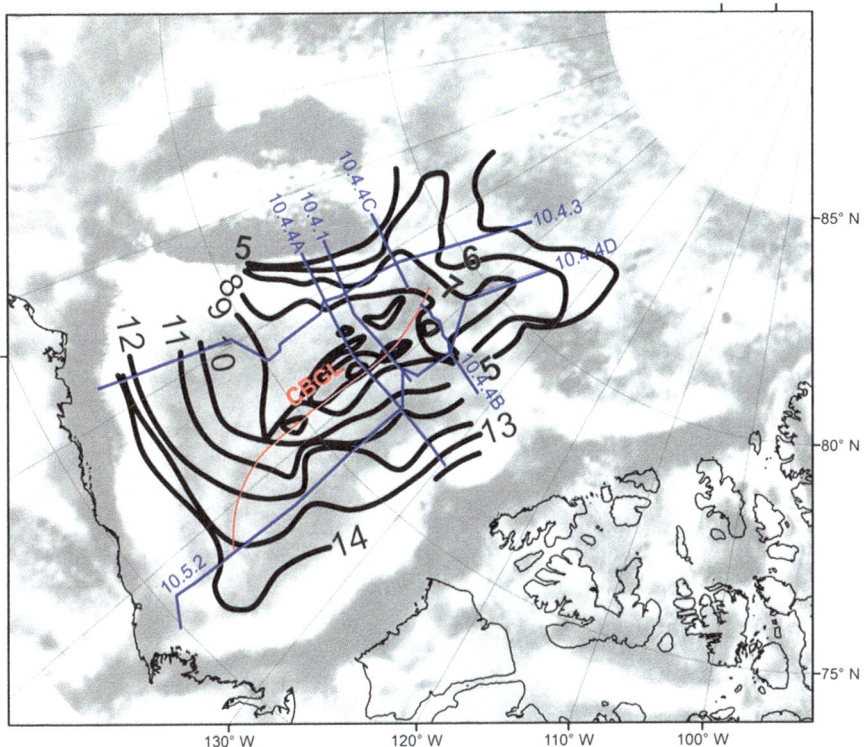

Fig. 10.7 Depth to basement (km) from modern sea level for Canada Basin. Contours represent depth from modern sea level to acoustic basement in 1 km intervals, overlain on the Bouguer gravity anomaly map shown in Fig. 10.3. The central gravity low (CBGL) is shown in red. To generate this map, seismic travel time to depth conversion was based on sediment velocities determined by Chian and Lebedeva-Ivanova (2015) and Shimeld et al. (2016). Locations of seismic profiles shown in figures in this chapter are shown in blue

is in excess of 500 m shown on the contours. These depths compare with those of the modern Gakkel Ridge (Jokat and Micksch 2007), and are suggestive of an ultra-slow spreading ridge. Additionally, the structure correlates precisely with the region mapped by Chian et al. (2016) as layer 3 oceanic crust. This region also encom-passes the magnetic lineations that are interpreted to represent the landward limit of oceanic crust, as shown above in Fig. 10.5.

Examination of acoustic basement reflections shows four broad types of seismic/ echo characteristics: smooth, blocky/rugose, faulted (imbricated, and horst and gra-ben), and hummocky (Fig. 10.9). Seismic coverage is sparse at the scale of the basin, however, these four types were mapped regionally. Smooth basement lies at the periphery of the margin and generally corresponds to the region identified by velocity analysis as continental or transitional crust along the flanks of the basin (Chian et al. 2016), particularly in the southern half of Canada Basin (Fig. 10.9a). Blocky/rugose crust is prevalent in the central Canada Basin, in the region identified

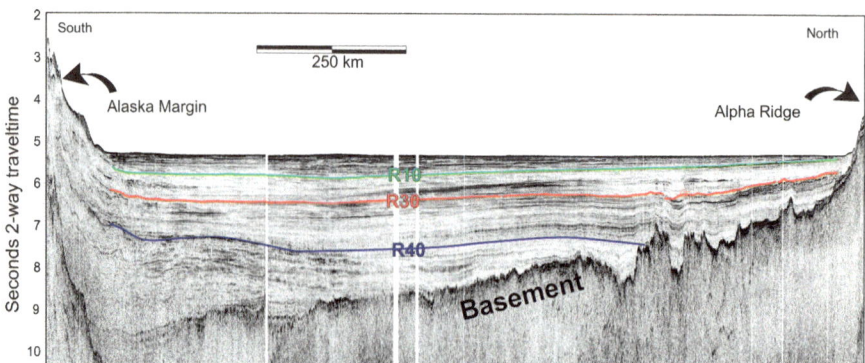

Fig. 10.8 Seismic reflection profile transecting Canada Basin from the Alaska margin in the south to Nautilus Spur in the north. Aside from basement and seafloor, three other horizons were picked that transect throughout much of Canada Basin (see Mosher et al. 2012a). The profile location is shown in Figs. 10.7 and 10.10

as oceanic crust by Chian et al. (2016). A deep graben is apparent in several profiles close to the Canadian Arctic Archipelago margin (Fig. 10.9b) showing imbricated and faulted basement. Another area of shingled (faulted?) structures occurs northeast of Chukchi Borderland in the north-central region of Canada Basin (Fig. 10.9c), as well as further east along Sever Spur. These imbricate and graben fault structures are believed to have resulted from crustal extension, but it is possible that they may include some volcanic sills. Hummocky topography of the basement reflection horizon is prevalent in the region where the HAMH underlies the basin. In several cases, in fact, buried and even emergent seamounts are obvious (Fig. 10.9d).

10.5 Seismic Stratigraphy

The seismic stratigraphy for Canada Basin shown in this section was developed by Mosher et al. (2012a). The total sediment thickness distribution interpreted from seismic reflection profiles is shown in map form in Fig. 10.10. The sediment sequence thins dramatically over basement high areas but the overall pattern within the basin proper shows thinning from south to north and east to west. Not surprisingly, the thickest sediment section is centered about southeastern Canada Basin, in the area where the Mackenzie River enters Canada Basin to form the thick Mackenzie fan deposits. These patterns of sediment thickness are shown in basin-crossing seismic profiles of Figs. 10.6, 10.8, 10.9a and 10.11. The sediment sequence was divided into four units by mapping three regionally identifiable seismic reflection horizons, not including basement and the seafloor. These horizons are referred to as R40, R30 and R10.

The sedimentary unit between basement and the R40 horizon thins from south to north; it is about 5 km thick where it is first imaged in the south and pinches out to

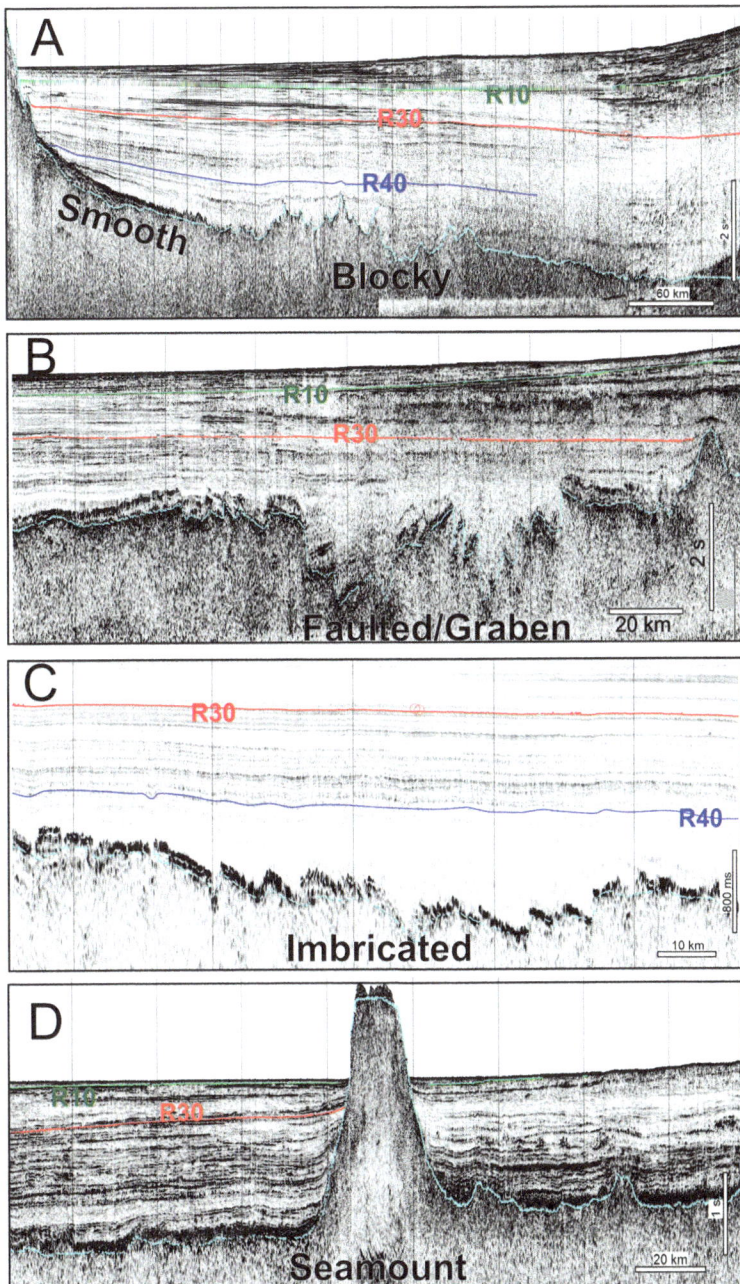

Fig. 10.9 The basement reflection is shown by the teal horizon. (**a**) Blocky basement in the central Canada Basin and smooth basement towards its flanks. (**b**) A graben structure in the northeast portion of the basin. (**c**) Shingled basement in the northwest. (**d**) Savaqatigiik Seamount, discovered in 2009. See Figs. 10.7 and 10.10 for the location of these profiles

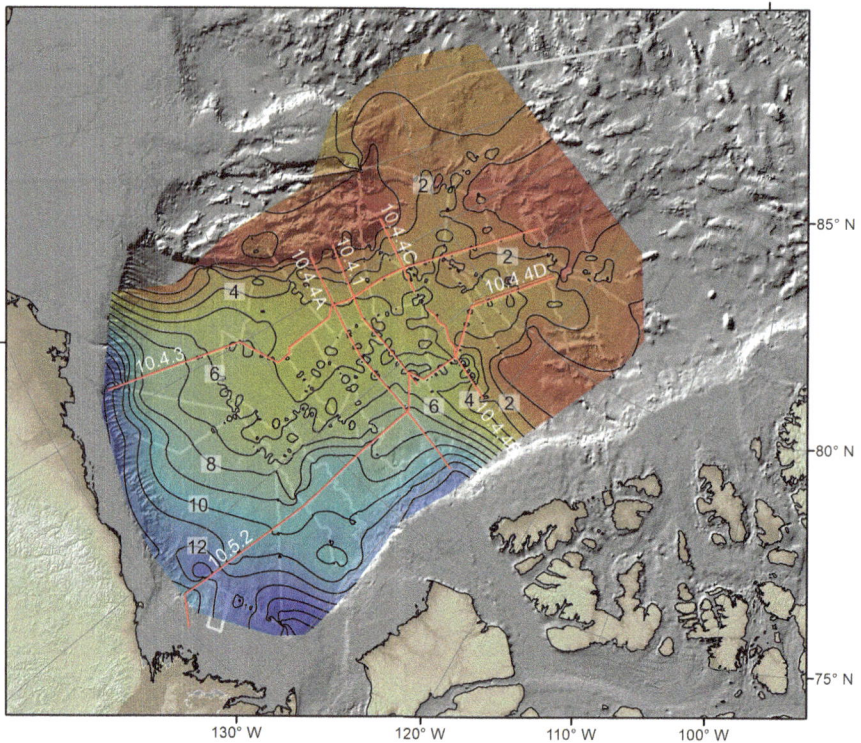

Fig. 10.10 Total sediment thickness, derived from seismic reflection profiles. Contours are in kilometres. Seismic tracks which represent the data control for the grid are shown in faded white lines. Red tracklines are locations of seismic profiles used as figures

the west against Northwind Escarpment and to the north at about 78°N where it onlaps an apparent vertical step in basement morphology (Figs. 10.8 and 10.11). This unit appears to represent the earliest phase of basin infill, prograding offshore from surrounding margins (Mosher et al. 2012a). If these sediments are synrift to early post-rift, then they presumably reflect sediments deposited during and possibly for some time after the opening of Canada Basin; that is, Upper Cretaceous to Paleocene (Døssing et al. 2013). Dixon et al. (1992) argued that offshore strata are principally post-Maastrichtian in age. Helwig et al. (2011) showed that synrift sediments beneath the Mackenzie fan are mid-Lower Cretaceous in age but admitted that these correlations are highly tentative. This deepest offshore sedimentary sequence is, therefore, potentially equivalent to the Fish River and Aklak sequences of the Beaufort Shelf, spanning late-Upper Cretaceous to end-Paleocene (Dixon et al. 1992).

Detailed interpretation of the seismic stratigraphy is provided in Mosher et al. (2012a) and Coakley et al. (2016), so a brief summary is provided here. Following rift opening of the Canada Basin, accumulation of sediment of sufficient volume to

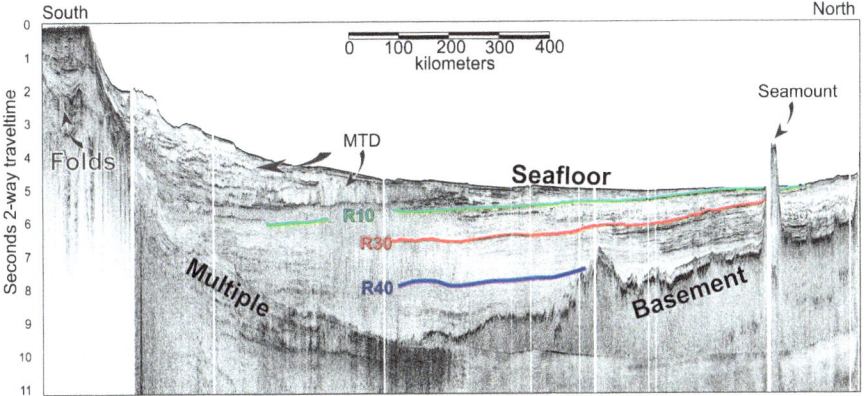

Fig. 10.11 A south to north profile transecting Canada Basin through the Mackenzie River prodelta (foreslope). Note the mass transport deposits (MTD) comprising the foreslope and folds underlying the shelf, part of the Beaufort foldbelt. Also note basement character and the Savaqatigiik Seamount to the north. The R40 horizon onlaps the basement high, shown here, which occurs at about 78°N. See the location of the profile in Figs. 10.7 and 10.10

be seismically resolvable was probably diachronous along the rifted margins and largely depended on the transport of sediment to and beyond the rift shoulders (Houseknecht and Bird 2011). In general, because the deepest sediment package in the basin is recognized to prograde northwards, then presumably basal sediments become younger to the north. Progressive onlap of reflections may reflect the early basin differential subsidence history. This pronounced subsidence-related phenomenon is better observed here than in other more complex ocean basins.

The R30 horizon is correlated throughout the abyssal plain from beneath the distal Mackenzie River fan in the south to onlap Alpha Ridge in the north (Figs. 10.8 and 10.11). In general, it dips slightly to the east and south. The sedimentary sequence between R30 and R40 largely consists of parallel coherent reflections that correlate for hundreds of kilometers. This R30 to R40 package is greater than 6 km thick to the east off M'Clure Strait and thins to the west and north (see Mosher et al. 2012a; Fig. 10.9a).

Horizon R10 is the shallowest horizon correlated for this study. It also correlates throughout the Canada Basin Abyssal Plain and traces the furthest north of the three horizons, where it onlaps Nautilus Spur, a southern extension of the Alpha Ridge and other basement highs (Mosher et al. 2012a; Fig. 10.8). It daylights at the seafloor before reaching the Stefansson and western Nautilus basins, however. It underlies the wedge-shaped sedimentary unit of the Mackenzie River prodelta in the south and shoals to the north (Fig. 10.10). The sedimentary package between R10 and R40, including R30, consists of parallel, coherent, laterally continuous reflections. The sequence thins significantly from east to west. Wedge-shaped units of coherent reflections interpreted as mass-transport deposits emanate from the Canadian Archipelago margin, accounting for the >3 km thickness of the R10 to R40 unit in this region (Mosher et al. 2012a, b).

Reflections of each unit mapped in Canada Basin terminate abruptly against the Northwind Escarpment in the west (Fig. 10.6); some upturning of reflections suggests onlap, but it is relatively minor. The R30 to R10 unit is about 400 m thick immediately basinward of Northwind Escarpment. This sequence may be equivalent to the Taglu and Richards Sequences of the Canadian Beaufort shelf and thus is Eocene in age. It may also extend up into the Oligocene and include the Kugmallit Sequence (Mosher et al. 2012a; Coakley et al. 2016).

The interval between R10 and the seafloor is more than 4 km thick in the Mackenzie River prodelta and thins basinward in a radial pattern well into northern Canada Basin. The sequence in this area consists of stacked mass-failure deposits as shown in cross section (Fig. 10.11). Farther offshore and underlying the Canada Basin abyssal plain, reflections become highly coherent, continuous and parallel for tens to hundreds of kilometers with varying amplitudes along strike. These reflections are interpreted to represent interbedded unconfined turbidite and hemipelagic deposits (Mosher et al. 2012a). The seafloor horizon shows no deflection at the interface with these bedrock highs, but rather abrupt termination. Upward deflection of reflections at the onlap surface increases with depth, however, suggesting minor syndepositional basin subsidence.

The R10 to seafloor interval correlates into the eastern part of Nautilus Basin, north of the Northwind Escarpment. It cannot be mapped into Stefansson Basin where the higher seafloor elevation suggests deposition was from a local source, likely the northern Canadian Arctic Archipelago.

Assigning an age to the R10 horizon is problematic. It correlates to Onlap Unconformity 3 (OU-3) of Grantz et al. (2011), which they placed as upper Eocene. In the study by Helwig et al. (2011), it correlates more closely to base Pliocene (Iperk) or possible base Miocene or base Oligocene reflections on their line 5600 offshore of the Mackenzie prodelta. Their Oligocene–Miocene section is thin, so precise correlation is difficult. The Iperk sequence was deposited as a large delta complex centered over the eastern Beaufort Sea (Dixon et al. 1992), as we see the R10 to seafloor sequence (Fig. 10.11). Helwig et al. (2011) showed that the Iperk sequence onlaps the Alaska Beaufort and Banks Island margin, which conforms with our observations of the R10 to seafloor unit. For these reasons, we prefer to correlate R10 to the Iperk sequence, although the offshore section may contain thin Mackenzie Bay and Akpak sequences dating back to the Oligocene; the majority of the section, however, is probably Pliocene of the Iperk sequence.

10.6 Morphology and Surficial Geology

The general geography and morphology of Canada Basin was described in the introductory section. In this section, we will describe in more detail the constituent parts of the basin, starting with shelf areas and proceeding with the various surrounding components including Canada Basin abyssal plain, Nautilus Basin, Nautilus Spur,

Stefansson Basin, the Canadian Arctic Archipelago margin, the Canadian Beaufort margin, the Alaska margin and Northwind Escarpment.

10.6.1   Shelf Areas

Canada Basin is flanked on only two sides by shallow shelf seas; along the Canadian Arctic Archipelago margin and the Alaska-Beaufort margin (Fig. 10.12). Along the Canadian Arctic Archipelago, the shelf is generally between 120 and 150 km wide. It terminates in a rounded shelf edge in about 500 m water depth. This shelf-slope transition is deepest seaward of shelf-crossing glacial troughs within MacLean and M'Clure straits and Amundsen Gulf (Fig. 10.12). In the southeast, Mackenzie-Beaufort region of the basin, the Mackenzie River delta has greatly influenced the shelf morphology. The shelf is over 150 km wide in this area and it terminates in about 100 m water depth with a sharp gradient change to the slope that comprises the Mackenzie fan. This Mackenzie-Beaufort region is terminated on its western

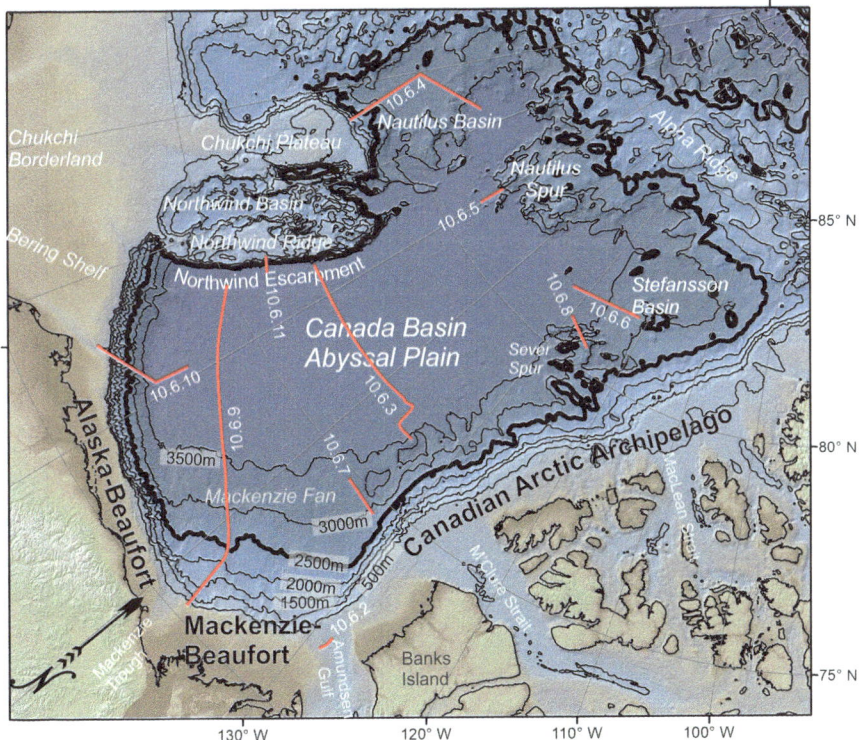

Fig. 10.12 Morphology of Canada Basin derived from the IBCAO v. 3 (Jakobsson et al. 2012) with 500 m isobaths. Red lines are locations of subbottom profiles shown in subsequent figures

edge by the Mackenzie Trough – a bathymetric trough that cuts back to the coast and terminates seaward with a rounded shelf edge (Fig. 10.12).

Along the Alaska-Beaufort margin, the shelf is less than 100 km wide and the shelf break occurs in less than 100 m water depth with a sharp gradient change to the slope. There is little bathymetric variability on the shelf in this area. The shelf of the Chukchi Borderland is more extensive; especially where it opens to the Chukchi Shelf. The Borderland extends northward and encompasses Chukchi Plateau and Northwind Escarpment and Northwind Basin. Although technically not considered shelf area, Chukchi Plateau forms a gradual ramp from the Chukchi Shelf that dips northwards and remains above 1000 m depth throughout its length. Its northernmost edge is rounded before dropping in irregular steps to the deep sea.

These shelf areas surrounding Canada Basin, including the Amundsen Gulf, Mackenzie delta, Alaska-Beaufort shelf, Bering Shelf and Chukchi Borderland, are treated in detail in various publications such as Polyak et al. (2001), Hill et al. (2007), Hill and Driscoll (2008, 2010), Blasco et al. (2013), Batchelor and Dowdeswell (2014), Dove et al. (2014), Jakobsson et al. (2015), King (2015) and King et al. (2017). The Canadian Arctic Archipelago shelf, however, is poorly studied due to perennial ice conditions. Ice camps and helicopter spot sampling programs are the only source of information in this region (e.g. Mosher et al. 1990; Hein et al. 1990).

In general, shelf areas throughout Canada Basin reflect the fact they have endured Quaternary shelf-crossing glaciations and subaerial exposure during corresponding sea level low-stands. These episodes have had a significant impact on the surficial geology. Erosion and ice scour of the seabed is apparent down to at least 500 m water depth (e.g., Fig. 10.13), although Jakobsson et al. (2014, 2015) suggest the ice shelf may have been more than 1000 m thick. Glacial diamict and well-laminated glacial marine sediments lie below this scoured interval. Shelf areas also had, and may still have, permafrost layers contained within the shallow subsurface sediments,

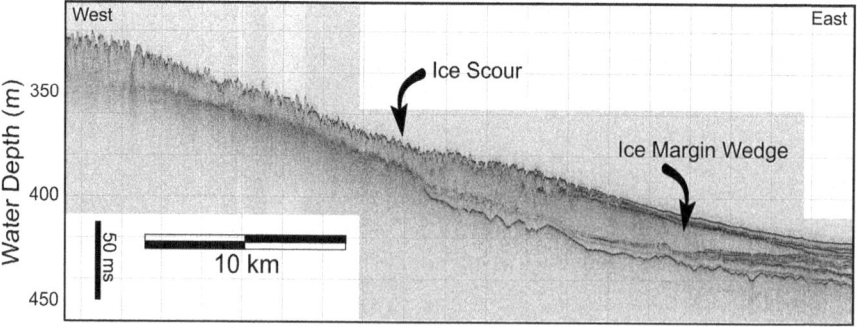

Fig. 10.13 Subbottom profile in the Amundsen Gulf. Note the rugose topography of the seafloor created by ice scour above 400 m depth, the ice-margin wedge (till tongue) that occurs downslope and then laminated glacial marine sediments further downslope. See Fig. 10.12 for the profile location

although along the Alaskan Beaufort shelf, permafrost is mapped only within the inner shelf to water depths less than 25 m (Brothers et al. 2016; Ruppel et al. 2016).

10.6.2 Canada Basin Abyssal Plain

Canada Basin abyssal plain is 494,000 km^2 in area, lying between 3500 and 3900 m water depth (Fig. 10.12). It is exceptionally planar and featureless in large scale and generally dips to the west and to the south. At 77° N, its eastern edge lies in 3700 m of water and in the west where it abuts Northwind Ridge, it is 3850 m deep. Grantz et al. (1996) and Mosher et al. (2011) interpreted that near surface sediments are turbidites based on sediment description of shallow sediment cores.

This thick sedimentary sequence in the basin has infilled the underlying topography and created the flat, relatively featureless modern seafloor. Subbottom profiler data show the basin may be divided into two parts (Fig. 10.14). The western part is flatter and featureless and interpreted as comprising turbidites, while the eastern portion appears to consist of low gradient sediment drifts suggestive of geostrophic currents rather than turbidites (Fig. 10.14). Although sediment delivery mechanisms may be hemipelagic and turbidity current in origin, it appears as though some degree of reworking has occurred.

10.6.3 Nautilus Basin

Nautilus Basin is about 124,000 km^2 (Jakobsson et al. 2003), surrounded to the south by the Chukchi Plateau, to the west by the Mendeleev Ridge and to the north and northeast by the Alpha Ridge and a promontory of the ridge known as Nautilus Spur. There are two areas to the basin (eastern and western components) largely divided by the 3750 m isobath (Fig. 10.12). The eastern portion is contiguous with

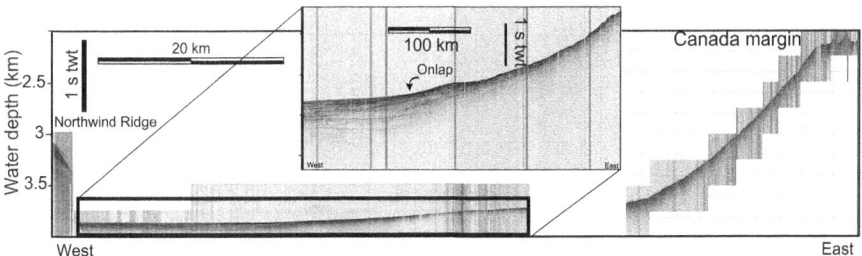

Fig. 10.14 Subbottom profile across Canada Basin (see Fig. 10.12 for location). Although exceptionally flat, vertical exaggeration shows deep sea sediments of Canada Basin are subjected to reworking and resedimentation creating bedforms and drift deposits

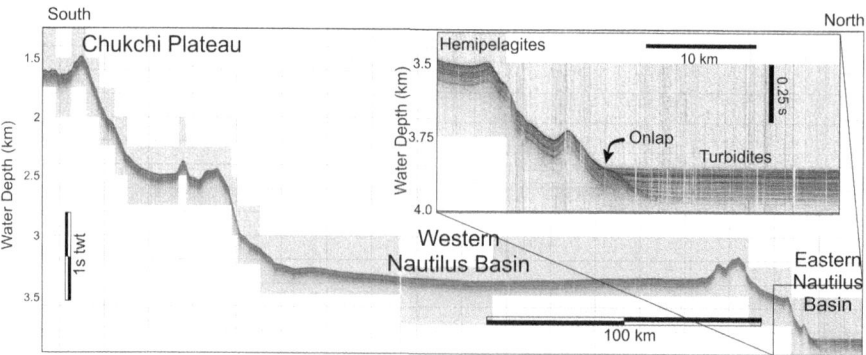

Fig. 10.15 Subbottom profile transecting Nautilus Basin. Note the elevation change between the eastern and western portions of the basin. Also note the onlap of turbidites sediments from the eastern Nautilus Basin onto the higher portion of the western part of the basin, as shown in the inset enlargement. See Fig. 10.12 for the profile location

Canada Basin and the sediment stratigraphy is readily correlated. The western portion, however, is elevated and appears to host hemipelagic sediment rather than turbidites that characterize western Canada Basin (Fig. 10.15). Flat-lying turbidites of the eastern portion onlap the elevated western side creating a sharp contrast in geology and gradient (Fig. 10.15).

10.6.4   Nautilus Spur

Nautilus Spur lies east of Nautilus Basin and is a southern promontory of Alpha Ridge that protrudes into Canada Basin. Much of it lies deeper than the 2500 m isobath, so it is included within Canada Basin in this context. The promontory is about 64,000 km² in area. The sedimentary sequence on top of the spur, equivalent to that on top of Alpha Ridge, dips beneath sedimentary sequences of Canada Basin and forms the seismic unit just above acoustic basement. The youngest sediment reflections of the abyssal plain of Canada Basin onlap the spur at a relatively precise contact at the 3750 m isobath, similar to Nautilus Basin (Figs. 10.8 and 10.16). This sharp contact juxtaposes different geologic units and forms a distinctive gradient change between the spur and Canada Basin abyssal plain.

10.6.5   Stefansson Basin

East of Nautilus Spur is Stefansson Basin; a 120,000 km² embayment in the northeast of Canada Basin. It is bordered to the north by Alpha Ridge and to the east by the shelf seaward of the northern Canadian Arctic Archipelago (Fig. 10.1). The

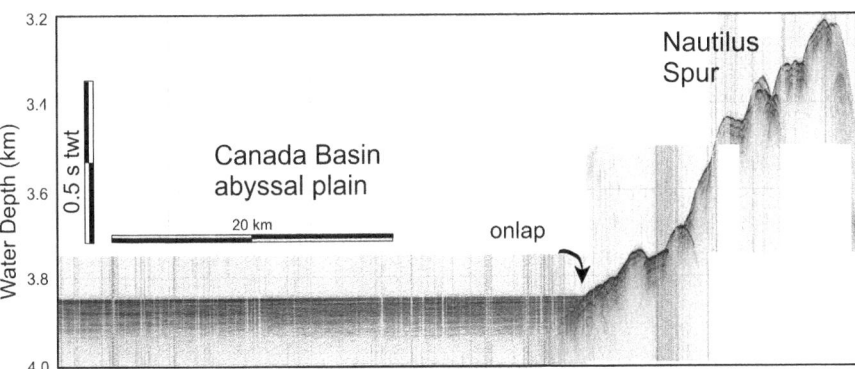

Fig. 10.16 Subbottom profile showing onlap of turbidites of Canada Basin abyssal plain onto Nautilus Spur. See Fig. 10.12 for the profile location

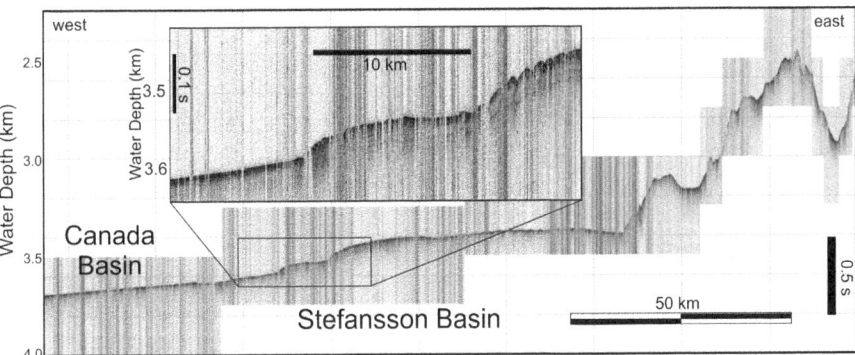

Fig. 10.17 Subbottom profile crossing from Canada Basin into Stefansson Basin. The data are noisy because of heavy ice-breaking required to operate in this area at any time of year. There is a marked higher elevation and hummocky topography of Stefansson Basin caused by debris flows. The inset enlargement shows bedform-like features on the seafloor that are asymmetric upslope. See Fig. 10.12 for the profile location

basin is elevated above the adjacent Canada Basin Abyssal Plain and the sediment stratigraphy does not correlate. Subbottom profiler data show surficial debris flows form the seafloor of Stefansson Basin, likely sourced from the Canada Arctic Archipelago (Fig. 10.17). Subbottom profiles show bedform-type features on the seafloor that are interpreted as debris flow ridges that are known to occur at the termination of debris flow lobes (e.g., Frey-Martinez et al. 2006).

10.6.6 Canadian Arctic Archipelago Margin

Stefansson Basin is bordered to its east by the Canadian Arctic Archipelago margin. This segment of Canada Basin continues to the south to Amundsen Gulf and includes regions seaward of MacLean Strait and M'Clure Strait. Morphologically, it appears to be a classic passive rift continental margin. The shelf is 130 to 150 km wide with a slope break in about 500 m water depth. There are a number of deep glacial troughs along the Canadian Archipelago margin that emanate from the aforementioned straits and extend out to the shelf edge. Gradients along the continental slope range from 2 to 4° on average (Fig. 10.13). The margin is poorly surveyed due to perennial sea ice cover. Subbottom profiler data, however, show surficial glacigenic debris flow deposits dominate the slope environment (King 2015; Mosher 2016; Baldwin et al. 2017; Fig. 10.18). This evidence suggests the slope is similar to other northern margins that hosted shelf-crossing glaciers (e.g., Batchelor and Dowdeswell 2014). These glaciers produced trough-mouth fans outboard of shelf troughs that consist of stacked glacigenic debris flows (Vorren et al. 1989; Vorren and Laberg 1997).

Sever Spur (Fig. 10.12) is a morphological promontory that comprises a series of marginal bedrock ridges. The exact morphology of the ridges is poorly known because there is not much measured data in the area. On subbottom profiler, these ridges are steep-sided with little sediment cover (Fig. 10.19). This structure seems to be a series of horst and graben (or half-graben) structures that may be indicative of extension related to rift formation of the margin. These features may, therefore, be similar in origin to the graben structure further seaward (Fig. 10.9b).

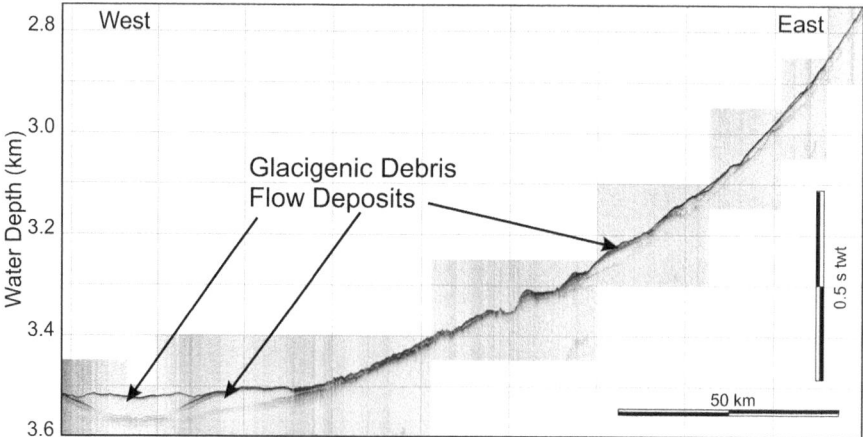

Fig. 10.18 Subbottom profile across the lower slope of the Canadian Archipelago margin showing glaciogenic debris flow deposits form the surficial sediment type along much of the margin. See Fig. 10.12 for the profile location

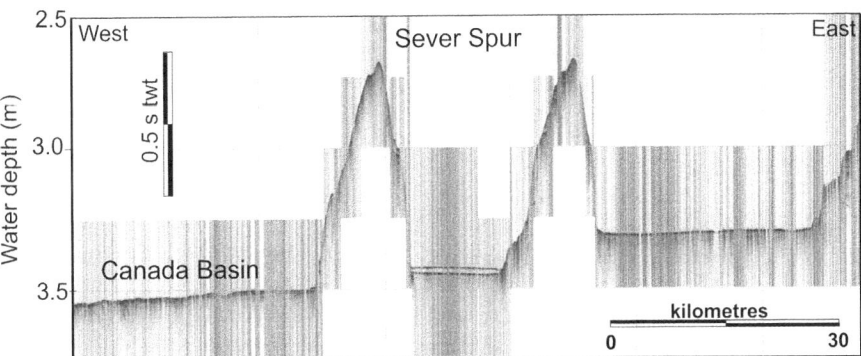

Fig. 10.19 Subbottom profile transecting several ridges of Sever Spur along the Canadian Arctic Archipelago margin. Note the successive change in elevation from Canada Basin to the landward basins between ridges. See Fig. 10.12 for the profile location

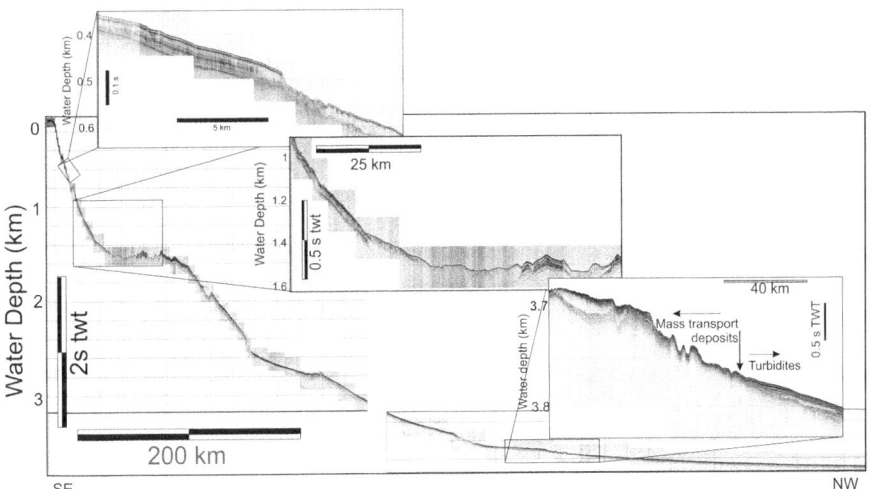

Fig. 10.20 Subbottom profile transecting the Mackenzie fan. It shows headscarps with mass transport deposits along the upper slope, debris flows on the mid and lower slope and debris flows transitioning to turbidites at the lower slope. See Fig. 10.12 for the profile location

10.6.7 Mackenzie-Beaufort Margin

The Mackenzie fan dominates the seafloor in the southeastern Canada Basin (Fig. 10.12). Since Miocene time, the Mackenzie River and its predecessors constructed a 200,000 km^2 submarine fan, with a foreslope that traverses the shelf break in 100 m water depth to the abyssal plain (Fig. 10.20). At large scale, the modern slope morphology is relatively smooth and dips gently (<1°) northward. Multibeam bathymetric and high-resolution subbottom data, however, show multiple large headscarps, slumps, slides and debris flows on the seafloor (Mosher et al. 2012b; Cameron et al. 2017) (Figs. 10.11 and 10.20). These features pinch out downslope and evolve into unconfined turbidites that distribute throughout the Canada Basin Abyssal Plain (Mosher et al. 2012a, b). This transition from mass-transport deposits (MTD) to turbidites reflects slope versus abyssal processes.

10.6.8 Alaska-Beaufort Margin

The Alaska-Beaufort margin is distal to the main influences of the Mackenzie River. In this area, the modern slope morphology shows a relatively consistent 3 to 4° regional gradient, lessening to 2° along the lower slope. Downslope-trending channels and valleys with relatively straight axes dominate the morphology of the slope, hence the seafloor is considered erosional. These channels incise the slope up to 500 m deep. Lack of sediment at the base of slope suggests the margin is sediment starved. Lobes of sediment deposits with symmetric ridges on their surfaces are apparent at the bases of some of these channels. These forms may be indicative of reworking by geostrophic currents creating bedforms or may be debris flow compressional ridges resulting from sediment mass failure. In the eastern region of the margin, the lower slope (~3500 m) is overlain by onlapping debris flow deposits of the Mackenzie fan. Further west, the margin is onlapped by flat-lying reflections that represent turbidites sourced from the fan (Fig. 10.21).

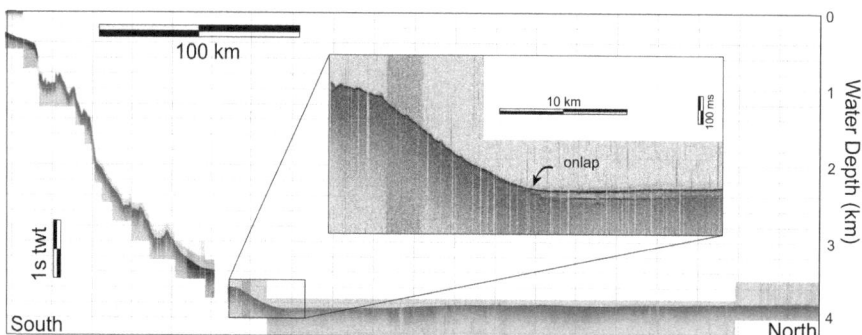

Fig. 10.21 Subbottom reflection profile from the Alaska margin, showing how distal turbidites of the Mackenzie fan onlap the Alaska slope. See Fig. 10.12 for the profile location

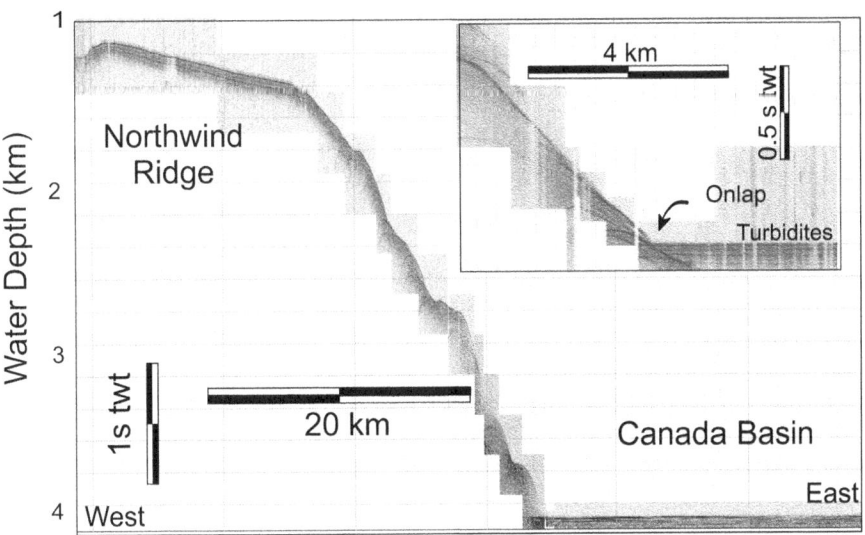

Fig. 10.22 Subbottom profile showing turbidites of Canada Basin onlapping Northwind Escarpment. See Fig. 10.12 for the profile location

10.6.9 Northwind Ridge

Northwind ridge is 120,000 km^2 in area and is separated from the Chukchi Plateau by the intervening Northwind Basin. The seafloor rises from 3860 m in the Canada Basin abyssal plain at the base of Northwind Ridge to less than 1000 m at its top forming the Northwind Escarpment. Slope angles are in excess of 10°. Little sediment cover on this steep slope indicates it is erosional or non-depositional. Turbidites of the abyssal plain truncate abruptly against the ridge, forming a sharp change in seafloor gradient and a distinctive change in surficial geology (Figs. 10.6 and 10.22). At its northern edge, Northwind Ridge forms a ramp down to the abyssal plain that terminates at about 79°30'N. North of this is a similar extension of the Chukchi Plateau as it drops into Nautilus Basin.

10.7 Tectonics

Tectonic reconstruction of Canada Basin is not trivial; in part due to lack of data and also because there is no clear geomorphological fit to the shapes of its margins. There is also no readily apparent evidence (such as transform faults) to constrain seafloor spreading direction or deformation that might indicate subduction within the basin to account for its formation or destruction. Furthermore, the presence of Alpha Ridge and Mendeleev Rise and the associated High Arctic Large Igneous Province (Døssing et al. 2013; Saltus et al. 2011; Oakey and Saltus 2016) obscures any pre-existing structures of northern Amerasia Basin that may have been related to its tectonic origins. Without data to constrain interpretations, many models have been proposed to explain basin formation (e.g., Carey 1955; Tailleur 1973; Grantz et al. 1979, 1998, 2011; Lawver and Scotese 1990; Embry 1990; Laxon and McAdoo 1994; Lane 1997; Lawver et al. 2002). More recent models, based on modern data, include those of Pease et al. (2014); Doré et al. (2016), Miller et al. (2017), and Hutchinson et al. (2017). It is not the purpose of this chapter to review these plethora of models, but Hutchinson et al. (2017) propose a complex tectonic scenario that accounts for most of the observations from the diversity of data sets available and largely presented in this chapter (Fig. 10.23).

Hutchinson et al. (2017) recognized three sub-parallel, north-east trending structures in Canada Basin, based on geomorphological and geophysical data sets (Fig. 10.23d). They interpreted these structures as evidence of strike-slip, or transtensional tectonic deformation in the early formation of Canada Basin. These features include Northwind Escarpment, features along the orientation of Sever Spur and the Alaska-Prince Patrick magnetic lineament (APPL) that appears to limit the southeastern extent of oceanic crust mapped by Chian et al. (2016) (Fig. 10.5). Strike-Slip tectonics are also proposed by Evangelatos and Mosher (2016) for the Amerasia flank of Lomonosov Ridge.

In addition to these transform features, the CBGL has long been interpreted as a fossil spreading axis (Grantz et al. 1979; Taylor et al. 1981; Coles and Taylor 1990; Laxon and McAdoo 1994). Only with modern seismic reflection data, however, has this gravity anomaly been shown to have a basement expression consistent with that of a slow- or ultra-slow spreading center (Fig. 10.6, 10.7 and 10.9a), and refraction data have shown oceanic and transitional crust nearly symmetrical about the anomaly (Fig. 10.5). These data, in addition to geologic similarities along the Alaskan and Canadian margins, have led to a consensus model in which Alaska and Arctic Canada are conjugate margins in a rotational rifting scenario that separated the margins in Lower Cretaceous time after a prolonged period of extension (Carey 1955, Tailleur 1973; Grantz et al. 1979, 1990, 1998, 2011; Chian et al. 2016).

Hutchinson et al. (2017), therefore, proposed a two-phase opening model of Canada Basin. In the first phase of opening, depicted in Fig. 10.23b, rifting proceeded with strike-slip or trans-tensional motion along the three northeast-trending features that led to the development of pull-apart basins, such as that shown in Fig. 10.9b or Northwind Basin that separates Northwind Ridge from Chukchi Plateau. In the second phase, depicted in Fig. 10.23c, rotational opening completed the formation of oceanic crust that resulted in the configuration of crustal blocks as observed today (Fig. 10.23d). A third phase that affected northern Amerasia Basin and less so of Canada Basin proper may have involved extension, as proposed recently by Døssing et al. (2017), although earlier hypothesized by Lawver and Scotese (1990), Miller et al. (2006; 2017), Evangelatos and Mosher (2016) and Evangelatos et al. (2017). To account for opening of Makarov Basin and extensional structures observed on Alpha Ridge, Evangelatos and Mosher (2016) proposed that structures that began as strike-slip along the Amerasia flank of Lomonosov Ridge

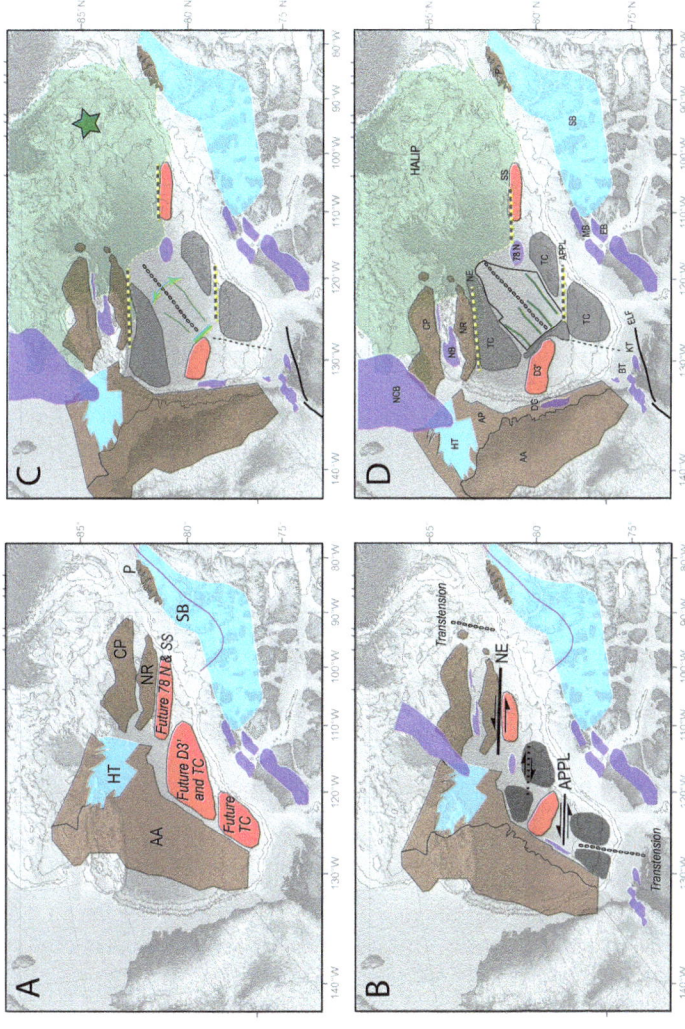

Fig. 10.23 Conceptual model for the evolution of the Canada Basin, from Hutchinson et al. (2017). (**a**) Closure. (**b**) Phase 1 rifting showing strike-slip exten- sion and transtension. (**c**) Phase 2 seafloor spreading showing rotation. (**d**) Present time. Abbreviations are AA, Arctic Alaska; AP, Arctic Platform; CP, Chukchi Plateau; D3′, crustal block; DG, Dinkum Graben;, Eskimo Lakes fault zone; HT, Hanna Trough; KT, Kugmallit Trough; NB, Northwind Basin; NCB, North Chukchi Basin; NE, Northwind Escarpment; NR, Northwind Ridge; P, Pearya; SB, Sverdrup Basin; SS, Sever Spur; TC, transition crust. The location of a eruption center associated with the HALIP (Døssing et al. 2013) is shown by the green star in panel C

became trans-tensional and then extensional in these northern areas. This phase in the northern Amerasia Basin may have been concurrent with the rotational opening of Canada Basin.

Acknowledgements The authors would like to thank the many individuals and organizations who have been involved in acquisition, processing and interpretation of the new data upon which this chapter is based. There are far too many people to mention individually, but we especially wish to express appreciation to Dr. Gordon Oakey for the potential field data compilations, Mr. John Shimeld for his exceptional efforts in seismic data acquisition and processing, and Mr. Kai Boggild for helping compile the subbottom reflection data.

References

Andersen OB, Knudsen P, Berry PAM (2010) The DNSC08GRA global marine gravity field from double retracked satellite altimetry. J Geodyn 84:191–199

Baldwin K, Mosher DC, Gebhardt C et al (2017) Surficial geology mapping of the Arctic Ocean. European geophysical union annual congress, Vienna, Austria, 24–28 April 2017, (poster)

Batchelor CL, Dowdeswell JA (2014) The physiography of high Arctic cross-shelf troughs. Quat Sci Rev J 92:68–96

Blasco S, Bennett R, Brent T et al (2013) 2010 State of knowledge: Beaufort Sea seabed geohazards associated with offshore hydrocarbon development. Geological Surv of Canada, Open File 6989, https://doi.org/10.4095/292616

Bown JW, White RS (1994) Variation with spreading rate of oceanic crustal thickness and geochemistry. Earth Planet Sci Lett 12:435–449

Brothers LL, Herman BM, Hart PE et al (2016) Subsea ice-bearing permafrost on the U.S. Beaufort margin: 1. Minimum seaward extent defined from multichannel seismic reflection data. Geochem Geophys Geosyst J 17:4354–4365. https://doi.org/10.1002/2016GC006584

Brumley K, Miller EL, Konstantinou A et al (2015) First bedrock samples dredged from submarineoutcrops in the Chukchi borderland, Arctic Ocean. Geosphere J 11:76–92

Cameron GDM, King EL, Murray D et al (2017) Relative timing of sediment failures within slide-valley complexes in the Kugmallit fan area of the Central Beaufort slope, offshore northwest territories. Geological survey of Canada, Scientific presentation 741 sheet, https://doi.org/10.4095/306013

Carey SW (1955) The orocline concept in geotectonics – part I, publication 28. The Papers and Proceedings of the Royal Society of Tasmania, 89, p 255–288

Chian D, Lebedeva-Ivanova N (2015) Atlas of Sonobuoy velocity analyses in Canada Basin. Geological Survey of Canada, Open File 7661, https://doi.org/10.4095/295857

Chian D, Keen CE, Reid I et al (1995) Evolution of nonvolcanic rifted margins: new results from the conjugate margins of the Labrador Sea. Geol J 23:589–592

Chian D, Jackson HR, Hutchinson DR et al (2016) Distribution of crustal types in the Canada Basin, Arctic Ocean. Tectonophysics J 691:8–30

Coakley B, Brumley K, Lebedeva-Ivanova N et al (2016) Exploring the geology of the Central Arctic Ocean – understanding the basin features in place and time. J Geol Soc 173:967–987

Coles RL, Taylor PT (1990) Magnetic anomalies. In: Grantz A, Johnson L, Sweeney JF (eds) The Arctic region. Geological Society of America, vol L. The Geology of North America, Boulder, Colorado, pp 119–132

Dixon J, Dietrich JR, McNeil DH (1992) Upper cretaceous to Pleistocene sequence stratigraphy of the Beaufort-Mackenzie and Banks Island areas, Northwest Canada. Geol Surv Can Bull 407

Doré AG, Lundin ER, Gibbons A et al (2016) Transform margins of the Arctic: a synthesis and re-evaluation. Geol Soc Lond Spec Publ 431:63–94

Døssing A, Jackson HR, Matzka J et al (2013) On the origin of the Amerasia Basin and the high Arctic large Igneous Province (results of new aeromagnetic data). Earth Planet Sci Lett J 363:219–230

Døssing A, Hansen TM, Olesen AV et al (2014) Gravity inversion predicts the nature of the Amundsen basin and its continental borderlands near Greenland. Earth Planet Sci Lett J 408:132–145

Døssing A, Gaina C, Brozena JM (2017) Building and breaking a large igneous province: an example from the high Arctic. Geophys Res Lett 44

Dove D, Polyak L, Coakley B (2014) Widespread, multi-source glacial erosion on the Chukchi margin, Arctic Ocean. Quat Sci Rev 92:112–122

Embry AF (1990) Geological and geophysical evidence in support of the hypothesis of anticlockwise rotation of northern Alaska. Mar Geol J 93:317–329

Evangelatos JE, Mosher DC (2016) Seismic stratigraphy, structure and morphology of Makarov Basin and surrounding regions: tectonic implications. Mar Geol J 374:1–13

Evangelatos JE, Funck T, Mosher DC (2017) The sedimentary and crustal velocity structure of Makarov Basin and adjacent alpha ridge. Tectonophysics J 696–697:99–114

Frey-Martinez J, Cartwright J, James D (2006) Frontally confined versus frontally emergent submarine landslides: a 3D seismic characterization. Mar Pet Geol J 23:585–604

Funck T, Jackson H et al (2011) The crustal structure of the alpha ridge at the transition to the Canadian polar margin: results from a seismic refraction experiment. J Geophys Res Solid Earth 116(B12):26

Gaina C, Werner S, Saltus R et al (2011) The CAMP-GM group. Circum-Arctic mapping project: new magnetic and gravity anomaly maps of the Arctic. In: Spencer AM (ed) Arctic petroleum geology, Geological Society of London, vol 35, pp 39–48

Grantz A, Eittreim S, Dinter DA (1979) Geology and tectonic development of the continental margin north of Alaska. Tectonophysics J 59:263–291

Grantz A, May SD, Taylor PT et al (1990) Canada Basin. In: Grantz A, Johnson L, Sweeney J (eds) The geology of North America. The Arctic Ocean region. Geological Society of America, Boulder, pp 379–402

Grantz A, Phillips RL, Mullen MW et al (1996) Character, paleoenvironment, rate of accumulation, and evidence for seismic triggering of Holocene turbidites, Canada abyssal plain, Arctic Ocean. Mar Geol J 133:51–73

Grantz A, Clark DL, Phillips RL et al (1998) Phanerozoic stratigraphy of Northwind ridge, magnetic anomalies in the Canada Basin, and the geometry and timing of rifting in the Amerasia basin, Arctic Ocean. GSA Bull 110(6):801–820

Grantz A, Hart PE, Childers VA (2011) Geology and tectonic development of the Amerasia and Canada basins, Arctic Ocean. Geol Soc Lond Mem 35:771–800

Hein FJ, van Wagoner NA, Mudie PJ (1990) Sedimentary facies and processes of deposition: Ice Island cores, Axel Heiberg shelf, Canadian polar continental margin. Mar Geol J 93:243–265

Helwig J, Kumar N, Emmet P, Dinkelman MG (2011) Regional seismic interpretation of crustal framework, Canadian Arctic passive margin, Beaufort Sea, with comments on petroleum potential. In: Spencer AM (ed) Arctic petroleum geology, Geological Society of London, vol 35, pp 527–544

Hill JC, Driscoll NW (2008) Paleodrainage on the Chukchi shelf reveals sea level history and meltwater discharge. Mar Geol J 254:129–151

Hill JC, Driscoll NW (2010) Iceberg discharge to the Chukchi shelf during the Younger Dryas. Quat Res 74:57–62

Hill JC, Driscoll NW, Brigham-Grette J et al (2007) New evidence for high discharge to the Chukchi shelf since the last glacial maximum. Quat Res 68:271–279

Houseknecht DW, Bird KJ (2011) Geology and petroleum potential of the rifted margins of Canada Basin. In: Spencer AM (ed) Arctic petroleum geology, Geological Society of London, vol 35, pp 509–526

Hutchinson DR, Jackson HR, Houseknecht DW et al (2017) Significance of northeast-trending features in Canada Basin, Arctic Ocean. Geochem Geophys Geosyst J 18

Jakobsson M, Kristoffersen GA et al (2003) Physiographic provinces of the Arctic Ocean sea floor. Geol Soc Am Bull J 115:1443–1455

Jakobsson M, Mayer L, Coakley B et al (2012) The International Bathymetric Chart of the Arctic Ocean (IBCAO) version 3.0. Geophysical research letters. Wiley 39(12)

Jakobsson M et al (2014) Arctic Ocean glacial history. Quat Sci Rev 92:40–67

Jakobsson M et al (2015) Evidence for an ice shelf covering the Central Arctic Ocean during the penultimate glaciation. Nat Commun 7(10365):10. https://doi.org/10.1038/ncomms10365

Jokat W, Schmidt-Aursch MC (2007) Geophysical characteristics of the ultraslow spreading Gakkel Ridge, Arctic Ocean. Geophys J Int 168:983–998

King EL (2015) Late glaciation in the eastern Beaufort Sea: contrasts in shallow depositional styles from Amundsen Gulf, Banks Island Shelf and M'Clure Strait [Abstract]. Natural Resources Canada, Earth Science Sector, Contribution Series 20150335. ArcticNet Conference

King EL, Li M, Wu Y et al (2017) A belt of seabed erosion along the Beaufort Sea margin, offshore northwest territories, governed by Holocene evolution of the Beaufort shelf-break jet; geological evidence, current measurements, and initial oceanographic modelling. Geological Survey of Canada, Open File 8198

Kovacs LC, Bernero C, Johnson GL et al (1985) Residual magnetic anomaly chart of the Arctic Ocean region Map and Chart Series (Geological Society of America). Vol. 53, 1 sheet, Geological Society of America

Lane LS (1997) Canada Basin, Arctic Ocean: evidence against a rotational origin. Tectonics J 16:363–387

Lawver LA, Scotese CR (1990) A review of tectonic models for the evolution of the Canada Basin. In Grantz A, Johnson L, & Sweeney J (eds) The Geology of North America. The Arctic Ocean region. Geological Society of America, Boulder, Chap 31, p 593–618

Lawver LA, Grantz A, Gahagan LM (2002) Plate kinematic evolution of the present Arctic region since the Ordovician. In: Miller EL et al (eds) Tectonic evolution of the Bering Shelf–Chukchi Sea–Arctic margin and adjacent landmasses, Geological Society of America

Laxon S, McAdoo D (1994) Arctic Ocean gravity field derived from ERS-I satellite altimetry. Science J 265:621–624

Maus S, Yin F, Luhr H et al (2008) Resolution of direction of oceanic magnetic lineations by the sixth-generation lithospheric magnetic field model from CHAMP satellite magnetic measurements. Geochem Geophys Geosyst 9.:Q07021. https://doi.org/10.1029/2008GC001949

Miller EL, others m (2017) Circum-Arctic Lithosphere Evolution (CALE) transect C: displacement of the Arctic Alaska–Chukotka microplate towards the Pacific during opening of the Amerasia Basin of the Arctic. In: Pease V, Coakley B (eds) Circum-arctic lithosphere evolution. Geological Society of London, London. https://doi.org/10.1144/SP460.19

Miller EL, Toro J, Gehrels G et al (2006) New insights into Arctic paleogeography and tectonics from U–Pb detrital zircon geochronology. Tectonics J 25(3)

Mosher DC (2016) Continental margins of the Arctic Ocean: implications for law of the sea. European Geosciences Union general assembly 2016, Vienna, Austria, April 17–22, 2016

Mosher DC, Mudie PJ, Thibaudeau SA et al (1990) Ice Island sampling and investigation of sediments: an environmental marine geology program: video documentary. Geological survey of Canada open file report 2261: 1 video/vidéocassette

Mosher DC, Shimeld J, Jackson R et al (2011) Sedimentation in Canada Basin. Geological Survey of Canada Open File 6759, Poster

Mosher DC, Shimeld J, Hutchinson D et al (2012a) Canada Basin revealed. Arctic Technology Conference paper OTC 23797, p 11

Mosher DC, Shimeld JW, Hutchinson D et al (2012b) Submarine landslides in Arctic sedimentation: Canada Basin. In: Yamada Y, Kawamura K (eds) Submarine mass movements and their consequences V, Advances in Natural and Technological Hazards Research, vol 31, pp 147–158

Mosher DC, Courtney RC, Jakobsson M et al (2015) Mapping the surficial geology of the Arctic Ocean: a layer for the IBCAO. Offshore Technology Conference paper OTC 22561

Nady D (1966) The gravitational attraction of a right rectangular prism. Geophys J 31:362–371

O'Brien TM, Miller EI, Benowitz JP et al (2016) Dredge samples from the Chukchi borderland: implications for paleogeographic reconstruction and tectonic evolution of the Amerasia Basin of the Arctic. Am J Sci 316:873–924

Oakey GN, Saltus R (2016) Geophysical analysis of the alpha-Mendeleev ridge complex: characterization of the high Arctic large Igneous Province. Tectonophysics J 691:65–84

Pease V, Drachev S, Stephenson R et al (2014) Arctic lithosphere – a review. Tectonophysics J 628:1–25

Pickup SLB, Whitmarsh RB, Fowler CMR et al (1996) Insight into the nature of the ocean–continent transition off West Iberia from a deep multichannel seismic reflection profile. Geol J 24:1079–1082

Polyak L, Edwards MH, Coakley BJ et al (2001) Ice shelves in the Pleistocene Arctic Ocean inferred from glaciogenic deep-sea bedforms. Nature J 410:453–457

Ruppel CD, Herman BM, Brothers LL, Hart PE (2016) Subsea ice-bearing permafrost on the U.S. Beaufort margin: 2. Borehole constraints. Geochem Geophys Geosyst J 17:4333–4353. https://doi.org/10.1002/2016GC006582

Saltus RW, Miller EL, Gaina C (2011) Regional magnetic domains of the circum-Arctic: a framework for geodynamic interpretation. In: Spencer AM (ed) Arctic petroleum geology, Geological Society of London, vol 35, pp 49–60

Shimeld J, Li Q, Chian D et al (2016) Seismic velocities within the sedimentary succession of the Canada Basin and southern Alpha-Mendeleev ridge, Arctic Ocean: evidence for accelerated porosity reduction? Geophys J Int 204:1–20

Tailleur IL (1973) Probable rift origin of Canada Basin, Arctic Ocean. AAPG Mem 19:526–535

Taylor PT, Kovacs LC, Vogt PR et al (1981) Detailed aeromagnetic investigation of the Arctic Basin. J Geophys Res 86:6323–6333

Verhoef J, Roest WR, Macnab R et al (1996) Magnetic anomalies of the Arctic and North Atlantic oceans and adjacent land areas. Geological survey of Canada, Open File 3125a, p 577

Vorren TO, Laberg JS (1997) Trough mouth fans - palaeoclimate and ice-sheet monitors. Quat Sci Rev J 16:865–881

Vorren TO, Lebesbye E, Andreassen K et al (1989) Glacigenic sediments on a passive continental margin as exemplified by the Barents Sea. Mar Geol 85:251–272. https://doi.org/10.1016/0025-3227(89)90156-4

Chapter 11
Pliocene-Pleistocene Sedimentation

Daria V. Elkina, Vera I. Petrova, Alexey L. Piskarev,
and Irina A. Andreeva

Abstract Paleomagnetic data, including the recent high-quality measurements, estimate the average mean sedimentation rate in the Mendeleev Ridge for the last 4 Ma as 1–1.5 mm/kyr, rising slightly towards the shelf seas of northeast Russia. The rates also increase towards the Lomonosov Ridge: in its near-Greenland sector, the Brunhes/Matuyama transition was identified in the sediment core at 330 cmbsf, giving rates of 4.4 mm/kyr for the Brunhes chron.

Recently established presence of volcanic material in bottom sediments indicates active, at times even catastrophic, the Pleistocene volcanic activity in the Arctic Basin. It could be safe to state that the Eurasian Basin in the Arctic Ocean was a scene for at least one such a powerful volcanic eruption with huge volumes of ejected material at ~1.1 Ma.

The study of hydrocarbon molecular markers and the Late Cenozoic sedimentation in the Amerasian continental margin allowed to examine the importance of different processes (terrigeneous denudation, glacial transport, turbidites, oceanic slope contouring currents, submarine erosion, and bedrock material re-deposition) in accumulation of the sedimentary cover.

Keywords Arctic Basin · Paleomagnetism · Sedimentation rate · Volcanic eruption · Hydrocarbon molecular markers

D. V. Elkina (✉) · V. I. Petrova · A. L. Piskarev
All-Russian Research Institute of Geology and Mineral Resources of the World Ocean
(VNIIOkeangeologia), Saint Petersburg, Russia

Saint Petersburg University, Saint Petersburg, Russia

I. A. Andreeva
All-Russian Research Institute of Geology and Mineral Resources of the World Ocean
(VNIIOkeangeologia), Saint Petersburg, Russia

© Springer International Publishing AG, part of Springer Nature 2019
A. Piskarev et al. (eds.), *Geologic Structures of the Arctic Basin*,
https://doi.org/10.1007/978-3-319-77742-9_11

11.1 Sedimentation

In the last decades, the processes of sedimentation in the abyssal Arctic Ocean attracted steadily increasing attention and new data are constantly being collected for underwater ridges and uplifts, especially - the Lomonosov and Mendeleev ridges. The ever increasing volume of information feeds active investigations and discussions of sources, composition and modes of transportation of sedimentary material. But despite all these developments, dating of the sediments deeper than 1.5–2 m still remains problematic.

Reliable estimate of sedimentation rates is needed for correct evaluation of thickness of the recent sedimentary horizons and its sources. However, in the Arctic Ocean, the sedimentation rates can differ by up to 3 orders of magnitude – from several meters per millennium (in fiords and estuaries) to less than 1 cm per millennium in the abyssal plains (Levitan 2015). Moreover, in the central Arctic Ocean, a significant discrepancy exists between the very high ratio of the source areas to the sedimentation area and very low sedimentation rates (Backman et al. 2004).

The sea bottom deposits are a very important source of information regarding paleoclimate and its role in forecasting the future climate of our planet. The last, but not the least, urgent problem of delineation of the Russian Federation continental shelf outer limit in the Arctic Ocean relies heavily on our detail knowledge of the geology of the submarine elevations, and the Mendeleev Ridge in particular. Recent geological and geophysical expeditions concentrated their efforts on collecting as much information as possible on age, composition, sources, modes and direction of transportations, separation of imported and "in situ" clastic material etc.

Lacking microfossil material, the utmost important task of determining absolute age of the sediments depends on radiocarbon method and paleomagnetic data. However, the former method can be applied with confidence only to the most recent (tens of thousands of years) formations, and the latter, as it will be shown later, is prone to ambivalent correlations to established geochronological scale.

A paleomagnetic record on sediments reversely magnetized can be interpreted as an apparent reversal of the Earth's magnetic field (Clark 1970; Steuerwald et al. 1968; Witte and Kent 1988; Piskarev et al. 2013; Piskarev and Elkina 2014), or excursions (Jakobsson et al. 2000; Nowaczyk et al. 2003; O'Regan et al. 2008; Spielhagen et al. 2004), or as a result of diagenetic chemically induced remanent magnetization (CRM) (Channell and Xuan 2009; Xuan and Channell 2010). Thus, different approaches may produce chronological models varying by the order of magnitude.

The first paleomagnetic investigations of the cores collected on the Mendeleev Ridge north from 83°N gave very low sedimentation rates, similar to the abyssal plains of the Pacific Ocean: 1–3 mm/kyr cited in (Lin'kova 1984). In some cores, position of the Brunhes/Matuyama transition gave rate 0.8–1.6 mm/kyr (Clark 1970; Steuerwald et al. 1968). However, in the recent publications these results for the central Arctic Ocean including the Mendeleev Ridge were revised upwards to more than 10 mm/kyr. Below we will discuss the reasons for such revision.

Witte and Kent (1988) questioned the results by (Clark 1970; Steuerwald et al. 1968) because these authors did not use demagnetizing procedures on the samples. However, when the magnetic reversals clearly manifest themselves (which was the case), stepwise demagnetization will not alter the results but make the reversely magnetized zones even more apparent. That was clearly demonstrated in (Witte and Kent 1988), where samples from two cores collected at the south-eastern slope of the Mendeleev Ridge confirmed the Brunhes/Matuyama boundary at 228 cmbsf (cm below sea floor) (core T3-67-6) and 85 cmbsf (core T3-67-12) after demagnetization applied. The sedimentation rates were calculated at 1 and 3 mm/kyr, accordingly (Fig. 11.1).

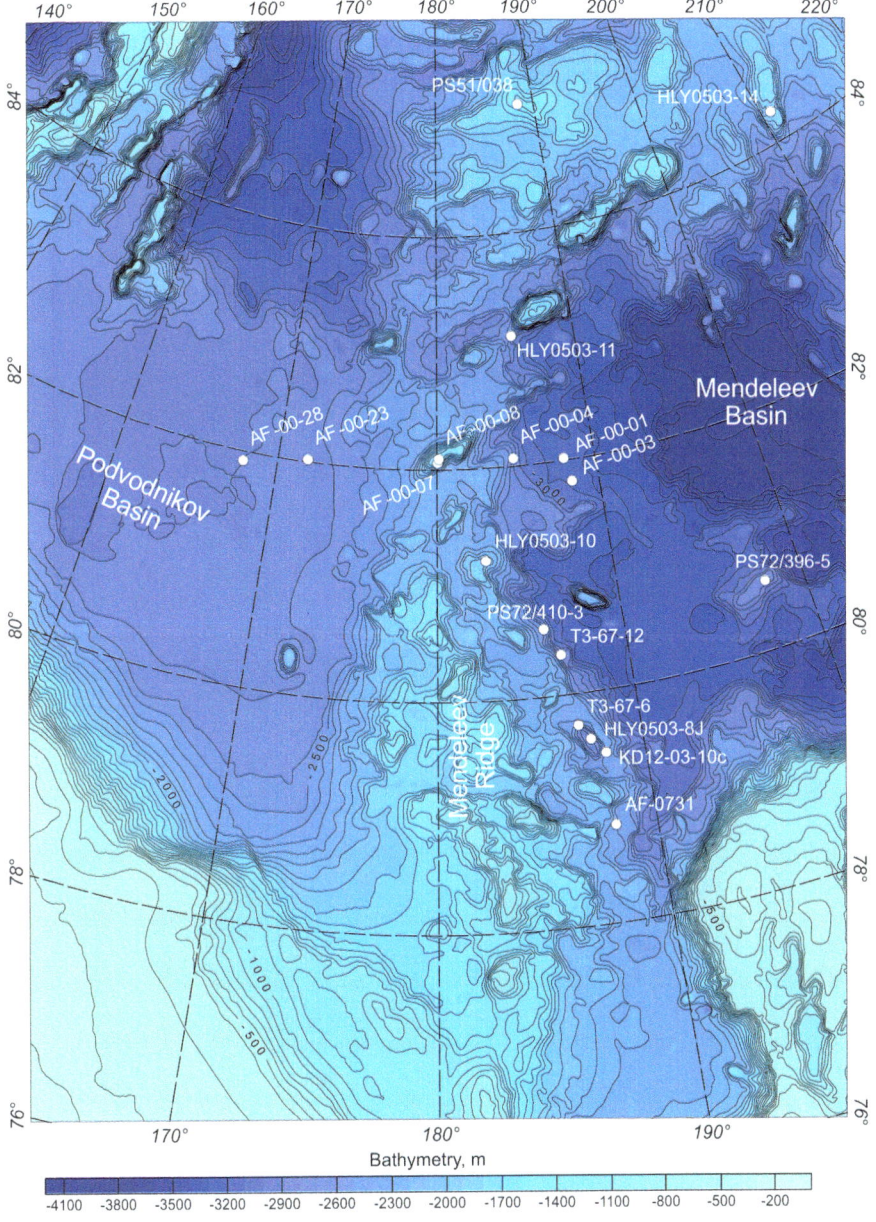

Fig. 11.1 Sediment coring sites (RV Akademik Fedorov in 2000, Arctic-2012 etc.)

Frederichs (1995) conducted a detailed paleomagnetic study of marine sediments from the Lomonosov Ridge, the Amundsen Basin, the Ermak and Morris-Jesup plateaus, and the Fram Strait. The author estimated that the sedimentation rates were 10 mm/kyr at the near-polar segment of the Lomonosov Ridge (core 2185–6) and 15 mm/kyr in the adjacent part of the Amundsen Basin (core 2171–4). The research did not provide any information regarding the Amerasian Basin.

Paleomagnetism of sedimentary cores up to 13 m long, also from the western part of the Eurasian Basin, was carefully studied by Schneider et al. (1996) using demagnetization by alternating magnetic field up to 60 mT strong. The derived Pleistocene sedimentation rate was estimated at several centimeters per thousand years. Despite some ambiguity in chronology of certain reverse polarity episodes (subchrons), the total duration of them comprises less than 10% of the total period of sedimentation covered by cores, which agrees with the world-wide numbers.

The paper by Jakobsson et al. (2000) discusses the analysis of the 772-cm-long core from the near-polar segment of the Lomonosov Ridge (station 96/12–1pc). The Brunhes/Matuyama boundary was fixed at 550 cmbsf. However, judging by its shape, this boundary, in our opinion, should be placed at 270 cmbsf. In any case, it is difficult to explain the NRM vector inclination curve below 270 cmbsf because this interval has more than 30% of the reversely magnetized samples, which is unusual for the Brunhes chron. It is quite possible that positive magnetic polarity in this and similar columns has chemical nature and caused by post-deposional diagenetic processes. The conclusions by Jakobsson et al. (2000) discarding the low sedimentation rates in the Mendeleev Ridge as erroneous were corroborated by later studies (Backman et al. 2004; Backman and Moran 2009; Moran et al. 2006). However, the cited works were based on the radiocarbon dating (efficient only for the very top of the sedimentary columns) or biostratigraphic correlation (handicapped by lack of fossil and micro-fossil material in the Mendeleev-Alpha region).

Nowaczyk et al. (2001) analyzed sedimentation processes in the Makarov Basin near the Lomonosov Ridge and presented the convincing correlation of three cores 831–1372 cm long, based on variety of data, including paleomagnetics. The authors could not reconcile the conflicting information and, keeping in mind possible substantial depositional hiatuses, ended up with two sedimentation rates for the Makarov Basin: 13 mm/kyr and 40 mm/kyr. Again, as in (Jakobsson et al. 2000), the high (>30%) concentration of the reversely magnetized samples in the Brunhes interval remains unexplained.

The attempt to reconcile the sedimentation rates in the Eurasian Basin (the Lomonosov Ridge) and the Amerasian Basin (the Mendeleev Ridge) was made by Spielhagen et al. (2004). The correlation proposed by the authors left some questions unanswered. The base of normal polarity zone in core PS51/038 (the Mendeleev Ridge, just north from 85°N, 1473 bsl (below sea level), Fig. 11.1) was fixed at 131 cm and, judging by prevalence (≈80%) of reversely magnetized samples in 70 cm-long interval below, can be confidently identified as the Brunhes/Matuyama boundary. Thus, it defines the sedimentation rate during the Brunhes epoch at 1.7 mm/kyr. A Completely different situation exists in the Eurasian Basin. Identification of the Brunhes/Matuyama boundary in the cores from the Lomonosov

Ridge (96-12-1pc and PS2185) and Morris Jesup Plateau (PS2200) is quite questionable as below 3 m depth normal- and reversely magnetized sediments in these cores interlace each other with 2.5:1 ratio. Nevertheless, the authors correlate the reliable boundary in core PS51/038 with the much less reliable one in cores 96–12-1pc, PS2185 and PS2200, making it younger.

Paleomagnetic analysis (Polyak et al. 2009; Adler et al. 2009) of seafloor material obtained by the 2005 Healy-Oden TransArctic Expedition (HOTRAX) has demonstrated that some localities can have much higher sedimentation rates, as in core HLY0503-8JPC from the south-eastern slope of the Mendeleev Ridge (Adler et al. 2009). In this core, the Brunhes/Matuyama boundary was identified at 420 cm, giving the sedimentation rate $\geq$ 5 mm/kyr. Such fluctuations when rates vary substantionally in proximal locations are not unique (Vishnyakov et al. 1992). Indeed, the Brunhes/Matuyama transition in cores T3-67-6 and T3-67-12 collected from the same region was located at 228 cm and 85 cm, accordingly.

Radiocarbon dating cited in (Polyak et al. 2009) and (Adler et al. 2009) are applicable only to the first 50–60 cm of sediment cores representing around 50 Ka, while correlation with the paleomagnetic data from the Lomonosov Ridge may raise the question. Nevertheless, we may accept the following locations of the Brunhes/Matuyama boundary in the studied cores: 230 cm (10JPC), 80 cm (11JPC), and 220 cm (14JPC). At the eastern slope of the Northwind Ridge, the Brunhes/Matuyama transition is confidently identified at 540 cm (92-P27), >500 cm (92-P25), 200 cm (92-P39), 120 cm (92-P21), and 40 cm (92-P23), estimating the mean sedimentation rate several times lower than found by the authors (Polyak et al. 2009; Adler et al. 2009).

Krylov et al. (2011) supported the suggestion to correlate the magnetic field reversal in the sediment cores from the Mendeleev Ridge not to the Brunhes/Matuyama boundary but to the Biva-II excursion (272 ka), also taking into account data on core 96/12-1pc from the Lomonosov Ridge.

The last decade was marked by increased attention to radioactive methods of age determination. Uranium-thorium dating of the sediments (Andreeva et al. 2007) and the age model determined by the paleomagnetic data seem to agree with each other. The ratio of ^{230}Th to ^{231}Pa in the HOTRAX cores from the Mendeleev Ridge (Not and Hillaire-Marcel 2010) gave the mean sedimentation rate of 1.5 mm/kyr, which is lower than the ones in the Eurasian Basin by the order of magnitude. The researchers noted that with such a low mean sedimentation rate the radiocarbon methods could be reliably applied only to the first centimeters of the sedimentary column.

Gusev et al. (2012) applied radiochemical analysis to cores AF-00-02, AF-00-07 and AF-0731 from the eastern Mendeleev Ridge. The first two columns showed rates of 1.5 mm/kyr and 1.1 mm/kyr, accordingly. The third (AF-0731) core, taken further southwards and closer to the continental slope, had sedimentation rate of 4.4 mm/kyr over the last 200,000 years, which is close to what we have predicted from the NRM inclination in core HLY0503-8JPC.

With such a variety of existing estimates and stratigraphic interpretations, fresh information from cruises Arctic-2000, *RV* Akademik Fedorov, and Arcric-2012,

Table 11.1 Sediment cores subjected to paleomagnetic studies

Core	Latitude	Longitude	Sea depth, m	Core length, m
AF-00-01	82°00.51'N	171°58.60'W	3110	2.73
AF-00-03	81°48.71'N	171°38,50'W	3321	3.10
AF-00-04	82°03.45'N	175°09.20'W	2704	2.40
AF-00-07	82°03.24'N	179°56.17'W	1555	2.45
AF-00-08	82°05.22'N	179°52.00'W	1490	2.65
AF-00-23	82°00.95'N	171°53.99'E	2750	3.30
AF-00-28	81°54.90'N	167°52.32'E	2814	3.34
KD12-03-10c	79°27.75'N	171°55.08'W	2200	6.00

icebreaker Kapitan Dranitsyn (Morozov et al. 2013), can be instrumental in improving our understanding of the sedimentation processes in the Amerasian Basin.

The *RV* Akademik Fedorov (year 2000) traverse ran from the Mendeleev Basin (cores AF-00-01 and AF-00-03, 82°N), across the eastern slope of the Mendeleev Ridge (AF-00-04), its central part (AF-00-08, collected from the local mound named after the ship, and AF-00-07, on its south-western slope) into the Podvodnikov Basin (AF-00-23 and AF-00-28). Core KD12-03-10c was obtained further south from this traverse on the eastern slope of the Mendeleev Ridge in 2012 (Fig. 11.1, Table 11.1).

Using the method recommended by Kochegura (1992), samples were carefully collected from only undeformed sediments. First, 30–50 mm thick segments were cut from a core along its axis then several properly oriented glass cylinders (in average, 22.5 mm dia.) were pressed in sediments forming the samples; the size of the cylinders determined the increment of observations. In total, 824 samples were subsampled from the AF cores (year 2000) and 244 - from KD12-03-10c (year 2012).

Of two parameters, defining the spatial orientation of the natural remanent magnetization (NRM) vector - inclination and declination – only inclination I can be used due to arbitrary horizontal orientation of cores.

Spinner magnetometers JR-4 and JR-6A, AGICO provided measurements of intensity J and inclination I for specimens of 2000 and 2012, respectively. Magnetic susceptibility k of the AF cores was measured by a kappabridge KLY-2c meter calibrated to ±3% with volume correction. Magnetic (volume) susceptibility of 1st and 2nd meters was obtained from discrete measurements of specimens by a Kappabridge MFK1-FA, AGICO. Starting from the 3rd meter magnetic susceptibility was measured by Kappameter KT-5 and MS2E surface sensor, Bartington.

Inclination I of the NRM vector as a function of depth is one of the most important paleomagnetic parameters needed for dating and correlation of bottom sediments at the high geographical latitudes. We used the following geomagnetic polarity time scales for referencing the above function to the principal and secondary geomagnetic events (chrons, subchrons, and excursions) (Pospelova 2004; Gee and Kent 2007):

Excursion Biwa-III (B), reverse polarity – 0.37 Ma
Excursion Elunino (Elun), reverse polarity – 0.71 Ma

Base of Brunhes (Br) chron, normal polarity - 0.78 Ma;

Subchron Jaramillo (Jar), normal polarity – 0.90-1.06 Ma
Subchron Gilsa (Gil), Cobb Mountain-(Cobb M), normal polarity   1.21 1.24 Ma;
Subchron Olduvai (Old), normal polarity −1.78–2.00 Ma;
Subchron Reunion (R), normal polarity −2.08–2.14 Ma;
Base of Matuyama (M) chron, reverse polarity – 2.58 Ma;

Base of Gauss (G) chron, normal polarity – 3.58 Ma (boundary with the Gilbert (Gi) reverse chron);

Subchron Kaena (K), reverse polarity – 3.05-3.12 Ma.

Fig. 11.2 presents plots of the NRM inclination for cores from *RV* Akademik Fedorov, and ChRM (Characteristic remanent magnetization) versus NRM inclination for core KD12-03-10c, illustrating the transition from mainly large positive values to smaller and negative ones.

All the studied sediment cores reflect the transition from Brunhes to Matuyama, confirming the possibility of its identification even at very low mean sedimentation rates (Lin'kova 1984; Tretyak et al. 1989; Vishnyakov et al. 1992). Some peaks of negative I in the Brunhes interval, and positive - in the Matuyama, can be explained by reverse events, typical for both chrons.

In some cores, the transition from the domain of positive I values into that of negative is less apparent. It could be contributed to the viscous remanent magnetization (VRM) which, if present, will increase values of normally magnetized sediments and decrease it in reversely magnetize ones. Core AF-00-01 illustrates the influence of the VRM component. Here the sediments deposited during normal and reverse polarity of the Earth's Magnetic field differ not only by the inclination, but also by its values: in normally magnetized intervals average $J \approx 2 \times 10^{-3}$ nT, while in reversely magnetized - $\approx 1 \times 10^{-3}$ nT. Because magnetic susceptibility k remains relatively constant in both chrons, the difference in J can only be explained by influence of the VRM component.

The same phenomena can be observed in practically all the presented AF cores. Reversely magnetized intervals were experimentally confirmed by demagnetization of the samples with alternating magnetic field up to 12.3 mT. Demagnetization process affects the inclination of normally and reversely magnetized sediments differently: in the former, the vector only slightly deviates from the original inclination, no more than 10°, as a rule, while in the latter, the deviation approached 30°. Such a difference is due to the fact that the superimposed secondary VRM magnetization is less stable than primary magnetization acquired during sedimentation. Magnetic susceptibility k, varying between $0.2–0.4 \times 10^{-3}$ SI (save for some anomalous peaks) and, naturally, reflecting concentration of ferromagnetic fraction in different layers, is often used for independent confirmation of otherwise identified boundaries. For instance, the very strong localized spikes are recorded at the top of the first reversely magnetized layer in cores AF-00-08 and AF-00-23, hence confirming identification of the Brunhes/Matuyama boundary.

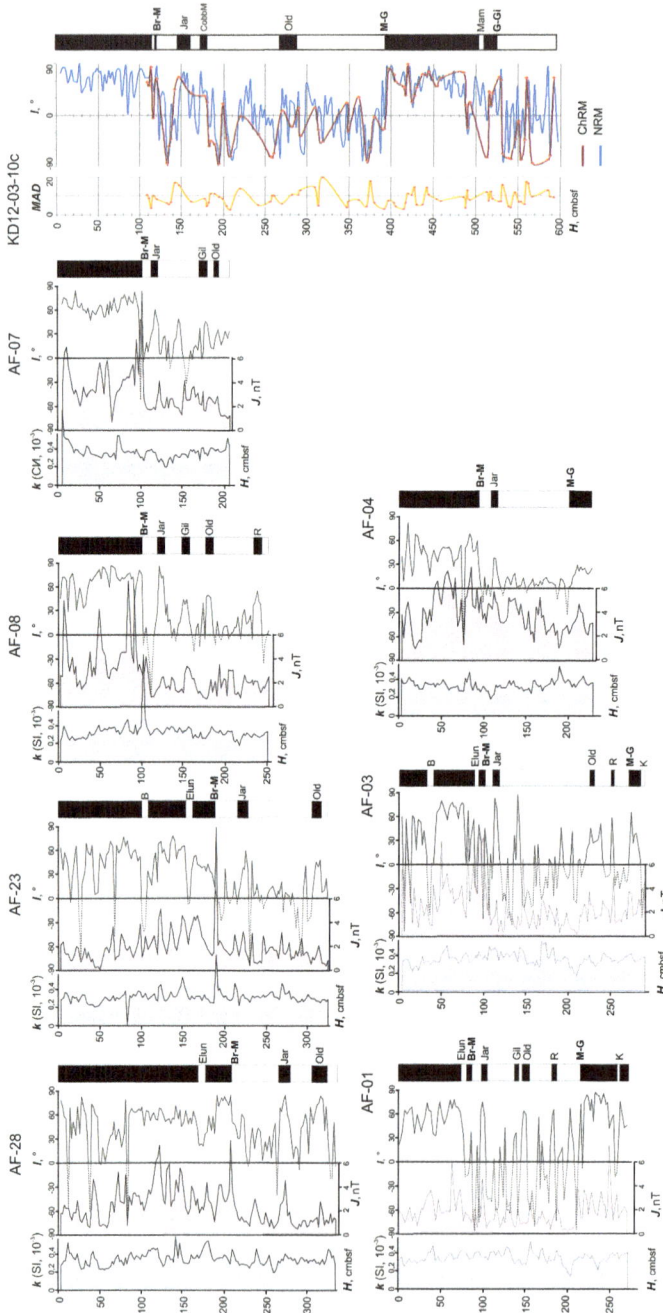

Fig. 11.2 Pliocene-Pleistocene paleomagnetic and stratigraphic correlation of the Mendeleev Ridge cores (AF cores — cruise of *RV* Akademik Fedorov in 2000, KD12-03-10c — Arctic-2012), where k -magnetic susceptibility; J – intensity of NRM; I- inclination; MAD — maximum angular deviation; black - normal polarity

Concentrating attention on the finer details, we can see that cores AF-00-01 and AF-00-03 from the Mendeleev Basin (eastern end of the traverse) display identical patterns of magnetization parameters (Piskarev et al. 2013). Large inclinations dominate the upper parts of both cores. Several sharp and short negative deflections are unidentifiable falling beyond the resolution of the method. The Biwa (-III) excursion was identified by its most likely position in the middle of the Brunhes interval. Transition of the inclination values into negative domain, accompanied by noticeable drop in J, marks onset of to the Matuyama chron. The reverse transition to the Gauss chron of the normal polarity is more apparent in the AF-00-01 core than in AF-00-03 and supported by analysis of the demagnetization data.

Cores AF-00-04, AF-00-07 and AF-00-08 from the slope and top of the Mendeleev Ridge are not long enough to penetrate the Gauss chron and they have reached only horizons identified as subchrons Olduvai or Reunion (AF-00-08). The Brunhes here has the same thickness as in the Mendeleev Basin. The Brunhes/Matuyama boundary is additionally manifested by spikes of k and J, pointing on short-lived change of the sedimentation regime leading to higher concentration of ferromagnetic material. The intensity J in these cores is also somewhat higher than in sediments from the Mendeleev Basin.

Cores from the Podvodnikov Basin, AF-00-23 and AF-00-28 display the thicker Brunhes interval, and the Brunhes/Matuyama boundary is also marked by peak values of k and J. The intervals of positive polarity in the lower parts of the columns are identified as Jaramillo and Olduvai.

The considerably longer (600 cm) core KD12-03-10c provided paleomagnetic information for sediments as old as ca. 4 Ma (Gilbert chron, Early Pliocene) – Fig. 11.2.

Positive inclinations are prevailing to 123.5 cmbsf with an abrupt change to the negative inclination at this mark. The inclination remains mostly negative down to 394 cm with some short lived positive deviations.

As the J (Fig. 11.2) curve demonstrates, magnetization of the interval with normal polarity is higher than with reverse: 4×10^{-3} A/m in 0–121 cmbsf vs 1.8×10^{-3} A/m in 123–394 cmbsf (some positive anomalies at 78.5 cm, 93 cm, and 118.5 cm and below correlate with simultaneous positive surges of magnetic susceptibility k). After a slight increase in the Gauss chron (394–470 cm), J does not exceed $1.4 \cdot 10^{-3}$ A/m in the underlying Gilbert chron.

The ChRM curve based on the data produced by the demagnetization procedure (described in Piskarev and Elkina (2014), Elkina (2014)) demonstrates the alternating intervals of normal and reverse polarity along the whole length of KD12-03-10c more clearly (Fig. 11.2).

Assuming, as first approximation, the constant sedimentation rate (quite reasonable assumption, in the light of available data), we calculated the forecasted positions of the chrons for marine sediments of the Mendeleev Ridge using average rates for Brunhes and Matuyama chrons - Table 11.2. These positions can be used for the most probable identification of interfaces separating domains with different polarity.

Table 11.2 The Mendeleev Ridge - depth of chrons and mean sedimentation rates

	Boundary depth (cmbsf.)			Mean Sedimentation Rate mm/kyr		
Cores	Brunhes-Matuyama (0.78 Ma)	Matuyama-Gauss (2.58 Ma)	Gauss-Gilbert (3.58 Ma)	Brunhes	Matuyama	Gauss
AF-00-01	86	213		1.10	0.70	
AF-00-03	102	271		1.31	0.99	
AF-00-04	94			1.21	1.06[a]	
AF-00-07	102			1.31	0.84[a]	
AF-00-08	101			1.29	1.01[a]	
AF-00-23	188			2.41	1.15[a]	
AF-00-28	208			2.67	0.95[a]	
KD12-03-10c	123.5	394.5	531	1.58	1.5	1.36

[a]For 2000 cores, the mean sedimentation rate was calculated by subchrons' positions

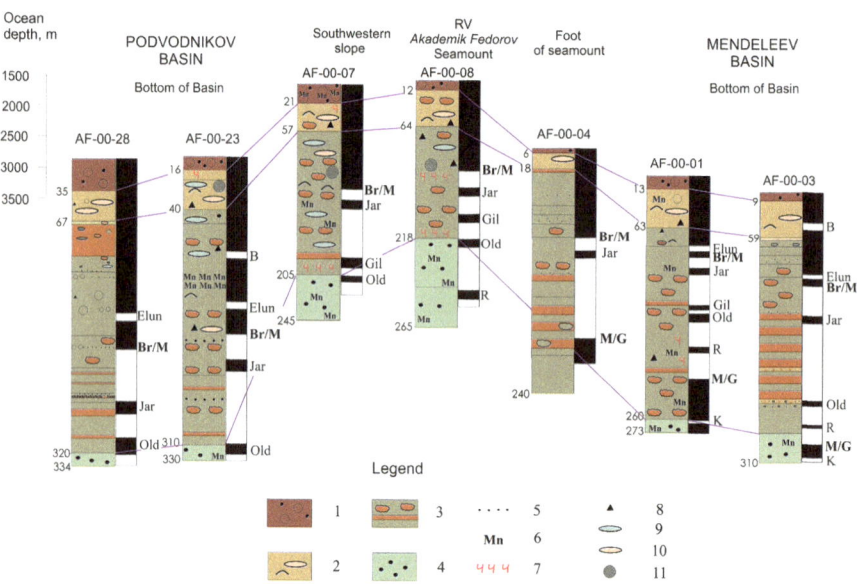

Fig. 11.3 Lithology and paleomagnetic correlation of the cores, collected onboard RV Akademik Fedorov (2000) in the Mendeleev Ridge, where 1 — "oxidized" pelites, 2 — calcareous pelites, 3 – variegated pelites, 4 – spotted pelites, 5 – sand interlayers, 6 – Mn microconcretions and crusts, 7 – Fe hydroxides, 8 – inclusions of rock fragments, 9 – clayed lenses, 10 – calcareous lenses, 11 – clayed pellets

Lithological analysis of sediments collected in 2000 has provided additional information to support the established boundaries. In the cores, we identified the following units (from top to bottom): "oxidized" pelites (or, equivalently, clays hereinafter), calcareous pelites, variegated pelites, and spotted pelites (see Fig. 11.3).

"Oxidized" pelites are composed of brown to dark brown sandy aleuropelites (or, equivalently, silty-clays hereinafter) with abundant plankton and benthic microfauna, and shell fragments (12–21 cmbsf on crest and slope, 13–17 cmbsf in local depressions) in the upper part of all the cores. The first 2 cm of cores is usually in the semi-liquid state and the rest has viscous, "runny" texture. The more compacted dark brown 2–7 cm thick layer marks the well-defined base of the unit.

Calcareous pelites (36–52 cm) forms the unit most consistently present on the Mendeleev Ridge and Mendeleev Basin. Its thickness varies between 36–52 cm (stable 50 cm on the eastern slope and inside the depression). It consists of olive brown or light brown aleuropelites with lenses of sand, loose calcareous material and bluish clay. Some layers are rich ferrous hydroxide, manganese concretions, and plankton/bentic microfauna. The coarse-grained clastic material is present in shallowest parts of the Mendeleev Ridge; rounded clayish-calcareous fragments are found in the Podvodnikov Basin. The transition to the underlying unit is gradual and identified only by change of colors and texture.

The units of variegated pelites consist mainly of intercalating olive green or brown aleuropelites, spotty and bead-layered pelites and sand lenses (some containing abundant microfauna). The unite thickness varies from 144 cm in the uplifts and 270 cm in depressions. Sediments are bioturbated and saturated by microconcretions, manganese crusts and clumps, streaks of iron hydroxides; rounded beads, lenses or thin beds of the calcareous clays are common in the top of the unit. General texture of all sediments can be described as fluid-ductile, less compacted than overlaying unit. The base is clearly identified by a change of colors and a thin basal layer of iron hydroxides.

Spotted pelites form the basal unit of penetrated section. The 111 cm thick unit consists of greenish-grey and light olive-green aleuropelites freckled with tiny manganese inclusions.

The boundaries of the units mentioned above coincide, or nearly coincide with polarity boundaries. This is especially true for the distinct basal section of variegated pelites in cores AF-00-28 and AF-00-23 where the unit base coincide with the Olduvai subchron as well as for AF-00-07 and AF-00-08. The same boundary in core AF-00-03 is clearly below the Olduvai subchron and in AF-00-01 it is found in the Gauss chron. Therefore, the change in the sedimentation regime, signifying the end of accumulation of spotted pelites and the beginning - of variegated, happened several hundred thousand years later in the western side of the Mendeleev Ridge than in the eastern.

Gradual transition between the base of the calcareous pelites and the top of the variegated pelites takes place inside the Brunhes. Thickness of combined "oxidized" and calcareous pelites changes only slightly along this transect, then increase of the Brunhes sediments thickness in the Podvodnikov Basin is due to increased thickness of the variegated pelites.

The Brunhes – Matuyama boundary was confidently identified at 86–124 cmbsf in all cores from the Mendeleev Ridge. In two cores from the Podvodnikov Basin, west from the Mendeleev Ridge, it was fixed at 188 and 208 cmbsf. This increase of the sedimentation rate can be explained by their location closer to the East-Siberian shelf - a major source of terrigenous material.

During the recent years, even more long cores from the Arctic Basin was sub-jected to paleomagnetic studies under the joint collaborative research of the VNIIOkeangeologia, St. Petersburg State University, Alfred Wegener Institute for Polar and Marine Research (AWI), and University of Bremen. Paleomagnetic study on four sediment cores, retrieved from the Mendeleev Ridge, and the Lomonosov Ridge, have provided the opportunity to compare sedimentation regimes on these two profound structures of the Arctic Basin. Cores PS72/396-5 and PS72/410-3 (Fig. 11.1), and PS87/023-1 and PS87/030-1 were collected during the cruises of *RV* Polarstern from the Mendeleev Ridge (Stein et al. 2010), and the Lomonosov Ridge (Stein 2015), respectively.

Measurements of NRM with the following alternating field (AF) demagnetiza-tion were carried out on u-channel samples, obtained from the cores, at the Center for Geo-Environmental Research and Modeling (GEOMODEL) of the Research Park in St. Petersburg State University, and University of Bremen.

According to the preliminary results, core PS72/396-5 from the Mendeleev Ridge has shown a change from positive to negative inclinations at ca. 90–120 cmbsf prevailed at least up to ca. 350 cmbsf, following which inclinations revert to positive ones. This trend is comparable with our previous results on all the cores from the Mendeleev Ridge mentioned above. In contrast, for core PS87/023-1 from the Lomonosov Ridge, a relevant drop to negative inclinations has been observed only after ca. 340 cmbsf. That could signify a dramatic difference in sedimentation rates between the sites, acting during the Pliocene and Quaternary. Nevertheless, a rather complicated picture of the AF data assumes effects of secondary overprints, having influenced the initial magnetization (Elkina and Piskarev 2017). Undoubtedly, additional careful studies of magnetization, geochemistry and lithology are needed to remove existing ambiguity.

It would be safe to conclude that for the last 3.5–4 Ma the mean rate of sedimen-tation in and around the Mendeleev Ridge did not exceed 1–1.6 mm/kyr, increasing only in the direction toward the continental shelves of north-eastern Russia and to the Lomonosov Ridge.

Estimation of the Pliocene-Pleistocene mean sedimentation rates in the Eurasian Basin and at the Lomonosov Ridge is more complicated due to some ambiguity in fixing the position of the Brunhes/Matuyama boundary. According to analysis of cores 96-12-1pc and PS2185 (Lomonosov Ridge), published by (Jakobsson et al. 2000; Spielhagen et al. 2004), and (Frederichs 1995), the Brunhes rate was set at 10–15 mm/kyr, while their published paleomagnetic data may be interpreted differ-ently with resulting sedimentation rate of only 4–5 mm/kyr.

11.2   Volcanic Activity

The modern stage of Pliocene-Pleistocene sedimentation process in the abyssal Arctic Ocean began after huge volume of bathymetry seismic and lithology data became available. Inside this volume there are indications that volcanic activity, periodically, and sometimes with catastrophic consequences, introducing volcanogenic material, must be considered as one of many factors affecting sedimentation process. Therefore, comprehensive analysis of the bottom sediments is needed in order to reconstruct the volcanic activity in the abyssal basins, namely, the Eurasian Basin.

In the previous Chapters, we presented the modern interpretation of the Eurasian Basin as a result of Cenozoic ocean floor spreading due to addition of the oceanic crust in the Gakkel Ridge axial zone (Brozena et al. 2003; Gleboskiy et al. 2006). Spreading was predated by the Late Cretaceous – Early Paleocene continental rifting. The Lomonosov Ridge drifted eastward as a part of the Eurasian Plate until arriving at the present position. The Gakkel Ridge is the slowest segment of the global system of the mid-ocean ridges, with some specific features peculiar to ultra-slow spreading centers.

According to the classical model, the evolution of the Eurasian Basin went through several stages (Khain 2001). During the first stage, between 55–56 and 33–34 Ma (Eocene), the spreading speed was around 1.2 cm/yr. At the onset of Oligocene, in the time of the Fram Strait opening, spreading speed in the Gakkel Ridge dropped to 0.5 cm/yr (ultra-slow spreading). Large intraplate volcanic plateau in the Late Oligocene was split by spreading into the northern Morris-Jesup Plateau attached to the Lomonosov Ridge, and southern Ermak Plateau attached to Svalbard Archipelago. At this time (Late Oligocene), spreading of the Gakkel Ridge south from 78°N practically ceased and the Ridge became buried under the sediments. Currently, however, we know many facts contradicting this simplified model and have to admit that the true history of the Eurasian Basin evolution is yet to be written.

Seismic surveys of 2011–2015 in the Eurasian Basin were one of the first steps in this direction. Especially important information was collected in 2014 when MCS seismic and multibeam sonar line 2014–05 shot across the axial zone of the Gakkel Ridge displayed the caldera centered at 81°31' N, 120°00' E .

The caldera was identified for the first time on the International Bathymetric Chart of the Arctic Ocean (IBCAO), v. 1 (2.5 × 2.5 km grid) compiled by group of international experts and presented to the American Geophysical Union in 1999. The latest version of the caldera bathymetry map after several refinements is presented in Fig. 11.4.

The caldera is 40 km wide, 80 km long and 1.2 km deep; numerous peaks, mounds and ridges up to 0.5–1 km high are visible along its rim. The expected total volume of volcanic material ejected during formation of the caldera is at least 3000 km^3 as one can calculate from caldera's depth and size. This puts it at par with the largest Quaternary calderas on Earth: the Yellowstone and Toba.

Seismic and multibeam echosounding data acquired in 2014 (see Fig.11.1b for profiles location) brought to light the important information on recent tectonic history of the Gakkel Ridge and Eurasian Basin. The data highlight the details of caldera slopes and demonstrate that the contemporary tectonically active rift valley (divergent plate boundary) 10 km wide and 500 m deep is visible at the caldera floor 4800 m bsl (Fig.11.5).

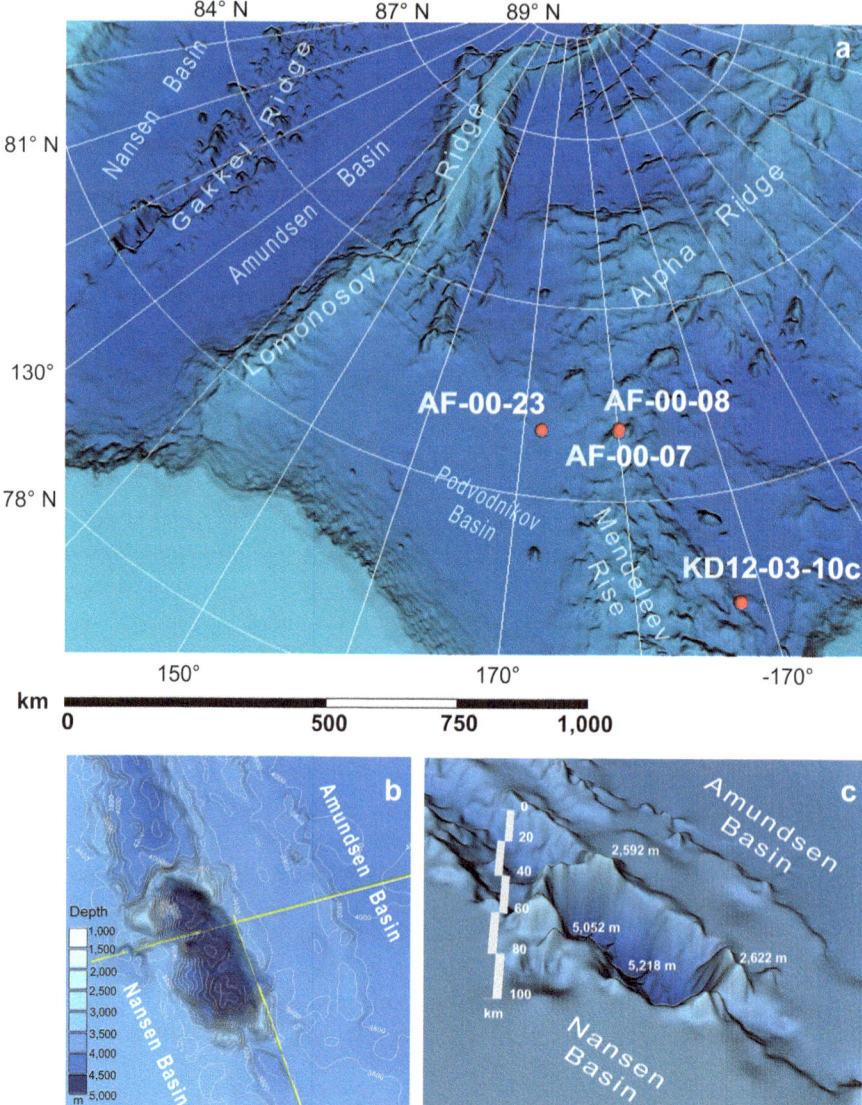

Fig. 11.4 (**a**) View of the Arctic Basin structures and some coring sampling sites; (**b**) caldera on the Gakkel Ridge rift valley (yellow lines - MCS seismic lines and multibeam survey, 2014); (**c**) 3D view of caldera and surrounding ocean floor topography (Geocap (2015) software on IBCAO grid v. 3.08)

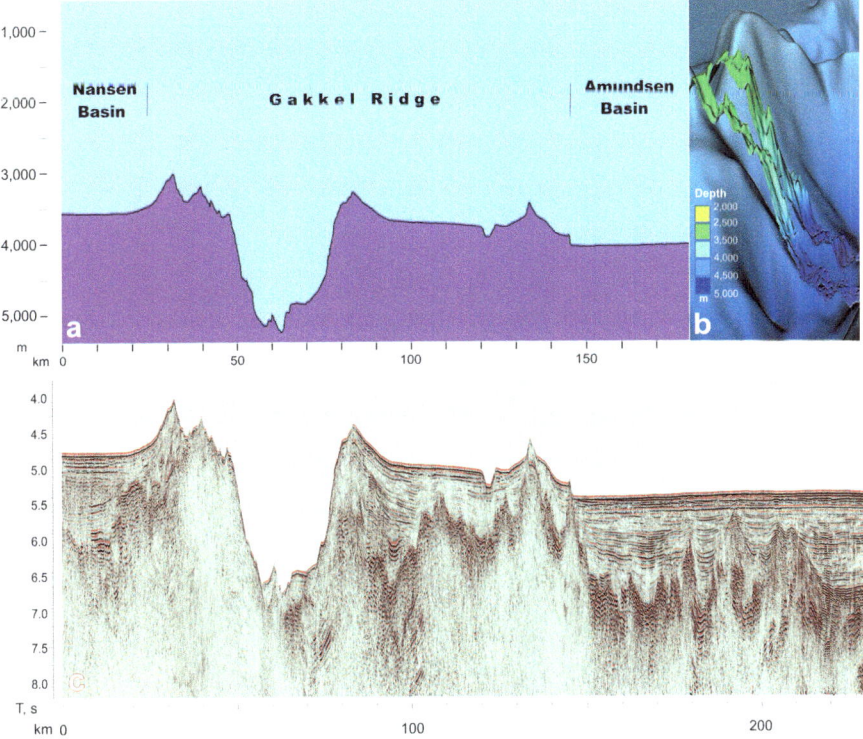

Fig. 11.5 (**a**) Cross section of the caldera according to multibeam echosounding, line 2014–05; (**b**) View at the western slope of the caldera with multibeam echosounding lines superimposed on IBCAO grid v. 3.08 (Geocap (2015)); (**c**) seismic section, line 2014–05

Assuming the average rate of spreading in the Eurasian Basin about 1 cm per year, it took approximately 1 million years to form the 10 km-wide rift valley at the bottom of the caldera. Therefore, the very important stage of tectonic evolution of the Eurasian Basin tectonic evolution – intensive magmatic activity and creation of giant caldera – came to a close at around 1 million year ago.

Introduction of a huge amount of volcanic material in the waters of the Arctic Ocean should have left a noticeable impact on composition and properties of sediments formed during and after the eruption. Therefore, wide distribution of omnipresent sedimentary layers of volcanogenic origin is expected. Such layers can be identified by increased concentration of monoclinic pyroxenes and opaque ore minerals (Owens et al. 2016; Buchs et al. 2015; Gorbarenko et al. 2002) and by higher values of magnetic susceptibility and residual magnetization caused by higher concentration of magnetite and titanomagnetite.

Several such layers, with anomalously high values of magnetization and magnetic susceptibility were identified in sedimentary cores collected in the Mendeleev Ridge,

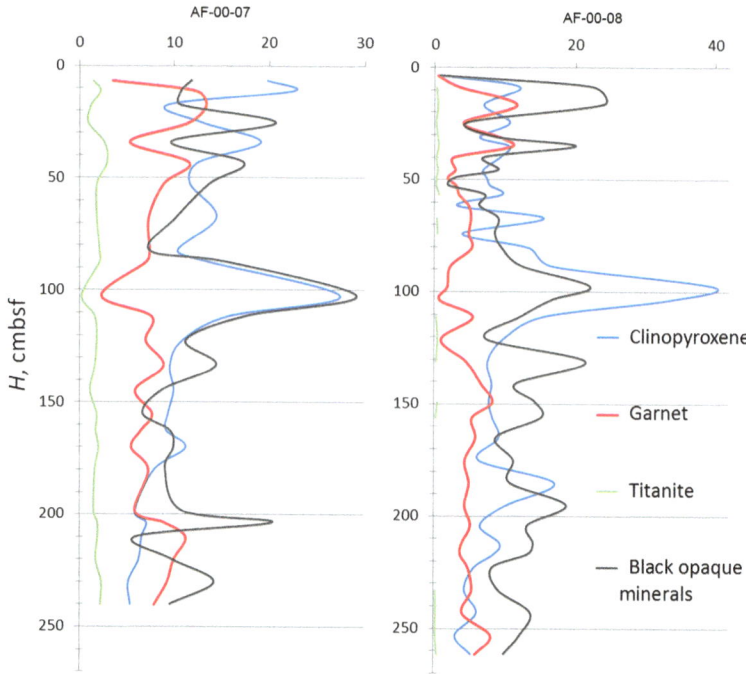

Fig. 11.6 Mineralogical analysis of heavy fraction (cores AF-00-07 and AF-00-08, Mendeleev Ridge)

about 1000 km away from the caldera (Piskarev et al. 2013), and dozens of kilometers from each other. One of them stratigraphically is very close to the Brunhes / Matuyama boundary and could be dated with high degree of confidence at about 750 Ka (Fig. 11.2, 11.7). The mineralogical analysis of this and other layers with high k and NRM demonstrates sharp increase of pyroxene and ore minerals and depletion of garnet and titanite (Fig. 11.6).

Increased concentrations of ore minerals and pyroxene indicate a probable volcanogenic nature of the studied thin sedimentary layer. At the same time, a significant decrease in concentration of such undoubtedly clastic minerals as garnet and titanite means an abrupt surge of the sedimentation rate during the formation of this volcanogenic layer in the sedimentary column. Apparently, for a short period of time there was 500% increase of sedimentation rates compare to its average value.

Even more valuable information on the specifics of Pliocene-Quaternary sedimentation was obtained from core KD12-03-10c. The core consists of relatively homogenous aleuropelite with some differences in colors and bedding configuration. The magnetic parameters were measured along the core with 2.5 cm (Fig. 11.7).

Five intervals with peak values of k and NRM identified in Fig. 11.7 were then dated with high degree of confidence by correlation with the paleomagnetic chrons:

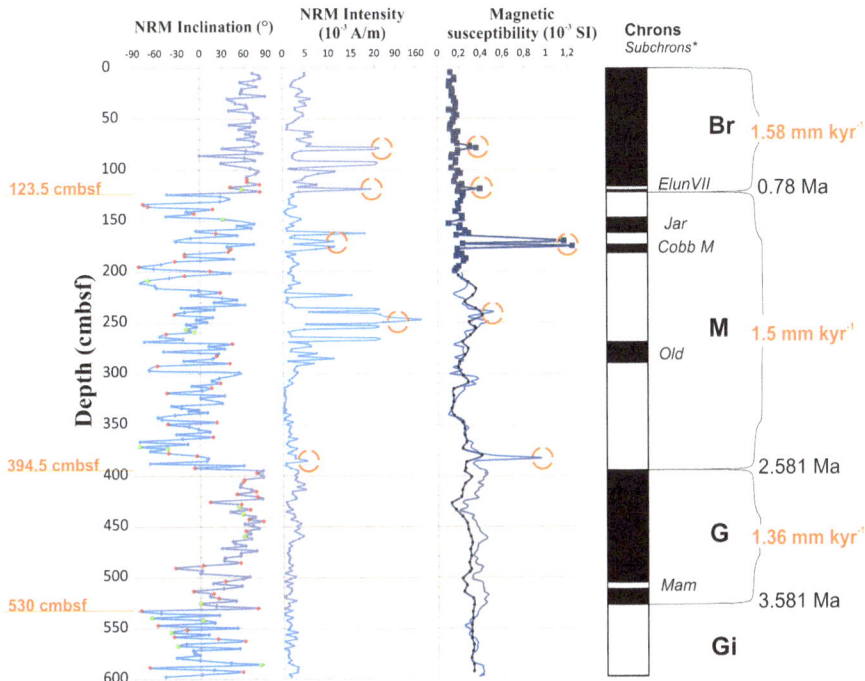

Fig. 11.7 Paleomagnetic correlation with the Geomagnetic Polarity Time Scale (Gee and Kent 2007; Zhamoida et al. 2000) on sedimentary core KD12-03-10c. The mean sedimentation rates have been determined for each paleomagnetic chron based on the established paleomagnetic boundaries, and related age (Piskarev and Elkina, 2014). Yellow circles mark abnormally high values of the magnetic susceptibility, and natural remanent magnetization (NRM). See Fig.11.1 for location

775 mmbsf — 0.47 Ma;
1185 mmbsf — 0.727 Ma;
1700–1750 mmbsf — 1.09 Ma;
2400 mmbsf — 1.62 Ma;
3850 mmbsf — 3.52 Ma.

Weight percentage distribution of the coarse fraction (> 500 microns) along the core demonstrated some correlation between intervals with a high coarse fraction content and magnetic susceptibility: three out of five peaks of the coarse fraction content (70–80 cm, 110–130 cm and 145–180 cm) coincide with spikes of magnetic susceptibility.

In order to identify the lithological differences between the bulk of the core and above mentioned anomalous intervals, 26 thin sections from 13 sampled locations of both were prepared and studied. The samples were prepared from the sediments collected into glass cylinders 20 mm dia., pressed into the sediments and impregnated by Canada balsam. After solidification the vertical and horizontal slices were

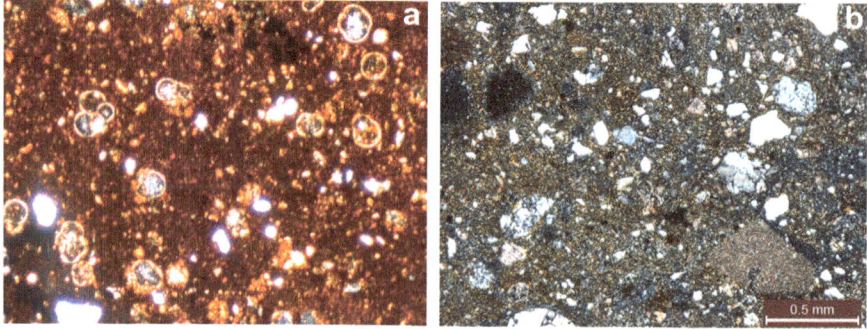

Fig. 11.8 Polarizing microscope. Nicoli X. Vertical thin sections: (**a**) Horizon 114 cmbsf, bulk of the core; (**b**) Anomalous horizon 175 cmbsf, high magnetization values

cut, grinded into thin sections and studied under binocular and polarized microscopes.

In thin section under microscope the bulk of the core can be described as a biogenic sediment with heavily ferruginous clay-carbonate matrix, saturated by fragments of quartz, plagioclase, carbonates and planktonic foraminifera. Iron hydroxides tint the clay-carbonate basic mass in yellow-brown colors and form a small spotted brown impregnation in cement and small halos around of some foraminifera (Fig. 11.8a).

Sediments of the marked anomalous intervals look differently, as illustrated by thin section of material at 175 cmbsf (Fig. 11.8b). Clastic fraction is completely unsorted neither by size, nor by degree of roundness; the color of this horizon is noticeably lighter and in general it has an appearance of tuffite; fauna fragments are remarkably rare, ash flyers are present, there are noticeably more sharply angular volcanic glass fragments up to 0.5 mm in size then in the bulk of the core. Also typical for these horizons are fragments of hornblende, ore minerals, iron hydroxides, glauconite, calcite, and, the most importantly, higher content of clinopyroxene.

This comprehensive analysis of core KD12-03-10c proves that inside the prevailing mass of the biogenic pelites there are thin tuffite-like layers enriched by volcanogenic material (coarse angular fragments of quartz, feldspars, volcanic glass and clinopyroxene) – an irrefutable testimony of the periodic explosive volcanic activity. The most pronounced of these layers is located at 170–175 cmbsf and dated at 1.09 Ma, which is very close to 1 Ma age derived from the time needed to create 10 km-wide rift valley at the bottom of the caldera with spreading rate of 1 cm/yr.

The caldera is asymmetrically located with regards to the Gakkel Ridge as clearly demonstrated by MCS line 2014–05 crossing the entire width of the Gakkel Ridge from flank to flank (Fig. 11.5).

Thus, we can conclude that the Eurasian Basin of the Arctic Ocean in the Pleistocene was a scene for a unique and powerful volcanic eruption with huge volumes of ejected material. Apparently, this eruption dated at ~1.1 Ma was not the only powerful eruption in the Arctic Basin during the Pliocene and Pleistocene.

The presence of large volcanic structures was found earlier in the western part of the Gakkel Ridge during the expedition AMORE 2001 (Thiede et al. 2002; Jokat and Schmidt-Aursch 2007) Ultraslow-spreading ridges have a unique feature - amagmatic rifts that expose mantle peridotite (with only traces of basalt and gabbro) directly on the seafloor thus forming a new (fourth) type of plate boundaries (Snow and Edmonds 2007). These discoveries raise new questions about violent processes in ultra-slow-spreading systems (Sohn et al. 2008). We consider the described caldera as an evidence of some new form of volcanism related to this type of plate boundary. Probably it was the most powerful volcanic eruption which left the significant marks on the topography and sedimentation within the Arctic Ocean. It also might affect the recent (Pleistocene) spreading geometry of the eastern part of the Eurasian Basin by triggering a jump of the Gakkel Ridge spreading axis which, instead of typical central position, is located in its south-western flank (Fig. 11.5).

The size of caldera, which we suggest to call "Gakkel Caldera" due to its location, put it in category of supervolcanoes with Volcanic Explosive Index 8. This category includes volcanoes with eruptive release of more than 1000 km^3 of lava and volcanic ash. In this case, the latter contaminated not only atmosphere but hydrosphere as well, as proven by presence of volcanic material in sediments thousands of kilometers away from the volcanic sources. The scope and duration of climatic changes related to the catastrophic Pleistocene volcanic events and their impact on all living organisms including humans are still under discussion, as a debate on ecological impact of the Toba eruption 75 Ka may attest (Lane et al. 2013).

The tectonic position of Gakkel Caldera is a remarkably interesting feature. As far as we know, Gakkel Caldera is a unique example of a supervolcano formed in the rift zone of a mid-ocean ridge. All other known supervolcanoes are located above subduction zones or in the immediate vicinity as that creates the favorable conditions for generating giant magma chambers. It should be noted that, in fact, Gakkel Caldera is located at the termination of the mid-ocean ridge where its rift valley turns to a shallow graben several hundred meters deep verging towards the Laptev Sea shelf. Faults forming the graben's walls are traceable deep into the thick sedimentary sequence with the Cretaceous sediments at its base, according to (Kim and Ivanova 2000, Piskarev et al. 2017).

Previously we mentioned that the magnetic anomalies in this region are based on 50 years old surveys with notoriously low accuracy. Nevertheless, even on these unreliable maps one can draw a line separating the central part of the Eurasian Basin where magnetic anomalies form a distinct linear pattern, from the south-eastern part, close to the Laptev Sea, where the pattern is drastically different. It was earlier suggested (Piskarev 2004) that this line represents the tectonic suture of the first order cross-linking the Cenozoic and Mesozoic ocean floor (Fig. 3.2, 3.36). Now we may suggest that this suture might act as a magma conduit for the volcanic episode leading to explosive birth of the caldera, geographically associated with it. However, geological and geophysical data do not allow precise definition of the nature of this zone so far. Evidently, Gakkel Caldera appears to be one of its characteristic features.

11.3   Hydrocarbon Molecular Markers as Indicators of Late Cenozoic Sedimentation in the Amerasian Continental Margin

The sediments of the abyssal Arctic Ocean as a terminal sedimentation basin contain the stratified material allowing to estimate the contribution of various sources (river run-off, glacial transport, turbidity flows, oceanic slope contouring currents, subaqueous erosion, and re-deposition of bedrocks) to the formation of bulk sediments.

According to (Yamamoto and Polyak 2009; Stein et al. 2009), composition of the dispersed organic matter (DOM) of the abyssal sediments in the Amerasian continental margins is determined by two principal sources:

- aquatic transport of terrigenous humic organic matter (OM) during glacial periods;
- aquatic transport of terrigenous humic OM and ice/icebergs transport of rocks containing a thermally mature (lithic) OM during deglaciations.

Subaqueous erosion and re-deposition of bedrock, usually buried under considerable volume of the Meso - Cenozoic sedimentary sequences, are considered to play a negligible role in the formation of seafloor sedimentary cover. However, the seismic data from the HEALY expedition demonstrate an elevation of the acoustic basement up to the seafloor surface (Bruvoll et al. 2012). An MCS survey, run on the southeastern slope of the Shamshura seamount and western and eastern slopes of the Trukshin seamount during the Arctic-2012 expedition has also registered the tectonic uplift of the bedrock to the seafloor surface (Gusev et al. 2014; Gusev et al. 2017).

Here we present our investigation of composition and distribution of DOM as indicators of sources and accumulation conditions of the bottom sediments in the Amerasian continental margin.

11.3.1   Material and Methods

We used sediment cores collected from two meridional transects approximately along 169°E and 180°E between 76° N and 85° N during *RV* Akademik Fedorov (2000, 2005, 2007, 2008) and icebreaker "Kapitan Dranitsin" (2012) expeditions (Fig. 11.9). Marine sediments were sampled from 6 m gravity cores with plastic inserts into sterile boxes and kept at −18 °C. Analysis of DOM included determination of the elemental (total organic carbon – TOC, carbonate carbon – C_{carb}, organic nitrogen – ON) and group composition of bitumoid, chromatographic separation of fractions of saturated and aromatic hydrocarbons (HCs), and their GC-MS study on a Hewlett Packard 5973/6850 setup with a quadrupole mass detector and a software for processing analytical data (Petrova et al. 2010, 2017).

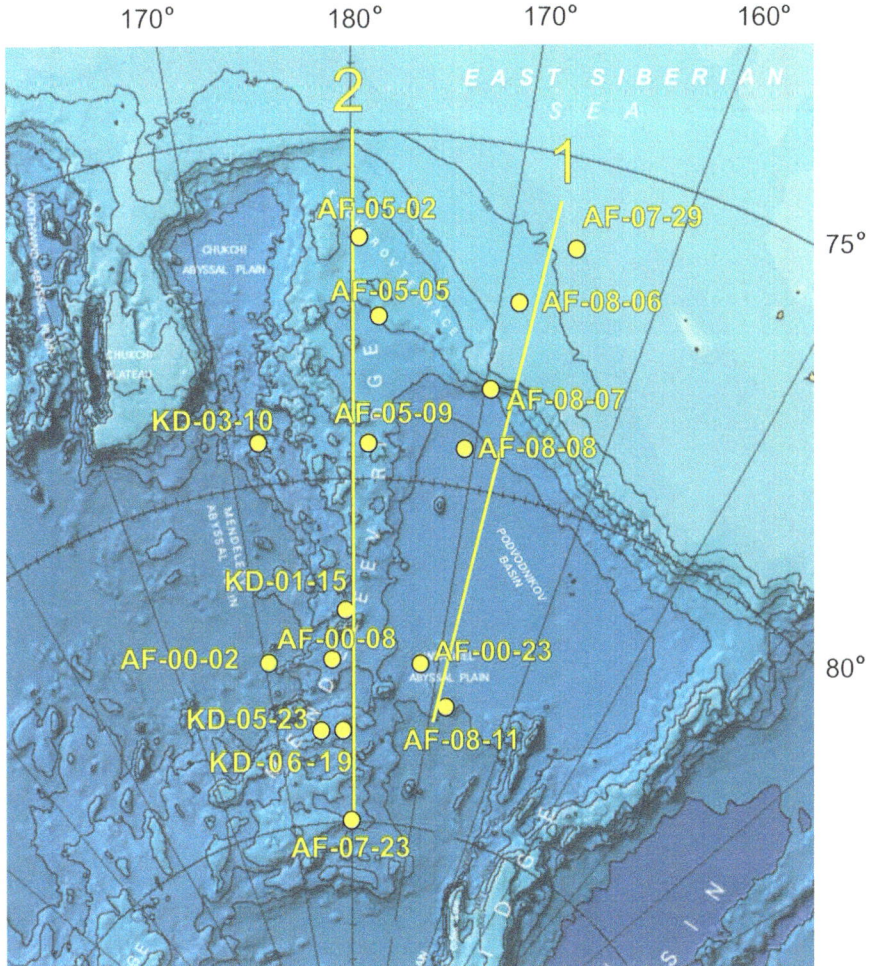

Fig. 11.9 Coring locations over bathymetry

11.3.2 Results and Discussions

Transect 1 starts over the outer shelf with cores AF-07-29, AF-08-06. Core AF-08-07 is on the upper continental slope, AF-08-08 – upper bench of the Podvodnikov Basin. The last core (AF-08-11) was taken at its northern end (The Arctic Basin (Geology and Geomorphology) 2017). Prevailing sediments in the cores (up to 3 m long) are none-carbonate aleuropelites with low organic carbon content (Corg = 0.3–0.8%) (Fig. 11.10). The highest TOC content is found in core AF-07-29, the lowest – in the abyssal sediments of AF-08-11, which also contains variable amounts of C_{carb} (0.05–1.34%) not detected in cores from the continental slope and rise. DOM group composition points out to the considerably high level of

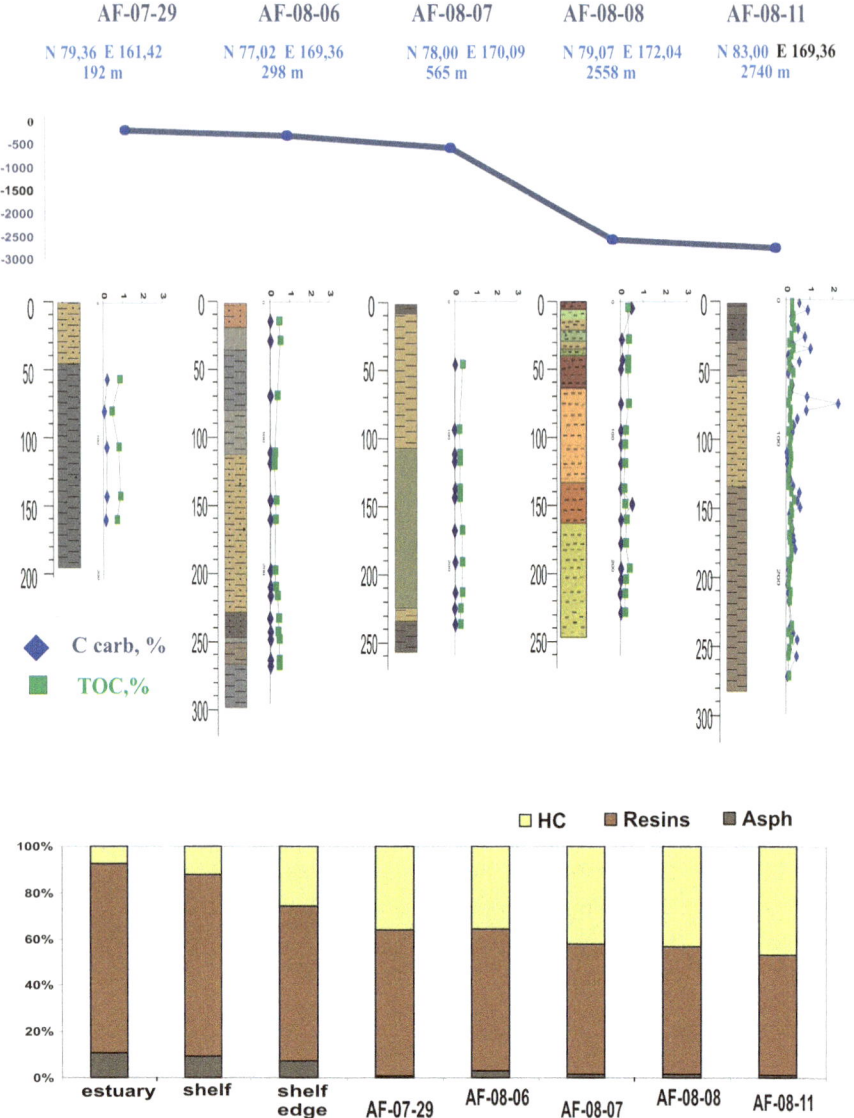

Fig. 11.10 Lithologic and geochemical composition of sediments along Transect 1

its transformation (residual organic matter (ROM) > 90%). The highest level of OM transformation (ROM ca. 97.2%) is detected in the southern part of the profile (core AF-07-29 from the upper part of the continental slope). However, this unexpected result agrees with the previously established the middle Pleistocene age of sediments from this section (~172 Ka (Gusev et al. 2013)) and determines the late diagenetic stage of DOM transformation.

It can be noted that the general tendency in change of composition of the soluble DOM from a shelf to a pelagic zone is similar to that of the World Ocean as a whole (Romankevitch 1977) and the Arctic Basin in particular (Romankevich 1982). Namely, the concentration of polar components (resins and ashpaltens) falls and concentration of the HC rises reflecting the general direction of diagenetic and post-diagenetic transformations of DOM (Fig. 11.10).

Transect 2 starts at the shelf - Kucherov Terrace boundary and then follows the Mendeleev Ridge (Fig. 11.11), which, geo-morphologically, can be described as a system of stepping down benches (Naryshkin 1995; Kaban'kov et al. 2004, 2008).

The cores consist mainly of clays and silty-clays with irregularly spaced inclusions of sands and gravels. Fine-grained sediments from the continental slope (AF-05-02, AF-05-05) are predominantly carbonate-free ($< 0.1\%$) with low TOC ($<0,5\%$, Fig. 11.11). The average C_{carb} values increase up to 0,6% and vary from 0.04% to 1.75% through the 6 m core from the southern slope of the Mendeleev Ridge (AF-05-09). But the correlation between the high content of C_{carb} and the coarse-grained fraction (>63 µm) – Fig. 11.11 is detected only in the lower part of the core, attesting to insignificant influence of glacial transport. Moreover, the "pink" layers suggested by Stein et al. (2009) as markers of the drifting carbonate material input are absent in this core. However, according to the named paper, which deals with the transect along 77° N across the Mendeleev Ridge, the pink layers were detected in sediments collected from its eastern slope (PS72/340-5) and to the north at 80°N. Comparative study of C_{carb} and coarse-grained fraction distribution in sediments from the western slope base (AF-00-23), central part (AF-00-08) and eastern slope base (AF-00-02) of the Mendeleev Ridge demonstrates their strong correlation in the central and eastern parts of the transect. C_{carb} content is drastically reduced and does not correlate with the coarse-grain fraction in the Podvodnikov Basin.

Correlation of the maximum C_{carb} (up to 4%) with the "pink" layers are detected in the sediment cores from the top of the Shamshura seamount (AF-00-08, KD12-01-15c) and from the eastern slope of the Mendeleev Ridge to the southeast of the Pochtarev Plateau (KD12-03-10c, Fig. 11.11. (Gusev et al. 2013)). Correlation between C_{carb} and coarse-grained fraction content is better distinguishable in core AF-00-08 (Fig. 11.11) than in KD12-03-10c. Moreover there is a significant discrepancy between the low content of coarse-grained fraction ($‘$ 10%) and the high abundance of dolomites and fossil fragments, which drives up the C_{carb} level in the upper part of the core (Rekant et al. 2015). The efforts to correlate this sedimentary core with the previously documented cores (HLY0503-08JPC (Yamamoto and Polyak 2009); PS72/404-3 (Stein et al. 2009)) are unsuccessful due to the difference in stratification descriptions of these neighboring cores. The more recent paleontological and paleomagnetic investigations (Taldenkova et al. 2016; The Arctic Basin 2017) also have shown the different Neo-Pleistocene and Pliocene age of the KD12-03-10c sediments, repectively.

Detail description of two cores - KD12-05-23c from the eastern (30°–40°) slope of the Trukshin seamount, and KD12-06-19c – from the western (35°–50°), was given in (Gusev et al. 2014), mentioning fractured bedrocks outcropping at the sea bottom. The cores consist of carbonate-free silty-clays ($C_{carb} < 0.1\%$) with

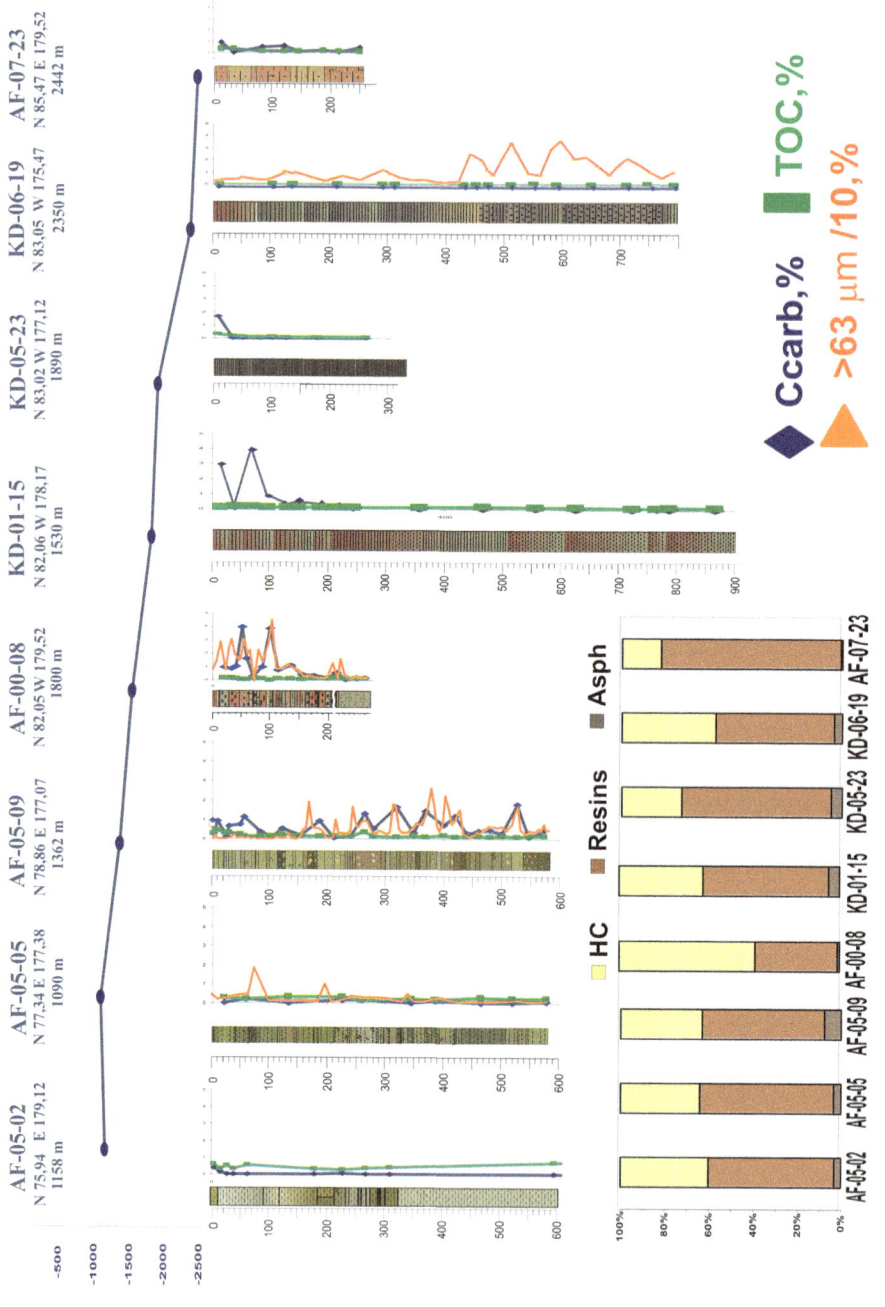

Fig. 11.11 Lithologic and geochemical composition of sediments along Transect 2

$C_{org} < 0.2\%$, with no apparent correlation between C_{carb} and coarse-grained fraction (> 63 μm). The group composition of DOM (ROM ≤ 98%) and its extractable part (bitumoid or extractable OM – EOM; Fig. 11.11) indicates considerable degree of post-diagenetic transformation.

In contrast to Transect 1 cores, the Transect 2 sediments do not display a pattern of regular changes in composition and distribution of DOM, typical for transition from shelf to abyssal sedimentation zone. Instead, every core has its own characteristics witnessing the substantial differences in source material and post-depositional transformations of accumulated sediments, and DOM molecular composition reflects this phenomena most clearly.

11.3.3 DOM Molecular Composition

Distribution of aliphatic HCs in sediment cores from Transect 1 attests to a mixed sapropelic-humic origin and significant level of DOM transformation, which has taken place in littoral and shallow-water marine environmental conditions (see Connan-Cassou diagram; Fig. 11.12).

n-Alkane profiles show a bimodal distribution, with predominance of low-molecular weight (LMW; C_{17}-C_{19}) and high-molecular weight (HMW; C_{25}-C_{31}) homologues, that is typical of mixed marine- terrigenous DOM. The content of terrigenous components gradually decreases (C_{17}/C_{27} from 0.33 to 0.96; Table 11.3) and DOM transformation level increases seawards that is corroborated by the decrease of CPI and OEP indices and relative increase of lithic (thermally matured) aliphatic HCs (Lithic/$\sum$Alk (Yamamoto and Polyak 2009)). An increase in LMW aquatic components content ($C_{17}/C_{27} = 0.96$) in core AF-08-11 is likely due to the drift or redeposited carbonates input. The similarity in *n*-alkane distribution in the upper and lower (up to 3 m) parts of the cores indicates stability of the Late Quaternary sedimentation.

Geochemical parameters of the Transect 2 sediments (along the Mendeleev Ridge) are specific and completely differ from Transect 1 *n*-alkanes and isoprenoid characteristics. Significant variations in origin, maturity degree and sedimentation conditions are reflected on the Connan-Cassou diagram as a scattered data field (Fig. 11.12). For example, sediments from the continental slope and southern part of the Mendeleev Ridge (AF-05-02, AF-05-05) have the highest concentration of aquatic components (Table 11.4).

Most prominent level of DOM transformation exhibit the sediment core from the eastern slope of the Mendeleev Ridge (KD12-03-10c). Low values of odd-even preference index (CPI = 1.07, OEP17–19 = 0.59, OEP27-31 = 1.47) are typical of post-diagenetic maturity stage of DOM and together with the high concentration of lithic *n*-alkanes attest to the significant contribution of sedimentary material from lithified rocks.

It should be noted that DOM of this particular core is rather specific. According to the Taldenkova et al. (2016), the stratigraphic scheme the lower part of the core

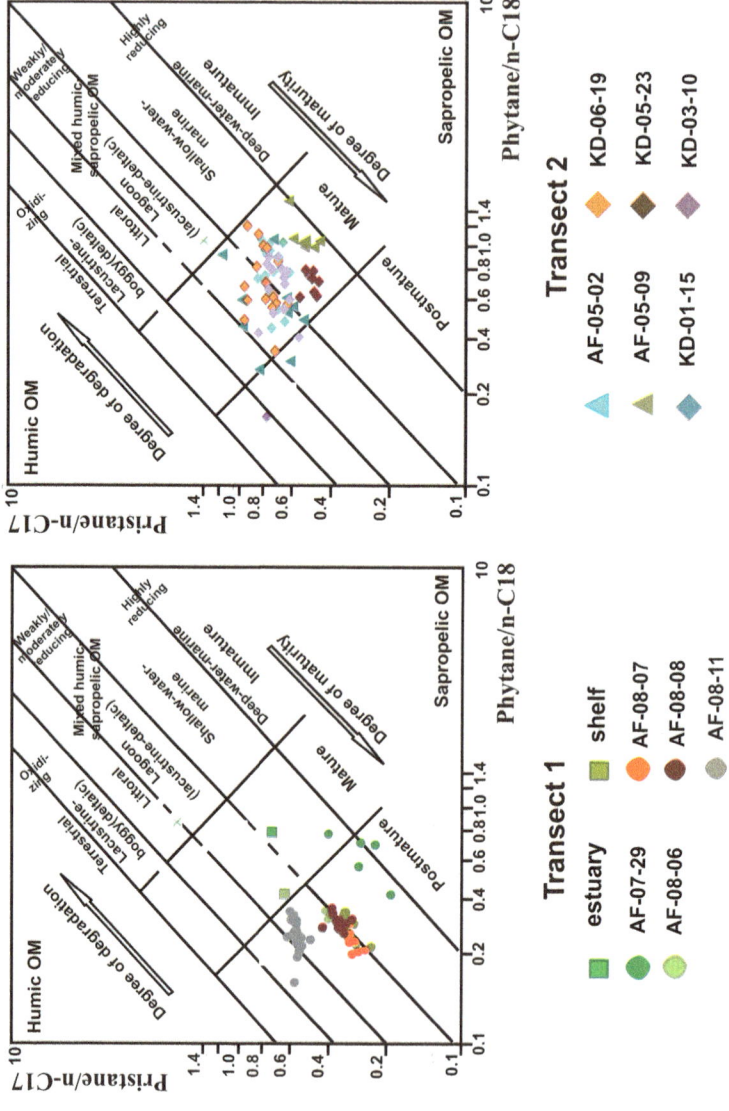

Fig. 11.12 Genetic types and facies of depositional environment

Table 11.3 *n*-Alkanes and isoprenoids in sediment cores from Transect 1

Core	CPI$_{23-33}$	C17/C27	OEP17–19	OEP27–31	Pr/Ph	K$_{iso}$	Lithic/$\sum$n-alk
AF-07-29	2.98	0.33	1.30	4.11	0.81	0.47	0.32
AF-08-06	2.39	0.67	0.81	4.65	0.89	0.32	0.22
AF-08-07	2.14	0.86	0.71	3.78	0.84	0.35	0.48
AF-08-08	2.38	0.58	0.74	4.70	0.95	0.37	0.39
AF-08-11	1.84	0.96	0.49	3.98	1.08	0.41	0.57

Table 11.4 *n*-Alkanes and isoprenoids in sediment cores from Transect 2

Core	CPI$_{23-33}$	C17/C27	OEP17–19	OEP27–31	Pr/Ph	K$_{iso}$	Lithic/$\sum$n-alk
AF-05-02	2.14	1.12	0.59	4.05	0.88	0.63	0.48
AF -05-05	1.94	1.49	1.03	3.36	0.82	0.64	0.42
AF -05-09	1.52	0.92	1.20	2.33	0.95	0.76	0.39
KD12-03-10c	1.07	0.88	0.59	1.47	0.90	0.63	0.89
KD12-01-15c	1.44	0.51	0.69	1.92	0.92	0.71	0.66
KD12 -05-23c	2.22	0.27	0.64	3.40	1.00	0.56	0.44
KD12 -06-19c	2.09	0.48	1.16	3.25	0.76	0.79	0.50

(450–570 cm) is older than MIS 16 and was deposited under moderate climate conditions with the seasonal ice cover. Interval 350–450 cm corresponds to MIS 16–13 and reflects an active ice transport of shelf material by Transpolar Drift Stream. The upper part of the section (MIS 12–1) corresponds to the extensive development of the ice cover and carbonates transport by the Beaufort Gyre.

Sedimentation rates obtained during the more recent paleomagnetic studies (The Arctic Basin 2017) are much slower than the previous data ($\leq$ 1.5 mm/kyr) and assess the age of the lower part of the core KD12-03-10c as 3.58 Ma (Pliocene). The three intervals of direct and inverse polarities are identified: 531–394.5 cm, 394.5–123.5 cm and < 123.5 cm, where 394 cm is the Matuyama/Gauss magnetic polarity boundary, and 123.5 cm – the Brunhes/Matuyama boundary.

Changes of *n*-alkanes and isoprenoids distribution during the detected sedimentation periods allow define three distinct units in core KD12-03-10c (Fig. 11.13):

- Sediments from the lower part of the core (>350 cm) contain DOM of humic-sapropelic origin with the uniform composition and level of OM transformation (C17/C27 = 0.9–1.2; CPI = 1.3–1.6; Litic/$\sum$ = 0.3–0.5). The high input of the slightly transformed OM enriched with the aquatic OM agrees with the supposed climatic optimum.
- An increase in thermally mature DOM content in the middle part of the section (<350 cm) may be due to both terrigenous discharge of the bedrock abrasion products and ice transport (CPI = 0.8–0.9; Lithic/$\sum$ = 0.8–0.9).

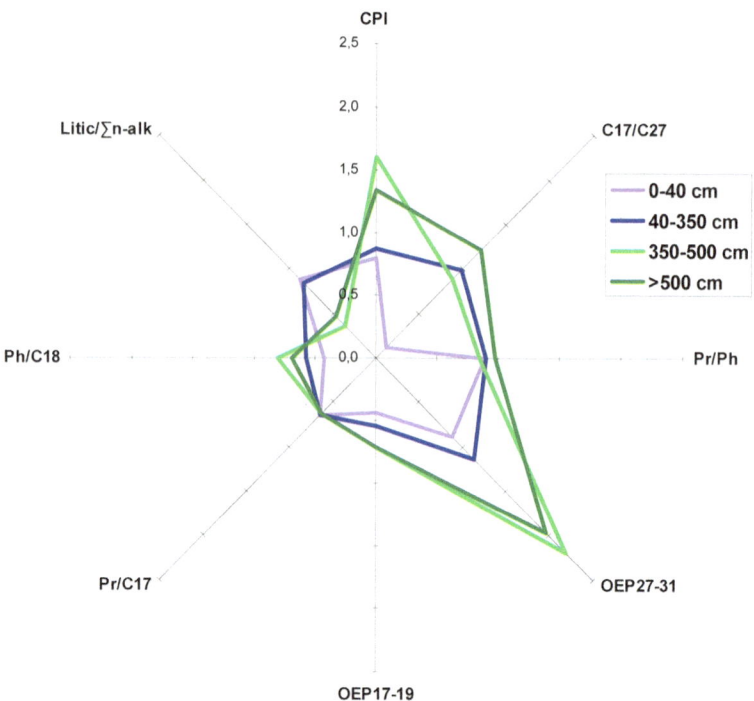

Fig. 11.13 Geochemical parameters of *n*-alkanes and isoprenoids in sediment core KD12-03-10c

- The upper section of the core (≤ 40 cm) shows the most surprising results due to the high degradation level of the low molecular weight *n*-alkanes (C17/C27 = 0.13) and the anomalous predominance of the even-chain components (OEP17-19 = 0.44).

The revealed specificity of *n*-alkanes distribution at the 0–40 cm interval may indicate the intensive bacterial transformation of DOM (Peters et al. 2004). This agrees with the geomorphological position of the station KD12-03-10c located in the NW oriented structural depression on the eastern slope of the Mendeleev Ridge (The Arctic Basin 2017) and confined to the fault zone with a highly probable hydrothermal activity. The iron-manganese concretions which are unusual for the abyssal Arctic were also discovered on the slope. The detailed analysis carried out by the group of Konstantinova et al. (2016) has demonstrated the hydrothermal origin of the inner part of the concretions and aquatic genesis of their surface layers. The obviously authigenic carbonate crusts which were detected by Pachalko et al. (2017) also evidence the hydrothermal activity in the area. The possible source of the hydrothermal fluid can be located within the normal large-amplitude discharge indicated on the HLY 0521 seismic profile (Gusev et al. 2014).

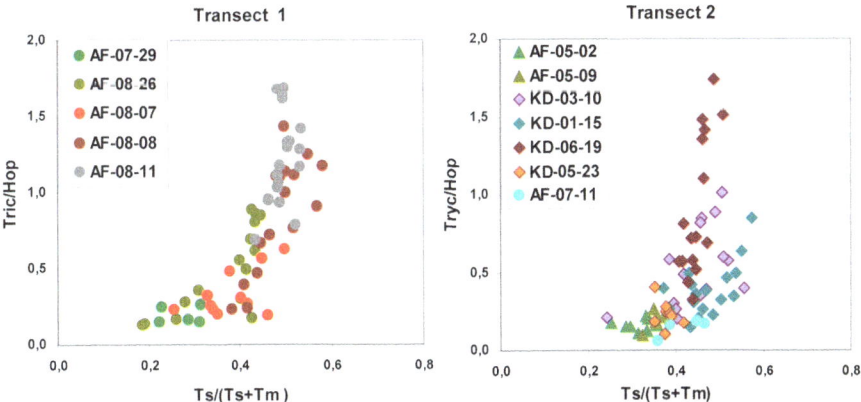

Fig. 11.14 Comparison of maturity parameters of hopanes in the DOM of sediments

Abnormal concentration of terrigeneous components with the decreased transformation degree (which is nevertheless higher than the values in the Podvodnikov
Basin – Transect 1, C17/C27 = 0.3–0.5) is detected in cores from Shamshura seamount in the northern part of the Mendeleev Ridge.

Analysis of triterpanes and steranes can be used as an additional evaluation tool
of the DOM thermal maturity and its origin. Significant influence of the transformed
geological material on the DOM composition in sediments of Transect 1 is reflected
in hopanes maturity index (Fig. 11.14). An increase of tricyclanes content (Tric/
Hop) northward, which may indicate an input of eucaryotic OM (Kontorovich et al.
2009), is unlikely due to the general seaward decline of humic OM content.
Tricyclans are known to be more thermodynamically stable and more persistent
during DOM transformations than hopanes (Peters et al. 2004). There is a linear
correlation between the tricyclic ratio and values of the Ts/(Ts + Tm) index (up to
0.3) which is typical of the catagenetic stage close to the main stage of the oil and
gas generation (Phomin 2011) and confirms the high level of DOM transformation.
The similar values of the DOM thermal maturity, which are detected in sediments
of Transect 2, does not show the same northward increase most likely due to the
difference in origin and composition of sedimentary material (the triterpanes maturity index for sapropelic OM is lower than that for terrigenous OM (Phomin 2011)).

Steranes maturity parameters are the indicators of OM catagenetic transformation and include epimeric ratios of bio- (αααR) and more thermodynamically stable
geo-structures (αββR and αββS) (Peters et al. 2004). Their values are limited by the
thermodynamic threshold of isomerization which coincides with the catagenetic
main stage of the oil and gas generation. Correlation between the different maturity
parameters of steranes of the Transect 1 sediments are shown on the Fig. 11.15a.
Sediments from the southern (AF-07-29) and northern (AF-08-11) part of the transect exhibit the highest degree of DOM transformation consistent with the values

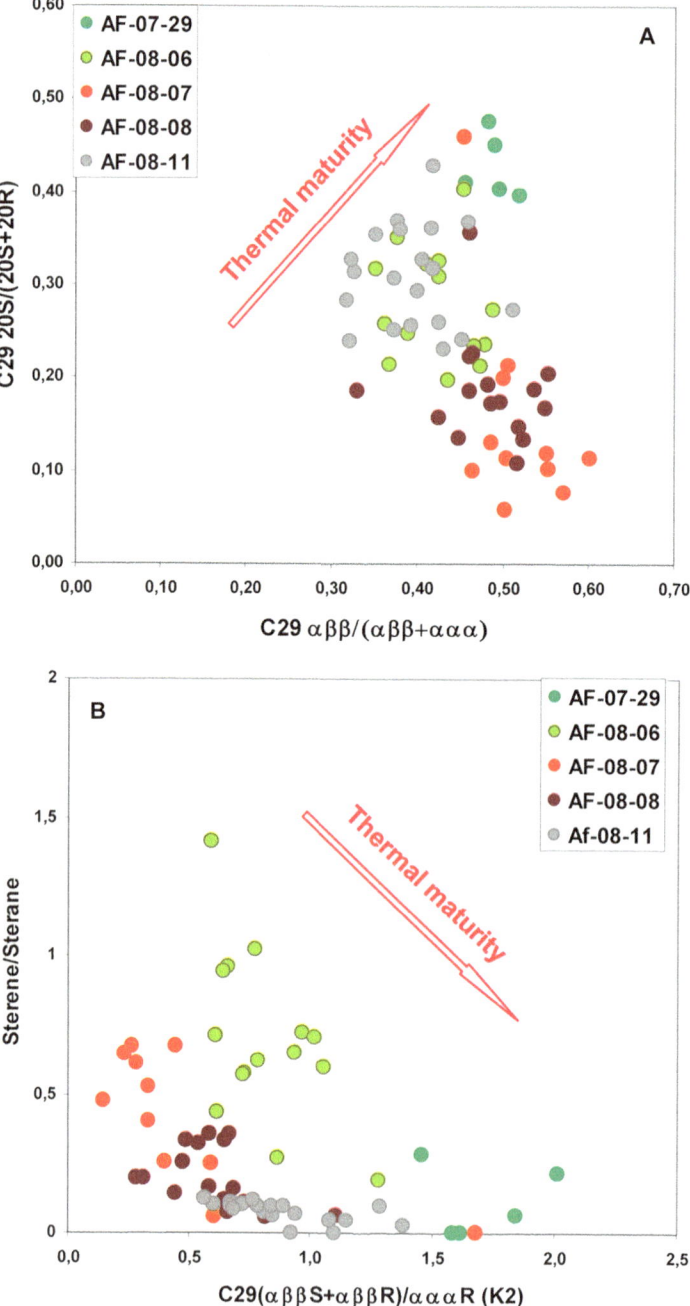

Fig. 11.15 Comparison of maturity parameters of steranes in the DOM of sediments

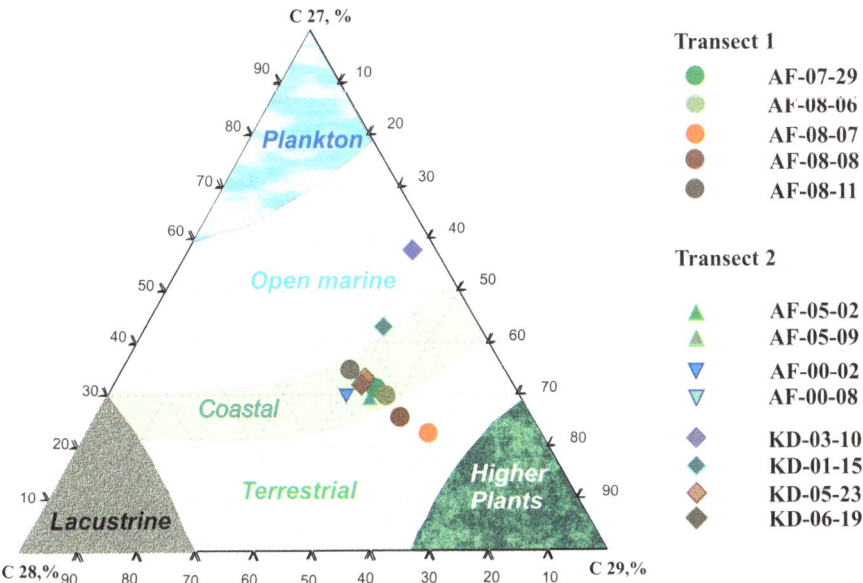

Fig. 11.16 Ternary diagram showing relative abundances of regular steranes and facial conditions of sedimentary DOM formation

obtained for the other groups of molecular markers. The ratio between the geologically stable forms of steranes and their biological precursors from higher land plants – sterenes (oleanen and friedooleanen) can also be used as a maturity parameter, which traces the input of modern terrigenous flow and shows the transformation degree of the humic OM. The seaward decrease of the sterenes relative content (Fig. 11.15b) indicates both the decline of the humic material input and the increase of OM transformation degree. This also agrees with the sterane maturity index K2.

Trace concentrations of biogenic epimers of steranes in sediments of Transect 2 hinder the estimation of the maturity degree but not affect the definition of the DOM genesis.

The facial conditions of sedimentation and DOM formation can be distinguished according to the relative concentration of steranes C27-C29 – markers of sapropelic and humic OM (Fig. 11.16). The ternary diagram demonstrates that Transect 1 sediments were accumulated mostly in coastal and shallow-marine conditions, and only the northern part of the Transect gets closer to open marine conditions. The sediments from the continental slope and the southern part of the Mendeleev Ridge of Transect 2 also exhibit predominantly coastal accumulation conditions.

However, the steranes composition in sediments from the eastern slope of the Mendeleev Ridge (KD12-03-10c) indicates the aquatic nature of DOM. This agrees with the characteristics of the other HC markers (*n*-alkanes, isoprenoids, terpanes) and emphasizes the specific characteristics of DOM of this sedimentary section.

The Fig. 11.17 demonstrates the input of procariotic and eucariotic biomarkers and attests to the predominantly terrigenous nature of DOM of Transect 1 due to the

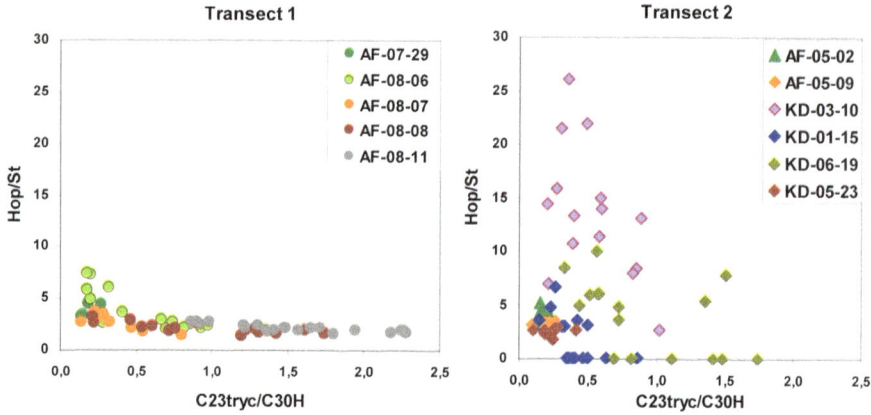

Fig. 11.17 Comparison of cyclanes indexes as indicators of DOM origin

increase of steranes and C23tricyclic terpane concentration. The distribution of these molecular markers in sediments of Transect 2 reflects significant variation in the origin of DOM. Thus, the extremely high content of procariotic molecular markers (Hop/St up to 27) in core KD12-03-10c is likely due to the significant contribution of the bacterial OM and agrees with the previously supposed hydrothermal activity in the region (Pachalko et al. 2017).

And, finally, the degree of DOM thermal maturity can be estimated using the ratios of aromatic compounds with different thermodynamic stability limits. For instance, the methylphenanthrene index (MPI) is based on ratio of phenanthrene and isomers of its monomethyl homologues (Radke 1988, Peters et al. 2004). The distribution of these compounds varies according to the DOM catagenetic maturity degree in the wide range of temperatures and has a linear correlation with the vitrinite reflectance (Phomin 2011).

MPI values vary between 0.2–0.6 in sediments of Transect 1 indicating the DOM transformation stage from the early catagenetic up to the main stage of the oil and gas formation (Fig. 11.18). The lowest MPI values are detected in northernmost core AF-08-11. The high content of carbonates detected in the mineral composition of sediments of this core may suppress the index value (Peters et al. 2004). The maximum thermal maturity degree of the DOM that agrees with the previously defined age of sediments is detected in the core from the upper part of the continental slope (AF-08-06).

The ratio of phenanthrene to its tetraalkylated homologue – retene (Ret/(Ret+Pn)) defines the process of aromatization and dealkalation of abietic acid (biogenetic precursor of phenanthrene) during the diagenesis and early catagenesis (Bastow et al. 2001). Furthermore retene also indicates the input of the land plant gymnosperms. Seaward decline of this parameter is shown on the diagram (Fig. 11.18).

The drastic variations in the polycyclic aromatic HC (PAH) characteristics in the Mendeleev Ridge sediments (MPI = 0.2–0.8; Ret/(Ret+Ph) = 0.01–1.0) attest to the

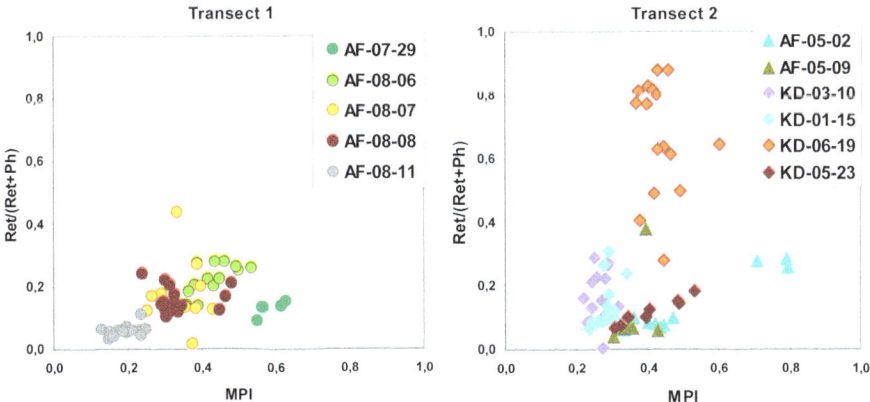

Fig. 11.18 Polycyclic aromatic HCs in the DOM of sediments from transects 1 and 2

wide diversity of the sediment sources and DOM nature. The most specific composition of PAHs is detected in the core KD12-06-19c, where an increased concentration of eucariotic molecular markers (C23 tricyclicT/30H up to 1.7) and a high concentration of retene (Ret/$\sum$PAH >0.5; Ret/(Ret+Pn) up to 0.9) is detected. Moreover, the correlation between retene distribution and the lithic n-alkanes C23-C33 – the components of thermally mature terrigenous DOM is revealed (Fig. 11.19).

The high concentration of retene as a molecular marker for terrigenous OM has been previously detected in the abyssal parts of the Arctic Ocean (Yamamoto et al. 2008; Petrova et al. 2013). But its association with the slightly transformed DOM of the gray clays and correlation with the other terrestrial markers (oleanenes, tetrahydrochrysenes, perylene), not detected in this core, makes unreliable the data comparison. The most probable analogue of the DOM molecular composition of core KD12-06-19c is the Cretaceous formation of Indigiro-Zyrianskaya depression in the northeastern Yakutia (Kashirtsev et al. 2012; Fig. 11.19). The position of the sediment core is on the side wall of the big underwater canyon with the Cretaceous unconformity located upper than the sampling point (Rekant et al. 2015). This suggests the significant influence of the processes of underwater weathering and re-deposition of the Cretaceous bedrock on sedimentation in this point.

The main results of our investigations are:

- The analisys of seventeen cores (273 samples) collected along the two meridional transects is carried out. The main geochemical parameters including the composition and distribution of HC molecular markers (n-alkanes, isoprenoids, terpanes, steranes, PAHs) are studied. The analysis of the main genetic and maturity parameters of DOM based on these molecular markers allow characterize the sources and sedimentation conditions of the region.
- The input of terrigenous sedimentary material enriched in abrasion products of the metamorphic rocks from the eastern source province determines the Holocene-Pleistocene sedimentation at the continental slope and Podvodnikov

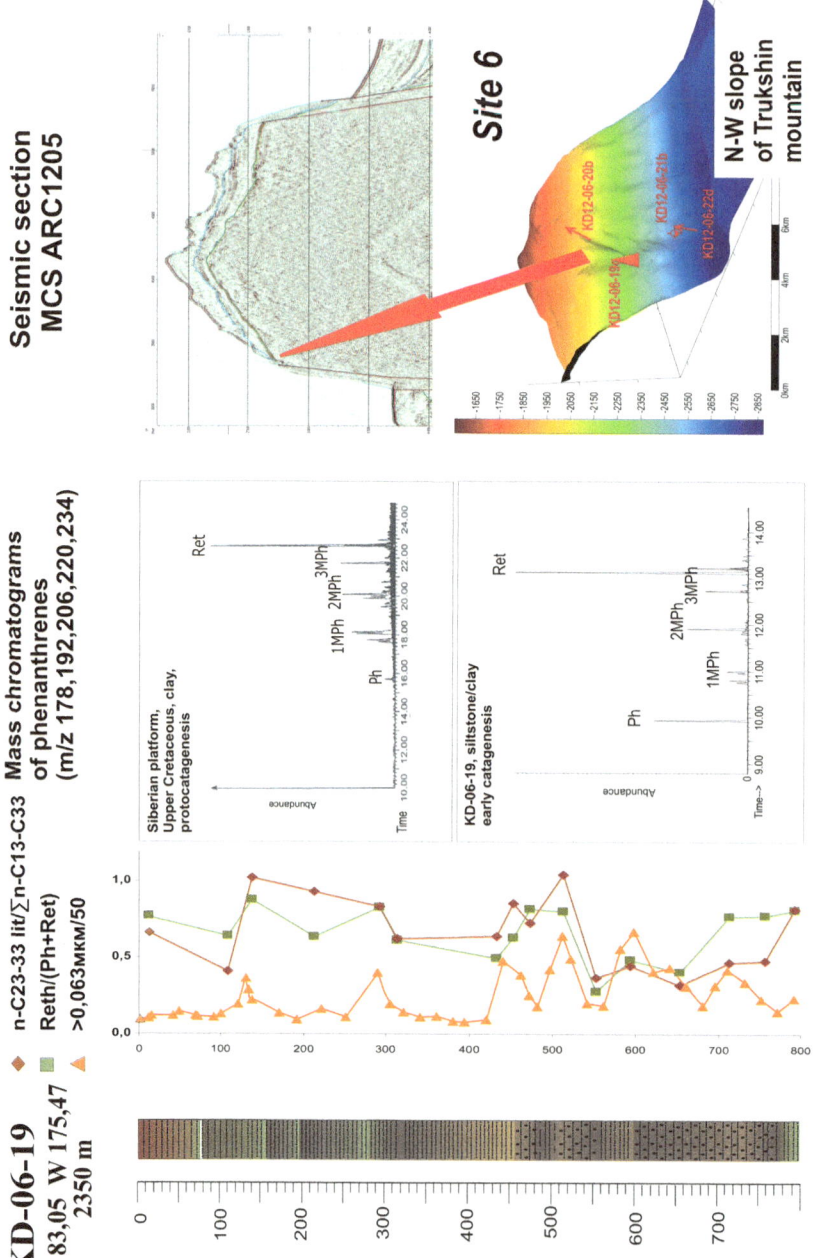

Fig. 11.19 Geomorphology, lithology and geochemistry of core KD12-06-19c

Basin (Transect 1). The revealed similarities between the main geochemical parameters and the high thermal maturity degree of the DOM corroborates this assumption
- Individual characteristics of the DOM of the Late Cenozoic sediments from the Mendeleev Ridge (Transect 2) reflect the variety of sedimentation conditions and sources. Moreover, the process of the pre-Holocene bedrock underwater denudation and redeposition should be considered of the same importance as the terrigenous flow and ice-transport in the region.

References

Adler RA, Polyak L, Ortiz JD et al (2009) Sediments record from the western Arctic Ocean with an improvement late quaternary age resolution: HOTRAX core HLY0503-8JPC, Mendeleev ridge. Glob Planet Chang J 68:18–29

Andreeva IA, Basov VA, Kupriyanova NV, Shilov VV (2007) Age and depositional environment of bottom sediments in the Mendeleev Rise (Arctic Ocean). Materials on the polar regions and the central part of the mid-Atlantic ridge in the Phanerozoic. Fauna, flora and biostratigraphy, vol 211. VNIIOkeangeologia, pp 131–152

Backman J, Moran K (2009) Expanding the Cenozoic paleoceanographic record in the Central Arctic Ocean : IODP Expedition 302 Synthesis. Cen Euro J Geosci 1(2):157–175

Backman J, Jakobsson M, Lovlie R et al (2004) Is the Central Arctic Ocean a sediment starved basin? Quat Sci Rev J 23:1435–1454

Bastow TP, Singh RK, van Aarssen BG et al (2001) 2-Methylretene in sedimentary material: a new higher plant biomarker. Org Geochem J 32:1211–1217

Brozena JM, Childers VA, Lawver LA et al (2003) New aerogeophysical study of the Eurasia Basin and Lomonosov ridge: implications for basin development. Geol J 31(9):825–828

Bruvoll V, Kristoffersen Y, Coakley B et al (2012) The nature of the acoustic basement on Mendeleev and northwestern alpha ridges, Arctic Ocean. Tectonophysics J 514:123–145

Buchs DM, Cukur D, Masago H et al (2015) Sediment flow routing during formation of forearc basins: constraints from integrated analysis of detrital pyroxenes and stratigraphy in the Kumano Basin. Jan J Earth Planet Sci Lett 414:164–175

Channell JET, Xuan C (2009) Self-reversal and apparent magnetic excursions in Arctic sediments. Earth Planet Sci Lett 284(1):124–131

Clark DL (1970) Magnetic reversals and sedimentation rates in the Arctic Ocean. Geol Soc Am Bull 81(10):3129–3124

Elkina D, Piskarev A (2017) Comparative paleomagnetic study of the Quaternary-Pliocene sedimentation rates in the Arctic Basin: first results, GP33B-0978, Presented at 2017 Fall Meeting, AGU, New Orleans, LA, 11–15 Dec

Frederichs T (1995) Regional and temporal variations of rock magnetic parameters in Arctic marine sediments. Ber zur Polarforsch J 164:1–212

Gee JS, Kent DV (2007) Source of oceanic magnetic anomalies and the geomagnetic polarity timescale. Treatise on Geophys J 5:455–507

Geocap 7.1.1 for Windows http://www.geocap.no/ (2015)

Glebovsky VY, Kaminsky VD, Minakov AN et al (2006) Formation of the Eurasia Basin in the Arctic Ocean as inferred from geohistorical analysis of the anomalous magnetic field. Geotectonics J 4:21–42

Gorbarenko SA et al (2002) Magnetostratigraphy and tephrochronology of the upper quaternary sediments in the Okhotsk Sea: implication of terrigenous, volcanogenic and biogenic matter supply. Mar Geol J 183:107–129

Gusev EA, Maksimov FE, Novikhina ES et al (2012) On stratigraphy of bottom sediments from Mendeleev rise (the Arctic Ocean) region. Vestnik Saint-Petersburg University 7(4):102–115

Gusev EA, Maksimov FE, Kuznetsov VY et al (2013) Stratigraphy of bottom sediments in the Mendeleev ridge area (Arctic Ocean). Dokl Earth Sci J 450:602–606

Gusev EA, Lukashenko RV, Popko AO et al (2014) New information on slope structure of the Mendeleev ridge seamounts (the Arctic Ocean). Dokl RAN J 455(2):184–188

Gusev E, Rekant P, Kaminsky V et al (2017) Morphology of seamounts at the Mendeleev rise, Arctic Ocean. Polar Res J 36:2–10

Jakobsson M, Lølie R, Al-Hanbali H et al (2000) Manganese and color cycles in Arctic Ocean sediments constrain Pleistocene chronology. Geol J 28:23–26

Jokat W, Schmidt-Aursch M (2007) Geophysical characteristics of the ultraslow spreading Gakkel ridge. Arctic Ocean Geophys J 168:983–998

Kashirtsev VA, Gayduck VV, Zueva IN (2012) Geochemistry of biomarkers and organic matter catagenesis of cretaceous and Cenozoic formations in the Indigir-Zyriyansk depression (northeastern Yakutyia). Geol Geofiz 53(8):1027–1039

Khain VE (2001) Tectonics of continents and oceans (year 2000). Scientific World, Moscow

Kim BI, Ivanova NM (2000) On the age of seismic units revealed on the Laptev Sea continental slope and adjacent part of the Eurasian Basin. Geological-geophysical features of the lithosphere of the Arctic. Region 3:82–92

Kochegura VV (1992) Implementation of Paleomagnetic methods for geological survey of a shelf. VSEGEI, Saint Petersburg

Konstantinova NP, Cherkashev GA, Novikov GV et al (2016) Ferro-manganese plaques of the Mendelelv Ridge: composition and genesis Artika. Ecologia i ekonomika 3(23):16–28

Kontorovich AE, Kashirtsev VA, Danilova VP et al (2009) Molecules-biomarkers in fossilizedn organic matter and naphtides of the Siberian PreCambrian and Phanerozoic formations, SPb, VNIGRI, p 108

Krylov AA, Shilov VV, Andreeva IA et al (2011) Stratigraphy and accumulation of upper quaternary sediments in the northern part of the Mendeleev rise (Amerasian Basin, Arctic Ocean). Probl Arctic Antarctic J 2(88):7–22

Lane CS, Chorn BT, Johnson TC (2013) Reply to Roberts et al.: A subdecadal record of paleoclimate around the Youngest Toba Tuff in Lake Malawi. Proceedings of the National Academy of Sciences 110(33):E3048. doi:https://doi.org/10.1073/pnas.1309815110

Levitan MA (2015) Sedimentation rates in the Arctic Ocean during the last five marine isotope stages. Oceanology J 55(3):425–433

Lin'kova TI (1984) Paleomagnetism of upper Cenozoic sediments of the World Ocean. Nauka, Moscow. (in Russian)

Moran K, Backman J, Brinkhuis H et al (2006) The Cenozoic palaeoenvironment of the Arctic Ocean. Nature J 441:601–606

Morozov AF, Petrov OV, Shokalsky SP et al (2013) New geological evidence justifying the nature of the continental area of the Central Arctic elevations. Reg Geol Metallogeny J 53:34–55

Naryshkin GD (1995) Orographic map of the Arctic basin (GUNiO), VNIIO, St. Petersburg

Not C, Hillaire-Marcel C (2010) Time constraints from [230]Th and [231]Pa data in late quaternary, low sedimentation rate sequence from the Arctic Ocean: an example from the northern Mendeleev ridge. Quat Sci Rev J 29:3665–3675

Nowaczyk NR, Frederichs TW, Kassens H et al (2001) Sedimentation rates in the Makarov Basin, Central Arctic Ocean: a paleomagnetic and rock magnetic approach. Paleoceanography J 16:368–389

Nowaczyk NR, Antonow M, Knies J et al (2003) Further rock magnetic and chronostratigraphic results on reversal excursions during the last 50 ka as derived from northern high latitudes and discrepancies in precise AMS14C dating. Geophys J 155:1065–1080

O'Regan M, King J, Backman J et al (2008) Constraints on the Pleistocene chronology of sediments from the Lomonosov Ridge. Paleoceanography J 23:PA1S19

Owens PN et al (2016) Fingerprinting and tracing the sources of soils and sediments: earth and ocean science, geoarchaeological, forensic, and human health applications. J Earth Sci Rev 162:1–23

Pachalko AG, Krylov AA, Mirolyubova ES et al (2017) First findings of Pleistocene Autogenetic Carbonate Plaques (ACP) on the Mendeleev Ridge, the Arctic Ocean. V International conference of young scientists and specialists in memory of academitian Karpinsky, VSEGEI, Saint Petersburg, 28 February – 3 March 2017

Peters K, Walters C, Moldowan J (2004) The biomarker guide. Cambridge University Press, p 364

Petrova VI, Batova GI, Kursheva AV et al (2010) Geochemistry of the bottom sediments organic matter in Central Arctic Uplifts Province, the Arctic Ocean. Geologiya i geofizika J 51(1):113–125

Petrova VI, Batova GI, Litvinenko I et al (2013) Organic matter in the Lomonosov Ridge Holocene-Pleistocene bottom sediments – biomarkers record 26th International Meeting on Organic Geochemistry (IMOG), 15–20 September 2013, 2:275–276

Petrova VI, Batova GI, Kursheva AV et al (2017) Molecular geochemistry of the north-eastern Barents Sea Triassic formations and affects of tectonics and magmatizm. Geologiya i geofizika J 58(3):398–409

Phomin AN (2011) Catagenesis of organic matter and petroleum potential of Mesozoic and Paleozoic deposits of the western Siberian Megabasin. INGG SO RAN, Novosibirsk. (in Russian)

Piskarev AL (2004) The basement structure of the Eurasia Basin and central ridges in the Arctic Ocean. Geotektonics J 38(6):443–448

Piskarev AL, Elkina DV (2014) Pliocene quaternary sediment accumulation rate at Mendeleev high, the Arctic Ocean from paleomagnetic data on bottom sediment columns. Karotazhnik 5:3–16

Piskarev AL, Andreeva IA, Gus'kova EG (2013) Paleomagnetic data on sedimentation rate in the Mendeleev rise region (Arctic Ocean). Oceanography J 53(5):694–704

Piskarev AL, Kireev AA, Poselov VA et al (2017) Areas of Pre-Cenozoic Basement in the Eurasian Basin (Arctic Ocean). 79th EAGE Conference and Exhibition 2017. doi: https://doi.org/10.3997/2214-4609.201701311

Polyak L, Bischof J, Ortiz JD et al (2009) Late quaternary stratigraphy and sedimentation patterns in the western Arctic Ocean. Glob Planet Chang 68:5–17

Pospelova GA (2004) Geomagnetic excursus in brief history and current conditions of geomagnetic studies carried out in the Institute of Earth's physics, Russian Academy of Sciences. Inst Fiz Zemli RAN, Moscow pp 44–55

Radke M (1988) Application of aromatic compounds as maturity indicators in source rocks and crude oils. Mar Pet Geol J 5:224–236

Rekant PV, Petrov OV, Kashubin SN et al (2015) History of formation of the sedimentary cover of Arctic basin. Multychannel seismic approach. Reg Geol Metallogeny 64:11–27

Romankevich EA (ed) (1982) Arctic seas – biogeochemistry of organic matter. Nauka, Moscow

Romankevitch EA (1977) Geochemistry of oceanic organic matter. Nauka, Moscow

Schneider DA, Backman J, Curry WB et al (1996) Paleomagnetic constraints on sedimentation rates in eastern Arctic Ocean. Quat Res J 46:62–71

Snow JE, Edmonds HN (2007) Ultraslow-spreading ridges. Rapid paradigm changes. Oceanography J 20(1):90–101

Sohn RA, Willis C, Humphris S et al (2008) Explosive volcanism on the ultraslow-spreading Gakkel ridge, Arctic Ocean. Nature J 453:1236–1238

Spielhagen RF, Baumann K-H, Erlenkeuser H et al (2004) Arctic Ocean deep-sea record of northern Eurasian ice sheet history. Quat Sci Rev J 23(11–13):1455–1483

Stein R, Matthiessen J, Niessen F (2009) Towards a better (litho-) stratigraphy and reconstruction of quaternary paleoenvironment in the Amerasian Basin (Arctic Ocean). Polarforschung J 79(2):97–121

Stein R, Mattheissen J, Niessen F et al (2010) Towards a better (Litho-) stratigraphy and reconstruction of quaternary Paleoenvironment in the Amerasian Basin (Arctic Ocean). Polarforschung J 79:97–121

Stein R (2015) The expedition PS87 of the research vessel polarstern to the Arctic Ocean in 2014. Berichte zur Polar-und Meeresforschung = Reports on polar and marine research, vol 688. Alfred Wegener Institute for Polar and Marine Research, Bremerhaven, p 273

Steuerwald BA, Clark DL, Andrew JA (1968) Magnetic stratigraphy and faunal patterns in Arctic Ocean sediments. Earth Planet Sci Lett 5:79–85

Taldenkova EE, Nikolaev CD, Stepanova AY et al (2016) Stratigraphy and paleogeograaphy of the Arctic Amerasian Basin in neo-Pleistocene in light of lithology and paleontology information. Vestnik Moskovskogo Universiteta (ser. 5). Aust Geogr 6:3–17

The Arctic Basin (Geology and Geomorphology) (2017) Kaminsky VD (ed) VNIIOkeangeologia, Saint Petersburg

Thiede J, the Shipboard Scientific Party (2002) Polarstern Arktis XVII/2 cruise report: AMORE 2001 (Arctic Mid-Ocean ridge expedition). Report polar mar. Res, vol 421

Tretyak AN, Vigilyanskaya LI, Dudkin VP et al (1989) Fine structure of geomagnetic field in late Cainozoe. Naukova Dumka, Kiev

Vishnyakov A, Piskarev A, Cherkashev G et al (1992) Detailed mapping of deep water bottom sediments by towing geophysical complex. Vestn Ross Akad Nauk 324(1):77–80

Kaban'kov VY, Andreeva IA, Ivanov VN (2004) On bottom sediments sampled on geo-traverse "Arktika-2000" in the Arctic Ocean (the Mendeleev ridge region). Dokl RAN J 399(2):224–226

Kaban'kov VY, Andreeva IA, Krupskaya VV et al (2008) New information on on composition and origin of seafloor sediments in the southern part of the Mendeleev ridge (the Arctic Ocean). Dokl RAN J 419(5):653–655

Witte WK, Kent DV (1988) Revised magnetostratigraphies confirm low sedimentation rates in Arctic Ocean cores. Quat Res J 29:43–53

Xuan C, Channell JET (2010) Origin of apparent magnetic excursions in deep-sea sediments from Mendeleev-alpha ridge, Arctic Ocean. Geochem Geophys Geosyst 11(2)

Yamamoto M, Polyak L (2009) Changes in terrestrial organic matter input to the Mendeleev ridge, Arctic Ocean during the late quaternary. Glob Planet Change 68:30–37

Yamamoto M, Okino T, Sugisaki S et al (2008) Late Pleistocene changes in terrestrial biomarkers in sediments from the Central Arctic Ocean. Org Geochem 39(6):754–763

Zhamoida AI et al (2000) Supplements to the stratigraphic code of Russia. VSEGEI, Saint Petersburg

Chapter 12
General Features of the Crust and the Sedimentary Cover

Sergey M. Zholondz, Andrey A. Chernykh, Vladimir Yu. Glebovsky,
Natalia E. Leonova, Anatoly D Pavlenkin, Victor A. Poselov,
and Lidia G. Poselova

Abstract The wide spectrum of the geophysical information was used for compilation of the "Sediment Thickness Map of the Arctic Ocean". The sedimentary cover includes all un-deformed Meso-Cenozoic (abyssal Arctic Ocean) or Paleo-Cenozoic (shelves) coilogenic stratified sedimentary complexes from the seafloor to the top of the acoustic basement (TAB) was presented at this Map. The belt of deep depressions located either inside shelves or in continental margin-shelf transition zones ringing the abyssal part of the Arctic Ocean, is clearly visible on the Map.

The demonstrated contemporary crustal thickness map of the Arctic Ocean was compiled using not only existing seismic information and bathymetry but also regional digital gravity dataset. The map vividly illustrates the radical difference in crustal thickness between clearly oceanic regions of the Arctic Ocean (Eurasian and Canada Basins) and remaining parts with continental or transitional types of crust.

Keywords Arctic Basin · Sedimentary cover · Earth's crust

S. M. Zholondz (✉) · V. Y. Glebovsky · N. E. Leonova · A. D. Pavlenkin
V. A. Poselov · L. G. Poselova
All-Russian Research Institute of Geology and Mineral Resources of the World Ocean
(VNIIOkeangeologia), Saint Petersburg, Russia
e-mail: szh@vniio.nw.ru; gleb@vniio.nw.ru; pav@vniio.nw.ru

A. A. Chernykh
All-Russian Research Institute of Geology and Mineral Resources of the World Ocean
(VNIIOkeangeologia), Saint Petersburg, Russia

Saint Petersburg University, Saint Petersburg, Russia
e-mail: a.a.chernykh@vniio.ru

© Springer International Publishing AG, part of Springer Nature 2019   365
A. Piskarev et al. (eds.), *Geologic Structures of the Arctic Basin*,
https://doi.org/10.1007/978-3-319-77742-9_12

12.1 Sedimentary Cover

After a slow start of more than half a century ago, the last decades saw a dramatic increase of the geophysical activity, especially seismic exploration, in the Arctic Ocean. These sustained multinational efforts in the Arctic Ocean consisted of seismic surveys run in different modes during various research expeditions.

The Russian programs were acquired by:

- drifting ice stations «Severny Polyus (SP)» («North Pole», NP);
- High Latitude Airborne Expeditions «Sever» (HLAE «North»);
- «Transarktika» geotraverse program;
- «Arktika-2000», «Arktika-2005», «Arktika-2007», «Arktika-2010», «Arktika-2011», «Arktika-2012» and «Arktika-2014» expeditions.

Other country programs:

- all-the-year-around and drifting ice stations;
- polar expeditions, icebreakers («Oden», «Polarstern», USCGC Healy, CCGS Louis S. St-Laurent) and some other polar expeditions;
- projects «ARTA» and «LORITA» (Canada, Denmark).

Fig. 12.1 shows location of the above mentioned surveys.

The Fig. 12.1 shows uneven distribution of the seismic data with shelves, especially western, being specifically targeted for many years, generally better covered

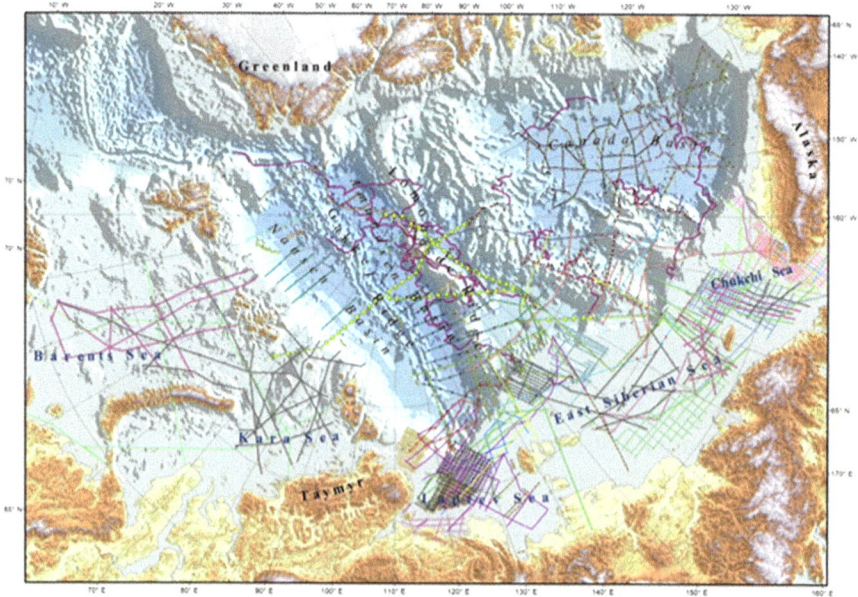

Fig. 12.1 Some lines of the seismic surveys in the Arctic Ocean used in the sediment cover map compilation

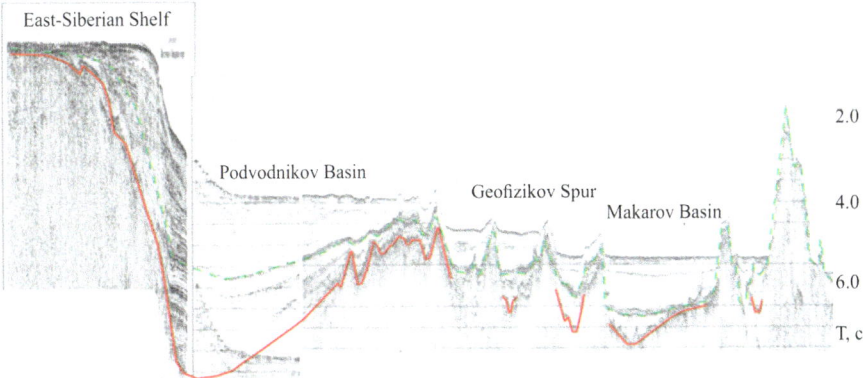

Fig. 12.2 TWT section along NP-28 East-Siberian shelf - Makarov Basin drift track (close to DSS traverse «Transarktika 1989–91»)

than the abyssal parts of the Arctic Ocean. Barents Sea shelf, for one, was subject of many areal seismic surveys summarized in (Verba 2008; Verba et al. 2001, 2005; Bogolepov et al. 1991; Sharov 2005; Shipilov and Mossur 1990). In the east, the Laptev and Chukchi Seas are better studied than Kara and East Siberian Seas.

The geophysical information used for compilation of the Sediment cover map can be grouped in three categories:

- MCS lines– modern high quality data with long record length and full representation of stratified sequences above acoustic basement; examples – see Chap. 2, Fig. 2.2, 2.6 etc.
- Older single channel seismic reflection data acquired from drifting «North Pole» ice stations; example - 0. 12.2
- Published seismic information and sediment cover maps from various sources - Fig. 12.3, (Shimeld et al. 2015).

We include into the sedimentary cover all undeformed Meso-Cenozoic (abyssal Arctic Ocean) or Paleo-Cenozoic (shelves) coilogenic stratified sedimentary complexes from the seafloor to the top of the acoustic basement (TAB) composed, in turn, by dislocated and/or metamorphosed to some degree formations of intermediate metasedimentary layer. The scattered depth data were interpolated into regular 5x5 km grid (format GRID ArcInfo). The Sediment Thickness Map was then created by contouring this grid with contour interval of 1 km for the complexes 0–4 km thick and 2 km - for thicker formations. The map carries limited geographical information – coastal lines, toponymics and latitude-longitude grid (Fig. A 12.4) but represents the important step forward compare to the previous versions because it is based on much bigger and accurate database.

The belt of deep depressions located either inside shelves, or in continental margin-shelf transition zones ringing the abyssal part of the Arctic Ocean – Barents-North Kara, Vilkitsky-North Chukchi, Beaufort-Mackenzie Delta, Sverdrup, Linkoln Sea etc. – is clearly visible on the presented Sediment Thickness Map. These depressions contain up to 18–20 km of sediments with main portion of which

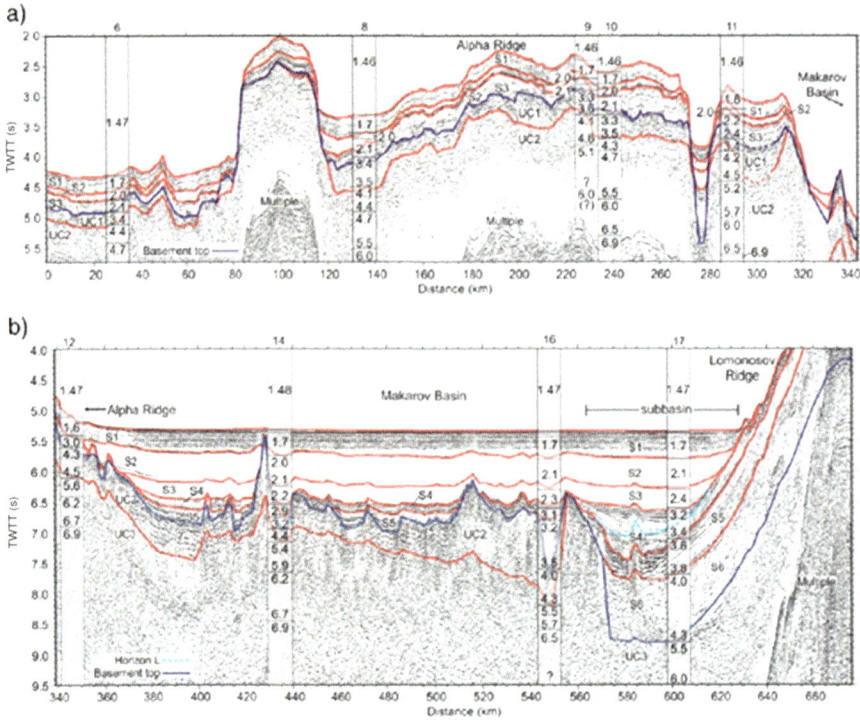

Fig. 12.3 Seismic section from (Evangelatos et al. 2017)

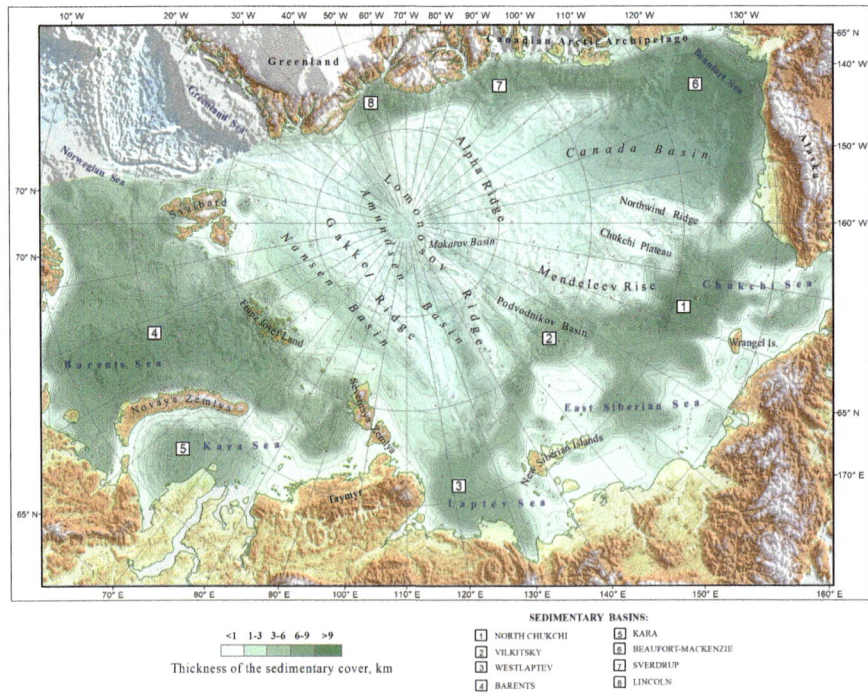

Fig. 12.4 Sediment Thickness Map of the Arctic Ocean

belong to Paleo-Mesozoic complexes overlain by 10–11 km of Late Mesozoic-Cenozoic formations. Most likely, these structures are palinspastic, formed by several epochs of continental rifting (Permian-Triassic, Late Mesozoic and younger) superimposed on each other.

Younger sub-meridional Eurasian-Laptev Sea system of shallower (6–10 km of the Late Mesozoic- Cenozoic sediments) depressions is superimposed on the Paleo-Mesozoic system of shelf-margin depression. Within separating uplifts (Lomonosov-New Siberian and Alpha-Mendeleev-Wrangel) the sedimentary cover reduced to 1 km.

In the abyssal Eurasian Basin the orientation of the prevailing depocenters follows the Gakkel Ridge, the active and the youngest spreading structure where thickness of sediments is drastically reduced (the local uplifts of the Gakkel Ridge are completely denuded of sediments). Barents-Kara shelf was the main source of clastic material for the Nansen Basin while the Amundsen Basin was supplied from the Amerasian sector of the Arctic Ocean. The thickness of the sedimentary cover in both basins seldom exceed 4 km. Supported by young Cenozoic oceanic crust, the depocenters of the Eurasian Basin are also of Cenozoic age and become progressively younger toward the axial zone of the Gakkel Ridge. In the Amerasian Basin the pattern of sedimentary cover distribution is similar to that of the Eurasian Basin – up to 1 km on the uplifts, and around 4 km - in the separating depressions. The thickness of the sedimentary cover in the Canada Basin grows towards the Mackenzie Delta from 3–4 km at the vicinity of the Mendeleev-Alpha system to more than 15 km in the Beaufort Sea.

At the Amerasian Basin - Eastern-Siberian shelf borderline the sedimentary cover becomes 10 km thick at the junction of the Lomonosov Ridge with the periphery of the Laptev Sea shelf (Vilkitsky Basin). The axial zone of the Vilkitsky Basin depocenter can be traced from the Podvodnikov Basin through the continental slope into shelf were it joins the super-deep North Chukchi Basin with 18–20 km of Upper Paleozoic (Lower Ellesmere) – Cenozoic (Upper Brooks) sediments.

Special thing to note is presence of isolated depressions with up to 5 km of sediments inside the Central Arctic Uplifts, more than 1000 km away from the shallow coastal waters. Such distance and isolation preclude any possibility of transporting the shelf-derived clastic material into those depressions, leaving the uplifts of the Amerasian Basin as the only source. These facts also support classification of the Amerasian Basin as an extensional structure on attenuated continental crust.

Another important conclusion can be made from the analysis of the Sediment Thickness Map – the commonality of some geotectonic structures of the shelf and the abyssal region. This manifests itself by continuity of the principal morphological features of the Central Arctic Uplifts (the Lomonosov Ridge, Mendeleev-Alpha Ridges and Makarov-Podvodnikov Basins systems) from the abyssal parts of the Arctic Ocean into its shelves, clearly visible on the Sediment Thickness Map.

We understand that this map is only one more step in the continuous analysis and compilations of the Arctic Ocean seismic data and, undoubtfully, it will be subject of further corrections and improvements coming along with influx of the new information. But we do not expect the radical changes of the principal conclusions we drew from this map, especially ones related to the continuity of the shelf structures into the abyssal parts of the Arctic Ocean.

12.2   The Earth's Crust of the Abyssal Arctic Ocean

Since discovery of the Mohoroviĉić discontinuity (Moho, for short) as widely accepted base of the Earth crust, enormous amount of worldwide information exists regarding its depth and, consequently, the thickness of the crust. The very first seismic experiments in the World Ocean demonstrated the sharp difference between oceanic and continental crustal thickness, the latter being several times thinner than the former (Demenitskaya 1975) – the fact confirmed later as irrefutable fact.

One of the first contemporary crustal thickness map of the Arctic Ocean was compiled under the international project «TeMAr» («Tectonic map of the Circum Polar Arctic») (Kashubin et al. 2011). The compilation used not only existing seismic information and bathymetry (Jakobsson et al. 2008) but also new updated grid of free-air gravity anomalies (Glebovsky et al. 2012) dataset. The latter was especially helpful in the regions which are not studied by seismic methods and where the crustal thickness was estimated, based on its empirical dependence with Bouguer gravity anomalies.

The deep structure of the abyssal Arctic Ocean was later specified by results of 3-D gravity modeling using procedure developed in VNIIOkeangeologia (Glebovsky et al. 2013). The specification became possible after compilation of new digital map (Kaminsky et al. 2012) and grid of the sedimentary cover thickness.

The actual process of 3-D modeling consisted of two main stages. During the first stage combined gravity effect from «water-seafloor» and «seafloor-acoustic basement» interfaces was calculated (software package developed in Geological Survey Of Canada (Verhoef et al. 1995) built around the FFT-based Parker algorithm (Parker 1974)) and subtracted from the observed gravity field. The second stage consisted of finding depth to causative interface by solving the inverse problem of gravity interpretation under the assumption that the residual gravity, free from the influence of the shallower boundaries, reflects topography of the Moho interface. The procedure was refined later by taking into consideration increase of sediments density with depth, decompaction of the upper mantle under the Gakkel Ridge and adjacent basins (Glebovsky et al. 2013) and regional trends established earlier by regionalization of the potential field anomalies (Glebovsky et al. 2011, 2012). The resulting map is shown on Fig. 12.5.

Comparing this model with DSS or seismic-gravity modeling data we found that it is accurate within ±1.7 km RMS, falling to ±2–3 km in poorly studied areas, such as the Eurasian Basin.

The map on Fig. 12.5 vividly demonstrates the radical difference in crustal thickness between clearly oceanic regions of the Arctic Ocean (Eurasian and Canada Basins) and remaining parts with continental or transitional types of crust.

The Eurasian Basin displays apparent axial symmetry in crustal thickness relative to the axial minimum of the mid-oceanic Gakkel Ridge. The Lomonosov Ridge with 18–26 km thick crust, separating the Eurasian and the Amerasian Basins, stands out as an abyssal continuation of the Laptev Sea shelf.

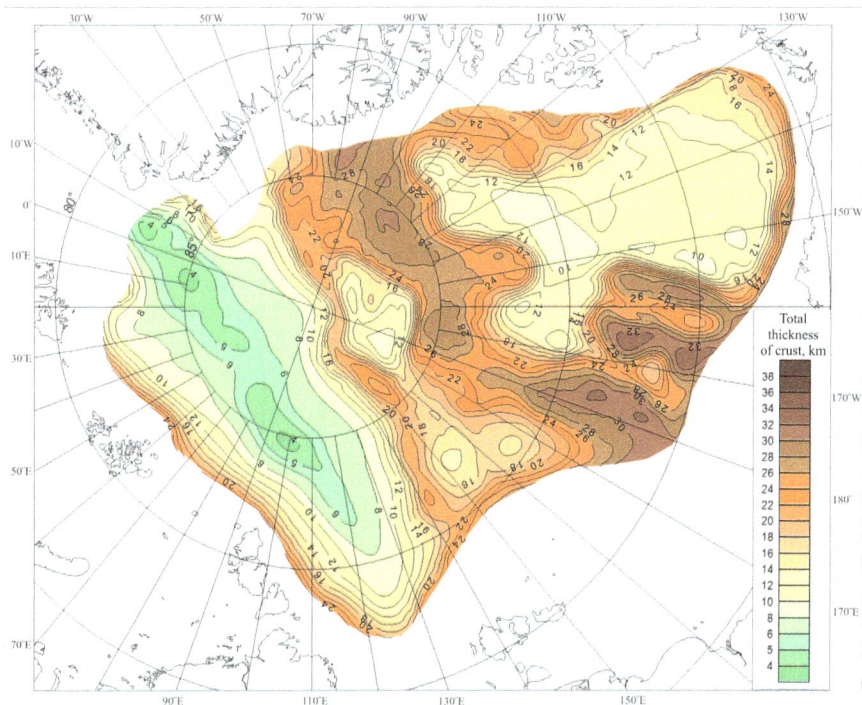

Fig. 12.5 Crustal thickness in the abyssal Arctic Ocean

The Central Arctic Uplifts in the Amerasian Basin, within the Alpha and Mendeleev Ridges forms the unified block with enlarged crustal thickness of 24–30 km, reaching 30–32 km at the Russian and Canadian continental shelves. The maximum crustal thickness in the Amerasian Basin (up to 34 km) is found at the Northwind Ridge and the Chukchi Plateau. Within the Sever Spur and Pearya Spur situated northward the Canadaian Archipelago at 120°W, the crustal thickness varies within 18–26 km. Within the abyssal Canada and Makarov Basins it is reduced to less than 10–16 km and increased to 14–24 km in the Podvodnikov Basin.

References

Bogolepov AK, Zhuravlev VA, Shipilov EV et al (1991) Deep structure of the Eurasian Arctic Ocean-continent transition zone (western sector). In: Belousov VV (ed) Deep structure of the USSR landmass, Nauka, Moscow, pp 31–41. (in Russian)

Demenitskaya RM (1975) The Eart crust and mantle. NEDRA, Leningrad. (in Russian)

Evangelatos JE, Funck T, Mosher DC (2017) The sedimentary and crustal velocity structure of Makarov Basin and adjacent alpha ridge. Tectonophysics J 696–697:99–114

Glebovsky V, Chernykh A et al (2011) Integrated analysis of updated potential field database: implications for the compilation of the new Circum Arctic Tectonic Map. VI International Conference on Arctic Margins (abstracts), Fairbanks, Alaska, USA 31 May-2 June 2011

Glebovsky VY, Chernykh AA et al (2012) Structural-tectonic regionalization of potential fields in the Arctic Ocean for the latest compilation of circumpolar tectonic map of the Arctic. In: Collected articles. Geology and geophysics of the Arctic region lithosphere, vol 8. VNIIOkeangeologia, Saint Petersburg, pp 20–29

Glebovsky VY, Astafurova EG et al (2013) Thickness of the Earth's crust in the deep Arctic Ocean: results of a 3D gravity modeling. Russian geology and geophysics J 54(3):247–262

Jakobsson M, Grantz A, Kristoffersen et al (2003) Physiographic provinces of the Arctic Ocean sea floor. Geological Society of America Bulletin J 115:1443–1455

Jakobsson M, Macnab R, Mayer L et al (2008) An improved bathymetric portrayal of the Arctic Ocean: implications for ocean modeling and geological, geophysical and oceanographic analyses. Geophysical Research Letters J:35. https://doi.org/10.1029/2008GL033520

Kaminsky VD, Suprunenko OI, Lazurkin DV et al (2012) On investigation of the abyssal potentially hydrocarbon–bearing sedimentary basins of the Eurasian continental margin and the Arctic Ocean. Min J 3:66–71

Kashubin SN, Petrov OV et al (2011) Crust thickness in the circumpolar Arctic. Regionalnaya geologiya i metallogeniya J 46:5–13. (in Russion)

Parker RL (1974) A new method for modeling marine gravity and magnetic anomalies. J Geophys Res 79:2014–2016

Sharov NV, Mitrofanov FP (2005) Lithosphere of the Russain segment of the Barents region Karelian scientific Center RAN, Petrozavodsk. (in Russian)

Shimeld J, Li Q, Chian D et al (2015) Seismic velocities within the sedimentary succession of the Canada Basin and southern alpha-Mendeleev ridge, Arctic Ocean: evidence for accelerated porosity reduction? Geophys J Int 204:1–20

Shipilov EV, Mossur AP (1990) Deep structure of the Arctic Region sedimentary cover. Izv. AN SSSR, Geology Ser. 1:90–97. (in Russian)

Verba ML (2008) Comparative geodynamics of the Eurasian Basin. Nauka, Saint Petersburg. (in Russian)

Verba ML, Ivanova NM, Katsev VA et al (2001) Regional seismic traverses AP-1 and AP-2 in the Barents and Kara seas. Razvedka I Okhrana Nedr J 10:3–7. (in Russian)

Verba ML, Roslov YV, Sakulina TS (2005) Novaya Zemlya and Ob-Barents sedimentary megabasin in light of interpretation of the seismic traverse AP-2. Razvedka I Okhrana Nedr J 1:6–9. (in Russian)

Verhoef J, Usov K, Oakey G et al (1995) Potential field data processing. Geol Survey of Canada Atlantic & Blue Vajra Computing

Chapter 13
Conclusions

Valery D. Kaminsky, Alexey L. Piskarev, Victor A. Poselov,
and David C. Mosher

Concerted international efforts researchers from Russia, Canada, USA, Germany, Denmark and Norway at the beginning of the XXI century greatly improved our knowledge of the Arctic Ocean. Multi-discipline geological and geophysical programs operated by specialized RVs, often supported by nuclear ice-breakers, included reflection and refraction seismic, Side Scan Sonar bathymetry, bottom sampling, and airborne geophysical surveys. Regional compilations of existing and newly acquired airborne and satellite data brought to light revised modern maps of potential field anomalies. All this activity created prerequisites of new analysis and revision of existing regional geological concepts.

The optimal, to the best of our knowledge, seismic stratigraphy model was created for the Arctic Ocean sedimentary cover. Based on the ACES well core (the Lomonosov Ridge, 2004) and ages of the regional unconformities, the Cenozoic portion of the sedimentary cover was subdivided in two seismic-stratigraphic complexes continuously traced from the Eurasian shelf, through the transitional zone, to the Lomonosov Ridge and, through the Vilkitsky Basin – into the Podvodnikov

V. D. Kaminsky · V. A. Poselov
All-Russian Research Institute of Geology and Mineral Resources of the World Ocean
(VNIIOkeangeologia), Saint Petersburg, Russia
e-mail: okeangeo@vniio.ru

A. L. Piskarev (✉)
All-Russian Research Institute of Geology and Mineral Resources of the World Ocean
(VNIIOkeangeologia), Saint Petersburg, Russia

Saint Petersburg University, Saint Petersburg, Russia

D. C. Mosher
Geological Survey of Canada, Natural Resources Canada, Bedford Institute of Oceanography,
Dartmouth, NS, Canada
e-mail: dmosher@ccom.unh.edu

© Springer International Publishing AG, part of Springer Nature 2019   373
A. Piskarev et al. (eds.), *Geologic Structures of the Arctic Basin*,
https://doi.org/10.1007/978-3-319-77742-9_13

Basin and Mendeleev Ridge. Information from Burger and Popcorn wells drilled on the Chukchi shelf helped to identify major unconformities in the Pre-Cenozoic sequences, date and trace them from the North Chukchi Basin to the Mendeleev Ridge and, through the Vilkitsky Basin, into the Podvodnikov Basin. In total, four seismic-stratigraphic complexes were identified in the Pre-Cenozoic sedimentary sequences from Upper Ellesmere to Lower Brooks (Aptian-Albian and younger). In 2011–2014, the Russians concentrated their seismic surveys on the Lomonosov Ridge, Podvodnikov Basin, Alpha and Mendeleev Ridges and Chukchi Plateau. Collected data convincingly demonstrate the rift-related stretching of the basement, attenuation of the continental crust and uninterrupted continuation of the principal structural complexes from the Eurasian shelf to the Greenland - Ellesmere continental margins on the opposite side of the Arctic Ocean, bridged by the Lomonosov Ridge and the Central Arctic Uplifts.

The extensional structures of the Russain Eastern Arctic Seas shelves and the Central Arctic Uplifts are typical for the stretched continental margins where a structural style is controlled by systems of grabens and half-grabens created by normal faulting, displacing acoustic basement and lower parts of the overlying sediments. The Podvodnikov Basin itself resides on the apparently stretched margin-type continental crust. Its flanks with the Lomonosov and Mendeleev Ridges are formed by sub-meridional systems of stepped normal faults. The orientation of stretching forces might have changed with time, being originated either from the Makarov Basin, or from the Eurasian side. Seismic signature of the acoustic basement in this region is drastically different from the regions with oceanic type of crust (the Amundsen Basin).

Rift-related stretching of the continental crust, apart from intensive normal faulting, was also accompanied by increasing magmatic activity inside the continental crust and on surface, attested by strong seismic reflectors at the upper part of the acoustic basement interpreted as seismic expression of the interbedding basalt flows, pyroclastics and sediments. Chemicaly different from tholeitic basalts of mid-oceanic ridges, basalts from the slopes and seamounts of the Mendeleev -Alpha system confirmed active magmatizm of the Amerasian Basin.

The structural and morphological homogeneity of the Central Arctic Uplifts region is broken by presence of isolated enclave of the oceanic crust – the Makarov Basin, formed inside the Central Arctic Uplifts as, perhaps a "pull-apart" depression, approximately synchronous to the opening of the Canada Basin. The recent seismic survey shows rift-like basement structure interpreted as possible Late Cretaceous or Paleogene spreading center.

According to the new geophysical information, large peripheral parts of the Eurasian Basin - southern parts of the Nansen Basin and Near-Laptev Sea margin (except one narrow zone) - do not contain any linear anomalies parallel to the Gakkel Ridge. The Gakkel Ridge central rift valley came to the present location in the eastern part of the Eurasian only in Pliocene, 2–5 Ma ago. Between that time and ≈ 35 Ma (magnetic anomaly 13) anomalies remain parallel to the ridge telling that during this period spreading direction did not differ much from the present. But in the epoch preceding the magnetic anomaly 13, the spreading vector was at an angle to the present.

Seismic evidence of extensive non-oceanic basement beneath Canada Basin is an important recent discovery. It coincides with proof of a spreading axis with associated limited oceanic crust within Canada Basin. Despite these constraints on its tectonic development, there are still significant challenges and disagreement over specific tectonic models. These differences result from the fact that the basin did not from by a simple rift/extension scenario, but rather likely formed by a complexity of events that included variably oriented extension, trans-tension and transform tectonics. Additionally, emplacement of the high arctic magnetic domain (Alpha Ridge and Mendeleev Ridge) masks underlying tectonic structures and lack of age control inhibits correlation with global events.

Extremely low Pliocene-Pleistocene sedimentation rates in the Mendeleev Ridge and the Podvodnikov Basin are indicative of tectonic stability of this region. On the other hand, volcanogenic material established in the cores of seafloor sediments proves that there was Pleistocene volcanic activity. It is safe to state that at least one such event, catastrophic by its scale and amount of ejected material, took place in the Eurasian Basin. This event (≈ 1.1 Ma) was not, by all means, the only one in during Pliocene-Pleistocene sedimentation cycle, but it was one of the strongest, powerful enough to cause short-lived climate change not only in the Arctic but worldwide. It also might trigger the spreading axis "jump" - the latest tectonic rearrangement in the Eurasian Basin.

The last 10–15 years have created a solid foundation on which we can build our progressive understanding of the origin and evolution of the Arctic Basin. The MCS surveys with long streamers providing reliable depth-velocity functions reaching deep inside the upper crust; modern high resolution aeromagnetic and aerogravity surveys with highly accurate positioning; deepwater drilling – are the major tools at our disposal from which we expect the new discoveries related to the history of the major structures of the Arctic Ocean.

Printed by Printforce, the Netherlands